High Purity Silicon VII

Proceedings of the International Symposium

Editors

C. L. Claeys
IMEC
Leuven, Belgium

M. Watanabe
SEZ Japan
Bunkyou-ku, Tokyo, Japan

P. Rai-Choudhury
Saltronics International
Bellingham, Washington, USA

P. Stallhofer
Wacker Siltronic AG
Burghausen, Germany

Sponsoring Division:
Electronics

Proceedings Volume 2002-20

THE ELECTROCHEMICAL SOCIETY, INC.
65 South Main St., Pennington, NJ 08534-2839, USA

Published by:

The Electrochemical Society, Inc.
65 South Main Street
Pennington, New Jersey 08534-2839, USA

Telephone 609.737.1902
Fax 609.737.2743
e-mail: ecs@electrochem.org
Web: www.electrochem.org

Library of Congress Catalogue Number: 2002116856

ISBN 1-56677-344-X

Printed in the United States of America

PREFACE

The previous international symposia on High Purity Silicon were held at the Electrochemical Society Meetings in respectively 1990 in Seattle, 1992 in Toronto, 1994 in Miami, 1996 in San Antonio, 1998 in Boston and in 2000 in Phoenix. Due to the strong growing interest in the field, it was decided to start from the fourth symposium editing a Proceedings Volume.

This Proceedings Volume includes papers that were presented at the Seventh Symposium on High Purity Silicon held in Salt Lake City, Utah at the 202nd Meeting of the Electrochemical Society, October 20-25, 2002. The Symposium, sponsored by the Electronics Division of the Electrochemical Society, was international in scope and included authors from Belgium, China, Denmark, Finland, France, Germany, Ireland, Italy, Japan, Poland, The Netherlands, Ukraine, United Kingdom, and the United States of America. This Proceedings contains 34 papers, including 10 invited papers, providing reviews and recent advances related to silicon crystal growth aspects, material properties, device performance and applications, and novel or improved characterization techniques for high purity silicon.

The Section on **Crystal Growth and Defect Control** starts with an invited paper by Fukuda *et al.* on the impact of wafer topography on the next generation processes. For obtaining high performance silicon oxide polishing it is essential to improve the nanotopography and to use polishing pads with the appropriate stiffness. It is important to differentiate between single and double sided polished wafers. The invited paper by Voronkov and Falster elaborates on the grown-in microdefects as a guide to the properties of point defects. A model is discussed whereby the concentration field of silicon self-interstitials and vacancies depends on 10 parameters, of which 7 can be calculated and the other 3 have to be fitted to the experimental data. The contributed paper by Kulkarni *et al.* further investigates the dynamics of point defects and the formation of microdefects in Czochralski silicon. The proposed simulation model is used to make predictions, which are validated by analyzing experimental results. Both the vacancy and interstitial type defects are approximated as spherical clusters. Wang and Brown discuss in their invited paper the role of the oxide morphology on the simulation of oxide formation and point defect dynamics in silicon. The model takes into account the importance of stress effects associated with the growing precipitates. The stress is directly linked to the morphology of the oxide precipitates. Predictions are given for crystals grown under both vacancy-rich and vacancy-poor conditions. The former gives spherical precipitates while the latter leads to disk-shaped precipitates.

The status of the simulation of silicon Czochralski growth is addressed by Anttilla *et al.* The melt flows are modeled in a large scale cylindrical growth system and good time dependencies in agreement with experiments are found, even when not taking into account turbulence effects. Different parameters influencing the melt flow patterns, the melt stability and the transport of impurities are accounted for. The evolution in crystal growth simulation has strongly improved during the last decade. Float zone crystals can be grown with a higher pulling rate than Czochralski crystals because of the better heat dissipation, making such crystals of interest for applications such as photovoltaics requiring low cost materials. This issue is studied in the paper by Luedge *et al.* Their work pointed out that dislocation-free FZ crystals can be grown with pull rates of 85% of the theoretical maximum value, whereas this was limited to 60% in the past. Since the discovery of Voronkov's critical growth law, several investigators have reported the existence of a ring-like stacking fault region (ring-OSF) in the crystals. These rings have a detrimental impact on the gate oxide quality. Nakashima *et al.* report on a new highly selective ion etching method to detect the nuclei of these ring-OSFs. These nuclei are small oxygen precipitates present already in the as-grown crystals. Vanhellemont *et al.* developed a model for the nucleation of stacking faults at oxide precipitates by taking into account the role of the stress associated with the precipitates. The strain release leads to the emission of self-interstitials which condense in order to nucleate a stacking fault. Beside the size of the precipitates, the involved strain energy controls the process.

The second Section is concentrating on **Impurities in Silicon**. The first invited paper by Yang and Yu addresses a nowadays more common Cz crystal growth technique, namely in-situ nitrogen doping of the crystal. Nitrogen doping reduces the size and density of crystal originated particles (COPs), enhances oxygen precipitation, and at the same time results in the formation of nitrogen-related grown-in defects. The detection of nitrogen by infrared absorption techniques is the subject of the contributed papers by Hashimoto *et al.*, Inoue, and Funao *et al.*, respectively. To increase the resolution of the technique it is necessary to eliminate interference absorption. The obtained nitrogen profiles are in good agreement with those resulting from SIMS measurement. At the melting point a correlation exists between the distribution coefficient and the solid solubility and the measured nitrogen concentration increases with the solidified fraction. Hydrogen in silicon is playing an important role as in first instance it enhances the thermal donor formation rate. The invited presentation by Job *et al.* clearly illustrated how that a low temperature hydrogen plasma treatment, leading to such an enhanced thermal donor formation, can be used for the fabrication of deep p-n junctions with a low leakage current as required for high voltage applications. Both single and two-step low thermal budget processes are discussed. As pointed out by Kurita *et al.*, the surface structure after hydrogen annealed wafers can be preserved by using an ozonized water and diluted HF cleaning. Both AFM and TEM analyses have been used to study the surface roughness. A model is proposed to explain the protrusions observed in the TEM micrographs. For several decades it is well known that oxygen atoms in silicon may act as dislocation pinning centers. However, the invited review by Sendaker *et al.* presents new insights into this dislocation mechanism that consists of five different steps. The temperature dependence of the unlocking stress allows to determine the binding energy of the oxygen to the dislocations. At lower temperatures enhanced oxygen diffusivity has been observed. An old technique for determining the oxygen concentration in Cz silicon is based on infrared absorption spectroscopy. Based on theoretical considerations De Gryse *et al.* have refined the analysis of the FTIR spectra in order to determine the chemical structure of oxide precipitates in heavy boron doped silicon. For the first time it is reported that in these samples the measured spectra of the precipitates can be explained by a mixture of SiO_2 and B_2O_3, with for heavily doped sample a B_2O_3 volume fraction as high as 0.41.

The third Section is devoted to **Process- and Irradiation-Induced Defects in Silicon** and contains 11 contributed and 3 invited papers. The Section starts with a paper on backside grinding which may introduce surface and sub-surface stress. The stress levels can be investigated by Raman spectroscopy as discussed by Watanabe *et al.* The detrimental influence of metals on the device performance is well known since the early 60ties. Polignano *et al.* performed an extensive study on the detection of metals in internally gettered wafers. The influence of iron and nickel in internal gettered CZ and epitaxial silicon is pointed out by studying the possible degradation of the generation lifetime. The potential and limitations of both SPV measurements and DLTS are addressed. Lemke elaborates on analytical distributions of substitutional transition metals in FZ silicon. Theoretical and experimental results are discussed for Cu, Ni, Co, Ag, Pd, Rh, Au, Pt and Ir. The metal behavior in both interstitial- and vacancy-rich crystals is discussed. The invited paper by Martinuzzi and Palais gives an overview of gettering and lifetime engineering in silicon. The preferred gettering technique strongly depends on the metal involved. An adequate method to evaluate the gettering efficiency is based on lifetime measurements by means of a contactless technique such as microwave photoconductance decay of the microwave phase shift. Microwave photoconductive decay (μ-PCD) has also been used by Väinölä *et al.* to study the impact of Cu in p-type silicon. While for low concentrations the copper mainly diffuses out of the silicon, for higher concentrations the Cu precipitates in the bulk. Oxygen precipitates are relatively inefficient as sinks for copper. Using high intensity illumination might increase the efficiency. Hojo *et al.* report on the topographical change at the SiO_2/Si interface after multiple oxidations, by using AFM. The oxidation temperature plays an important role in the stress relaxation. As demonstrated by Poyai *et al.* the analysis of the diode current-voltage characteristics is a very powerful tool to study process-induced defects such as e.g.

the p-well ion implantation damage. A higher leakage current and yield losses are observed for a lower implantation energy, while the role of the implant dose is not so important. The diode leakage current depends on the electrical field. Czerwinski *et al.* are proposing new methods to accurately determine the electrical field enhancement factor in junctions and apply the technique to diodes with a shallow trench isolation and high electrical fields. The techniques are based on the analysis of the activation energy. The local electrical field near STI diodes may be much larger than the field related to ionized impurties. Defect reactions of copper in silicon are reported by Knack et al. A photoluminescence line has been identified as belonging to the Cu_iCu_s pair. The invited paper by Buzkowski *et al.* illustrates the use of photoluminescence analysis as a powerful tool for contactless characterization of silicon wafers. PL can be used for imaging extended defects such as dislocations and precipitates. AFM and MOS capacitors analysis were used by Tokuda *et al.* to study the degradation of intentionally contaminated SiO_2 films. TXRF investigations point out that a high density of Cu atoms on the SiO_2 surface induces dielectric breakdown, while a low density of Cu atoms inside the SiO_2 film induces a high leakage current for low electrical fields. The last three papers of the Section are dealing with irradiation-induced defect. The invited paper by Schulze *et al.* reviews the impact of irradiation-induced defects on the electrical performance of power devices. The irradiation-induced degradation of the carrier lifetime and the effective doping concentration can efficiently be used for optimizing the devices. The lifetime modifications are used to control the device switching speed, while the doping effect is used to adjust the overvoltage protection. One has to differentiate between the impact of electron and proton irradiations, respectively. Neimash *et al.* use a special experimental design in order to differentiate between the impact of electron irradiation induced defects and thermally induced defects on the thermal donor generation at 450ºC. The main conclusions are that electron irradiations have a strong impact on the oxygen-related thermal donor generation rate, related to the vacancy-assisted enhancement of the effective oxygen diffusivity. Higher order (V_x-O_y) complexes may generate additional OTD nuclei. In the last paper of the Section, Stoddard *et al.* report on observations of point defect generation and complexing during electron beam irradiation of nitrogen doped Czochralski silicon. TEM studies are used to characterize the nucleation and growth of extended defects such as voids, stacking faults and oxygen precipitates, whereby nitrogen plays an important role.

The last Section is on **Alternative Silicon Substrates and Structures**. In their invited paper Maleville *et al.* discuss the Smart-cut approach for the fabrication of high volume silicon-on-insulator wafers. McCann *et al.* report on the electrical and structural characterization of SOI bonded interfaces, taking into account parameters such as the bonding conditions, the annealing temperature and differences in crystal orientation and growth. Direct bonded silicon surfaces exhibit a low leakage current, low defect densities and controllable doping redistribution. Reiche *et al.* use hydrophobic wafer bonding for the fabrication of epitaxial wafers that are designated for optical applications. The final paper of the symposium deals with the fabrication of 3-D structures on p-type silicon relying on electrochemical etching. The morphology of the micro porous silicon is affected by the silicon resistivity and the composition of the electrolyte, while the sidewall width of silicon between trenches increases with the Si resistivity. The fabrication of an accelerometer structure is demonstrated.

As a general conclusion it can be stated that there is a strong interest in High Purity Silicon. The research activities of the material suppliers are triggered by the future requirements of the device manufacturers, with emphasis on the fabrication of both higher quality and purer material and 300 mm diameter wafers. Beside standard polished and epitaxial wafers, there is also a growing interest in Silicon-on-Insulator. The higher purity of the starting material in combination with the increased sensitivity of the devices necessitates the availability of appropriate diagnostic tools. This either requires a refinement of existing techniques or the implementation of new ones.

November 2002

C.L. Claeys
M. Watanabe
P. Rai-Choudhury
P. Stallhofer

TABLE OF CONTENTS

* Invited Paper

IMPURITIES IN SILICON

PROCESS- AND RADIATION-INDUCED DEFECTS IN SILICON

* Invited Paper

Facts about ECS

The Electrochemical Society (ECS) is an international, nonprofit, scientific, educational organization founded for the advancement of the theory and practice of electrochemistry, electrothermics, electronics, and allied subjects. The Society was founded in Philadelphia in 1902 and incorporated in 1930. There are currently over 7,000 scientists and engineers from more than 70 countries who hold individual membership; the Society is also supported by more than 100 corporations through Contributing Memberships.

The technical activities of the Society are carried on by Divisions and Groups. Sections of the Society have been organized in a number of cities and regions. Major international meetings of the Society are held in the spring and fall of each year. At these meetings, the Divisions and Groups hold general sessions and sponsor symposia on specialized subjects.

The Society has an active publications program that includes the following:

Journal of The Electrochemical Society — The *Journal* is the peer-reviewed leader in the field of electrochemical and solid-state science and technology. Articles are posted online as soon as they become available for publication. This archival journal is also available in a paper edition, published monthly following electronic publication.

Electrochemical and Solid-State Letters — *Letters* is the first and only rapid-publication electronic journal covering the same technical areas as the ECS *Journal*. Articles are posted online as soon as they become available for publication. This peer-reviewed, archival journal is also available in a paper edition, published monthly following electronic publication. It is a joint publication of ECS and the IEEE Electron Devices Society.

Interface — *Interface* is ECS's quarterly news magazine. It provides a forum for the lively exchange of ideas and news among members of ECS and the international scientific community at large. Published online (with free access to all) and in paper, issues highlight special features on the state of electrochemical and solid-state science and technology. The paper edition is automatically sent to all ECS members.

Meeting Abstracts *(formerly Extended Abstracts)* — Meeting Abstracts of the technical papers presented at the spring and fall Meetings of the Society are published in serialized softbound volumes.

Proceedings Series — Papers presented in symposia at Society and topical meetings are published as serialized Proceedings Volumes. These provide up-to-date views of specialized topics and frequently offer comprehensive treatment of rapidly developing areas.

Monograph Volumes — The Society sponsors the publication of hardbound Monograph Volumes, which provide authoritative accounts of specific topics in electrochemistry, solid-state science, and related disciplines.

For more information on these and other Society activities, visit the ECS website:

www.electrochem.org

Crystal Growth and Defect Control

The Impact of Wafer Topography Issues on The Next Generation Processes

Tetsuo Fukuda[1)], Masaharu Watanabe[2)], Seiichiro Kobayashi[3)], Masanori Yoshise[2)], Satoshi Akiyama[4)], Yasuhiro Shimizu[5)], and Masayuki Hashimoto[6)]

The Next-generation Wafer Technologies Subcommittee, Japan Electronics and Information Technology Industries Association (JEITA)
Mitsuikaijo Bekkan Bld., Surugadai 3-11, Kanda, Chiyoda-ku, Tokyo 101-0062, Japan

[1)]Fujitsu, 50 Fuchigami, Akiruno, Tokyo, 194-0833, Japan
[2)]SEZ Japan, 1-24-1 Hongo, Bunkyo-ku, Tokyo 113-0033, Japan
[3)]Eastern Japan Semiconductor Technologies, 1-280 Higashi-koigakubo, Kokubunji, Tokyo, 185-8601, Japan
[4)]Raytex, Suite 403 Bellebs Nagayama, 1-5 Nagayama, Tama, Tokyo, 206-0025, Japan
[5)]Sumitomo Mitsubshi Silicon, Seavans North, 1-2-1 Shibaura, Minato-ku, Tokyo, 105-8634, Japan
[6)]Applied Materials Japan, Technology Center, 14-3 Shin-izumi, Narita, Chiba, 286-8516, Japan

The impact of nanotopography of silicon wafer on the polishing of silicon-oxide films on the wafers was quantitatively evaluated. Oxides remaining after the polishing have the thickness variations that depend on both nanotopography features and pad stiffness. It is concluded that high performance in the STI-CMP process can be achieved by improving nanotopography and using pads of optimal stiffness. To obtain the satisfactory patterning performance in the lithography process step, it is shown that chucked-wafer flatness is one of key factors to be analyzed and improved. We succeeded to measure wafer flatness, chuck flatness, and chucked-wafer flatness, separately, and found that the chucked-wafer flatness was affected by the interaction between wafer backside surface and pin-chuck surface. It is shown that the interaction mechanism should be clarified.

1. Introduction

One of essential reasons why greater circuit-density of ultra large scale integrated circuits (ULSI) has been achieved is to employ the shallow trench isolation (STI) method [1), 2)]. The STI method can release the difficulties associated with the conventional isolation method, local oxidation of silicon (LOCOS) [3), 4)]. In advanced device processes, STI requires the employment of chemical mechanical polishing (CMP) [5), 6)] technique to reduce the thickness of gap-fill silicon-oxide (SiO_2) layers deposited on the front-side surface of wafers.

Nanotopography is the surface topography of wafers placed on a flat stage without chucking or clamping. The peak-to-valley height varies between several nanometers and several hundreds nanometers, and the spatial wavelength range is approximately up to 20 mm[7)]. In typical CMP machines, a front-side surface reference is employed, and the backside surface of a wafer touches soft carrier films or airbags, both of which absorb the topography variations of the backside surface. Thereby, wafers are in a condition without chucking or clamping in the CMP machines.

Using a stiff polishing pad to achieve high planarization efficiency causes lower polish pressure in the lower areas of nanotopography than in the raised areas. That results in the oxide films remaining in the lower areas. Both nanotopography and stiff polishing pads reveal the thickness variations of residual silicon-oxide films in the STI-CMP process, and ultimately incomplete circuit patterns in ULSI devices.

Most advanced processes adopt 300mm diameter wafers commonly produced in double side polished (DSP) process. The DSP wafers are expected to have smaller nanotopography features than those of single side polished (SSP) wafers. However, using 200mm DSP wafers, we point out that there is still small but critical impact of DSP wafers' nanotopography on CMP performance.

As critical dimensions (CD) shrink, the specification of CD variations across the wafer is tightening and the lithographer has to work with almost vanishing process margins. One of the important factors to improve CD variations is to adopt a near perfect flat wafer vacuumed on an as-flat-as-possible exposure tool chuck, resulting in excellent chucked-wafer flatness. Although local flatness is compensated by dynamical leveling mechanism of the scanner, the more the flatness deteriorates, the higher the risk of larger resulting CD variations becomes.

In some device manufacturers, 200 mm SSP wafers have been mainly used in the mass production of leading edge devices. This results in another topography issue, the impact of chuck flatness on chucked-wafer flatness. The resulting chucked-wafer flatness comes from the complex interaction between wafer backside surface and chuck structure because the backside surface of SSP wafers has some topography features[8)]. However, chuck flatness and chucked-wafer flatness have not been of such a major interest for the public as wafer flatness has been.

In this study, we evaluate wafer flatness, pin-chuck flatness, and chucked-wafer flatness, separately, and consider the impact of chuck parameters on chucked-wafer flatness. It is shown that the interaction mechanism should be studied further.

2. Experiment

2.1. Nanotopography

2.1.1. Sample Preparation

200 mm DSP and SSP wafers were prepared by a wafer vendor [SUMCO].

2.1.2. Quantitative Measurement of Nanotopography

A sample wafer was placed on a flat stage of NanoMapper [ADE Phase Shift], in which an interferometer is employed. In measuring the height of the wafer front-side surface, white light with the wavelength of 0.6 μm is used and the interference converts small changes in a test beam's path to the wafer surface into measurable changes in optical intensity at the detector.

The long spatial wavelength of a wafer shape (bow, sori, and warp) was effectively removed from the raw height values using a spatial high-pass filter. We employed a single Gaussian type filter that transmits 50% of input at the cutoff wavelength (20 mm). Since measurements made near a wafer edge can affect the height values in the fixed quality area (FQA)[7], a large edge exclusion (20 mm) was used to eliminate the edge effect. We finally obtained the height value map (nanotopography features) all over the sample wafer except the edge exclusive area.

One-dimensional profile of nanotopography was calculated in the wafer diameter direction from the height value map. This data is necessary to analyze the impact on oxide polishing (shown in chapter 3).

2.1.3. CMP Process

After the measurement of nanotopography, 1 μm thick silicon-oxide films were grown using a HDP-CVD chamber, Ultima [Applied Materials Japan] (HDP: High Density Plasma, CVD: Chemical Vapor Deposition), and the initial film thickness variations were measured with AcuMap [Japan ADE] using 1 mm grids.

Oxide polishing on a rotary polisher, Mirra [Applied Materials Japan], was performed using a Cabot SS-25 slurry at 200 cm^3/min for 78 to 87 s. In this work, three kinds of polishing pads with different degrees of stiffness were used: a soft pad (IC-1000/suba-IV), a medium-soft pad (IC-1400), and a stiff pad (IC-1000 solo). The polish pressure was 5 psi, the rotation rates of the head and polish platen were 146 and 109 rpm, respectively, and the head sweep was between 4.2 and 5.6 inch with a sweep length 1.4 inch at 10 sweeps/min. The targeted polish removal thickness was 500 nm for each polishing pad. Since we used a large edge exclusion, CMP conditions at the wafer periphery were not optimized.

After oxide polishing, the thickness variations of residual films were measured again with AcuMap using a 35 mm single Gaussian filter. This was to eliminate the non-uniformity in oxide polishing, which was not caused by nantopography but by the polisher.

2.2. Flatness

2.2.1. Sample Preparation

200 mm DSP and SSP wafers were prepared by three different silicon vendors. All SSP wafers were selected from mass-produced wafers currently employed to fabricate devices over three generations.

2.2.2. Flatness Measurement

Currently, three chuck manufacturers (C_1, C_2, and C_3) provided us with their pin-type chucks. DynaSearch [Raytex] was employed to measure optically both chuck flatness and chucked-wafer flatness, using provided chucks and wafers. The chucked-wafer flatness is considered equal to the flatness of the wafer vacuumed on the exposure tool chuck.

Wafer flatness was measured, using an electrostatic capacity sensing technique of Ultragage 9900 [Japan ADE].

Flatness values of wafers, chucks, and chucked-wafers were quantitatively expressed as SFQR values (SFQR: Site Front least sQuares Range) at the site size of 25x8 mm^2. This expression results from the consideration that scanners will be mainly used in the next generation lithographic process to fabricate ultra-microscopic gate electrode structures of transistors.

3. Results and Discussion

3.1. Impact of Nanotopography on Oxide Polishing

The pre-CMP oxide thickness was almost uniform across the wafer diameter of the interest: its total thickness variation was less than 6 nm, indicating that the uniformity of initial oxide films was 0.6%.

One-dimensional nanotopography profiles across the wafer diameter are superimposed on the corresponding inverted thickness profiles of remaining oxide films for DSP wafers in Figures 1(a), 1(b), and 1(c), and for SSP wafers in Figures 2(a), 2(b), and 2(c). In all figures, the inverted film thickness profiles were obtained after CMP with a soft pad in Figures 1(a) and 2(a), a medium-soft pad in 1(b) and 2(b), a stiff pad in 1(c) and 2(c).

For the combination of a DSP wafer and a soft pad (Figure 1(a)), no impact of nanotopography was found on the thickness variations of the films remaining after CMP. This was concluded because the thickness variations are only a few nanometers and the positions of peaks and valleys have nothing to do with those of nanotopography features. The results shown in Figure 1(b) confirmed a similar lack of the impact after oxide polishing of a DSP wafer with a medium-soft pad.

However, for the combination of a DSP wafer and a stiff pad, somewhat larger thickness variations (nearly 10 nm) of the residual film were found, as shown in Figure 1(c). This indicates that nanotopography features affect oxide polishing.

We assumed that a typical value for the allowable thickness difference of residual oxides is 10 nm or less because the 10 nm dishing is only a small percentage of the depth of the trench. Figure 1(c) show that the nanotopography features of the DSP wafer have the risk to cause problem (pattern failures in STI) in the oxide polishing with a stiff pad since the thickness variations of the residual film are almost 10 nm.

Figures 2(a) and 2(b) show correlations between the nanotopography and the thickness variations of residual films for the combination of a SSP wafer and a soft pad (2(a)), and that of a SSP wafer and a medium-soft pad (2(b)). Both results show that the nanotopography of the center of SSP wafers have an impact (10nm thickness variation) on oxide polishing with either polishing pads. SSP wafers have the risk to cause problem in oxide polishing with a soft pad. According to the same analysis, it is risky also in the combination of SSP wafers and a medium-soft pad.

Figure 2(c) shows the thickness variations of the oxide film remaining after polishing for the combination of a SSP wafer and a stiff pad. The nanotopography matches perfectly when a stiff pad was used, as was previously reported [9), 10)].

The results obtained in our experiment are qualitatively summarized in Table 1, where the remaining film thickness variation of a few nanometers is described as "Almost no effect", and that of 10 nm is "Slight effect".

3.2. Impact of Chucking on Flatness

The SFQR vs. accumulation percentage curves (S-curves) in Figure 3 show that the sample wafers were classified into groups of four different flatness degrees. In each group, the amount of sample wafers was five for DSP, I-SSP, and C-SSP, respectively and two for U-SSP (I: Improved, and U: Ultra).

S-curves of the chuck flatness were given in Figure 4, showing that all chucks were found to be very flat because 95% of measured sites have SFQR values less than 65 nm.

Figure 5, 6, and 7 show the chucked-wafer flatness for all wafer groups, where wafers were vacuumed on the pin-chuck C_1 (Figure 5), on the C_2 (Figure 6), and on the C_3 (Figure 7), respectively. In Figure 7, each S-curve contains SFQR values of all chucked-wafers in each group.

From Figures 3, 5, 6, and 7, S-curves of the chucked-SSP wafer flatness are quite different from those of SSP wafer flatness. In the combinations of SSP wafers and chuck C_1 or C_3, four different flatness groups were observed, while in the combination of SSP-wafers and chuck C_2, only three different flatness groups were found. This results from the complex interaction between backside surface of SSP wafers and chuck C_2. Table 2 shows parameters of all chucks used in this study. The most remarkable difference among them is pin-to-pin distance, which is 2 to 2.2 mm in chuck C_1 and C_3, and 1 mm in chuck C_2. This suggests that backside topography of SSP wafers are characterized by some wavelength which is closely related with typical pin-to-pin distance. However, the mechanism should be studied further.

In the combination of SSP wafers and pin-chucks, the chucked-wafer flatness was better than wafer flatness itself (however, the chucked-wafer flatness of DSP wafers was still better than that of SSP wafers). Pin-chucking apparently improves the flatness of SSP wafers, probably due to the interaction between backside surface of SSP wafers and pin chucks.

Since, in Ultagage 9900, wafer flatness is mathematically calculated from the thickness variation of the wafer, the resulting flatness of SSP wafers is affected by the backside surface topography. In other words, in SSP wafers, the topography features of backside surface tend to be printed through to the all-over front-side surface, then affect the wafer flatness. On the other hand, the backside surface of a vacuumed wafer on the pin-chuck contacts only with pin top surfaces. Since a typical pin-to-pin distance and a pin-top dimension are 1 to 2.2 mm and 0.15 to 0.2 mm, respectively (Table 2), the contact ratio defined by

Contact ratio (%) = (contact area between pin tops and wafer backside surface / the total area of wafer backside surface) $\times$ 100, (1)

is less than 3%. Therefore, the topography features of the backside surface is by far less printed through to the front-side surface, less affect the wafer flatness, then it causes that the chucked-wafer flatness is better than wafer flatness.

4. Conclusion

Quantitative measurements of nanotopography and flatness were performed. The impact of nanotopography on CMP performance was simply evaluated after polishing blanket wafers deposited with CVD silicon-oxide films. The impact of chucking on the chucked-wafer flatness was evaluated through adopting an actual exposure tool chuck and wafers of different flatness degrees. Our conclusions are as follows:

(1) On DSP wafers polished with a soft or a medium-soft pad, the thickness variations of residual films are a few nanometers. So, the nanotopography of DSP wafers cause almost no problem in the oxide polishing of the STI, regardless of which pad is used.

(2) On DSP wafers polished with a stiff pad, the thickness profile was found to be around 10 nm, and slightly correlates to nanotopography. So, the combination of DSP wafers and a stiff pad has the risk to cause pattern failures in the oxide polishing of STI process.

(3) On SSP wafers polished with a soft or a medium-soft pad, the thickness profile was found to be around 10 nm, and clearly correlates to nanotopography features. So, the combination of SSP wafers and a soft or medium-soft pad is risky in resulting in pattern failures in the oxide polishing of STI process.

(4) On SSP wafers polished with a stiff pad, the nanotopography heavily affects the thickness variations of residual films, as reported previously.

(5) Chucked-wafer flatness is affected not only by wafer flatness but also by some chuck parameters, probably pin-to-pin distance. This indicates that chucked-wafer flatness depends on the interaction between wafer backside surface and chuck surface.

(6) Generally speaking, chucked-wafer flatness of SSP wafers is better than that of wafer flatness itself, as far as the flat pin-type chuck (SFQR<65 nm at 95% of sites) is employed.

(7) DSP wafers have the most excellent wafer flatness and chucked-wafer flatness.

Acknowledgement

The authors would like to thank many people, in particular, Mr. Masayuki Hashimoto [Applied Materials Japan], Mr. Takashi Kumagai [Applied Materials Japan], Dr. John F. Valley [ADE Phase Shift], Dr. Chris L. Koliopoulos [ADE Phase Shift], Mr. Atsuya Shinoda [Japan ADE], Mr. Hideo Takano [Canon Sales] for their great and devoted contribution to the CMP experiment, and nanotopography and flatness measurements.

Reference

1. P.C. Fazan and V. K. Mathews: IEDM Tech. Dig. (1993) 57.
2. A. Bryant, W. Hansch, and T. Mii: IEDM Tech. Dig. (1994) 671.
3. J. W. Lutze, A. H. Perera, and J. P. Krusius: J. Electrochem. Soc. **137** (1990) 1867.
4. S. Deleonibus, F. Martin, M. heitzmann, J. Ch. Guigert, and A. M. Papon: J. Electrochem. Soc. **144** (1997) L164.
5. J.-Y. Cheng, T. F. Lei, T. S. Chao, D. L. W. Yen, B. J. Jin, and C. J. Lin: J. Electrochem. Soc. **144** (1997) 315.
6. J. M. Boyd and J. P. Ellul: J. Electrochem. Soc. **144** (1997) 1838.
7. SEMI M43-0301 (Semiconductor Equipment and Materials International, San Jose, CA, 2001), p. 1.
8. M. M.-Roy, T. C. Huan, T .Y. Kwang, and G. S. Samudra: Prod. SPIE **4404** (2001) 14.
9. J.-G. Park, T. Katoh, H.-C. Yoo, and J.-H. Park: Jpn. J. Appl. Phys. **40** (2001) L857.
10. J.-G. Park, T. Katoh, H.-C. Yoo, D.-H. Lee, and U.-G. Park: Jpn. J. Appl. Phys. **41** (2002) L17.

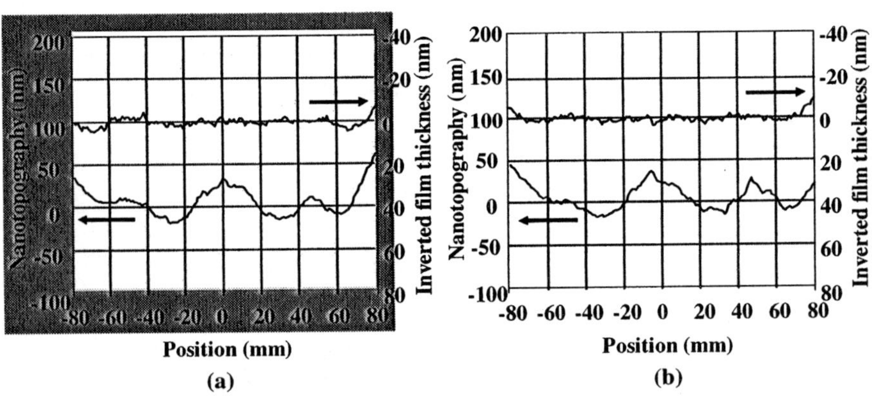

Figure 1. One dimensional nanotopography (lower curve) vs. inverted film thickness variation (upper curve) after oxide polishing. DSP and soft pad, (a), and DSP and medium-soft pad, (b).

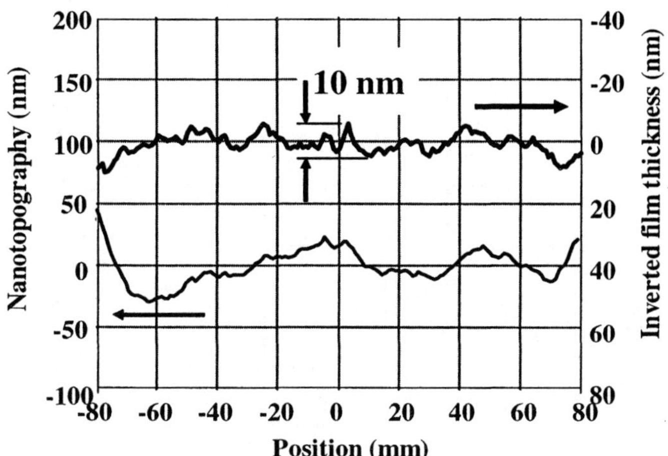

Figure 1 (c). One dimensional nanotopography (lower curve) vs. inverted film thickness variation (upper curve) after oxide polishing. DSP and stiff pad.

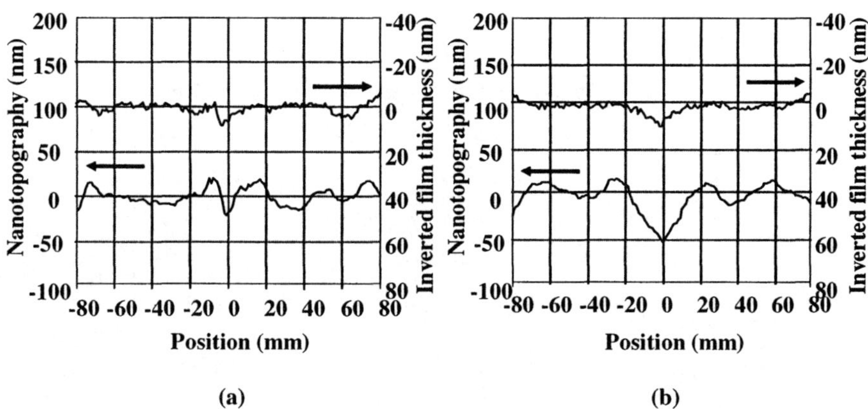

Figure 2. One dimensional nanotopography (lower curve) vs. inverted film thickness variation (upper curve) after oxide polishing. SSP and soft pad, (a), and

SSP and medium-soft pad, (b).

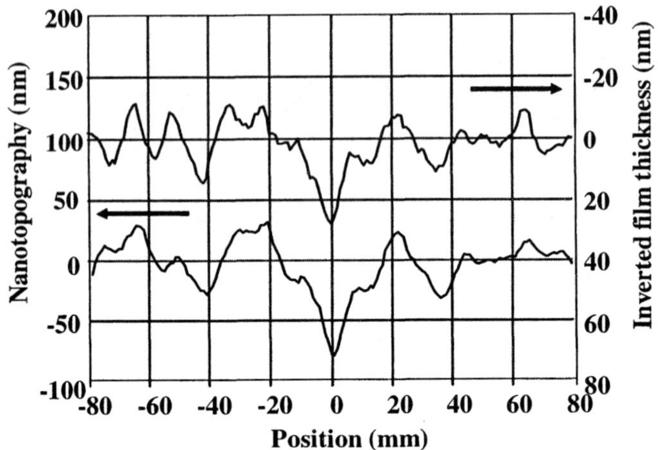

Figure 2 (c). One dimensional nanotopography (lower curve) vs. inverted film thickness variation (upper curve) after oxide polishing. SSP and stiff pad.

Table 1. The impact of nanotopography on oxide polishing

	DSP	SSP
Soft pad	Almost no effect	Slight effect
Medium-soft pad	Almost no effect	Slight effect
Stiff pad	Slight effect	Heavy effect

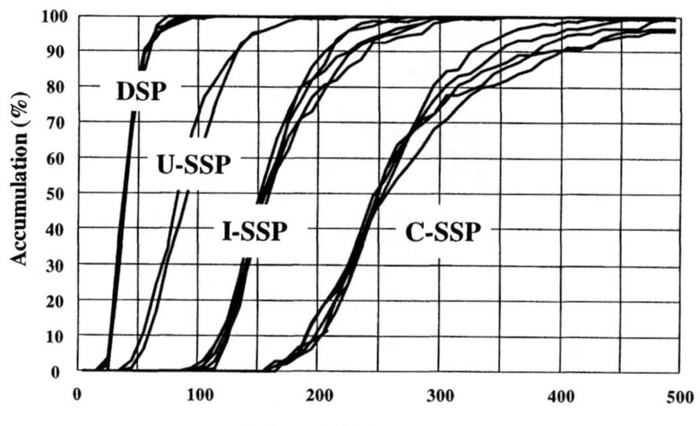

Figure 3. Wafer flatness.

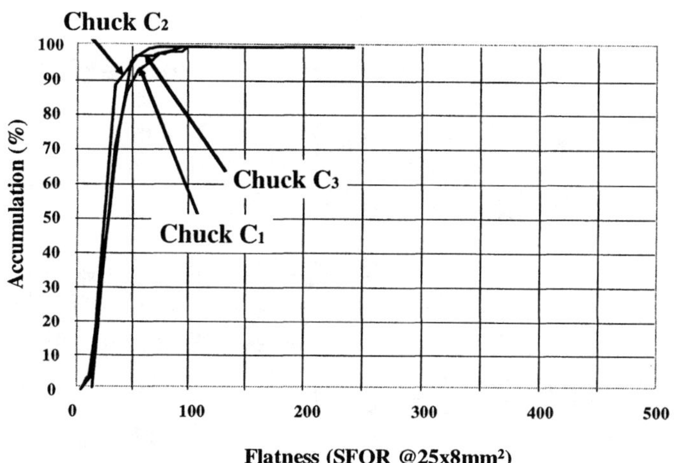

Figure 4. Chuck flatness.

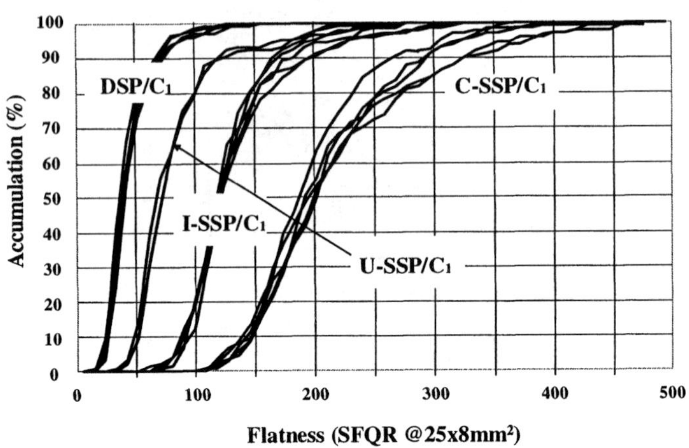

Figure 5. Chucked-wafer flatness (wafer / chuck C_1).

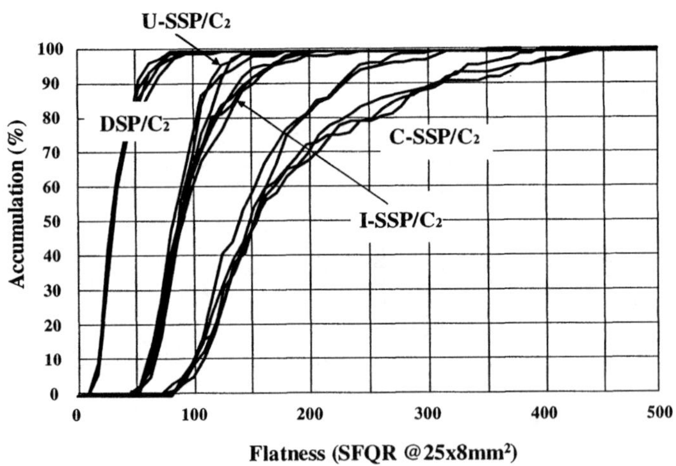

Figure 6. Chucked-wafer flatness (wafer / chuck C_2).

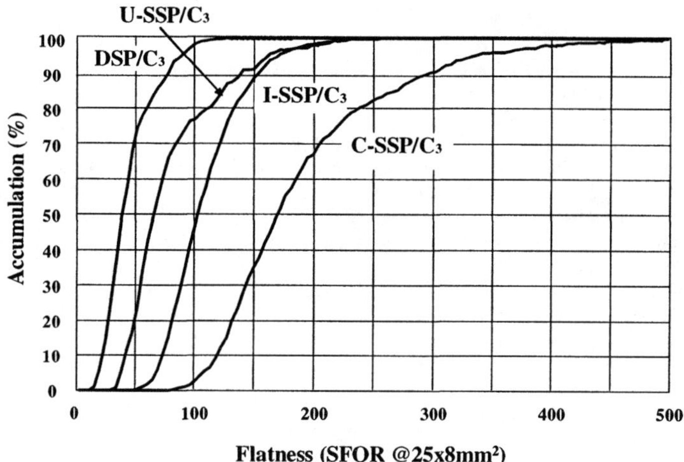

Figure 7. Chucked-wafer flatness (wafer / chuck C_3).

Table 2. Chuck parameters

Chuck	C_1	C_2	C_3
Type	Pin	Pin	Pin
Pin-to-pin distance (mm)	2	1	2.2
% of contact area (%)	0.56	3	0.6
Pin-top diameter (mm)	0.15	0.2	0.2
Flatness [@95% accumulation, SFQR(nm)]	65	50	50

GROWN-IN MICRODEFECTS IN SILICON AS A GUIDE TO THE PROPERTIES OF POINT DEFECTS

V.V.Voronkov and R.Falster*
MEMC Electronic Materials, via Nazionale 59, 39012 Merano BZ, Italy
*MEMC Electronic Materials, viale Gherzi 31, 28100 Novara 1, Italy

The concentration field of self-interstitials and vacancies in silicon depends on at least 10 parameters. But for the specific case of the growth of crystals deep in the interstitial mode, it is primarily sensitive to only the 3 parameters of the self-interstitial transport: the melting point diffusivity D_{im}, the migration energy E_{id} and the drift energy ε_i. A fit for these parameters can be obtained via the shape of A-swirl regions in quenched crystals. Using this approach, the resulting value for D_{im} is found to be about 1.5×10^{-4} cm^2/s (which considerably lower than a commonly assumed value) while ε_i is about 4 to 8 eV (a surprisingly strong uphill drift). This conclusion is still tentative and subject to some uncertainty related to a possible effect of carbon.

INTRODUCTION

Simulation of vacancy and self-interstitial distributions in silicon crystals is an important modern field of activity (1-3). In order to perform such simulations, values for many parameters of the intrinsic point defects are required. There are 5 basic constants for the self-interstitials: two for the equilibrium concentration (the melting point value C_{im} and the formation energy E_i), two for the diffusivity (the melting point value D_{im} and the migration energy E_{id}) and one for the drift velocity along the temperature gradient (the drift energy ε_i). Also, there are 5 similar constants for vacancies: C_{vm}, E_v, D_{vm}, E_{vd} and ε_v. The problem, in its simplest (and conventionally accepted) version – when a fast recombination of vacancies and self-interstitials is assumed – contains 10 parameters. None of them is well defined, in the current state of knowledge. If one chooses to consider a finite recombination rate, then two more parameters are required in order to describe the temperature dependence of the rate constant. We stick to the model of fast recombination, which is consistent with the V/G rule (3,4) i.e. that the type of intrinsic point defect incorporated into crystal locally – near the crystal-melt interface – is controlled by the ratio of the growth rate (V) to the local temperature gradient (G). Vacancies are incorporated if V/G is greater than some critical ratio $(V/G)_{cr}$ and self-interstitials are incorporated at $V/G < (V/G)_{cr}$.

In principle, the ten parameters listed above can be fit to experimental maps of grown-in microdefects produced by agglomeration of either vacancies or self-interstitials, using the constraints resulting from other available data, like self-diffusivity. However, the number of fitting parameters is too large to obtain definite and reliable values. The fitting problem is much simplified if one deals with a 'deep' interstitial growth mode, $V/G << (V/G)_{cr}$. The crystal is then in-flooded by self-interstitials that out-diffuse to the lateral crystal surface. The interstitial concentration field $C_i(r,z)$, where r and z are the radial and axial coordinates, respectively, is sensitive primarily to the interstitial transport constants which are D_{im}, E_{id} and ε_i. The number of fitting parameters is effectively reduced to three. To our knowledge, there is only one reported example of microdefect maps at very low V/G – in the paper by Roksnoer (5). The crystals of a small diameter (23 mm) were pedestal-pulled at a constant low V, and then quenched. In the course of cooling, the A-swirl-microdefects (dislocation loops of interstitial type) were formed in a region where the self-interstitial concentration was sufficiently high for that - above some critical value C_{cr}. In other words, the A-swirl region is confined by an iso-concentration contour, $C_i(r,z) = C_{cr}$. The fitting task is to simulate $C_i(r,z)$ for different sets of D_{im}, E_{id} and ε_i and select the set that gives the best description of the A-swirl regions.

CONSTRAINTS IMPOSED ON THE COMPLETE PARAMETER SET

Although the concentration field $C_i(r,z)$, in a deep interstitial growth mode, is controlled primarily by the three above-mentioned transport parameters, the 7 other parameters are also needed to perform the simulation. There are several constraints that yield these values approximately. And for the present purposes approximately is enough, due to a low sensitivity of the computed concentration field to all the parameters but D_{im}, E_{id} and ε_i. These constraints are listed below.

1. There is an approximate (but quite precise) analytical expression for $(V/G)_{cr}$ through the defect constants (3):

$$(V/G)_{cr} = (1/kT_m^2)\,[D_{im}C_{im}\,(E - \varepsilon_i) - D_{vm}C_{vm}\,(E - \varepsilon_v)]\,/\,(C_{vm} - C_{im}) \quad , \qquad [1]$$

where T_m is the melting temperature and $E = (E_i + E_v) / 2$ – the averaged formation energy. The experimental estimates for $(V/G)_{cr}$ are within a range 0.12 to 0.2 $mm^2/minK$ (1-4). We adopt the value of 0.14 $mm^2/minK$ (see the section discussing the temperature gradient in pedestal-pulled crystals).

2. The concentration difference, $C_{vm} - C_{im}$, can be determined (3) using the reported data (6) on the total amount of vacancies stored in voids (for a deep vacancy growth mode). There is some scatter in this estimate, and we choose the lower limit (which is most suitable to reconcile it with expression [1]) :

$$C_{vm} - C_{im} = 1.6\text{x}10^{14}\ \text{cm}^{-3} \quad . \qquad [2]$$

3. The equilibrium concentration difference, $C_{ve}(T) - C_{ie}(T)$, was measured (7), in a range of high T, in samples that received a Rapid Thermal Annealing. With the constraint [2], the experimental curve may be fitted using any value of E_i within an expected range from 3 to 5 eV (8), by selecting a proper value for the vacancy formation energy E_v. The relation between the two formation energies is well approximated by a parabolic law:

$$E_v = a + b\,E_i + c\,E_i^2 \quad , \qquad [3]$$

with a = - 0.482, b = 1.463, c = - 0.0914. The difference $E_i - E_v$ is relatively small, in the order of 0.2 eV, and it increases upon increasing E_i .

4. Self-diffusion and impurity diffusion data (9-11) give information about the diffusivity-concentration products, D_iC_{ie} and D_vC_{ve} (the subscript e indicates to the equilibrium value of the concentration). Most of the data refer to the temperature range around 1000°C, where there is an agreement about the values of the products: $D_iC_{ie} \approx D_vC_{ve} \approx 4x10^6$ $cm^{-1}s^{-1}$. However there is a remarkable discrepancy between the works (9,10) and (11) concerning the activation energies E_{is} and E_{vs} of the two self-diffusion products, D_iC_{ie} and D_vC_{ve} – and therefore a remarkable difference in the values of the two products extrapolated to the melting point, $D_{im}C_{im}$ and $D_{vm}C_{vm}$.

According to (9,10), the activation energy is high for self-interstitial (E_{is} = 4.95 eV) and considerably smaller for vacancy (E_{vs} = 4.11 eV) which results in a low $D_{vm}C_{vm}/D_{im}C_{im}$ ratio at the melting point ($D_{im}C_{im} = 2.4x10^{11}$ $cm^{-1}s^{-1}$, $D_{vm}C_{vm} = 1.85x10^{10}$ $cm^{-1}s^{-1}$). Since $C_{vm} > C_{im}$, the vacancy diffusivity D_{vm} is much lower than the self-interstitial diffusivity D_{im}. This concept was generally adopted in simulation works (1-3), though somewhat larger values of $D_{vm}C_{vm}$ and E_{vs} were used.

According to (11), the two activation energies are close one to the other: E_{vs} = 4.86 eV, E_{is} = 4.68 eV. The activation energy is slightly higher for the vacancy product, quite contrary to the previous result. Accordingly, the $D_{vm}C_{im}/D_{im}C_{im}$ ratio is larger than unity ($D_{vm}C_{vm} \approx 1.9x10^{11}$ $cm^{-1}s^{-1}$, $D_{im}C_{im} \approx 1.1x10^{11}$ $cm^{-1}s^{-1}$). The vacancy diffusivity D_{vm} is then slightly higher than the self-interstitial diffusivity D_{im}. However, $D_v(T)$ becomes lower than $D_i(T)$ upon lowering T, since the vacancy migration energy E_{vd} turns out to be larger than E_{id}.

The reason for the vacancy to self-interstitial change-over, upon reducing V/G, is quite different for the two above cases. The vacancy mode at higher V/G results from the inequality $C_{vm} > C_{im}$, in both cases. But the reason for the interstitial mode at lower V/G is different.

With $D_{im}C_{im} >> D_{vm}C_{vm}$ (BSM-type relation, by the initial letters of the authors (9)), the self-interstitial mode is realized due to a fast diffusion of self-interstitials, from the interface into the crystal bulk. The expression [1] is consistent with the experimental values for $(V/G)_{cr}$ – but only if the drift energy ε_i is zero – or at least considerably smaller than E (or if $\varepsilon_i < 0$ which means downhill drift of self-interstitials).

With $D_{vm}C_{vm} > D_{im}C_{im}$ (UGP-type relation (11)), the self-interstitial mode can be realized only due to a strong uphill drift of vacancies (ε_v is positive and larger than E). Then the vacancies, though faster diffusers than self-interstitials, can not penetrate into the crystal bulk at low V/G, because they drift back to the interface. The expression [1] for $(V/G)_{cr}$ can produce the correct number at any value of the self-interstitial drift energy ε_i , by a proper choice of the vacancy drift energy ε_v. For this reason, this case is more flexible than the previous one.

We will try both types of settings for the diffusivity-concentration products, using slightly corrected values (consistent with the above-mentioned value for D_iC_{ie} and D_vC_{ve} at 1000°C and with the reported self-diffusivity (10)) :

$D_{im}C_{im} = 2.3x10^{11}\ cm^{-1}s^{-1}$, $E_{is} = 4.95$ eV; $D_{vm}C_{vm} = 4x10^{10}\ cm^{-1}s^{-1}$, $E_{vs} = 4.25$ eV , [4a]

$D_{im}C_{im} = 1.0x10^{11}\ cm^{-1}s^{-1}$, $E_{is} = 4.6$ eV; $D_{vm}C_{vm} = 2x10^{11}\ cm^{-1}s^{-1}$, $E_{vs} = 4.9$ eV . [4b]

For any specified values of the three fitting parameters D_{im}, E_{id} and ε_i , and for the chosen self-diffusivity parameters (either [4a] or [4b]) the other seven parameters are defined by the following procedure:

1) The interstitial formation energy E_i equals $E_{is} - E_{id}$.
2) The vacancy formation energy E_v is a function of E_i defined by Eq.[3].
3) The vacancy migration energy E_{vd} equals $E_{vs} - E_v$.
4) The self-interstitial concentration (at T_m) is $C_{im} = (D_{im}C_{im}) / D_{im}$.
5) The vacancy concentration (at T_m) is $C_{vm} = C_{im} + (C_{vm} - C_{im})$.
6) The vacancy diffusivity (at T_m) is $D_{vm} = (D_{vm}C_{vm}) / C_{vm}$.
7) The vacancy drift energy, ε_v, is calculated from the Eq.[1] using the adopted value for the critical ratio, $(V/G)_{cr} = 0.14\ mm^2/minK$.

TEMPERATURE FIELD IN PEDESTAL-PULLED CRYSTALS

Before turning to simulation of the concentration fields, one should know the temperature field in the crystal, T(r,z). At low V, one can neglect the convection heat flux and apply an equation

$$\mathrm{div}(\chi \mathrm{grad} T) = 0 \quad , \qquad [5]$$

where $\chi(T)$ is the heat conductivity; it is approximately proportional to 1/T and can be presented as $\chi = \chi_m T_m / T$ where χ_m refers to the melting point. It follows from Eq.[5] that the function $U = \log(T_m/T)$ satisfies the Laplace equation $\Delta U(r,z) = 0$ where Δ stands for the Laplace operator. For this reason, the 'temperature potential' U is useful to represent the temperature field. It is also convenient to use the coordinates normalized by the crystal radius R ($\rho = r/R$ and $\zeta = z/R$). A simple analytical form for $U(\rho,\zeta)$ is obtained if the potential profile at the crystal surface (at $\rho = 1$), is represented by a polynomial

$$U(1,\zeta) = \Sigma P_k \zeta^k \quad , \qquad [6]$$

where k is from 1 to n. As it turns out, four terms (n=4) are enough for a good approximation. The axial distance z is counted from the interface edge so that U(1,0) = 0.

Next, we construct a polynomial $U_k(\rho,\zeta)$ that satisfies the Laplace equation and coincides with ζ^k at the crystal surface. It is easily done be starting with the ζ^k term, then adding a term ζ^{k-2} $(\rho^2 - 1)$ to compensate for $\Delta(\zeta^k) = k(k-1)\zeta^{k-2}$, and so on. A useful relation is $\Delta(\rho^k) = k^2 \rho^{k-2}$. The result is

$$U_1(\rho,\zeta) = \zeta, \quad U_2(\rho,\zeta) = \zeta^2 - (\rho^2-1)/2, \quad U_3(\rho,\zeta) = \zeta^3 - 3\,\zeta\,(\rho^2-1)/2,$$
$$U_4(\rho,\zeta) = \zeta^4 - 3\,\zeta^2\,(\rho^2-1) + 3\,(\rho^4-1)/8 - 3\,(\rho^2-1)/2. \quad [7]$$

In this way, we obtain a particular solution for the temperature potential:

$$U_p(\rho,\zeta) = \Sigma\, P_k\, U_k(\rho,\zeta) \quad , \quad [8]$$

that coincides with the profile [6] at the crystal surface.

At the crystal surface, the radial heat flux, $-\chi\, \partial T/\partial r$, should be balanced by the irradiated heat, σT^4, where σ is the Stefan-Boltzmann radiation constant multiplied by the gray-body emissivity. In terms of the potential U, and for the normalized radial coordinate, this boundary condition reads

$$\partial U/\partial \rho = [\sigma\, T_m^3\, R / \chi_m]\, \exp(-4\, U) \quad . \quad [9]$$

The parameter combination within the square brackets is equal to 0.11, for the thermal parameters conventionally used in temperature simulations (12) – with the emissivity 0.7 and the conductivity $\chi_m \approx 0.2$ W/cmK.

The polynomial solution [8] is not yet the final result since it is not completely consistent either with the boundary condition [9] or with the experimental shape of the crystal/melt interface. To achieve these requirements, an additional potential should be introduced. This potential, denoted by $U_b(\rho,\zeta)$, should be zero at the crystal surface - where the polynomial part $U_p(\rho,\zeta)$ already reproduces the surface axial profile [6] – and thus it can be expanded into the Bessel series:

$$U_b(\rho,\zeta) = \Sigma\, B_k\, J_o(\mu_k \rho)\, \exp(-\mu_k \zeta) \quad . \quad [10]$$

where J_o is the Bessel function of zero order and μ_k are the roots of this function (μ_1 = 2.40482; μ_2 = 5.52009).

The whole temperature potential U is the sum of the polynomial part U_p and the Bessel part U_b. The interface is defined as the surface $\zeta(\rho)$ at which $U(\rho,\zeta) = 0$. We use only two first terms (k=1 and k=2) in the series [10] to meet the two most important requirements:

1) to get the strict flux balance [9] at the interface edge (ρ=1, ζ=0).

2) to get the central interface deflection $\zeta(0)$ the same as that observed experimentally, about 0.1. The actual interface was exposed in ref.(5) by detaching the crystal from the melt; it is of a concave shape, and one parameter (central deflection) is enough to reproduce the shape.

The coefficients B_1 and B_2 are defined by these two requirements, at any specified coefficients P_k. In the final stage of temperature computation, the polynomial coefficients P_k were selected in such a way as to best satisfy the boundary condition [9]. The criterion was the averaged squared difference between the left-hand part and the right-hand part of the balance equation [9]. The smallest difference, less than 0.1%, was achieved at the following set of the coefficients: $P_1 = 0.3191$, $P_2 = -0.08655$, $P_3 = 0.0187$, $P_4 = -0.001781$, and these values were used to compute the temperature potential $U(\rho,\zeta)$, and thus the temperature itself, $T(\rho,\zeta) = T_m \exp(-U)$. The most important feature of the temperature field is the variation of the axial gradient G along the interface; this is shown in Fig.1. The average value of G is about 47.4 K/mm. It is close to the value calculated within the 1D model where T is considered as a function of z only, and the axial heat flux, $-\pi R^2 \chi\, dT/dz$, changes due to the heat loss from the lateral surface, $2\pi R\, \sigma T^4$. The gradient is then $G = (\sigma T_m^5 / R\, \chi_m)^{1/2} = 48.4$ K/mm. The 1D approximation allows also to take into account the convection heat flux, $V\, cT\, \pi R^2$ (where c is the heat capacity per unit volume). The corresponding correction to the gradient was found to be $-V\, cT_m / (3\, \chi_m)$. The interstitial to vacancy change-over in pedestal-pulled crystals occurs at V = 6 mm/min (5). At this relatively high growth rate, the correction to G is appreciable, - 5.6 K/mm, and the gradient is expected to be about 42.8 K/mm. The resulting estimate for the $(V/G)_{cr}$ is then 0.14 mm^2/minK – the value adopted in the present work.

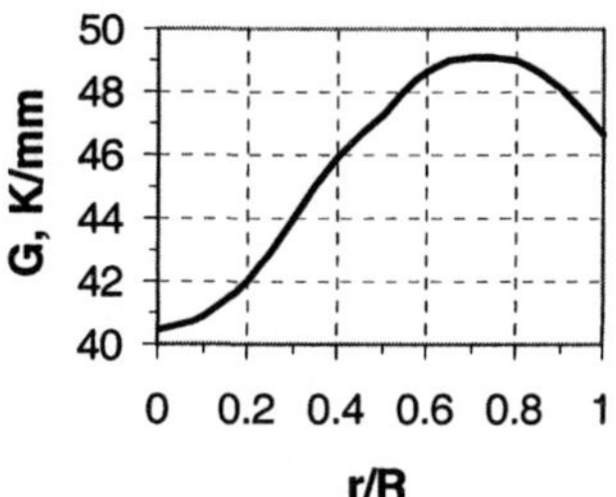

Fig.1 Calculated variation of the axial temperature gradient G along the interface, in pedestal-pulled crystal grown at low rate (crystal radius is R = 1.15 cm).

STEADY-STATE CONCENTRATION FIELD

Due to fast recombination, the self-interstitial concentration C_i and the vacancy concentration C_v are related by the equilibrium mass-action law:

$$C_i(r,z)\, C_v(r,z) = C_{ie}(T)\, C_{ve}(T) \quad , \qquad [11]$$

where the temperature T is a function of r and z that was defined above. First the crystal is grown in a steady-state mode, and the steady-state concentration fields are established. Then the crystal is quenched, and the two concentrations, C_i and C_v, decrease due to further recombination; the remaining concentration of self-interstitials will be identical to the initial (steady-state) concentration difference $C = C_i - C_v$. This difference is thus a quantity of interest, and it will be used as the primary concentration field. It is the iso-concentration contour of C(r,z) – rather than of $C_i(r,z)$ – that should be computed to delineate the A-swirl region. The self-interstitial concentration C_i is expressed through C from Eq. [11] :

$$C_i = C/2 + [C^2/4 + C_{ie}\, C_{ve}]^{1/2} \quad , \qquad [12]$$

and C_v is then expressed through C_i.

Generally – in a non-steady-state growth mode – the concentration C is changing with time due to the self-interstitial and vacancy fluxes, $\mathbf{J}_i$ and $\mathbf{J}_v$:

$$\partial C/\partial t = -\,\mathrm{div}(\mathbf{J}_i) + \mathrm{div}(\mathbf{J}_v) \quad . \qquad [13]$$

Each flux is composed of the three basic terms: Fickian diffusion, transportation by a moving crystal (convection) and drift along the temperature gradient. The self-interstitial flux vector is

$$\mathbf{J}_i = -\,D_i\,\mathrm{grad}\,C_i + \mathbf{V}\,C_i - D_i\,C_i\,\varepsilon_i\,\mathrm{grad}\,(1/kT) \quad . \qquad [14]$$

The last term (the drift flux) is based on a generally accepted assumption: the drift velocity is proportional to the gradient of the inverse temperature and to the diffusivity. The drift energy ε_i is just a kinetic coefficient that can be, in principle, of any value and sign. The positive sign means an uphill drift (from a colder to a hotter crystal part). The vacancy flux vector $\mathbf{J}_v$ is of the same form (only the index i should be replaced for v). The velocity vector $\mathbf{V}$ is along the axis z (the pulling direction). It is convenient, for subsequent use, to define the temperature-gradient vector $\mathbf{g} = \mathrm{grad}(1/kT)$. It can be expressed through the gradient of the temperature potential: $\mathbf{g} = (1/kT)\,\mathrm{grad}\,U$.

We are interested here in the steady state case, $\partial C/\partial t = 0$. The steady-state distribution can be obtained by relaxing some arbitrary initial field C(r,z) until the derivative $\partial C/\partial t$ becomes reasonably low.

The right-hand part of the decay equation [13] is convenient to express through the derivatives of the concentrations. The term $\mathrm{div}\mathbf{J}_i$ takes the form

$$\mathrm{div}\mathbf{J}_i \;=\; -\,D_i\,\Delta C_i \;+ [\mathbf{V} + D_i\,(E_{id} - \varepsilon_i)\,\mathbf{g}]\,\mathrm{grad}\,C_i + \varepsilon_i\,(E_{id} - kT)\,D_i\,C_i\,\mathbf{g}\,\mathbf{g} \quad . \qquad [15]$$

In derivation of this expression, a relation $\Delta(1/kT) = kT\,\mathbf{g}\,\mathbf{g}$ was used; it follows from the Laplace equation for $U = \log(T_m/T)$. The term $\mathrm{div}\mathbf{J}_v$ has a form similar to Eq.[15]. The first and second derivatives of C_i and C_v are expressed through the discrete values of the two concentrations, at the nodes of a rectangular grid.

The equilibrium boundary condition ($C_i = C_{ie}$, $C_v = C_{ve}$) were assumed both at the interface and at the lateral crystal surface. This corresponds to an ideal sinking of the point defects at these surfaces.

COMPUTATION RESULTS

The near-interface A-swirl regions were observed (5) in two crystals grown at V_1=0.5 mm/min and at V_2=0.2 mm/min (crystal 1 and crystal 2, respectively). These regions are reproduced by the lines in Fig.2. Their shape qualitatively corresponds to the out-diffusion of self-interstitials to the lateral crystal surface. At very low pull rate (Fig.2b) the A-defect region is narrow; it strongly indicates to an appreciable uphill drift of the self-interstitials that keeps them close to the interface. At a higher pull rate (Fig.2a), the self-interstitials are transferred by a moving crystal further into the crystal body, and the A-region becomes essentially larger. For the crystal 1, the boundary of the A-swirl region is somewhat diffuse in the central part (at ρ from 0 to 0.5); a tentative shape of this portion is shown by the dashed line in Fig.2a.

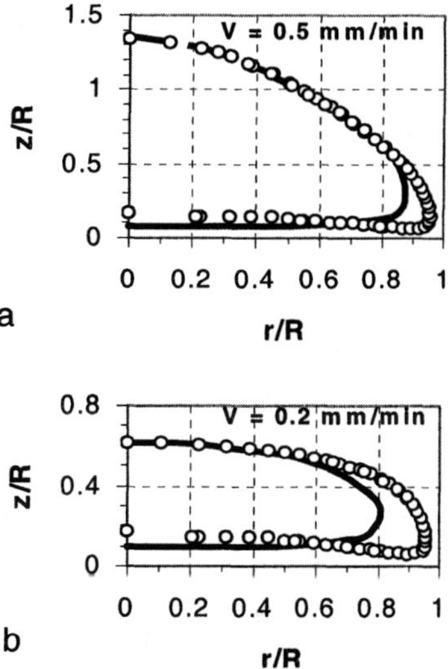

Fig.2 Experimental (lines) and simulated (open circles) boundary of A-swirl-region in pedestal-pulled and quenched crystals. The parameters correspond to the set 1 of Table 1.
a: in crystal 1 (0.5 mm/min);
b: in crystal 2 (0.2 mm/min)

The adopted fitting sequence was as follows. The interstitial migration energy, E_{id}, was first fixed, within a reasonable range from 0.2 to 1 eV. Then the interstitial drift energy ε_i was scanned within some range. For every ε_i, different values for the diffusivity D_{im} were tried. For each of them, the iso-concentration contour was drawn through a representative point at the A-swirl boundary in the crystal 1. This point was chosen at ρ=0.537, ζ=1.007 (just outside the diffuse portion of the boundary). The contour shape is very sensitive to D_{im}, and practically a unique value of D_{im} could be selected to best describe the most reliable portion of the A-swirl boundary (between ρ=0.5 and ρ=0.8). This value of D_{im} was then used to compute an iso-concentration contour in the crystal 2; this one was drawn through the point located at the center of the boundary (ρ=0, ζ=0.61). By this procedure, the concentration C for each contour is defined: C_1 for the crystal 1, C_2 for the crystal 2. Each of these two values should be identical to the critical concentration C_{cr} for the formation of A-swirls in the course of quenching. Therefore the parameter set is consistent with the experiment only if it leads to $C_1 = C_2$. As an example, the computed values of C_1 and C_2 (normalized by C_{im}) are plotted in Fig.3 in dependence of the assumed value of ε_i, for the UGP-type setting [4b] and for the low value of the migration energy, E_{id} = 0.2 eV. The criterion $C_1 = C_2$ is fulfilled at ε_i = 4.6 eV which means a strong uphill drift of self-interstitials; the corresponding value of the vacancy drift energy is still larger, ε_v= 9 eV.

The complete parameter set for this case is presented in Table 1 (indicated by the number 1). The computed iso-concentration contours in Fig.2 (shown by open circles) correspond just to this case. They reproduce the A-swirl regions satisfactorily, except for the portions near the crystal surface. A reason for this discrepancy may be a deviation of the actual sinking law at the surface from the assumed ideal sinking, $C_i = C_{ie}$.

Table 1. Some best-fit parameter sets for the self-interstitial and vacancy.

Set	C_{im} / 10^{14} cm^{-3}	E_i , eV	D_{im} / 10^{-4} cm^2/s	E_{id} , eV	ε_i , eV	C_{vm} / 10^{14} cm^{-3}	E_v , eV	D_{vm} / 10^{-4} cm^2/s	E_{vd} , eV	ε_v , eV
1	7.3	4.4	1.37	0.2	4.6	8.9	4.19	2.25	0.71	9.01
2	6.67	4	1.5	0.6	6.3	8.27	3.91	2.42	0.99	9.69
3	6.25	3.6	1.6	1	8	7.85	3.6	2.55	1.3	10.36
4	2.3	4	10	0.95	1.5	3.9	3.91	1.28	0.34	10.91

Similar fitting procedure for other values of E_{id} gives similar results: a strong uphill interstitial drift is required to meet the condition $C_1 = C_2$, and the computed contours are very similar to those shown in Fig.2. However, the deviation of the computed contours from the experimental ones increases upon increasing E_{id}. The parameter sets for E_{id} = 0.6 eV and 1 eV are shown in Table 1 (indicated by number 2 and 3, respectively). Though the set 1 of the Table 1 seems to be preferable (it results in a somewhat better fit) the present procedure can not define a certain value for the interstitial migration energy. However, quite narrow ranges for the melting point diffusivity D_{im} (around 1.5×10^{-4} cm^2/s) and for the drift energy ε_i (around 6 eV) follow from such a fit based on the UGP-type setting [4b].

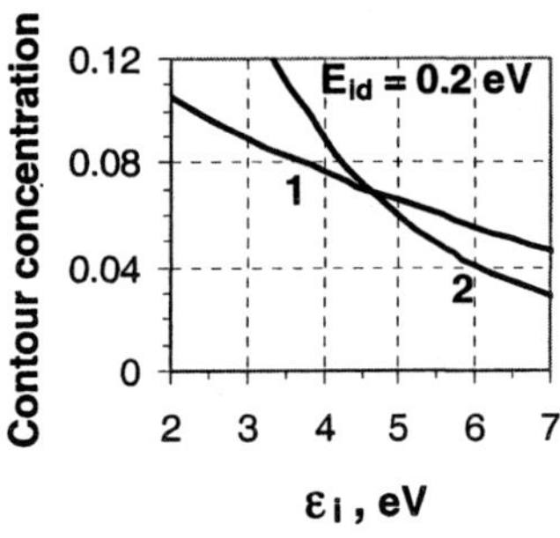

Fig3 Computed dependence of the contour concentrations C_1 and C_2 (for the crystals 1 and 2, respectively) on the assumed drift energy ε_i. The setting for the self-diffusivity products is that by Eq.[4b]; the interstitial migration energy is E_{id} = 0.2 eV.

For the BSM -type setting [4a], a similar fitting procedure fails: though a good fit to the shape of the A-swirl boundaries can be found - by a proper choice of ε_i and D_{im} at specified migration energy E_{id} - the contour values C_1 and C_2 are essentially different. It does not necessarily mean that the BSM setting should be rejected. The point is that self-interstitial agglomeration was reported to be sensitive to the carbon impurity (5). The critical concentration C_{cr} for A-swirl formation may then depend on the carbon content. The two crystals 1 and 2 may be of not identical carbon content, and the requirement of $C_1 = C_2$ is then in doubt. An example of parameter set consistent with only the contour shapes is listed in Table 1 as number 4. The corresponding contours are shown in Fig.4; here $C_1 = 0.123$ while $C_2 = 0.265$. The vacancy migration energy should be not lower than 0.35 eV (3). This condition is violated at $E_{id} < 0.95$ eV. On the other hand, much higher values for E_{id} are not likely, and the present choice for E_{id} is thus justified. With this value, the interstitial drift energy is practically fixed at 1.5 eV: lower values lead to an unreasonably high diffusivity D_{im} - much higher than 10^{-3} cm^2/s - while larger values of ε_i lead to unreasonably large vacancy drift energy - much larger than 10 eV. For these reasons, the parameter set 4 of the Table 1 is, practically, the only one consistent with the BSM setting for the self-diffusivity products.

Note that if the constraint $C_1 = C_2$ were ignored also in the previous consideration of UGP-type setting, the quality of fit would be improved, in a range of moderate drift energy, 1.5 to 4 eV. The computed contours would be similar to those shown in Fig.4. The diffusivity D_{im} would be then increased.

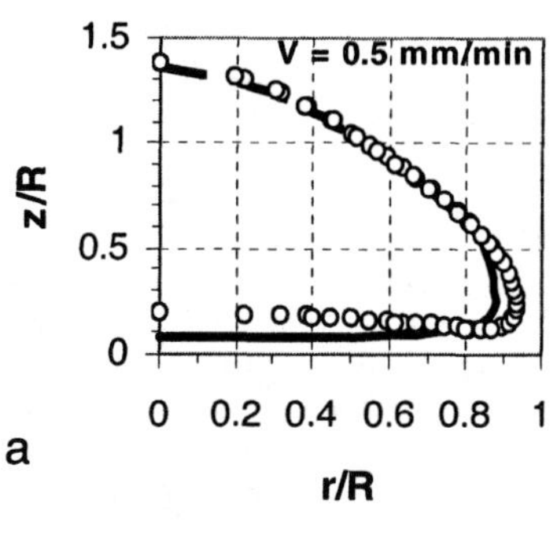

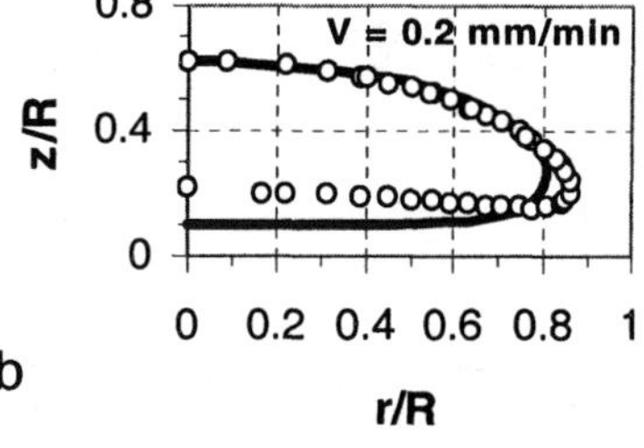

Fig.4 Experimental (lines) and simulated (open circles) boundary of A-swirl region, for the parameter set 4 of Table 1.
a: in crystal 1 (V = 0.5 mm/min)
b: in crystal 2 (V = 0.2 mm/min)

CONCLUSIONS

A 'deep' interstitial growth mode - realized at $V/G << (V/G)_{cr}$ – is useful to deduce the parameters of intrinsic point defects in silicon. By quenching a crystal - grown in a steady-state mode - a particular iso-concentration contour of the concentration field C(r,z) (the difference of the interstitial and vacancy concentrations) is delineated as a boundary of A-swirl-microdefects formed in the course of quenching. The observed shapes of this boundary impose a strong constraint on the choice of the parameters.

The three most relevant parameters are the melting point diffusivity D_{im}, the migration energy (the activation energy for diffusion) E_{id} and the drift energy ε_i. These were the fitting parameters of the problem while the other (less relevant) seven parameters of point defects were calculated using available data on the point defect properties. Particularly, the diffusivity-concentration products, $D_{im}C_{im}$ and $D_{vm}C_{vm}$ (for self-interstitial and for vacancy, respectively) are important for that purpose. Two different settings for these products were tried. The first is based on the metal (especially, Zn) diffusion data (9). The second is based on the self-diffusion and diffusion of phosphorus and antimony under non-equilibrium conditions (11). The essential difference between these two settings is that the $D_{vm}C_{vm}/D_{im}C_{im}$ ratio is very small in the former case but larger than 1 in the latter case.

Assuming the universal value for the critical concentration C_{cr} (for formation of A-swirls), the quenched A-swirl regions can be satisfactory explained only for the second setting - with $D_{vm}C_{vm}/D_{im}C_{im} > 1$. The migration energy E_{id} could not be definitely selected by this fit though lower values (like 0.2 eV) seem preferable. However definite (and surprising) conclusions can be drawn concerning the diffusivity D_{im} and the drift energy ε_i. The fitted diffusivity D_{im} turns out to be close to $1.5x10^{-4}$ cm^2/s - remarkably smaller than the commonly assumed value of about $5x10^{-4}$ cm^2/s (1-3). The drift energy ε_i was often assumed to be negligible (1-3), but the present work indicates to a relatively large value, around 6 eV - greater than the formation energy. The corresponding drift velocity, ε_i D_i grad(1/kT), is not however very large: it is in the order of 0.8 mm/min in pedestal-pulled crystals of small diameter (with very high temperature gradient) and in the order of 0.07 mm/min in modern Czochralski crystals where the gradient is much smaller. The vacancy drift is somewhat faster, by a factor of about 2.

These conclusions, if correct, change significantly the commonly adopted view of the high-temperature properties of point defects in silicon. However, a conventional view (small ε_i , large D_{im} – consistent with the setting of ref.(9)) can not be discarded, considering possible complications due to the carbon impurity. To make a final choice, more microdefect maps should be examined using the representative parameter sets of the type 1 and 4 of Table 1.

REFERENCES

1. T.Sinno, R.A.Brown, E.Dornberger and W. von Ammon, *J.Electrochem.Soc.*, 145, 303 (1998).
2. K.Nakamura, T.Saishoji, J.Tomioka, in *Semiconductor Silicon 2002, vol.2*, H.R.Huff, L.Fabry and S.Kishino, Editors, PV 2002-2, p.554, The Electrochemical Society Proceeding Series, Pennington NJ (2002).
3. V.V.Voronkov and R.Falster, *J.Appl.Phys.*, 204, 463 (1999).
4. V.V.Voronkov, *J.Crystal Growth*, 59, 625 (1982).
5. P.J.Roksnoer, *J.Crystal Growth*, 68, 596 (1984).
6. T.Saishoji, K.Nakamura, H.Nakajima, T.Yokoyama, F.Ishikawa and J.Tomioka, in *High Purity Silicon V*, C.L.Claeys, P.Rai-Choudhury, M.Watanabe, P.Stallhofer and H.J.Dawson, Editors, PV 98-13, p.28, The Electrochemical Society Proceeding Series, Pennington NJ (1998).
7. R.Falster, V.V.Voronkov and F.Quast, *Phys.Stat.Sol. (b)*, 222, 219 (2000).
8. T.Sinno, in *Semiconductor Silicon 2002, vol.1*, H.R.Huff, L.Fabry, S.Kishino, Editors, PV 2002-2, p.212, The Electrochemical Society Proceeding Series, Pennington NJ (2002).
9. H.Bracht, N.A.Stolwijk and H.Mehrer, *Phys.Rev. B*52, 16542 (1995).
10. H.Bracht, E.E.Haller and R.Clark-Phelps, *Phys.Rev.Letters*, 81, 393 (1998).
11. A.Ural, P.B.Griffin and J.D.Plummer, *Phys.Rev.Letters*, 83, 3454 (1999).
12. E.Dornberger, E.Tomzig, A.Seidl, S.Schmitt, H.J.Leister, C.Schmitt and G.Muller, *J.Crystal Growth*, 180, 461 (1997).

DYNAMICS OF POINT DEFECTS AND FORMATION OF MICRODEFECTS IN CZOCHRALSKI CRYSTAL GROWTH: MODELING, SIMULATION AND EXPERIMENTS

Milind S. Kulkarni[a], Vladimir Voronkov[b] and Robert Falster[c]

MEMC Electronic Materials, [a]St. Peters, MO 63376 USA, [b]39012 Merano BZ, Italy, [c]28100 Novara 1, Italy

ABSTRACT

Most common microdefects in Czochralski silicon, voids and dislocation loops, are formed by agglomeration of point defects, vacancies and self-interstitials, respectively. Dynamics of formation and growth of the microdefects along with the entire crystal pulling process is simulated. The Frenkel reaction, the transport and the nucleation of the point defects and the growth of the microdefects are considered to occur simultaneously. The nucleation is modeled using the classical nucleation theory. The diffusion-limited growth of the nucleated precipitates is assumed. The microdefect distribution at any given location is captured on the basis of the formation history of nuclei. The microdefect type and size distributions in crystals grown under steady state as well as unsteady state are predicted. The surface energies for voids and interstitial clusters are determined using experimental results. The model predictions agree very well with the experimental results. Various predictions of the model are presented and the results are discussed.

INTRODUCTION

The crystals grown by both the *Czochralski* process (CZ-silicon), in which a crystalline ingot is continuously pulled from the melt in a quartz crucible, and the *Float Zone* process (FZ-silicon), in which a molten zone traverses through a polycrystalline ingot to allow formation and growth of monocrystalline silicon, inherently contain many crystallographic imperfections or defects, popularly known as *microdefects* or *grown-in* defects. Prior to 1960s, crystals contained dislocations induced by thermo-mechanical stresses near the vicinity of the melt/crystal interface. A major breakthrough by Dash (1958, 1959) allowed crystal growth without the thermo-mechanically induced dislocations (1,2).

In crystals free of thermo-mechanically induced dislocations, various other microdefects can form by agglomeration of point defects that exist as solutes in the silicon matrix. There are two basic types of point defects in silicon — vacancies, each of which is formed by a missing silicon atom from the lattice, and self-interstitials (or simply, interstitials), which are interstitial silicon atoms not bonded with the atoms forming the lattice (3). Voronkov (1982) described the conditions leading to the formation of microdefects in both the CZ- and the FZ- crystal growth, on the basis of interplay between the transport of the point defects from the melt/crystal interface and the Frenkel pair reaction (4). The temperature inside a growing crystal sharply drops with the increasing distance from the melt/crystal interface, which decreases the equilibrium concentrations of both the point defects. Under such conditions, the Frenkel reaction drives the concentration of both the point defects to lower values. The point defect concentration gradients drive the diffusion of the point defects into the crystal. The convection of the crystal also contributes to the flux of the point defects. The net flux of the point defects, defined as the difference in the flux of, say vacancies and interstitials, very close to the interface, determines the difference between the vacancy concentration and the interstitial concentration in the crystal, at the end of a phase called the *initial incorporation*. The initial incorporation takes place within a short distance from the interface, termed the *recombination length*. In the absence of external sources and sinks, such as the crystal surface or the thermo-mechanically induced dislocations, the established point defect concentration difference remains constant. After the initial incorporation, beyond the recombination length, the established concentration difference determines the type of the prevailing point defect at a distance far away from the interface. As the pull-rate (V) of the crystal increases, the convection of the point defects dominates their diffusion into the crystal from the interface leading to the vacancy rich conditions, because the equilibrium concentration of vacancies at the interface is higher than the equilibrium concentration of interstitials. As the magnitude of axial gradient (G) near the interface increases, the sharp temperature drop near the interface dramatically increases the concentration gradient of the point defects, thus allowing the diffusion to dominate convection. Self-interstitials diffuse faster than vacancies and lead to the interstitial rich conditions. Thus, vacancies remain in excess at a higher V/G, self-interstitials remain in excess at a lower V/G and no point defect species dominates at a V/G closer to its critical value. The surviving point defect species precipitates at a lower temperature to form microdefects.

This model, although one-dimensional in nature, explains a 2-dimensional point defect distribution quite accurately, because the timescale of the initial incorporation is much smaller than the timescale of the radial diffusion of the point defects to and from the surface of the crystal. Typically, in a growing crystal, G increases along the radial position, which leads to the incorporation of vacancies in the central region, and of interstitials in the peripheral region. The two regions are separated by a so-called *v/i* boundary. Accordingly, there is a central region of the vacancy agglomerates and a peripheral region of the interstitial agglomerates, separated by a microdefect-free region. One of the early reports of the interstitial microdefects was made by Abe and coworkers (1966), although the origin of these defects was not clearly known (5). These defects were further studied and termed as either *A-clusters* and *B-clusters*, or *A-swirl* and *B-swirl* (5,6,7,8). Subsequent studies gradually identified A- and B-defects as the interstitial related dislocation loops and globular structures, respectively (9,10,11,12). The vacancy agglomerates or *voids* were first reported by Roksnoer and van den Boom (1981) and Roksnoer (1984), and termed *D-defects* (13,14). In addition to these microdefects, the CZ-crystals may contain a band at the periphery of the vacancy region, where stacking faults are formed after a high-temperature oxidation (so called OSF ring), which often defines the *v/i* boundary. In this band, the excess vacancy concentration during the crystal growth is too low to result in the formation and growth of D-defects at higher temperatures. Instead, the surviving residual vacancies interact with oxygen, introduced by the quartz crucible, leading to the formation of small oxide precipitates, which, upon appreciable growth, facilitate formation of OSFs. Voronkov's theory has been verified by data reported before and after its publication (13-27). A few attempts to explain the point defect dynamics in the absence of the Frenkel reaction were not successful and were later corrected (28-32).

The type of the microdefect formed depends on the initial incorporated point-defect concentration. The microdefect size distribution depends also on the cooling rate of the crystal. Voronkov and Falster (1998) quantified the void size distribution by decoupling the initial incorporation process and the subsequent nucleation process (18). A similar decoupling scheme, with a different approach based on oxide nucleation model developed by Esfandyari *et al.* (1996), has been successfully applied for prediction of the void size distribution (33). These decoupling schemes are based on the zero-dimensional approximation that does not account for the diffusion of the point defects in the vicinity of the nucleation front. Sinno (1998), and Sinno and Brown (1999) developed a 1-dimensional model for prediction of the void distribution in crystals grown under quasi-steady state (34,35). A 2-dimensional model incorporating simultaneous point-defect and microdefect dynamics was developed by Mori (2000) (36). However, these models assume a constant pull-rate, and the actual unsteady state crystal growth is not treated.

As the line widths of modern devices shrink, controlling the microdefect distribution in the CZ-crystal has become very important. These restrictions have resulted in the development of new CZ-growth techniques that require dynamic control and variation of the crystal pull-rate. Thus, the CZ-process of the modern era is essentially an unsteady state process, in which both the growth rate of the crystal and the temperature field change with time. Quantitative prediction of the microdefect formation and distribution in such transient processes requires simulation of the entire crystal growth with defect dynamics. This study focuses on the quantitative prediction and understanding of both the steady state and the unsteady state defect dynamics in a growing crystal.

DEFECT DYNAMICS

CZ-crystal growth is a dynamic process that involves continuous growth of a crystal from a hot-melt placed in a quartz crucible. Practical limitations imposed by the system dynamics require the growth of a conical section known as the *crown* in the beginning of the process, which is followed by the growth of the cylindrical portion known as the *crystal-body*, or simply, the *body*. The end of the process is marked by the growth of another long conical section termed the *end-cone*. Substrate wafers for the microelectronic devices are manufactured from the crystal-body, while the crown and the end-cone are either recycled or discarded.

Microdefect dynamics, or simply, defect dynamics is a collective term that describes the interplay between the point-defect transport, the Frenkel reaction, the nucleation of point defects and the growth of microdefects. In this study, effects of impurities are neglected and attention is focused on the formation of D-, A- and B-defects. The nucleated microdefects are approximated as spherical aggregates and termed *clusters*. The vacancy agglomerates are termed *v-clusters* and the interstitial agglomerates are termed *i-clusters*. D-defects are known to be octahedral voids (37,38,18). B-defects are believed to be globular interstitial clusters (11,22,36). Thus, Both D- and B-defects can be approximated to be spherical. A-defects are identified as the dislocation loops (9,10,11,12). However, it is believed that at the nucleation stage, 3-dimensional globular clusters are formed, and only later they can transform into A-defects (4,11,36). Thus, A-defect density can be predicted by considering the nucleation of the spherical clusters. However, the computed size distribution of *i*-clusters is just an indication of the real size of A-defects.

THE MODEL

The model describing the defect dynamics must include the Frenkel reaction kinetics, the nucleation and growth of clusters, the mass balance of the point defects, and the growth of the crystal.

The reversible mutual annihilation and formation of the point defect pair (vacancies, v, and self interstitials, i), are supposed to take place in the entire crystal at finite rates.

$$i + v \leftrightarrow Si \qquad [1]$$

where Si denotes a silicon lattice atom. The rate of annihilation of a point defect species is given by

$$-r_i = -r_v = k_{iv}\left(C_i C_v - C_{i,e} C_{v,e}\right) \qquad [2]$$

where r is the net rate of formation of a point defect per unit volume and time (atoms/cm^3.s), k_{iv} is the rate constant (cm^3/atoms.s), and C (atoms/cm^3) is the concentration of any species. The subscript i denotes interstitials, v vacancies, and e equilibrium conditions.

The nucleation of the point defects to form the microdefects takes place through a series of bimolecular reactions (39,40). For the sake of simplicity, the net rate of formation of stable nuclei or stable clusters is defined on the basis of the classical nucleation theory (41,18).

$$J_{cl,j}(m_j^*) = \left[4\pi R_{cl,j}(m_j^*) D_j C_j\right] \left[k_b T \ln\frac{C_j}{C_{j,e}} \left(12\pi F_j^* k_b T\right)^{-\frac{1}{2}}\right] \left[\rho_{site} e^{\left(-F_j^*/k_b T\right)}\right] \qquad [3]$$

where m^* is the critical size (the number of point defects in the critical cluster), $J_{cl,j}(m_j^*)$ (number/cm^3.s) is the net rate of the formation of critical clusters, $R_{cl,j}(m_j^*)$ (cm) is the radius of the critical clusters, D is the diffusivity of the nucleating point defect or the monomer (cm^2/s), F_j^* (eV) is the free energy change associated with the formation of the critical cluster, k_b (eV/atom) is the Boltzmann's constant, T (K) is the temperature, and ρ_{site} ($number/cm^3$) is the site density for the nucleation. The subscript cl denotes any cluster and j denotes any nucleating species. The superscript * denotes the stable critical nuclei. In equation (3), the first term in the square bracket is the attachment frequency of the monomers to a critical nucleus, the second term is the so-called Zheldovich factor, and the third term is the equilibrium concentration of the critical nuclei. The free energy change associated with the formation of a new cluster of size m is given by the contributions from the bulk free energy change associated with the supersaturation of the monomer and the surface energy of the new phase. The size of a critical nucleus at the maximum free energy change and the maximum free energy change are given by

$$m_j^* = \left[\left(2\lambda_{cl,j}\right) \Big/ \left(3k_b T \ln \frac{C_j}{C_{j,e}}\right) \right]^3 \qquad [4]$$

$$F_j^* = \frac{4}{27} \times \lambda_{cl,j}^3 \Big/ \left(k_b T \ln \frac{C_j}{C_{j.e}}\right)^2 \qquad [5]$$

where λ ($eV/atom^{3/2}$) is the surface energy coefficient of the cluster or the nucleus.

The diffusion-limited growth of the formed nuclei is assumed. During the crystal growth, clusters move at a rate V in the direction away from the melt/crystal interface, or in the z direction. The formation history of the nuclei is used to predict the size distribution of the nuclei. Let $R(z,\tau,t)$ be the radius of the clusters present at time t, at an axial location z, formed at elapsed time τ, at some location ξ (R , z and ξ are measured in cm and t is measured in seconds). The rate of consumption of the monomers at z, by the clusters formed before t, denoted by q ($atoms/cm^3.s$), is given by

$$q_j = 4\pi D_j \left(C_j - C_{j,e}\right) \int_0^t J_{cl,j}(\xi,\tau) R_{cl,j}(\xi,\tau,t)\, d\tau \qquad [6]$$

where $J(\xi,\tau)d\tau$ is the density of the nuclei formed at ξ, at the elapsed time τ.

The diffusion-limited growth rate of the clusters at any z, at any time t, which were formed at ξ, at the elapsed time τ is described by

$$\frac{\partial R^2_{cl,j}(z,\tau,t)}{\partial t} = \frac{2D_j}{\psi_{j,cl}}\left(C_j - C_{j,e}\right) - V\frac{\partial R^2_{j,cl}(z,\tau,t)}{\partial z} \quad [7]$$

where ψ is the density (atoms/cm^3) of the monomer atoms in the cluster.

The point defect mass balance includes the diffusion, the convection, the Frenkel reaction and their consumption by the clusters. In this study, the dynamic equilibrium for the Frenkel reaction is not assumed. However, it is assumed that the Frenkel reaction kinetics remains fast enough through the initial incorporation, and appreciable cluster growth, to allow supersaturation of the crystal with only one point defect species. Thus, the consumption of vacancies by i-clusters or the consumption of interstitials by v-clusters is ignored. Radial diffusion effects are also ignored. Based on these assumptions, the mass balance of the point defects can be written as

$$\begin{aligned}\frac{\partial C_j}{\partial t} &= \frac{\partial\left(D_j\frac{\partial C_j}{\partial z}\right)}{\partial z} - V\frac{\partial C_j}{\partial z} - k_{iv}\left(C_iC_v - C_{i,e}C_{v,e}\right) \\ &\quad - 4\pi D_j\left(C_j - C_{j,e}\right)\int_0^t J_{cl,j}(\xi,\tau)R_{cl,j}(\xi,\tau,t)d\tau\end{aligned} \quad [8]$$

The rate of increase in the height of the crystal, h (cm), is given by

$$\frac{dh}{dt} = V \quad [9]$$

The domain of the system is defined by the boundaries of the crystal. The initial height of the crystal is assumed to be zero. For a crystal of a finite length, the equilibrium conditions are assumed to prevail at all surfaces of the crystal including the melt/crystal interface. The initial size of the clusters is given by equation (4). The concentration of any clusters formed at some elapsed time τ, at some location ξ and present at a location z at some other time t are given by $J_{cl,j}(\tau,\xi)d\tau$. The equations described so far are solved simultaneously for both vacancies and interstitials.

The described system of equations must be solved with the equations describing the energy balance in the hot-zone. It is assumed that the dynamics of the heat-transport is very fast compared to the dynamics of the crystal-growth, and hence, the energy balance calculations are performed using the quasi-steady state approximations. Any time-dependent variation of the temperature field in a growing crystal is taken into account by computing the temperature profiles at various lengths of the crystal at various stages of growth. The quasi-steady state energy balance has been treated by many in the past and shown to be quite accurate for CZ-growth (42,43,44).

PROPERTIES OF POINT DEFECTS

Among many, Sinno *et al.* (1998), Voronkov and Falster (1999), and Falster *et al.* (2000) report the parameter sets that satisfy the *V/G* rule (19,21,24). In this study, the diffusivities and the equilibrium concentrations of the point defects given by Voronkov (2002) are used (45).

$$D_i(\text{atoms/cm}^2) = 0.19497 \times \exp(-0.9(eV)/(k_b T)) \quad [10]$$

$$D_v(\text{atoms/cm}^2) = 6.2614 \times 10^{-4} \times \exp(0.4(eV)/(k_b T)) \quad [11]$$

$$C_i(\text{atoms/cm}^3) = 6.1859 \times 10^{26} \times \exp(-4.0(eV)/(k_b T)) \quad [12]$$

$$C_v(\text{atoms/cm}^3) = 7.59982 \times 10^{26} \times (-4.0(eV)/(k_b T)) \quad [13]$$

The Frenkel reaction rate constant given by Sinno *et al.* (1998) is used (19). However, the enthalpic contribution is set to zero as suggested by Wang (2002) (46). The accurate estimation of this constant is not necessary as the reaction dynamics is very fast. The surface energy coefficients used for the *v*-clusters ($\lambda_{cl,v}$) and for the *i*-clusters ($\lambda_{cl,i}$), are 1.85 eV/atom$^{3/2}$ atoms and 2.95 eV/atom$^{3/2}$, respectively.

STEADY STATE SIMULATIONS

The discussed model was first solved using a fixed temperature profile in a growing crystal given by

$$1/T = (1/T_f) + (1/T_f^2)Gz \quad [14]$$

where the subscript *f* denotes the conditions at the interface. A value of 2.5 K/mm was used for *G*. In the discussion henceforth, units popular in the semiconductor industry are used. Various crystal growth processes at various fixed pull-rates were simulated. For a fixed temperature profile and a fixed pull-rate, the crystal growth occurs under a steady state. At a steady state, the point defect concentration profiles and the microdefect density profiles in the crystal represent the path followed by each crystal segment. Such profiles are shown in Figure 1 for a very high pull-rate. As is evident from the figure, at this pull-rate, vacancies are incorporated in preference to interstitials. After the initial phase of incorporation, the vacancy concentration remains practically unchanged resulting in the gradual vacancy supersaturation with the decreasing temperatures. Subsequent nucleation of vacancies and growth of v-clusters around 1112 ^{0}C depletes the vacancy concentration. The nucleation temperature is defined as the temperature at which the nucleation rate is the highest. The nucleation range, defined by the appreciable nucleation rates, is very short. The growth of the clusters continues much beyond the nucleation range. It can be observed that the Frenkel reaction increases the concentration of interstitials during the formation and growth phase of the v-clusters. Figure 2 shows the simulation results for a lower pull-rate, when interstitials remain the dominant incorporated species. The nucleation of interstitials takes place at a much lower temperature (around 900 ^{0}C).

The effect of the changing pull-rate at a fixed G, on the incorporated point defect concentration and the nucleation temperature, is shown in Figure 3. As the pull-rate decreases from a very high value, the incorporated vacancy concentration gradually decreases. The decrease in the incorporated vacancy concentration drives the nucleation temperature to a much lower value, as the supersaturation required for appreciable nucleation cannot be achieved at higher temperatures. As the pull-rate further decreases toward the critical V/G, the oxygen-vacancy interaction dominates the vacancy consumption, before the vacancy nucleation takes place to form v-clusters. This interaction is not predicted by the model. However, under these conditions, the developed algorithm ignores any spurious v-cluster precipitation. At the critical V/G, no point-defect species is dominant. The model prediction of the critical V/G (0.15 mm^2/K.min) matches almost exactly with the prediction of the analytical expression (Voronkov, 1982) (4). A further decrease in the pull-rate below the critical V/G allows the incorporation of interstitials. As the pull-rate further decreases, the incorporated self-interstitial concentration increases, which drives the interstitial nucleation temperature to a higher value. It is important to note that the incorporated interstitial concentration changes more dramatically than the incorporated vacancy concentration, with the same absolute change in the pull-rate. The developed algorithm ignores the formation of spurious undetectable precipitates under the interstitial-rich conditions very close to the critical V/G.

The cluster size distributions at various pull-rates are shown in Figures 4 and 5. The cluster size distribution is determined by the interplay between the formation of the new clusters driven by higher supersaturations of the dominant point defect and the growth of the already formed clusters, which consume the dominant point defect and decrease the supersaturation. The incorporated point defect concentration and the cooling rates through the nucleation range quantify this interplay. For higher cooling rates and for a fixed incorporated point defect concentration, before the formed clusters can appreciably consume the point defects, the nucleation proceeds with appreciably high rates as the temperature drops at a high rate. Thus, a large number of clusters are formed, which compete for the dominant point defect species. Hence, the nucleation through higher cooling rates produces a large number of smaller clusters. The concentration of the incorporated point defect species also influences the cluster size distribution. For a higher incorporated concentration, the concentration driving force for the cluster growth in the nucleation range is very high, allowing a faster growth of the formed clusters, which consume the dominant point defect and suppress further nucleation. Thus, as the dominant point defect concentration increases, for a fixed cooling rate, the density of the formed clusters decreases while the average size of the clusters increases.

The product of the pull-rate and the local negative axial temperature gradient gives the cooling rate of a crystal segment at any location. In the vacancy rich condition, as the pull-rate increases, the cooling rates increase through the nucleation range. However, the incorporated vacancy concentration also increases with the increasing pull-rate. Thus, an increase in the pull-rate generates two opposing effects on the cluster size distribution. Typically, the effect of a change in the cooling rate dominates. In the interstitial rich conditions, as the pull rate decreases, the cooling rates through the nucleation decrease. The incorporated interstitial concentration increases rather significantly with the decreasing pull-rate. Thus, with a change in the pull-rate, both the cooling rate and the incorporated interstitial concentration influence the *i*-cluster size distribution in the same way.

The nucleation period can be defined as the period during which the nucleation rates are greater than 36.8% of the maximum rate. The predicted nucleation periods at various pull-rates are shown in Figure 6. The vacancy nucleation takes place in a matter of minutes whereas the interstitial nucleation takes place over a longer period of time. The nucleation period is also determined by the interplay between the formation and the growth of clusters.

UNSTEADY STATE SIMULATIONS

The results of the numerical simulations performed to capture the defect dynamics and the microdefect distribution under unsteady state conditions are discussed in this section. The same temperature profile discussed in the previous section is supposed represent the simulated unsteady state crystal growth. However, an unsteady state pull-rate profile, shown in Figure 7, is used. The crystal growth is defined by the four sequential phases – the first constant pull-rate, the decreasing pull-rate, the increasing pull-rate and the second constant pull-rate. The crown is grown in the first constant pull-rate phase, the crystal body is grown in the decreasing and the increasing pull-rate phases, and the end-cone is grown in the second constant pull-rate phase. Figure 7 also identifies two pairs of *locations of symmetry* on the basis of the pull-rate profile – A and A′, and B and B′.

The predicted cluster type and cluster density profile in the crystal body grown in the two varying pull-rate phases, the decreasing pull-rate and the increasing pull-rate, is shown in Figure 8. Vacancies incorporated in the segments formed at higher pull-rates during the early periods of the decreasing pull-rate phase nucleate when the pull-rates are lower. In the first 13 cm long section of the body, although the incorporated vacancy concentration monotonically decreases with the length, the cooling rates through the nucleation ranges also continuously decrease. The stronger influence of the cooling rates results in a continuous decrease in the *v*-cluster density along the axial position. For the next 4 cm long section (13 cm - 17 cm), the vacancy incorporation takes place in the decreasing pull-rate phase but the nucleation takes place in the increasing pull-rate phase, which increases the *v*-cluster density.

In the succeeding short section of the crystal (17 cm – 20 cm), no point defect species concentration is high enough to form the clusters. Since the developed algorithm ignores the formation of clusters having negligible growth potential, the predicted microdefect-free section is relatively wider. This region defines the *v/i* boundary. The critical *V/G* under these transient conditions is predicted to be 0.1644 mm^2/K.min. This value is higher than the critical *V/G* predicted at the steady state, which is equal to 0.15 mm^2/K.min. This increase is caused by an increase in the net flux of interstitials. In the decreasing pull-rate phase, a crystal segment formed at a given pull-rate moves with a continually decreasing rate through the recombination length, allowing increased diffusion time, which increases the concentration of interstitials in the segment. Also, at any pull-rate, at the interface, the interstitial concentration gradient is a little higher and the vacancy concentration gradient is a little lower compared to the steady state conditions, as a result of the vacancy rich conditions established next to the interface at preceding higher pull-rates. Thus, in a given crystal segment formed at a given pull-rate, the incorporated interstitial concentration is a little higher and the incorporated vacancy concentration is a little lower compared to a segment grown at the same pull-rate under the steady state. Based on this hypothesis, it can be concluded that the shift in the critical *V/G* from its steady state value must increase with the increasing magnitude of the rate of change of the pull-rate with the crystal length. For industrial applications, a linear approximation of this shift for the decreasing pull-rate conditions is given as

$$[V/G]_{x,-slope} = [V/G]_x - 7.85(\mathrm{K/mm^2}) \times [dV/dL](\mathrm{min^{-1}}) \qquad [15]$$

where L (mm) is the length of the crystal. The subscript *–slope* indicates the transient conditions marked by a continuously decreasing pull-rate. Equation (15) can be applied to any crystal growth process and treated as a generic rule.

Beyond the discussed cluster-free section, the interstitial rich conditions prevail in the next 16 cm (20 cm – 36 cm) long section of the body, as the pull-rate further decreases. A fraction of this section undergoes the initial incorporation at the end of the decreasing pull-rate phase, and the remaining fraction undergoes the initial incorporation in the increasing pull-rate phase. The *i*-cluster density typically decreases with the decreasing pull-rate and shows a strong dependence on the incorporated interstitial concentration. In this case, the entire section goes through the nucleation range in the increasing pull-rate phase. The cooling rates through the nucleation range at each location in the section vary as the pull-rate continuously varies and the nucleation temperature itself shifts. However, as the change in the incorporated interstitial concentration with the changing pull-rate is very high (Figure 3), the change in the *i*-cluster density is dominantly affected by the change in the incorporated interstitial concentration.

The second *v/i* boundary appears in the short section beyond the *i*-rich section discussed above. In this section, the point defects are incorporated in the increasing pull-rate phase. The transient critical *V/G* predicted for this phase is 0.1248 mm^2/K.min, which is much lower than its steady state value. This drift can also be explained by the shift in the excess point defect flux. In this phase, a freshly formed crystal segment moves with a continually increasing rate through the recombination length, which decreases the diffusion time through the initial incorporation. Also, the interstitial diffusion flux from the interface into the crystal decreases because the concentration driving force for interstitials is lower compared to that at the steady state. Thus, more vacancies survive through the incorporation. The shift in the observed critical *V/G* under the increasing pull-rate conditions is also given as a function of the rate of change of the pull-rate with the crystal length as follows,

$$[V/G]_{x,-slope} = [V/G]_x - 13.745(\text{K/mm}^2) \times [dV/dL](\text{min}^{-1}) \qquad [16]$$

The subscript *+slope* indicates the increasing pull-rate conditions.

The crystal section beyond the second cluster-free section shows a monotonically decreasing *v*-cluster density. The incorporated vacancy concentration monotonically increases in this section but the cooling rates through the nucleation ranges do not vary much. Thus, the *v*-cluster density profile in this section is determined predominantly by the profile of the incorporated vacancy concentration as shown in Figure 8.

The shifts in the critical *V/G* in the increasing and the decreasing pull-rate phases are better explained by tracing the evolution of the dominant point defect concentration in the two chosen symmetric locations denoted by points A and A′ (Figure 7). The vacancy incorporation happens in the decreasing pull-rate phase at point A and in the increasing pull-rate phase at point A′. Based on our arguments, it must be expected that the incorporated vacancy concentration at A must be lower. Figure 9 shows this to be true. Although the incorporated vacancy concentration at location A is lower, the average cluster density is also lower, as the cooling rates at location A are lower than the cooling rates at location A′, during appreciable nucleation. Figure 10 shows the *v*-cluster size distributions at these two locations.

Typically, in the industry, the critical *V/G* is determined by tracing the microdefect distribution in a crystal grown under varying pull-rate conditions, similar to the conditions described here. Thus, the results discussed in this section are of significant industrial importance.

COMPARISON WITH EXPERIMENTS

The developed model is further validated by comparing the model predictions with the microdefect distribution in two crystals grown under unsteady state conditions. The first experiment represents a well-known technique used in the process development and the second experiment represents a typical manufacturing process.

Experiment – I

An experimental crystal was pulled by the varying non-dimensional pull-rate profile shown in Figure 11. On the basis of the pull-rate profile, two phases of the crystal-growth are identified – the decreasing pull-rate and the increasing pull-rate. The length of the crown is around 12 cm. The crystal length, measured from the beginning of the crystal body, is approximately 70 cm long. The end-cone is more than 25 cm long. The microdefect distribution only in the crystal body is discussed in this section.

The axial profiles of the predicted v-cluster density and the volume-averaged size at the center of the crystal are shown in Figure 12. Incorporated vacancies in the sections formed earlier in the decreasing pull-rate phase undergo nucleation in the same phase at much lower cooling rates. Thus, the v-cluster density marginally decreases in the first 20 cm long section, along the axial position. Incorporated vacancies in the sections formed in the later stages of the decreasing pull-rate phase undergo nucleation in the increasing pull-rate phase. For these sections, both the effect of the incorporated vacancy concentration and the cooling rate contribute to increase the cluster density. The cooling rates through nucleation continually increase for all sections formed beyond the later stages of the decreasing pull-rate phase. However, the dramatic rise in the v-cluster density at 40 cm away from the origin of the body is dominantly caused by the extremely low incorporated vacancy concentration, and the appreciable cooling rates through the nucleation period. In reality, oxygen interaction with vacancies dominates this region. Hence, the model predictions are not very accurate near the critical V/G. This dramatic rise in the cluster density is followed by a dramatic drop as the incorporated vacancy concentration increases with the increasing pull-rate, in the increasing pull-rate phase. The profile of the volume averaged v-cluster size shows the expected inverse behavior with the cluster density.

The predicted cluster density is compared with the cluster density determined by two popular experimental techniques. The first method involves etching of silicon crystal by a defect decorating etchant. Secco etchant is popularly used for microdefect decoration (47). However, Secco etching does not expose all microdefects. The cluster density profile determined by Secco etching is typically multiplied by a process dependent constant to obtain a semi-quantitative measure of the microdefect density. The experimental results match very well with the model predictions as shown in Figure 12.

The second method of measuring the cluster density involves growing copper precipitates on the clusters and subsequent surface polishing followed by Secco etching. Copper precipitates on a cluster as copper silicide and creates a much larger region of tiny copper precipitates around the cluster, collectively called the copper colony (48). A more quantitative treatment of the decorating etching of the copper colonies is accomplished by Kulkarni *et al.* (2002) on the basis of a generic theory of etching developed by Kulkarni and Erk (2000) (49,50). The method of copper precipitation followed by a sequential surface polishing and Secco etching can be used as a better quantitative representation of the density of larger clusters. As shown in Figure 12, the predicted cluster density matches very well with the experimentally determined microdefect density, for larger clusters.

Experiment – II

In this experiment, a crystal was pulled to mimic the real process conditions. The used non-dimensional crystal pull-rate profile is shown in Figure 13. The variation of the temperature profile in the crystal as a function of the crystal length was incorporated in the model. Pseudo-steady state energy balance in the hot-zone at various crystal lengths provides the axial temperature profile in the crystal at various crystal lengths. The real process is mimicked by assuming that the temperature profile in the crystal is more accurately captured by the sequential jumps from one predicted profile to the next depending on the crystal length. Thus, for a more accurate prediction of the defect dynamics, multiple predicted temperature profiles at various crystal lengths must be used.

The model predicted and the experimentally determined *v*-cluster density profiles agree very well as shown in Figure 14. The profile of the volume averaged cluster size is also shown in Figure 14. The model can be reliably applied to capture the defect distribution in real systems.

CONCLUSIONS

Both steady state and unsteady state point defect and microdefect dynamics in Czochralski (CZ) crystal growth can be accurately modeled using the steady state nucleation of the point defects and the diffusion limited growth of the microdefects.

The model predictions agree with the theory of the initial incorporation of the point defects, according to which, in a growing crystal, the type and the concentration of the excess point defect species is established within a short distance from the melt/crystal interface, known as the *recombination length.* The interplay between the convection and the diffusion of both the point defects – vacancy and self-interstitials, and the Frenkel reaction, leading to establishing either the vacancy rich conditions in the crystal beyond the recombination length at a higher *V/G* (pull-rate/negative axial temperature gradient), or leading to establishing the interstitial rich conditions at a lower *V/G*, can be accurately captured by the model for both the steady state and the unsteady state crystal growth conditions. A comparison of the steady state crystal growth and a representative unsteady state crystal growth shows significant differences in the quality of the crystals grown under different modes. The shift can be explained and quantified on the basis of the change in the net point defect flux.

The developed model approximates both the vacancy type microdefects and the interstitial type microdefects as spherical clusters. The interplay between the nucleation of the dominant point defect to form new clusters at higher supersaturations and the growth of the already formed clusters that reduce the nucleation rate by consumption of the point defect, determines the size and the density of the grown clusters. The cooling rates through the nucleation range and the incorporated point defect concentration influence this interplay. The model captures cluster size distributions under variety of steady state and unsteady state conditions. It is shown that, under an unsteady state crystal growth at a varying pull-rate, the change in the size distribution of the interstitial agglomerates, or *i*-clusters, is strongly influenced by the variation in the incorporated interstitial concentration. The change in the size distribution of the vacancy agglomerates, or *v*-clusters, is more strongly influenced by the cooling rates. For the same absolute shift in V/G, the shift in the nucleation temperature of the interstitials is higher than the shift in the nucleation temperature of vacancies. This behavior is a result of a more sensitive variation of the incorporated interstitial concentration with V/G. The interstitial nucleation periods, during which nucleation rates are appreciable, are longer than the vacancy nucleation periods.

The model predictions agree very well with the experimental data. Data from two unsteady state experiments establish the validity of the model and the developed algorithm. The entire complex of defect dynamics is accurately predicted.

ACKNOWLEDGMENTS

Milind Kulkarni gratefully acknowledges Jeffrey Libbert, J. W. Ryu and Zheng Lu for performing the experiments. Milind is also grateful to Lee Ferry, who provided the computed temperature fields in various crystal-pullers, necessary for capturing defect dynamics.

REFERENCES

1. W. C. Dash, *J. Appl. Phys.*, **29**, 736 (1958)
2. W. C. Dash, *J. Appl. Phys.*, **30**, 459 (1958)
3. F. Shimura, *Semiconductor Silicon Crystal Technology*, p56, Academic Press, Inc., San-Diego (1989).
4. V. V. Voronkov, *J. Crystal Growth*, **59**, 625 (1982).
5. T. Abe, T. Samizo, and S. Maruyama, *Jpn. J. Appl. Phys.*, **5**, 458 (1966).
6. A. J. R. de Kock, *Appl. Phys. Letters*, **16**, 100 (1970).
7. A. J. R. de Kock, *J. Electrochem. Soc.*, **118**, 1851 (1971).
8. A. J. R. de Kock, P. J. Roksnoer and P. G. T. Boonen, *J. Cryst. Growth*, **22**, 311 (1974).
9. P. M. Petroff and A. J. R. de Kock, *J. Cryst. Growth*, **30**, 117 (1975).
10. P. M. Petroff and A. J. R. de Kock, *J. Cryst. Growth*, **36**, 4 (1976).
11. H. Föll, U. Gösele and B. O. Kolbesen, *J. Cryst. Growth*, **40**, 90 (1977).
12. A. J. R. de Kock and W. M. van de Wiljert *J. Cryst. Growth*, **49**, 718 (1980).
13. P. J. Roksnoer and M. M. B. van den Boom, *J. Cryst. Growth*, **53**, 563 (1981).
14. P. J. Roksnoer, *J. Cryst. Growth*, **68**, 596 (1984).
15. S. Sadamitsu, S. Umeno, Y. Koike, M. Hourai, S. Sumita, and T. Shigematsu, *Jpn. J. Appl. Phys.*, **32**, 3675 (1993).
16. R. Falster, V. V. Voronkov, J. C. Holzer, S. Markgraf, S. McQuaid, and L. Mulèstagno, *Electrochemical Society Proceedings*, **98-1**, 468 (1998).
17. E. Dornberger and W. von Ammon, *J. Electrochem. Soc.*, **143**, 1648 (1996).
18. V. V. Voronkov and R. Falster, *J. Cryst. Growth*, **194**, 76 (1998).
19. T. Sinno, R. A. Brown, W. von Ammon, and E. Dornberger, *J. Electrochem. Soc.*, **145**, 302 (1998).
20. V. V. Voronkov and R. Falster, *J. Cryst. Growth*, **204**, 462 (1999).

21. V. V. Voronkov and R. Falster, *J. Appl. Phys.*, **86**, 5957 (1999).
22. V. V. Voronkov and R. Falster, *J. Electrochem. Soc.*, **149**, G167 (2002).
23. V. V. Voronkov and R. Falster, *J. Appl. Phys.*, **87**, 4126 (2000).
24. R. Falster, V. V. Voronkov and F. Quast, *phys. Stat. Sol.* **222**, 219 (2000).
25. E. Dornberger, D. Gräf, M. Suhren, U. Lambert, P.Wagner, F. Dupret, and W. von Ammon, *J. Cryst. Growth*, **180**, 343 (1997).
26. E. Dornberger, W. von Ammon, J. Virbulis, B. Hanna, and T. Sinno, *J. Cryst. Growth*, **230**, 291 (2001).
27. W. von Ammon, R. Hölzl, J. Virbulis, E. Dornberger, R. Schmolke, and D. Gräf, *J. Cryst. Growth*, **226**, 19 (2001).
28. R. Habu, K. Kojima, H. Harada, and A. Tomiura, *Jpn. J. Appl. Phys.*, **32**, 1754 (1993).
29. R. Habu, T. Iwasaki, H. Harada, and A. Tomiura, *Jpn. J. Appl. Phys.*, **33**, 1234 (1994).
30. R. Habu and A. Tomiura, *Jpn. J. Appl. Phys.*, **35**, 1 (1996).
31. R. Habu, I. Yunoki, T. Saito, and A. Tomiura, *Jpn. J. Appl. Phys.*, **32**, 1740 (1993).
32. R. Habu, K. Kojima, H. Harada, and A. Tomiura, *Jpn. J. Appl. Phys.*, **32**, 1747 (1993).
33. J. Esfandyari, C. Schmeiser, S. Senkader, G. Hobler, and B. Murphy, *J. Electrochem. Soc.*, **143**, 995 (1996).
34. T. R. Sinno, "Defects in Crystalline Silicon: Integrated Atomistic and Continuum Modeling," *Ph. D. Thesis*, Massachusetts Institute of Technology, 1998.
35. T. Sinno and R. A. Brown, *J. Electrochem. Soc.*, **146**, 2300 (1999).
36. T. Mori, "Modeling the Linkages between Heat Transfer and Microdefect Formation in Crystal Growth: Examples of Czochralski Growth of Silicon and Vertical Bridgman Growth of Bismuth Germanate," *Ph. D. Thesis*, Massachusetts Institute of Technology, 2000.
37. M. Kato, T. Yoshida, Y. Ikeda, Y. Kitigawara, *Jpn. J. Appl. Phys.*, **35**, 5597 (1996).
38. G. Kissinger, G. Morgenstern, J. Vanhellemont, D. Gräf and U. Lambert, *Appl. Phys. Lett.*, **72**, 223 (1998).
39. D. Turnbull and J. C. Fisher, *J. Chem. Phys.*, **17**, 71 (1949).
40. A. S. Michaels, *Nucleation Phenomena*, American Chemical Society, 1966.
41. D. Kashchiev, *Nucleation, Basic Theory with Applications*, Butterworth- Heinemann, 2000.
42. P. A. Ramachandran and M. P. Dudukovic, *J. Crys. Growth.*, **71**, 399 (1985).
43. R. K. Srivastava, P. A. Ramachandran and M. P. Dudukovic, *J. Crys. Growth.*, **73**, 487 (1985).
44. A. Virzi, *J. Crys. Growth*, **112**, 699 (1991).
45. V. V. Voronkov, *Private Communication* (2002).
46. Z. Wang, *Private Communication* (2002).
47. F. Secco d'Aragona, *J. Electrochem. Soc.*, **119**, 948 **(1972)**
48. S. P. Murarka, I. V. Verner, and R. J. Gutmann, *Copper-Fundamental mechanisms for microelectronic Applications*; John Wiley and Sons, Inc.: New York , 2000.
49. M. S. Kulkarni, J. Libbert, S. Keltner, and L. Muléstagno, *J. Electrochem. Soc.*, **149**, G153 (2002).
50. M. S. Kulkarni and H. F. Erk, *J. Electrochem. Soc.* **147**, 176 (2000).

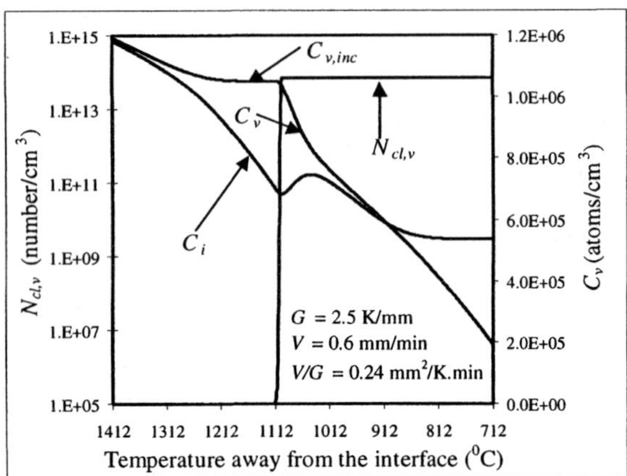

Figure 1. The profiles of concentrations of the point defects and the v-cluster density (as the functions of the crystal temperature) in a representative crystal grown under a steady state.

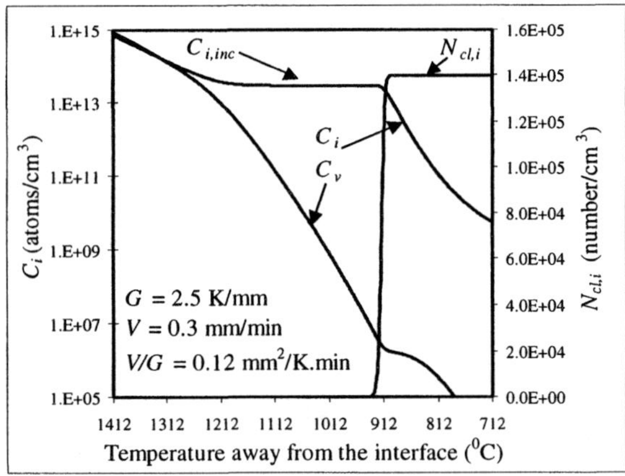

Figure 2. The profiles of concentrations of the point defects and the v-cluster density (as the functions of the crystal temperature) in a representative crystal grown under a steady state.

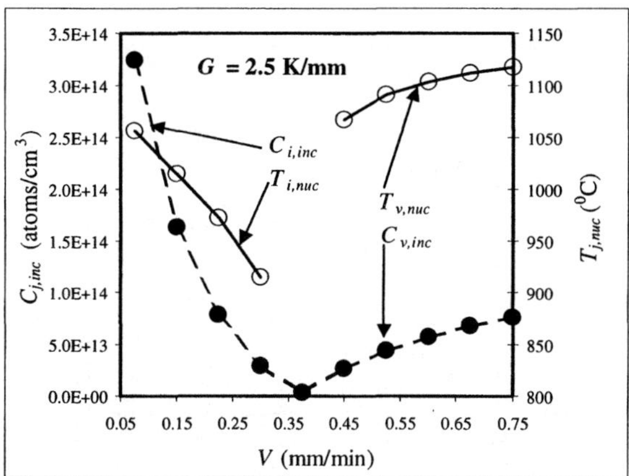

Figure 3. The dependence of the incorporated point defect concentration and the nucleation temperature on the pull-rate at a fixed G (Subscripts *inc* and *nuc* denote the incorporated concentration and the nucleation, respectively).

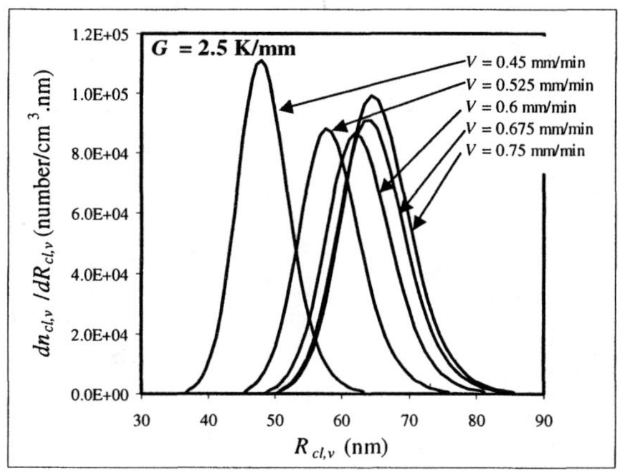

Figure 4. The size distributions of v-clusters in the representative crystals grown under steady states at various pull-rates (n (number/cm^3) is the density of the clusters).

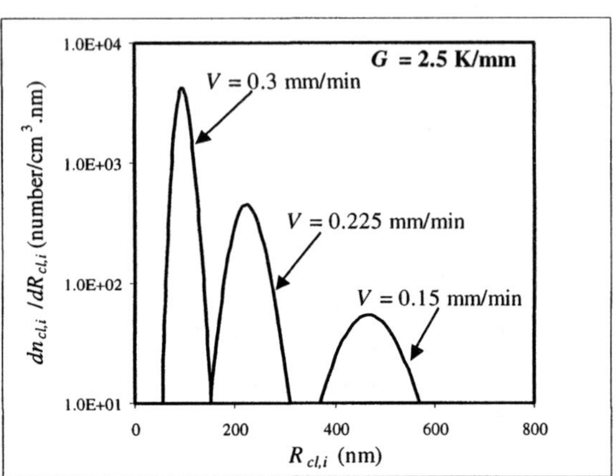

Figure 5. The size distributions of *i*-clusters in the representative crystals grown under steady states at various pull-rates.

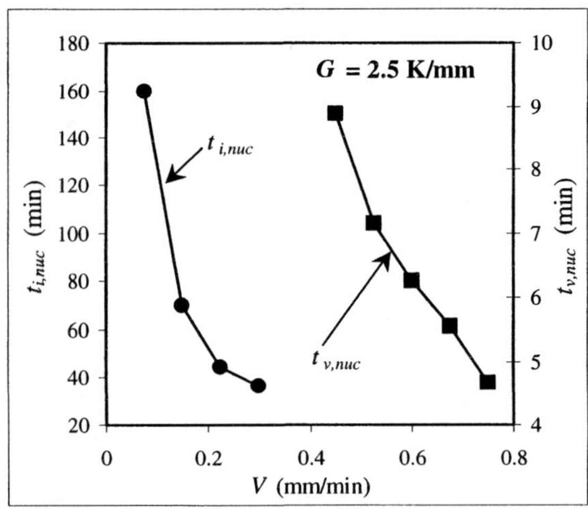

Figure 6. The self-interstitial and the vacancy nucleation periods as functions of the pull-rate.

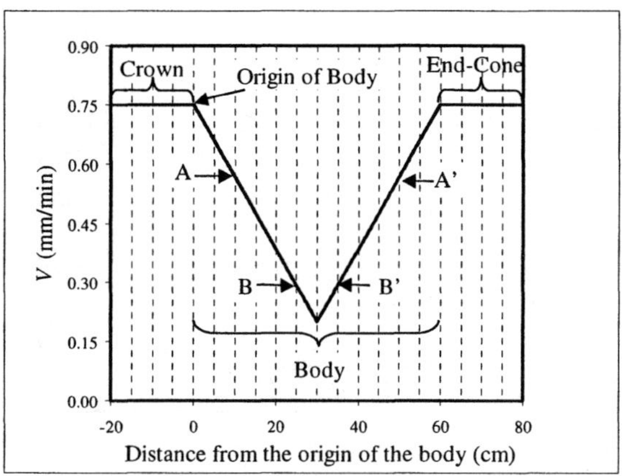

Figure 7. The crystal pull-rate profile used to understand the defect dynamics under unsteady state conditions.

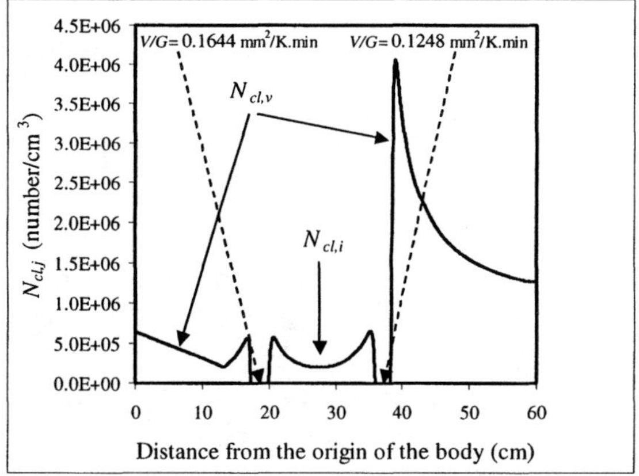

Figure 8. The cluster density profile in the crystal for the pull-rate profile shown in Figure 7.

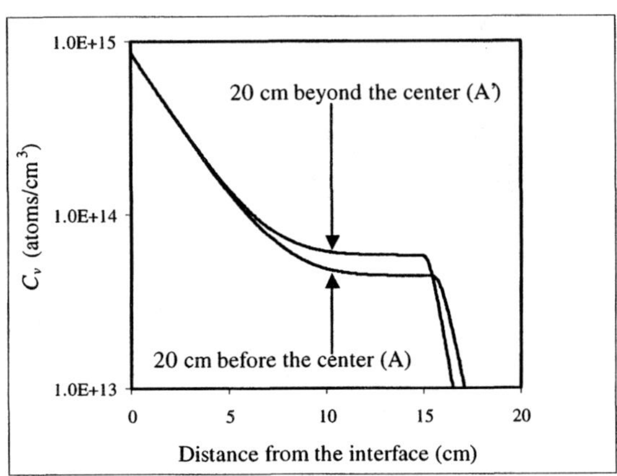

Figure 9. The evolution of the vacancy concentrations in the chosen symmetric locations, A and A′, as functions of their locations from the interface during the crystal growth.

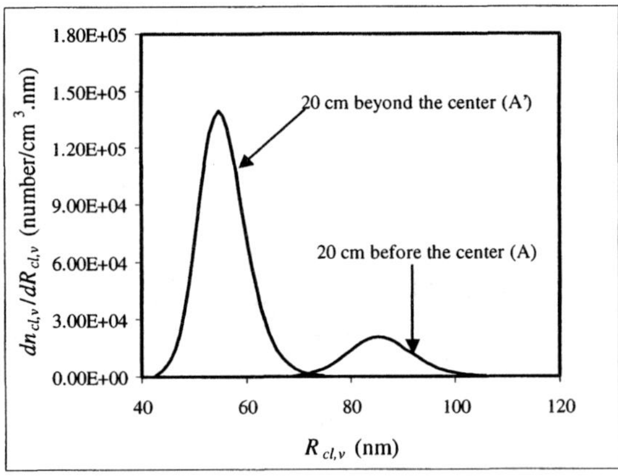

Figure 10. The v-cluster size distributions at the chosen symmetric locations, A and A′.

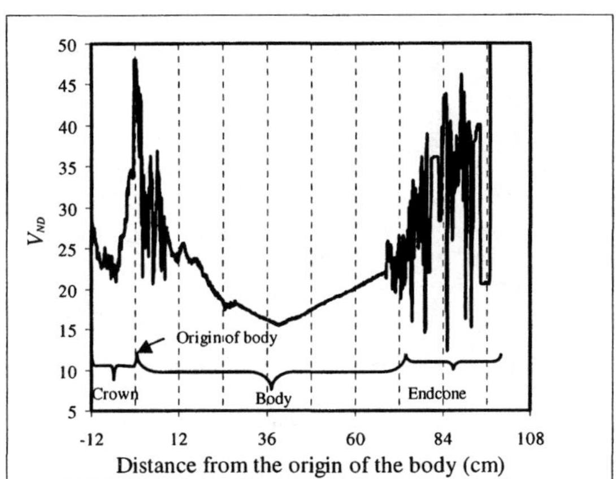

Figure 11. The non-dimensional crystal pull rate profile as a function of its length: experiment – I

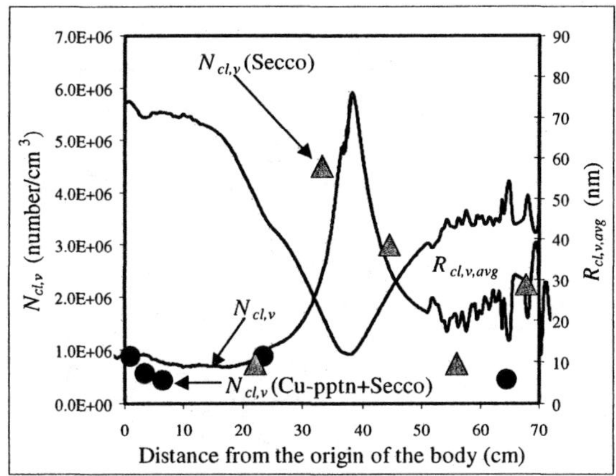

Figure 12. A comparison of the model predicted cluster density profile with the experimentally determined cluster density profiles: Experiment – I. The square legends indicate the cluster density profile determined by Secco etching, and the circles indicate the density profile determined by copper precipitation followed by Secco etching. The predicted profile of the volume-averaged size of the clusters is also shown.

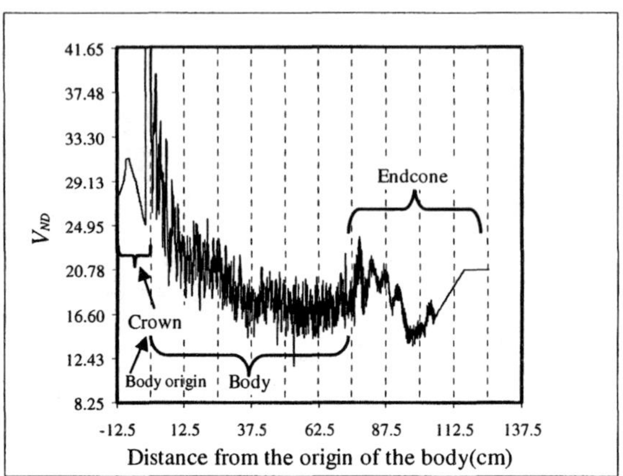

Figure 13. The non-dimensional crystal pull-rate profile as a function of its length: Experiment – II

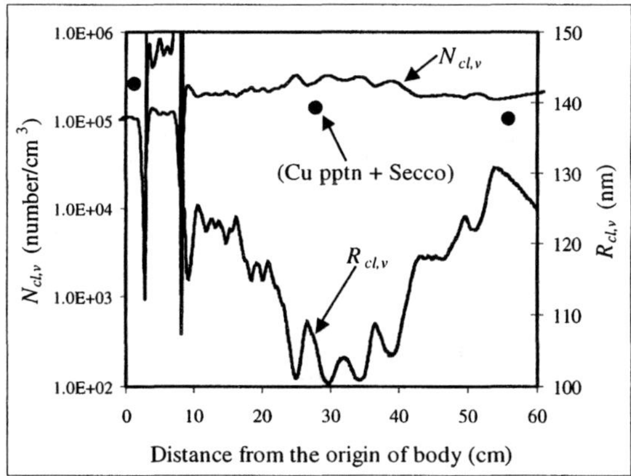

Figure 14. A comparison of the model predicted and the experimentally determined v-cluster density profiles: Experiment – II. The circular legends indicate the experimental data points. The volume averaged cluster size (radius) profile is also shown.

SIMULATION OF OXIDE FORMATION AND POINT DEFECTS DYNAMICS IN SILICON: THE ROLE OF OXIDE MORPHOLOGY

Zhihong Wang and Robert A. Brown
Department of Chemical Engineering
Massachusetts Institute of Technology
Cambridge MA 02139

In silicon wafers, oxygen precipitates with appropriate size and density serve as gettering sites for metallic impurities which are introduced during fabrication processes. A model for oxygen precipitation is presented that predicts the evolution of the precipitate size distribution. Precipitate growth models the important effect of stress by including both vacancy absorption into the precipitates and self-interstitial ejection into the silicon matrix as mechanism for alleviating stress. Most importantly, the effect of precipitate morphology is included by modeling the evolution of spherical and disk-shaped precipitates. The model successfully predicts experimental results for oxygen precipitate under both conditions when the crystal is vacancy-rich (spherical precipitates) or vacancy-poor (disk-shaped precipitates).

INTRODUCTION

The oxygen concentration that is inherent to CZ-grown silicon wafers is crucial to the mechanical and electrical properties of silicon wafers. The oxide precipitates that form from the oxygen super-saturation and their associated stacking faults and dislocation loops serve as getting sites for metallic impurities (1). The goal of this research is the development of quantitative models for oxide precipitation and the prediction of precipitate distribution as a function of thermal history. We present an internally consistent model for oxide precipitate growth that includes the dynamics of oxygen precipitation and the influence of point defects on this precipitation.

Point defect dynamics plays a critical role in oxide precipitation. Because the density of the precipitate is approximately half that of the matrix, elastic stress energy is created that can only be dissipated by vacancy absorption into the precipitate or self-interstitial ejection into the matrix. The model presented here builds on the models presented before (2-4) for point defect dynamics and aggregate growth.

Of particular interest is the prediction of the dependence of the oxide precipitate density on the temperatures and times used in a traditional Hi-Low-Hi cycle. The careful experiments reported by Kelton et al. (5,6) demonstrate the sensitive dependence of the density on the Low temperature used in the nucleation step. We aim to reproduce these experiments.

The morphology of the precipitate is a critical element of our model. Bergholz (7) has classified the morphologies seen in oxide precipitates for five ranges of nucleation temperature (T_{nucl}). These morphologies and the corresponding temperature ranges are: ribbon-shaped for $T_{nucl} \le 550°C$; a mixture of ribbon-shaped and disk-shaped for $550°C \le T_{nucl} \le 700°C$; disk-shaped for $700°C \le T_{nucl} \le 900°C$; octahedral for $900°C \le T_{nucl} \le 1100°C$; polyhedral or nearly spherical for $T_{nucl} \ge 1100°C$. Images of these precipitates are shown in Figure 1. We believe that these differences are driven by the effects of elastic stress and the point defect distributions. It has long been known that disk-shaped precipitates generate less elastic stress than spherical shapes. We model the precipitation of both spherical and disk-shaped precipitates and predict the transition in morphology and its impact on precipitate density. By including both the effects of morphology and point defect dynamics the model also successfully predicts the oxide densities for novel annealing processes, such as the MDZ (Magic Denuded Zone) process developed by MEMC (8), that effectively uses a high vacancy concentration to install a spatially varying precipitate distribution.

MODEL FORMULATION

Oxygen precipitation in silicon is expressed mechanistically as

$$P_n + (0.5 + \gamma_I)Si + O + \gamma_V V \leftrightarrow P_{n+1} + \gamma_I I + \textit{stress energy} \qquad [1]$$

where P_n is an oxygen precipitate of size *n* which forms by aggregating oxygen (*O*), silicon (*Si*), ejecting interstitials (*I*) and absorbing vacancies (*V*). The stoichiometric coefficients γ_I and γ_V give the number of interstitials ejected and vacancies absorbed, respectively, during the reaction. An internally consistently model for oxygen precipitation in silicon includes the dynamics of vacancies, interstitials and oxygen and the formation of clusters of each species. The monomers (vacancy, interstitial, and oxygen) are governed by convection (only for crystal growth), diffusion, recombination of vacancies and interstitials, and the source or sink due to the formation of all the clusters (voids, interstitial clusters, and oxygen precipitates). All the clusters are described by DRE's (Discrete Rate Equations) for small clusters and FPE's (Fokker-Planck Equations) for large clusters. The details of the model are referred to (2) where the dynamics of native point defects and their clusters are described. Only the models for oxygen precipitation and morphology are emphasized here.

The numerical simulations of vacancies, interstitials, oxygen and all clusters use the CC70 finite difference method to discretize the size space for all FPE's, DG method (Discontinuous Galerkin) for the spatial variables in the DRE's and FPE's, and LDG method (Local Discontinuous Galerkin) for the spatial variables in the governing equations for point defects and oxygen. Time integration involves an operating splitting method and implicit Eular integration in time. The accuracy and efficiency of the simulations are improved by using iteration and adaptive time steps for each operating splitting step. The details of the numerical methods used to solve the model are referred to (9,10).

Mesoscopic Growth Models

The diffusion-limited growth rate g and dissolution rate d are determined by mass balances about an individual microdefect (2,10,11). Consider a microdefect of arbitrary shape that grows by diffusion of a monomer from bulk silicon to the precipitate, as shown in Figure 2. The monomer concentration $C(\bar{r})$ in the bulk is governed by the species conservation equation

$$\nabla^2 C = 0 \tag{2}$$

where $\bar{r}$ is a position vector measured from the center of the precipitate. Equation [2] is solved with the boundary conditions

$$C \to C_\infty \quad as \quad |\bar{r}| \to \infty, \quad C = C_S \quad at \quad \bar{r} = \bar{r}_S \tag{3}$$

where $\bar{r}_S$ defines the surface of the precipitate where the concentration is C_S. This concentration is determined by a mass balance at the surface of the precipitate as

$$F(n) = g(n) - d(n) \tag{4}$$

where $F = F(n)$ is the monomer diffusion flux, and the growth $g(n)$ and dissolution $d(n)$ rates are

$$g(n) \equiv k_r(n) C_S, \quad d(n) \equiv k_r(n) C^{eq}(n) \tag{5}$$

In eq.[5] $k_r(n)$ is the surface reaction rate and $C^{eq}(n)$ is the equilibrium concentration for the monomer for a microdefect with n monomers. The equilibrium concentration $C^{eq}(n)$ is related to bulk monomer equilibrium concentration C^{eq} by

$$C^{eq}(n) = C^{eq} \exp(\frac{\Delta G^f_{n \to n+1}}{kT}) \tag{6}$$

where G^f is the non-configurational portion of the free energy of formation for the microdefect. A model form of the free energy is discussed in the next section. The reaction rate $k_r(n)$ is modeled as (11)

$$k_r(n) = V^i \frac{D}{\delta^2} \exp(-\frac{\Delta G_{n \to n+1}}{kT}) \tag{7}$$

where V^i is the volume of the interfacial layer, D/δ^2 is the attachment frequency, D is monomer diffusivity, δ is the thickness of the interfacial layer which is in the order of the lattice spacing (approximately 0.235 *nm*), and $\Delta G_{n \to n+1}$ is the total free energy of formation for a microdefect changing from *n* to *n*+1 monomers.

The formulations for the growth rate and dissolution rate are general for any defect shape. The monomer concentration at the surface of a spherical microdefect is obtained by solving eqs.[2] and [3] as

$$C_S = \frac{k_r(n)C^{eq}(n) + k_d(n)C_\infty}{k_r(n) + k_d(n)} \qquad [8]$$

where $k_d(n) \equiv 4\pi r_n D$, $V^i \equiv 4\pi r_n^2 \delta$ for $k_r(n)$ in eq.[7], and r_n is the radius of the spherical cluster of size *n*. Axisymmetric spheroid-shaped precipitates can be modeled by solving eqs.[2] and [3] in oblate spheroidal coordinates. An expression for the surface concentration C_S can be derived in the same form as eq.[8]. We consider only the limit of an oblate disk-shaped particle of radius r_n; for this particle shape C_S is given by eq.[8] with $k_d(n) \equiv 8r_n D$, and $V^i \equiv 2\pi r_n^2 \delta$.

Free Energy of Formation for Oxide Precipitates

For an oxide precipitate, the free energy of formation is modeled as

$$G(n) = -nkT \ln\frac{C_O}{C_O^{eq}} - \gamma_V nkT \ln\frac{C_V}{C_V^{eq}} + \gamma_I nkT \ln\frac{C_I}{C_I^{eq}} + E_V + E_S \qquad [9]$$

where the first three terms are configurational contributions from the super-saturations of oxygen, vacancies and interstitials, and (γ_I, γ_V) are the stoichiometric coefficients in eq.[1]. The term E_V represents the contribution to the free energy from the stress field created by the density mismatch of the particle and matrix, and E_S is the surface energy of the precipitate/matrix interface. These energies depend on the shape of the precipitate. For spherical oxygen precipitates, $E_V \equiv 6V_m \mu e_C^2 X$ and $E_S \equiv 4\pi r_f^2 \sigma$, where $X \equiv 1 + 4\mu/3K$ (12,13). In these expressions, V_m is the volume of the unconstrained cavity created by the precipitate in the matrix including the additional volume created by vacancy absorption and interstitial ejection. The volume of the unconstrained precipitate is V_p. For a silicon oxide precipitate $V_p > V_m$ and stress is generated in the precipitate and matrix to create the final precipitate volume V_f with compressive stress in the precipitate. The corresponding radii of spherical volumes are (r_m, r_p, r_f) and (μ, K, e_C, σ) are the (shear modulus of silicon, bulk modulus of silicon dioxide, constrained strain, surface energy between the precipitate and the matrix).

Minimizing the free energy [9] with respect to γ_I and γ_V shows that either one or the other of these coefficients is zero (10). Physically, growth proceeds either by vacancy adsorption ($\gamma_I = 0$, $\gamma_V \neq 0$) or interstitial ejection ($\gamma_I \neq 0$, $\gamma_V = 0$), but not simultaneously by both mechanisms. For self-interstitial ejection

$$e_C = \frac{kT}{4\mu V_{Si}} \ln \frac{C_I}{C_I^{eq}} + \frac{\sigma}{2\mu r_p}(1 - \frac{1}{X}) \quad [10]$$

$$\gamma_I = \frac{1}{2}\left((1 + Xe_C)^{-3} \frac{V_{SiO_2}}{V_{Si}} - 1 \right) \quad [11]$$

Similar equations are obtained for (e_C, γ_V) for the vacancy absorption mechanism.

These two regimes for spherical precipitate growth occur at very different rates. When there is a pre-existing vacancy concentration, as is installed during the MDZ process, precipitate growth occurs at high rates. For the growth by the interstitial ejection mechanism to proceed, the ejected interstitials must diffuse from the matrix. Otherwise, a super-saturation of self-interstitials is created leading to high free energies and very low growth rates. Low rates for spherical precipitate growth cause disk-shaped precipitates to dominate at low annealing temperatures.

The stress energy of a disk-shaped oxygen precipitate is modeled by Hu (14) as a circular dislocation loop with a giant Burgers vector which corresponds to the lattice displacement perpendicular to the disk. Because the stress energy of the disk-shaped precipitate is low, there is little need to release stress by vacancy absorption and interstitial ejection. Accordingly, it is assumed that $\gamma_I = \gamma_V = 0$. The free energy of formation for a disk-shaped oxygen precipitate is expressed simply as

$$G(n) = -nkT \ln \frac{C_O}{C_O^{eq}} + \frac{\mu r (hs)^2}{2(1-\upsilon)}\left(\ln\left(\frac{8r}{hs}\right) - 1 \right) + 2\pi r^2 \sigma \quad [12]$$

where υ is the Poisson's ratio. The second term is the energy of the dislocation loop with the Burgers vector *hs*. The last term is the surface energy for the two circular surfaces of the disk. The aspect ratio of the disk is determined by minimizing the free energy of formation described by eq.[12].

The free energies of formation for disk-shaped and spherical precipitates with varying super-saturations of interstitials and vacancies are shown in Figure 3 for *T*=650°C. Clearly, spherical precipitates have the lowest free energy when a super-saturation of vacancies exists. The free energy is always higher when the mechanism shifts to interstitial ejection and $C_I / C_I^{eq} > 1$. The disk-shaped precipitate has a lower free energy than all spherical precipitates except those with pre-existing super-saturations of vacancies. This observation is critical to the results presented below.

The oxygen diffusivity at high temperatures (*T*>750°C) is taken from (15) as

$$D_O = 0.23\exp(-\frac{2.56eV}{kT})\ cm^2/s \qquad [13]$$

However it is commonly believed (16) that the effective oxygen diffusivity at low temperatures is much higher than the values predicted by [13]. Although the mechanism of the faster oxygen diffusion at low temperature is still not clear, it has been established by various experiments that the effective activation energy of oxygen diffusion for T<750°C is 1.55 eV versus the value of 2.56 eV in [13]. The oxygen diffusivity for T<750°C is taken from (17) as

$$D_O = 2.16\times10^{-6}\exp(-\frac{1.55eV}{kT})\ cm^2/s \qquad [14]$$

The calculations described here include the temperature dependence of the precipitate/matrix surface energy σ, as would be expected from arguments presented in (18). Fitting oxygen precipitation data in silicon to the model for classical homogeneous nucleation also has predicted the temperature dependence of σ (6,19). We use the expression

$$\sigma = 0.51 + 3\times10^{-4}(T - 873K)\ J/m^2 \qquad [15]$$

in the simulations presented here, which gives $0.465\,J/m^2 \le \sigma \le 0.66\,J/m^2$ for $450°C \le T \le 1100°C$.

COMPUTATIONAL RESULTS

Hi-Low-Hi Wafer Annealing

The traditional Hi-Low-Hi wafer annealing cycle (20) can be used to produce a spatially non-uniform oxide precipitate distribution across the thickness of the wafer with precipitate-free or denuded zones (DZ) at the wafer surface. This process is simulated here and the results are compared with the experiments reported by Kelton et al. (5,6). The annealing cycle is shown in Figure 4. Physically, there is a finite ramp rate for the temperature between each set-point, although no ramping rate was given in the references (5,6). The importance of the ramp between T_{nucl} and T_{growth} is explored here.

The first high temperature step of the cycle dissolves grown-in oxygen precipitates and thus erases the thermal history of the wafer. When the duration of this step is short, little oxygen out-diffusion occurs near the surfaces of the wafer (which would be typical when a denuded region is formed during a long time, high temperature anneal). The low-temperature step at a preset temperature, $450°C \le T_{nucl} \le 750°C$ nucleates oxide precipitates which grow at the long-time, high-temperature third step.

In this simulation, the $T_{high} = 1000°C$ first step is not hot enough to establish the high vacancy concentration needed to grow spherical precipitates. As a result, most of the precipitates formed are disk-shaped. According to [12], there is essentially no interaction between the growth of these precipitates and the point defect densities. Results are presented for wafers with initial oxygen concentrations (C_O) of $6\times10^{17}cm^{-3}$, $7\times10^{17}cm^{-3}$ and $8\times10^{17}cm^{-3}$.

Predictions for the total density of precipitates are shown in Figure 5 for $C_O = 8\times10^{17}cm^{-3}$ and three ramping rates for varying temperature between steps in the annealing cycles. All three sets of results agree qualitatively with the experiments of Kelton et al. (5,6) and show a peak in precipitate density at approximately $T_{nucl} = 600°C$. This maximum represents a balance between the driving force of the oxygen super-saturation and inhibition of oxygen diffusion at low temperature where too few nuclei are formed in the set time.

The final oxide density also depends on the temperature ramp rate between T_{nucl} and T_{growth}. For a finite temperature ramp rate some small nuclei have a chance to grow that would otherwise dissolve if the temperature was more suddenly increased. The slower the ramping rate, the higher the predicted precipitate density. A typical industrial ramp rate is $5°C/\min$. For this rate, the experiments and simulations are in reasonable agreement for all three values of C_O, except for $T_{nucl} < 500°C$. The effect of the length of the nucleation step, defined as the time $t = t_{nucl}$ the wafer is hold at $T = T_{nucl}$ is shown in Figure 8 plotted both on linear (a) and logarithmic (b) scales. In the experiment and simulations the 1000°C high temperature anneal for 15 minutes is followed by $T_{nucl} = 650°C$ for the time t_{nucl}. The nuclei are grown by the final two-step high temperature cycle shown in Figure 4. The oxygen concentration was $C_O = 7\times10^{17}cm^{-3}$.

The simulations and experiments are in reasonable agreement. As clearly seen in Figure 8(a), the evolution is divided into an induction period in which no nuclei grow that survive the high temperature growth and a period of nearly constant nucleation rate. The induction time depends on both the nucleation temperature (T_{nucl}) and growth temperature (T_{growth}). Lower nucleation temperatures T_{nucl} lead to slower oxygen diffusion and longer induction times for the size of the nuclei to reach the critical size for the 800°C step. Higher values of T_{growth} give larger critical radius and increase the induction time.

As the results in Figure 5 – 8 indicate the model for oxide precipitation described here reasonably predicts precipitate densities and the dependence on nucleation time and nucleation temperature. The dominant morphology of these precipitates is an additional output of the model. For the conditions used in these simulation, no substantial vacancy concentration exists and the dominant morphology is disk-shaped. This point is emphasized in Figure 9 where the size distribution after a $T_{nucl} = 650°C$ step for 32 hours is shown for $C_O = 8\times10^{17}cm^{-3}$. Note that the density of disk-shaped precipitates is more than four orders-of-magnitude higher than the density for spherical precipitates. This result is easily understood by noting that $C_I^{eq}(T=650°C)=1.5\times10^5 cm^{-3}$, while $O(10^{11})cm^{-3}$ interstitials have to be ejected to create a density of $1\times10^{10} cm^{-3}$ precipitates of size $n=100$. Formation of interstitial clusters from interstitials ejected by spherical oxides is included in the model, but this process is not efficient enough to get low super-saturation of interstitial. To make this point, the size distribution also is shown for a hypothetical model in which the bulk interstitial concentration is held fixed at $C_I=C_I^{eq}(T)$, corresponding to a system where interstitials ejected from precipitates are instantly absorbed into an undefined defect structure. In this case, the density of spherical precipitates is slightly higher than the density for disk-shaped particles.

Vacancy-Assisted Oxygen Precipitation

The simulation described here can be used to explore a variety of annealing strategies. One very interesting approach is the MDZ technology developed by MEMC (8) to create wafers with precipitates in the middle of the wafer and surface denuded zones with a low thermal budget. In the MDZ process, the wafer is first heated rapidly (with a ramp rate of $50°C/s$) from room temperature to $1250°C$ and held there for 10 seconds before being cooled to room temperature at a cooling rate of approximately $100°C/s$. This rapid, high-temperature step is followed by a nucleation step at $T_{nucl}=650°C$ and a growth step at $T_{growth}=1000°C$ for 16 hours.

No out-diffusion of oxygen occurs during the very short first step; however, very small oxygen precipitates that may have existed in the crystal are dissolved. More importantly, the equilibrium vacancy concentration corresponding to $T=1250°C$ is created during this step, and diffuses to the surface only in a thin surface layer because of fast cooling after this step. As a result, there is a high vacancy concentration in the bulk of the wafer.

The predictions for precipitate density as a function of t_{nucl}, with and without vacancy assistance, are shown in Figures 10 and 11 and compared with experimental measurements (8) for $C_O=7\times10^{17} cm^{-3}$. Results are shown for the traditional Hi-Low-Hi process and for the MDZ annealing cycle with $T=1250°C$ for 10 seconds. In both cases, $T_{nucl}=650°C$ and $T_{growth}=1000°C$ for 16 hours.

The thermal ramp-rate between $650°C$ and $1000°C$ was not reported in the experiments and is crucial for prediction of the results. The sensitivity of the final oxide density to the ramp rate is shown in Figure 10 for classical Hi-Low-Hi annealing. The experimental results are best matched with a ramp rate of $20°C/\min$, which has been used in the remainder of the calculations. The match between experiments and simulations is reasonable, considering that the detection limit for measuring the oxide density is approximately $1\times10^7 cm^{-3}$.

The effect of the MDZ treatment on the oxide density in both experiments (8) and simulations is shown in Figure 11. The impact of the implanted vacancy distribution on the higher nucleation rate of spherical precipitates is clearly seen. Also the agreement between experiments and simulations is excellent.

CONCLUSIONS

We have presented a framework for modeling oxygen precipitation in silicon by including the details of the growth of oxygen precipitates and the important roles of self-interstitials and vacancies in this process. The framework includes the role of precipitate morphology by including micro-mechanical models for both spherical and disk-shaped precipitates. Spherical shaped precipitates grow preferentially when there is ample vacancy concentration in the crystal to supply the free volume required to release elastic stress. Disk-shaped precipitates are favor at low nucleation temperature when stress relief is only possible by interstitial ejection.

Simulation results successfully predict oxide densities and the effects of nucleation temperature and time during traditionally Hi-Low-Hi annealing. The importance of vacancy absorption in the MDZ annealing process is also captured. Finally the annealing experiments reported by Kelton et al. (5,6) are quantitatively predicted, except at very low nucleation temperature ($T_{nucl} < 500°C$) by a simple internally consistent simulation. The disagreement between simulation and experiment for $T_{nucl} < 500°C$ still remains unanswered, but may be related to further changes in the morphology of the precipitates as suggested by the micrographs of rod-shaped precipitates by Bergholz (7).

ACLNOWLEDGMENTS

We thank MEMC Corporation, and Wacker Siltronic for financial support. We also thank Prof. T. Sinno, Dr. T. Mori, and Dr. G. Sin for valuable discussions.

REFERENCES

1. T. Y. Tan, E. E. Gardner and W. K. Tice, *Appl. Phys. Lett.,* **30**, 175, (1977).
2. T. Sinno, R. A. Brown, W. von Ammon and E. Dornberger, *J. Electrochem. Soc.,* **146**, 2300, (1999).
3. T. Sinno, E. Dornberger, R. A. Brown, W. von Ammon and F. Dupret, *Material Science and Engineering Reports,* **243**, 1, (2000).
4. T. Mori, Z. Wang and R. A. Brown, *Electrochem. Soc. Proc.,* **17**, 118, (2000).
5. K. F. Kelton, R. Falster, D. Gambaro, M. Olmo, M. Cornara and P. F. Wei, *J. Appl. Phys.,* **85**, 8097, (1999).
6. P. F. Wei, K. F. Kelton and R. Falster, *J. Appl. Phys.,* **88**, 5062, (2000).

7. W. Bergholz, *Semiconductors and Semimetals,* **42**, 513, (1994).
8. R. Falster, V. V. Voronkov, J. C. Holzer, S. Markgraf, S. McQuaid and L. Mule'Stagno, *Electrochem. Soc. Proc.,* **98-13**, 135, (1998).
9. T. Mori, Ph.D thesis, Massachusetts Institute of Technology, (2000).
10. Z. Wang, Ph.D thesis, Massachusetts Institute of Technology, (2002).
11. M. Schrems, *Semiconductors and Semimetals,* **42**, 391, (1994).
12. F. R. N. Nabarro, *Proc. R. Soc. A,* **175**, 519, (1940).
13. W. J. Taylor, U. M. Gosele and T. Y. Tan, *Mater. Chem. and Phys.,* **34**, 166, (1993)
14. S. M. Hu, *Mat. Res. Soc. Symp. Proc.,* **59**, 249, (1986).
15. J. W. Corbett, R. S. McDonald and G. D. Watkins, *J. Phys. Chem. Solids,* **25**, 873, (1964).
16. R. C. Newman and R. Jones, *Semiconductors and Semimetals,* **42**, 289, (1994).
17. S. Senkader, P. R. Wilshaw and R. Falster, *J. Appl. Phys.,* **89**, 4803, (2001).
18. F. Spaepen, *Solid State Phys.,* **47**, 1, (1994).
19. J. Esfandyari, J. Vanhellemont and G. Obermeier, *Electrochem. Soc. Proc.,* **99-1**, 437, (1999).
20. F. Shimura, *Semiconductors and Semimetals,* **42**, 577, (1994).

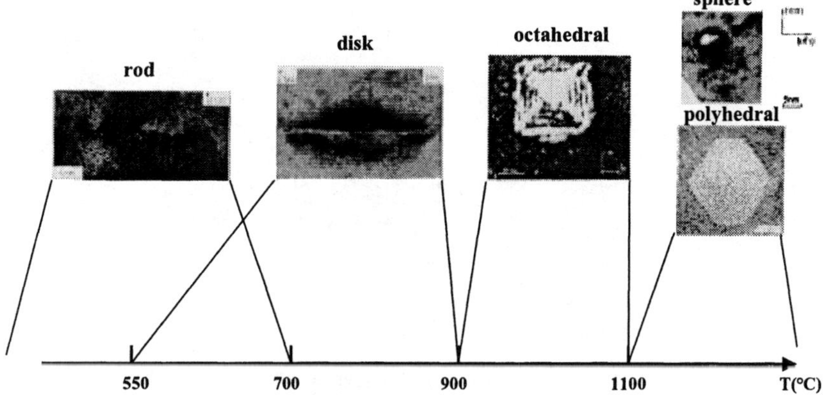

Figure 1. Morphology of oxygen precipitates under different nucleation temperature taken from ref.(7).

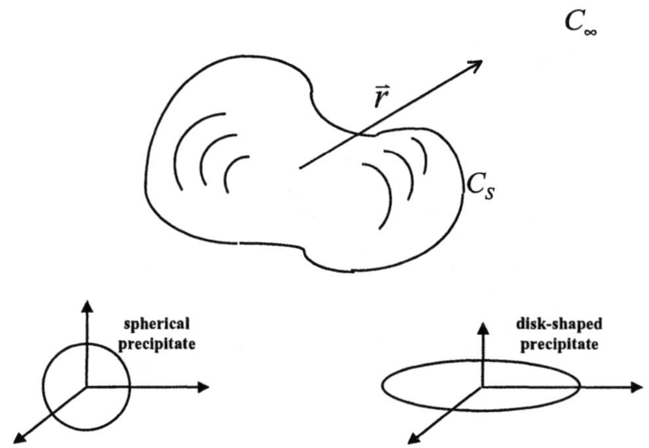

Figure 2. Coordinate system for arbitrary shape particle.

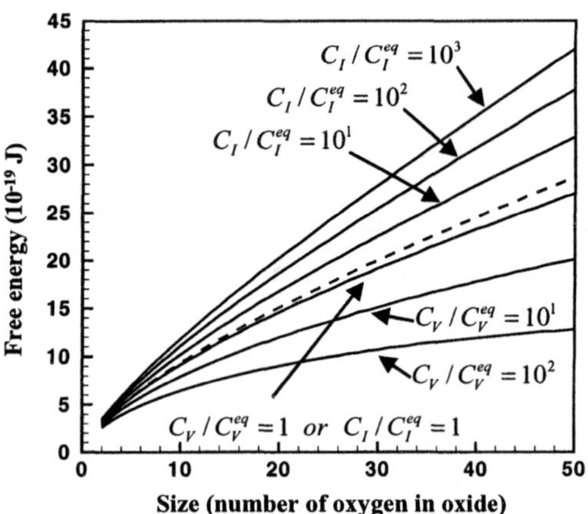

Figure 3. The free formation energy for the disk-shaped (- — —) and spherical (———) oxide precipitates with different supersaturation of interstitials and vacancies: $T = 650°C$.

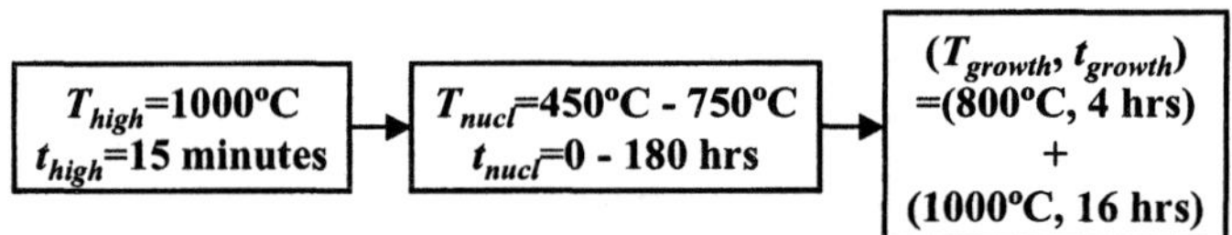

Figure 4. Annealing procedure used in simulations and taken from ref.(5).

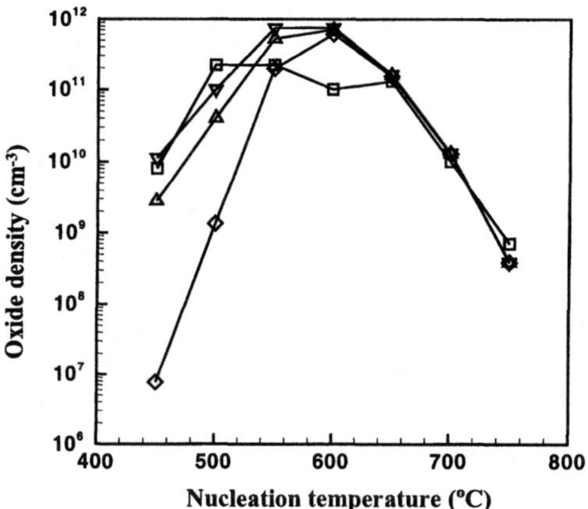

Figure 5. Oxide density as a function of nucleation temperature for $C_O = 8 \times 10^{17} cm^{-3}$. Results are shown for the experiments (□) and simulations with the thermal ramp rate of a step change (◇), $5°C/min$ (△), and $3°C/min$ (▽).

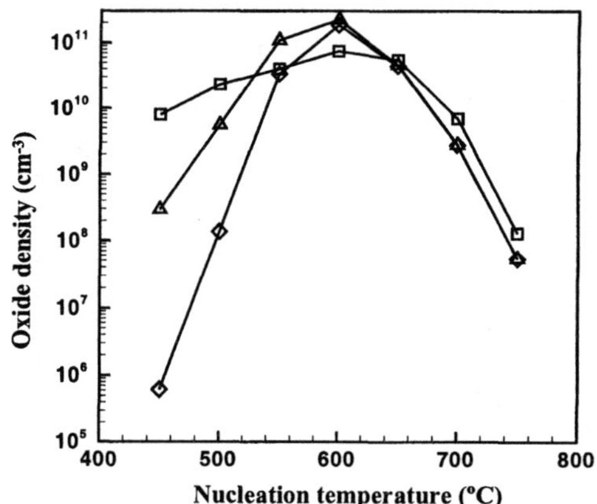

Figure 6. Oxide density as a function of nucleation temperature for $C_O = 7 \times 10^{17} cm^{-3}$. Results are shown for the experiments (□) and simulations with the thermal ramp rate of a step change (◇), and $5°C/min$ (△).

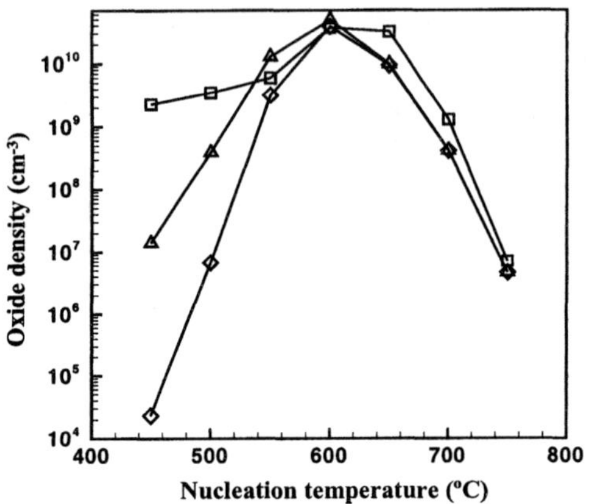

Figure 7. Oxide density as a function of nucleation temperature for $C_O = 6 \times 10^{17} cm^{-3}$. Results are shown for the experiments (□) and simulations with the thermal ramp rate of a step change (◇), and $5°C/min$ (△).

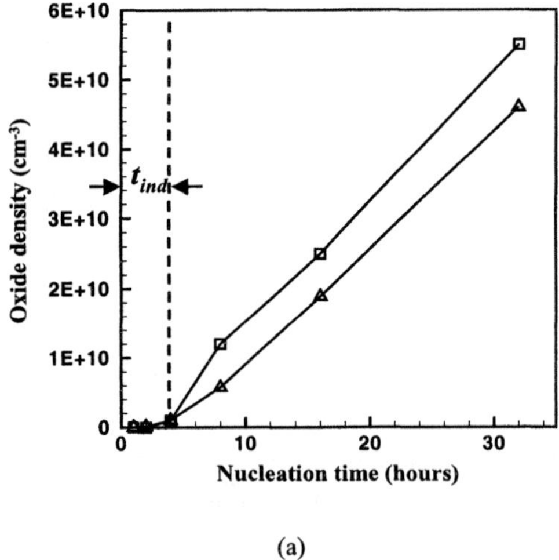

(a)

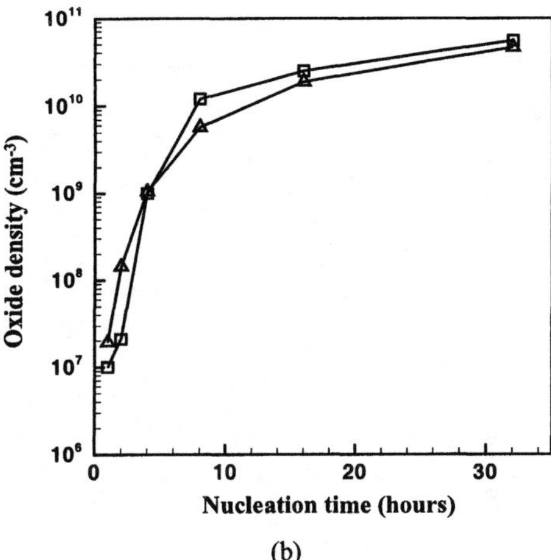

(b)

Figure 8. Oxide density as a function of nucleation time at $650°C$ for $C_O = 7 \times 10^{17} cm^{-3}$ plotted on linear (a) and logarithmic (b) scales. Experimental (□) and simulation (△) results are shown.

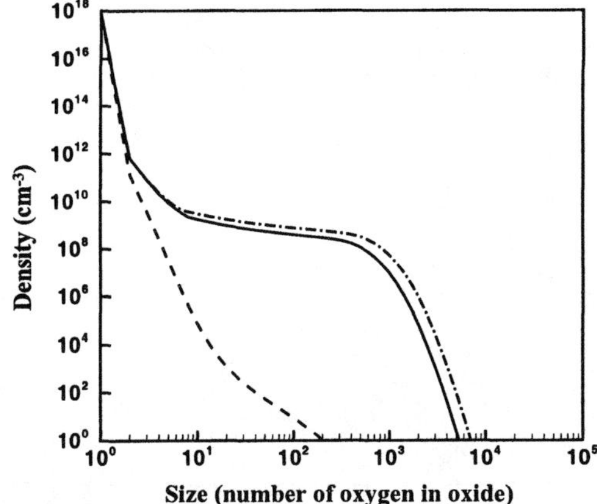

Figure 9. Size distributions of spherical oxides and disk-shaped oxides after 32 hours annealing at $650°C$ for $C_O = 8.0 \times 10^{17} cm^{-3}$. Results are shown for spherical (- — —), disk-shaped precipitates (———), and spherical precipitates under $C_I / C_I^{eq} = 1$ (– · — ·).

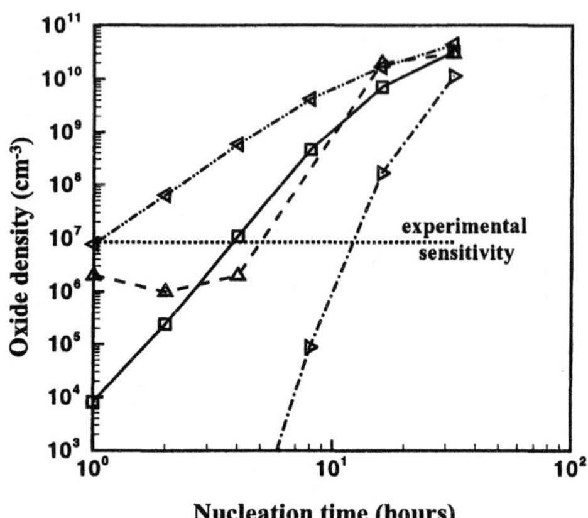

Figure 10. Sensitivity of oxide density on the ramping rate for traditional annealing process without vacancy assistance. Results are shown for the experiments (△) and simulations with the thermal ramp rate of a step change (▷), 20°C/min (□), and 5°C/min (◁).

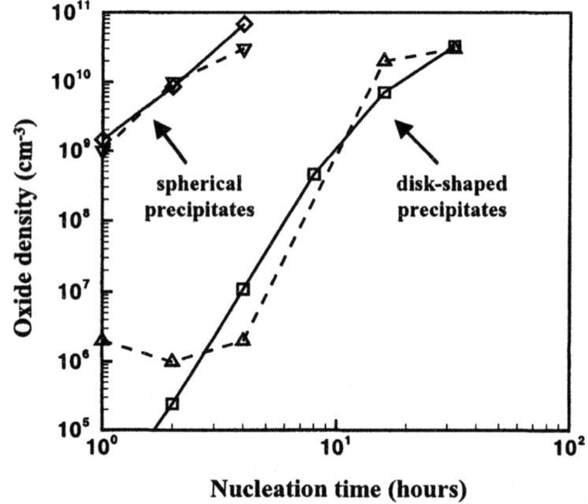

Figure 11. Oxide density as a function of the nucleation time with and without vacancy assistance. Results are shown for experiments (- — —), simulations (———), with MDZ treatment (◇, ▽), and without MDZ treatment (□, △).

SIMULATION OF SILICON CZ GROWTH: WHERE WE ARE NOW

Olli Anttila, Maria Laakso, and Jari Paloheimo
Okmetic Oyj, P.O. Box 44, FIN-01301 Vantaa, Finland

Jussi Heikonen, Antti Pursula, Juha Ruokolainen, Ville Savolainen, and Thomas Zwinger
CSC – Center for Scientific Computing, Ltd.
P.O. Box 405, FIN-02101 Espoo, Finland

Melt instabilities lead to crystal inhomogeneities as well as to occasional structural yield losses during silicon CZ growth. We have modeled melt flows in a large scale, cylindrically symmetric growth system. Time dependencies similar to what we see experimentally are reproduced in these simulations, without turbulence models. Validation and some widely known growth related aspects are discussed.

Introduction

Mathematical modeling of silicon CZ crystal growth was started with the phase change interface shape and temperature distribution calculations of the crystal about 30 years ago [1,2]. The early models faced challenges with radiative heat exchange and boundary conditions at the growth interface. Melt behavior was also included fairly early on but the numerical task was far too complex to be solved with the available computers and computational methods, except for extremely small melts. However, new insight in the growth was gained even from these simplified models, which helped to develop better hot zones for improved crystal quality and yields.

In CZ silicon wafer production the crystal growth is the first, and by far the most expensive process step. With increasing batch sizes, driven by the need for larger diameter wafers, the necessity of increasing crystal growth yields became more evident, and better stability of the melt became inevitable. Smarter hot zones that provide more suitable temperature distributions were developed, and gas flow patterns were modified to keep silicon monoxide out of critical areas. Magnetic fields of different shapes were also studied and used to slow down erratic melt flows.

The biggest challenge for modeling CZ growth is the large turbulent melt while the gas flow is normally somewhat easier to simulate. Temperature distribution calculations are relatively straightforward except for the impact of the melt flows. If the temperature distributions during different time steps of the growth are already known, the thermal history of the grown material as well as the thermal stresses can be easily evaluated. In addition, other phenomena in the crystal occurring after solidification are being modeled by several groups, most notably the behavior of vacancies and self-interstitials, as well as oxygen precipitation.

Melt flow simulations for industrial-sized crucibles reached a new level with the successful introduction of stabilized finite-element methods [3], with real silicon material parameters, and without the use of any turbulence model, as well as with other direct simulation tools [4]. Stabilized finite-element methods were first applied to cylindrically symmetric 2-D calculations [5,6]. In state-of-the-art PC's, realistic simulations can now be produced in a matter of few days. Simultaneously, advanced turbulence models have been developed to give results that seem to be fairly realistic with moderate computational effort [7]. These models need to be validated against experimental data and direct 3-D simulations. 3-D simulations without turbulence models have been performed, but their application to real life cases is still very limited: one calculation typically requires months on a powerful workstation cluster [8].

In this paper, the crucible size in the simulations was 20" and the melt depth 160 mm, corresponding to about 60 kg of silicon melt remaining in the crucible, except around Fig. 3. The crucible rotation rate was 5 rpm unless otherwise noted. The crystal diameter was 156 mm, counter rotating at 20 rpm. The crucible wall temperature was 30 K higher than the melting point 1685 K, and the external radiation temperature above the melt was 1600K. Most of the calculations were performed using a 2 GHz Pentium IV personal computer, with 2 GB of memory.

Evidence of time dependence

Experimental temperature measurements in the melt and simple visual observations of the melt surface portray time-dependence and 3-D melt behavior (see, e.g., refs. [4,8]). An example is shown in Fig. 1. This figure shows a minority carrier recombination lifetime map of a 125 mm wafer, cut out from a crystal, which was grown about ten years ago using a non-optimized hot zone design. The growth inhomogeneities have been made visible through a thermal anneal, which has created small oxygen precipitates. Higher density of oxygen precipitates is seen as lower lifetime. The ring like structure was born as the wafer was cut out perpendicular to the growth axis. A measurement of the dominant ring diameters, together with the knowledge of the freezing interface shape and the pull speed, reveals quasiperiodic time dependence of the growth rate or of the oxygen concentration [9], the dominant period being of the order of one minute in this case. The external pull speed, controlled by the crystal grower, was essentially constant.

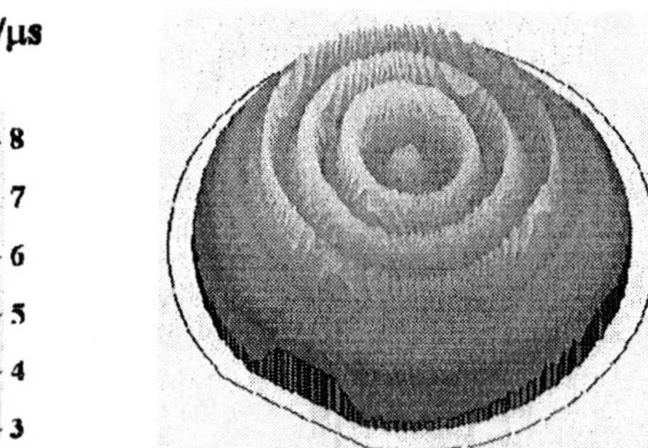

Fig. 1. This recombination lifetime map of a 125 mm wafer, after oxygen precipitate growth for 4 h at 850 °C followed by 12 h at 1000 °C, reveals quasiperiodic time dependence of the melt behavior.

Anybody who has the possibility to look into a CZ grower, while it is running, can easily see different shades of red, rotating approximately with the crucible, provided that enough free surface of the melt is visible for the observation. The elongated darker areas mark lower surface temperatures, and our conviction is that these areas are born at the boundaries of three dimensional convection shells, between which the melt probably stays close to the surface longer than average. The radiation then makes this more stagnant melt lose its temperature. With proper equipment, this kind of surface pattern may be transformed into a quantitative, time-dependent surface temperature map, as in ref. [4].

Examples of qualitative melt flow behavior as well as computed silicon flow patterns are relatively easy to find in the literature, but in most cases the data has been very unreliable or misleading. With typical crucible sizes and temperature differences between the crystal and crucible walls, the melt flow, in contrast to numerous calculated results, becomes turbulent, time dependent, and three-dimensional. However, well established turbulence models, used e.g. for the airflow outside an airplane, fit poorly to the fluid flows within a rotating crucible of a CZ crystal grower. Good experimental data of the melt behavior is scarce and it is hard and expensive to gain, mostly because of the reactivity, high temperature, and the opacity of the melt. The lack of validation of the various computational results has been a major deficiency.

Instead of one large regular convection roll or two, as often suggested in the literature, there are several of them, the outmost one still running upwards at the crucible wall, but there are also rolls rotating in the opposite direction (see, e.g., Fig. 4). However, it is impossible to show the flow behavior in a simple cross-sectional picture, as the situation changes constantly with time, and the three-dimensional nature of the flow makes visualization even more complicated. It is a major challenge to find good representative figures or numbers to make reasonable comparisons between different flows, and visual impressions become also important.

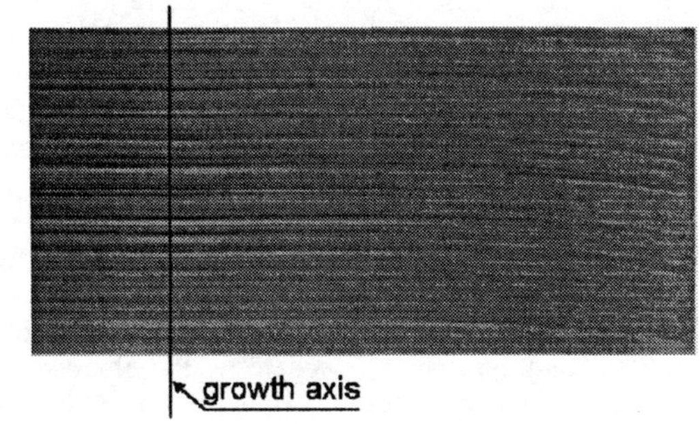

growth axis

Fig. 2. An LPS map of a 150 mm N(100) crystal. The sample has been cut out of a 50 mm high section of the crystal and etched.

One approach is to use test points within the melt, and calculate spectra for the temperatures at these points, as shown later in Fig. 3. Velocity components may also be represented in the same way. On some surfaces, like at the freezing interface or at the crucible wall, the temperatures or heat flux densities may be calculated for the whole surface. The thermal gradients in the melt, in the vicinity of the freezing interface, can be, together with the knowledge of heat flux densities in the crystal, converted into local growth velocities. These give us the momentary positions of the freezing interface, too. The time-dependent crystallization velocity leads to variations in the doping concentrations, which can be experimentally detected, e.g., by using Lateral PhotoScanning (LPS) technique, Fig. 2. In this method, a cross sectional sample out of a crystal is scanned using a focused laser beam [10]. The excess charge carriers are separated by internal electric fields in the sample, which are caused by dopant concentration variations, mainly due to crystal growth rate fluctuations. The separation of the charge carriers can then be detected, e.g., through ohmic contacts in both ends of the sample as small voltage signal. We expect that the comparisons between calculated growth rates and those observed experimentally using LPS will become a key means of validation.

Validation

As discussed above, results given by a computational model may be largely misleading unless the model has been properly validated. Validation is not easy and it is difficult to tell when it is sufficient. There are some crucial components though.

1) The model needs to be able to show the same essential features in the melt flow, which we experience in practice, e.g., characteristic temporal behavior (see Figs. 2 and 3). The accuracy of the modeling results may, however, not be any better than how well the relevant boundary conditions can be set, and this may be difficult too. As an example, when calculating melt flows in a crucible, the melt experiences radiative and conductive heat transfer through the free surface as well as through the silica crucible. The boundary conditions in the melt cannot be posed any more accurately than one can measure the temperatures in the melt, very close to the interfaces, and elsewhere in the system, a challenging task in itself. Or one can start modeling one step earlier, and get the thermal boundary conditions to the melt from a global thermal and flow model. At that point, there are uncertainties about the physical properties of the thermal insulators, structural materials, the melt itself, and the demanding task posed by the semitransparent, poorly conducting silica crucible. In addition, there are inadequately defined thermal contacts between different parts of the system and variations of the material properties with wear.

2) The model needs to show explanatory power. Oftentimes, a model cannot properly forecast individual parameters, like oxygen concentration in the crystal, without some tuning of the physical parameters adopted from the literature data. This should be accepted to a certain degree because of the uncertainties in the measured data. However, the model should be able to correctly forecast the effect of an intentional change in the system, and most of the times, give a reasonably good idea about the magnitude of the effect. Furthermore, careful study of the results should give some understanding about how, i.e., through which mechanism the imposed change influenced the observed parameters. As an example, we take the effect of crucible rotation rate on the oxygen concentration and on the required heater temperature during the growth. This will be treated later in a separate paragraph.

3) Most challengingly, the model needs to be able to predict experimental behavior not observed earlier. This requirement is more stringent than plain explanatory power. As one starts to model such behavior, which is already known, one tends to add mainly those limited sets of equations, which are needed to explain the known features. At the same time, important aspects, which were not crucial at the time of model building, may be neglected, making the model fail in its ability to forecast correctly the effects of completely new ideas. Furthermore, if one wants to explain observation, it is relatively easy to become biased in one's selection of the potentially important factors, which have an impact on the observed behavior.

4) The user needs to have confidence to the model. Validation is a very important part of this, but at the same time, the fewer is the number of adjustable parameters, and the closer the model is to generally accepted physical realm, the easier it is to build the confidence. As an example, if one would use turbulence models instead of solving the Navier-Stokes equations directly, a large body of independent and converging experimental data would be needed, as it exists in aviation research. However, even if this data would be available, and the model could be reliably used to present today's growth processes, its ability to forecast new developments could be poor.

In Fig. 3, we show a detail of a validation effort, where the temperature of one point in the melt has been tracked experimentally, and the same point is then followed during two different kinds of simulations, one our own. In this case the crucible size was 14", only, in the experimental setup, and the measured temperature profiles along the crucible walls were used in both simulations. The graphs show power spectral densities of the measured and simulated temperatures at the point located 30 mm from symmetry axis and 20 mm below the melt surface, tracked over a period of about 250 s. The largest amplitudes in the spectrum of the 3-D simulation [8] agree excellently with the experiments, whereas our 2-D simulation underestimates the amplitude, but the shape of the spectrum is very close to the experimental one. The probable reason why we underestimate the magnitude of the temperature variation is that our 2-D model ignores any positional wobble of the cool flow, which goes downward near the crucible center line. The relatively short time period for the calculation of the spectra does not allow us to resolve temperature fluctuations of about one minute or longer time scales.

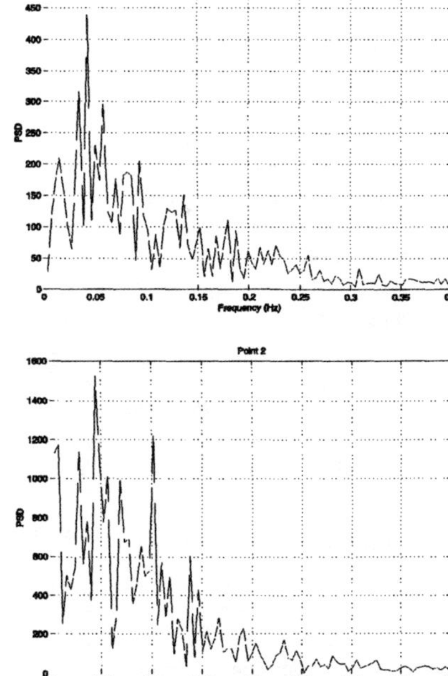

Fig. 3. Temperature power spectral densities (PSD) of a single point 30 mm from symmetry axis, 20 mm below the crystal. Right: Simulation with our FEM code. Below: Three-dimensional simulation from ref. [8]. Right and below: Experimental results, from ref. [8]. Crucible rotation rate was 5 rpm.

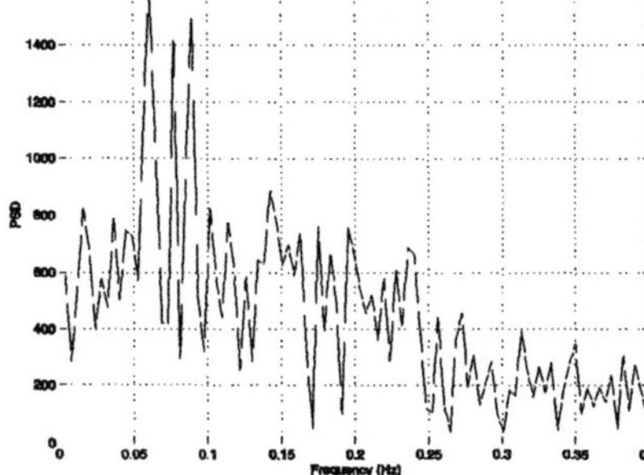

Turbulence and grid density requirements

Definition: *When describing complex evolution in space and time (sometimes called fully developed turbulence), it is applied where spatially localized behavior is chaotic and behavior at spatially separated points is decreasingly correlated as the separation increases. In fluids this appears as vortices and eddies of all sizes and at all length scales. Though originally used to describe fluid complexity, the term is now applied to any spatiotemporal complexity of this type.* (Academic Press Dictionary of Science Technology, 1996)

According to the above definition, turbulent flow has vortices and eddies of all sizes and at all length scales, which would make it impossible to resolve numerically the smallest details of the flow. We consider the silicon melt turbulent, but its behavior is definitely not of the type, which is called fully developed or hard turbulence. It is chaotic to a certain degree but it also shows signs of quasiperiodicity sometimes. At this point, the typical dimensions of the smallest significant eddies are still unclear to us. However, it might be possible to solve a turbulent flow problem numerically, even if the grid size and the time step are larger than the smallest eddies, without using turbulence models. A trustworthy solution would require e.g. that the overall result is largely independent on the discretization, i.e., on the spatial grid density and the size of the time step.

With coarse grid, the solution would lose the smallest details of the flow, but as long as the equations for mass and energy transfer are properly discretized, overall melt and heat flows could still be realistic. One of the requirements for this to take place is that the heat diffusivity is large, as it is with molten silicon. However, it would be dangerous to use the same kind of approach to calculate transport of slow diffusers, like oxygen and dopants. In order for us to model oxygen or dopant transport realistically, we need to address the issue of the smallest scale eddies, too.

The boundary layer thickness observed close to crucible walls, using dense grids there, suggests that 0.1-0.3 mm grid should be sufficiently fine, in accordance with the boundary layer theory [11]. Grid densities of this kind along the melt boundaries are not computationally prohibitive so oxygen dissolution as well as oxygen and dopant evaporation and incorporation into the crystal can already be tackled with the present numerical tools. This does not, however, tell us much about the grid density requirements in the bulk of the melt. For example, 0.3 mm dense grid in the bulk would still be beyond today's personal computers, even in 2-D. The few tests we have performed this far with locally denser grids, down to 0.2-0.3 mm, have, however, not given any clear indications that grid size smaller than 1-2 mm would be needed in the bulk of the melt. However, we are looking into large eddy simulation (LES), too, to characterize these sub-grid sized features more realistically, and to possibly give a better starting point for oxygen and dopant transfer in the bulk of the melt.

Effect of crucible rotation on heat transport

It is well known to the CZ silicon crystal growth society that if the crucible rotation rate is increased while keeping other crystal growth related parameters constant, specifically the average crystal growth rate, it is necessary to increase the heater power and temperature. The classical reasoning is straightforward. The average melt density is higher close to the melt surface, and higher rotation rate creates an additional force, which opposes the flow pattern caused by the buoyancy force. That is why the temperature along the crucible edge needs to be elevated, in order to transfer the same amount of heat towards the crystal as before. One can also find another factor, supporting this effect, related to Coriolis effect. However, this classic explanation is very qualitative, at it may not stand a more critical look.

The mechanism by which the crucible wall temperature is increasing with the rotation rate seems to be more complicated. Our simulations suggest that a higher rotation rate creates more elongated rolls in vertical direction, and their number is increased at the same time (Fig. 4). The radially more extended rolls, dominant at low rotation rates, transport heat more effectively over long distances.

Fig. 4. Momentary but representative velocity profiles for different crucible rotation rates. Right: 5 rpm. Below: 10 rpm. Right and below: 15 rpm. The perpendicular velocity component is shown, only. The displayed velocity scale is meters per second.

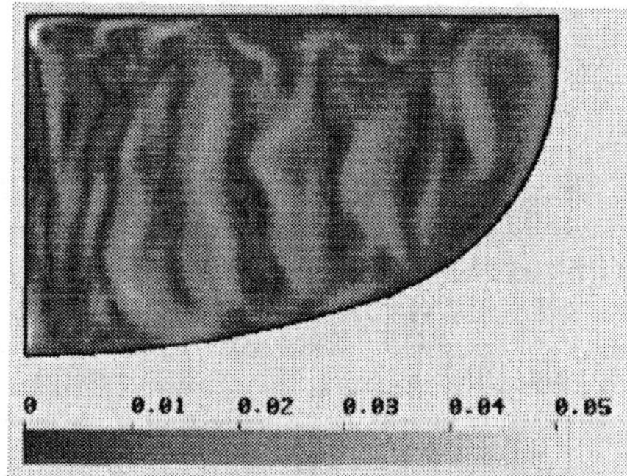

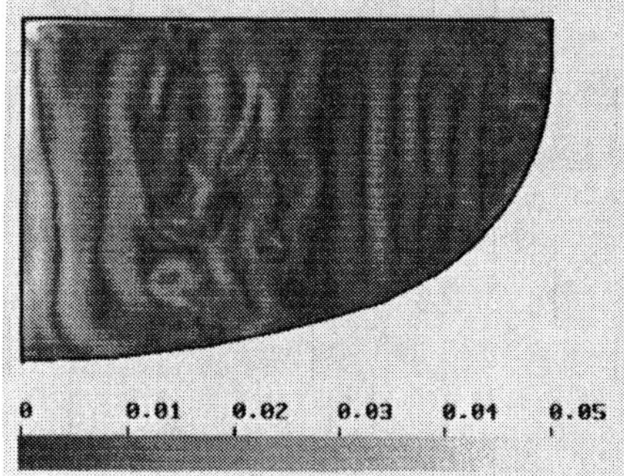

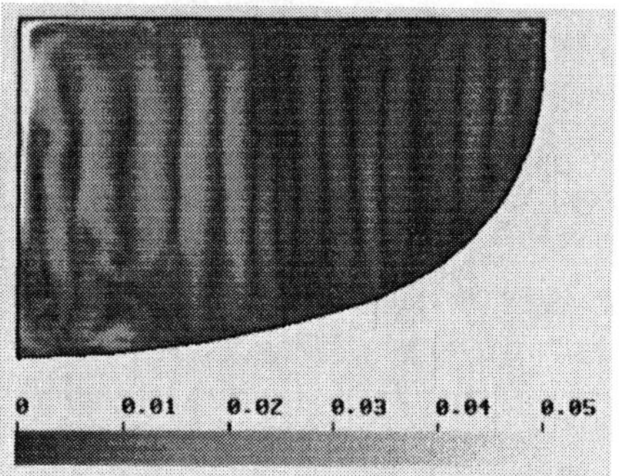

Now, why should larger crucible rotation rate create narrower, vertically elongated convection rolls. This is not fully clear, but our perception at this point is that the main reason is related to the Coriolis force. A broad convection roll is capable of effective transfer of azimuthal linear momentum in radial direction (which is in this case closely related to the angular momentum relative to the center axis), in addition to heat and mass. As the flow moving inwards hits another roll moving outwards, they exchange azimuthal linear momentum, and the outer flow's rotation rate slows down, as the other roll experiences acceleration. If the radial distance over which a single roll transfers linear momentum grows large, the amplitude of the momentum transfer to the neighboring roll is increased at the same time. And this creates additional turbulence, which is a good starting point for an additional convection roll to be born between the original rolls. So, in the first approximation, the radial width of the convection rolls should decrease inversely proportional with the crucible rotation rate, in fair agreement with our simulations. 3-D effects almost certainly complicate things further.

Strong wear at crucible corner area

The bulk of the melt rotates with almost the same angular velocity as the crucible, even though it takes several minutes for the melt to reach a stable rotation condition after the crucible rotation rate has been changed. However, there is a small difference, a few cm/s, in the azimuthal linear velocity of the melt, compared with the crucible linear velocity. This difference does not change much with the crucible rotation rate, if one looks the area outside about one third of the crucible radius. Most of the melt is rotating slightly faster than the crucible. The azimuthal velocity difference between the rotating melt and the crucible create a difference in the centrifugal forces between the melt layer in the immediate vicinity of the crucible and in the bulk of the melt. The tangential component of this centrifugal force difference is largest in the crucible corner area. In Fig. 4, this additional force, which makes the flow patterns stay close to the crucible wall a little longer than elsewhere, is barely visible, but its effect is very significant during abrupt changes in the crucible rotation rate. As the flow sweeps the corner section of the crucible more intimately, enhanced oxygen transport and wear follows. The location matches excellently with the wear pattern we typically see after long growth cycles. Normally, the crucible temperature is not uniform, though uniform wall temperature was used in these calculations, but it has highest values in the same area as where the flow dependent wear would be strongest. The temperature distribution is traditionally considered as the main culprit for the fast wear of the crucible in the corner area. The higher crucible wall temperature increases oxygen dissolution. However, solubility of oxygen into molten silicon, according to the literature data [12], should be only slightly dependent on the melt temperature. The conclusion here is that the temperature distribution is only partially responsible for the corrosion, and melt flow behavior must be controlled as well, to extend the crucible lifetime.

Oxygen distribution in the melt

It is a widely known phenomenon that the oxygen level in the growing crystal tends to decrease towards the tail end. The classical explanation has been to relate the surface area of the silica crucible, exposed to dissolution by the reactive melt, which decreases with recessing melt surface, to the free surface area of the melt, through which oxygen evaporates into the inert gas flow. This simple assumption, based on a notion of well mixed melt, has been questioned earlier, e.g., through electrochemical measurements of oxygen concentration in the melt, in which case higher concentrations have been observed far from the center axis [13]. In this case no crystal was growing. In Fig. 5, we show a calculated momentary oxygen concentration profile in the melt and in the gas phase just above the melt, scaled to the maximum solubility in both phases separately. The inert gas flow has been forced close to the melt surface using a baffle. There is a strong radial variation in the melt, as the low diffusivity of oxygen does not allow it to move but with the flow, and the dominant flow patterns are vertical rolls. Use of more realistic temperature boundary conditions may change the picture somewhat, as the Marangoni flow close to the crystal is underestimated here. Numerical diffusion caused by the relatively coarse grid here may also have an impact on the calculated oxygen distribution. Our confidence level concerning the calculated oxygen transport in the melt is still fairly low, e.g., the distribution presented in Fig. 5 is not compatible with the observed low oxygen concentrations in heavily antimony or arsenic doped crystals.

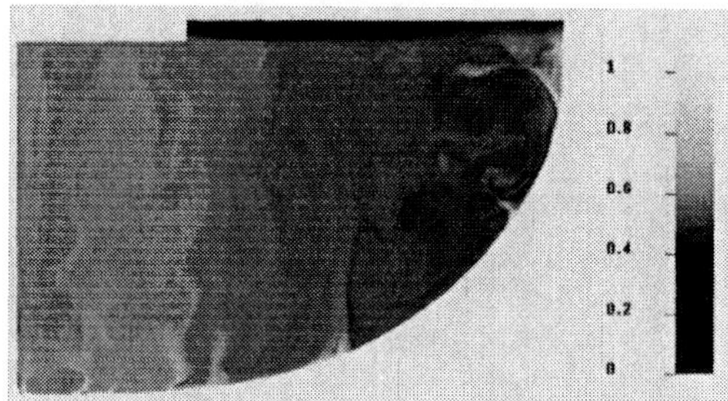

Fig. 5. Oxygen distribution in the melt, scaled to maximum solubility. Vertical convection rolls make it difficult for oxygen to move radially, and under the melt-gas interface the oxygen concentration is significantly lower through the whole depth of the melt.

Another well established phenomenon concerning oxygen in the crystal is that increased crucible rotation rate increases the oxygen level. At this point it is not yet clear to us, what the main mechanism is, through which the crucible rotation rate influences oxygen level. If it is the increased silica dissolution rate, especially at the crucible corner area or if it is the more difficult radial transport of oxygen in the melt, out from the region under the crystal.

Effect of magnetic cusp field

Out of the two types of direct current magnets used in the production now only the effect of a cusp magnet may be modeled in 2-D; the other commercial set-up of transverse field definitely breaks the symmetry. In ref. [6], the impact of increasing cusp field is shown to stabilize the flow patterns, as expected. Relatively weak fields damp the flows starting close to the crucible edge, i.e., where the field is strongest, but it has little effect close to the symmetry axis, until relatively strong fields, in the range of 100 mT are used.

Two-dimensional vs. 3-D calculations

Three-dimensional calculation is possible today but the computational cost is still prohibitive for an industrial application of this kind. In order to make proper use of a model within industry R&D, the software should be able to run on a powerful workstation, and preferably on high-end personal computers. The parallelization of the code, used in this work, has been preliminarily tested, and it has been running on a mainframe. However, in order for the modeling work to have a reasonable cost as well as proper impact, the actual use of the model should be made by the people within industry, using preferably their own reasonably priced hardware. A longer-term goal is to have the models run on existing high-end personal computers in the background, utilizing their idle time, and the communication between the processors taking place along fairly standard high speed intranets. The need for information exchange between the processors is large, because in 3-D simulations, with a large number of processors, the number of surface elements within each partition is significant compared with the number of volume elements. A smaller complication is that the thermal environment couples practically all surfaces together through radiative interaction.

The need for intensive information exchange between distributed processors poses serious challenges to the development of computational tools, too. However, we expect that the requirements for effective use of existing PC's and local networks where there is a need for heavy parallel processing will be satisfactorily solved within a few years. Until then, 3-D calculations are feasible mainly as references and in establishing knowledge about the limitations of 2-D calculations, as well as in gaining confidence about their applicability. They are definitely too expensive, at this point, for any practical process optimization or for screening the effects of new ideas.

Conclusions

Time dependencies similar to what we see experimentally are now produced using 2-D direct simulation without turbulence models on personal computers. The impact of various factors influencing the melt flow patterns, the melt stability as well as the transport of impurities can be foreseen. Computer simulations together with experimental work with production scale equipment help us understand the behavior of the melt as well as the influence of process changes such as hot zone modifications. In addition, completely new approaches may be found more easily. As we gain reasonable confidence that CZ growth models represent the real world with practical accuracy, such simulations will add significantly to the understanding of the growth processes. In addition, they will act as a very reasonably priced test bench for future developments. The development of computational methods as well as the even faster pace of computer technology has allowed us to address challenges which were well beyond our reach still ten years ago.

References

[1] T. Arizumi and N. Kobayashi, J. Crystal Growth, **13/14** (1972) 615.

[2] R.A. Brown, Am. Inst. Chem. Eng. J., **34** (1988) 881.

[3] L.P. Franca, S.L. Frey, and T.J.R. Hughes, Computer Methods in Applied Mechanics and Engineering, **99** (1992) 209.

[4] M. Tanaka, M. Hasebe, and N. Saito, J. Crystal Growth, **180** (1997) 487.

[5] J. Järvinen, J. Ruokolainen, V. Savolainen, and O. Anttila, A Stabilized Finite Element Analysis for Czochralski Silicon Melt Flow, Proc. for Mathematics In Applications, Novosibirsk (1999).

[6] V. Savolainen, J. Heikonen, J. Ruokolainen, O. Anttila, M. Laakso, and J. Paloheimo, J. Crystal Growth, **243** (2002) 243.

[7] I.Yu. Evstratov, V.V. Kalaev, A.I. Zhmakin, Yu.N. Makarov, A.G. Abramov, N.G. Ivanov, E.M. Smirnov, E. Dornberger, J. Virbulis, E. Tomzig, and W. von Ammon, J. Crystal Growth, **230** (2001) 22.

[8] S. Enger, Numerical Simulation of Flow and Heat Transfer in Czochralski Crucibles, Ph.D. Thesis, University Erlangen-Nürnberg, 2001.

[9] W. Zulehner, Origin and Effects of Inhomogeneous Impurity Distribution in CZ Silicon, Defect Control in Semiconductors, K. Sumino, ed., Elsevier (1990), p. 143.

[10] A. Lüdge and H. Riemann, Doping inhomogeneities in silicon crystals detected by the photoscanning (PS) method, Proc. DRIP VII, J. Donecker and I. Rechenberg, eds., Inst. Phys. Conf. Ser., **160** (1998), p. 145.

[11] H. Schlichting and K. Gersten, Boundary Layer Theory, 8th ed., Springer (2000).

[12] K. Hoshikawa and X. Huang, Si Melt Growth: Oxygen Transportation During Czochralski Growth, in Properties of Crystalline Silicon, R. Hull, ed., INSPEC (1999), p. 23.

[13] A. Seidl, R. Marten, and G. Müller, J. Crystal Growth, **166** (1996) 680.

HIGH-SPEED GROWTH OF FZ SILICON FOR PHOTOVOLTAICS

A. Luedge, H. Riemann, B. Hallmann, H. Wawra
Institute of Crystal Growth, Max-Born Str 2,
D-12489 Berlin
L. Jensen, T. L. Larsen, A. Nielsen
Topsil Semiconductor Materials A/S, P.O.Box 100, Linderupvej 4 DK-3600
Frederikssund, Denmark

Floating Zone (FZ) silicon can be grown with pull speeds comparably higher than for CZ silicon because the heat is dissipated by radiation into the colder surroundings. In 1955, Billig analyzed theoretically the heat balance of crystals growing under FZ-like conditions and computed approximately the maximum of the crystallization rate v_c as a function of the crystal radius regardless of the crystal structure. In this work we have experimentally tested how fast FZ silicon crystals of diameters up to 150mm can grow dislocation-free.

The main result is that typically at least 85% of the theoretical v_c value after Billig can be reached before the crystal dislocates, e.g. v_c = 3,8mm/min for a crystal of 125mm diameter. The deflection of the crystallization interface evaluated from the Lateral Photovoltage Scanning (LPS) striation pattern increases with v_c. When approaching the critical pull rate, the interface shows symmetry deviations indicating instability. Touching of non-molten core parts of the feed rod onto the growing interface could frequently be observed when the structure got lost.

The minority carrier lifetime τ is a main quality parameter for silicon solar cells. Ciszec found for thin crystals, that τ increases with a faster pull rate. On the other hand for high growth rates, the segregation coefficients are enhanced for impurities with recombination activity, too. We observed a stable or weakly rising τ with v_c. After the sample surfaces were carefully passivated, reasonably high lifetimes could be measured.

1. INTRODUCTION

In this work we have investigated the maximum growth rates for certain FZ processes for dislocation-free FZ crystals with different diameters up to 150mm. The study investigates reasons responsible for loosing mono crystallinity, if the pull rate is too high and how the minority carrier lifetime, being important for the solar cell efficiency, is influenced by high growth rates. The growth interface shape were derived from Lateral Photovoltage Scanning (LPS) measurements detecting the doping striations.

Today, solar cells made of FZ silicon show the highest efficiencies of silicon based cells, if the cell process is adapted to exploit advantages of the FZ material like large carrier diffusion length. On the other hand, FZ silicon developed for electronic devices is expensive compared to e.g. solar-grade mono crystalline CZ silicon. To overcome this disadvantage, one attempt is to develop a crystal growth process for low cost FZ silicon for photovoltaics. Because of the relatively high temperature gradient at the crystallization front, the FZ process has the potential of higher growth rates than those for CZ. Doing so some material parameters could degrade, like doping homogeneity or intrinsic defect structure, but they are less relevant for PV than for electronic applications.

2. EXPERIMENTAL

FZ silicon crystals with different diameters from 7 to 150 mm were grown to test the growth perfection at high pull rates. Depending on the crystal diameter different heating coils had to be used to allow for a dislocation-free crystal growth in particular for high growth rates. During the experiments, the pull speed was increased gradually until the crystal became dislocated. From some of those crystals samples were prepared corresponding to the different growth rates. Axially cut samples were used for the LPS visualization of the striations and radially cut samples for μPCD lifetime mapping measurements

3. RESULTS AND DISCUSSION

3.1 Maximum growth rates

Billig [1] determined the heat balance for the highest crystallization rate for non-dendritic growth theoretically and found the maximum growth rate to be

$$V_{\max} = \frac{1}{L * \rho} \sqrt{\frac{2\sigma\varepsilon K_m T_m^5}{3r}} \quad (1),$$

where	L	latent heat
	ρ	solid density
	σ	Stephan-Boltzmann constant
	ε	emissivity
	T_m	melting temperature
	K_m	heat conductivity in the solid at T_m
	r	crystal radius

under the following assumptions: cold surroundings, only radiative heat loss, flat interface, $K_m \sim 1/T$ and temperature-independent surface emissivity. Although Ciszek [3] introduced a correction factor of $\sqrt{(3/2)}$, we use equation (1) for the comparison with the experimental values according to Oh et al [2] who made similar comparisons for the Czochralski growth of silicon.

Table 1 shows the maximum growth rates of this work compared with the values after Billig and the experimental CZ values after Oh. Most of the maximum growth values of our FZ experiments reached 80-85% of the theoretical value, whereas the CZ values are between 40% and 64%. The variation of the FZ values, we interpret to be the consequence of different induction coils, feed rod diameters and rotation rates. The processes for the 147mm and 150mm crystal were not fully optimized to high growth rates. Here the lower maximum growth rate reflects the sensitivity of such large-diameter processes. However, the FZ growth experiments suggest that in any case maximum growth rates of 85% of the Billig's value are realistic. This would correspond to expected 3.48 mm/min for 150mm diameter and to 3 mm/min for 200mm.

diameter [mm]	max. growth rates [mm/min] FZ experiments	CZ experiments	Billig	V_{FZ}/V_{Billig}	V_{CZ}/V_{Billig}
7	16		18.9	0.85	
11	12		15.1	0.80	
25	9		10.1	0.89	
30		5.8	9.1		0.63
40		4.3	7.9		0.54
50	6		7.1	0.85	
67	5		6.1	0.82	
80		2.3	5.6		0.41
100	4.1		5.0	0.81	
120		2.2	4.6		0.48
125	3.8		4.5	0.85	
147	2.8		4.1	0.68	
150	2.7		4.1	0.66	
200	?		3.5		

Table 1 Maximum growth rates and relations to the theoretical values after Billig

The diagram in Fig. 1 shows both, the strong correlation between experimental and theoretical maximum growth rates and the differences between the expected and observed values.

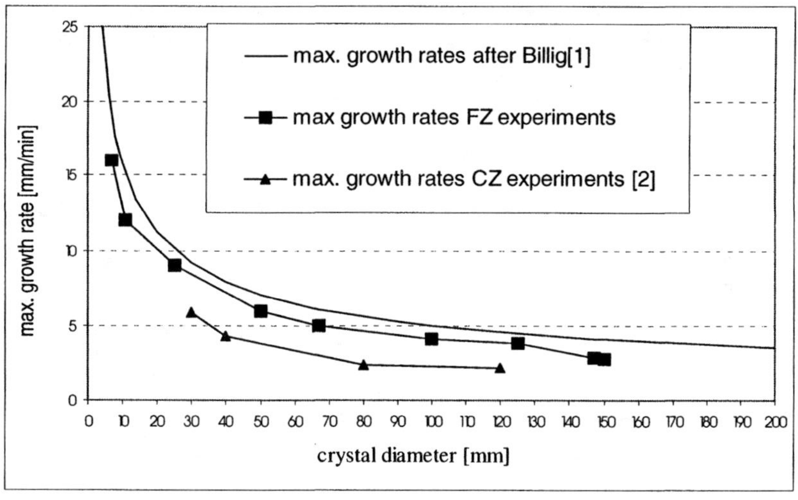

Fig. 1 Maximum growth rates: Calculated after Billig, from FZ experiments and CZ experiments [2]

3.2 Striation patterns and phase boundaries

Fig. 3 shows striation patterns for varying pull speeds for the 25mm crystal. At 9 mm/min, the structure was lost because of a remaining unmolten core of the feed rod, which was contacting the crystal interface.

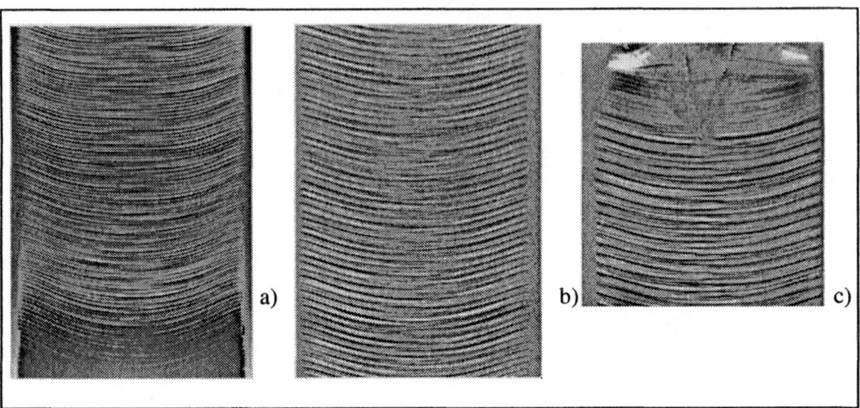

Fig. 3 Striation patterns for 3 different growth rates of the 25mm FZ crystal (7 rpm)
a) 2.8 mm/min, b)5 mm/min, c) 9mm/min : structure loss

In Fig. 4, the phase boundaries are compared for growth rates of 2.8, 4, 5, 6, 7, 8 and 9 mm/min. They are the more deflected the higher the pull rate is until 8 mm/min and for 9mm/min, the interface becomes unsymmetrical and less deflected.

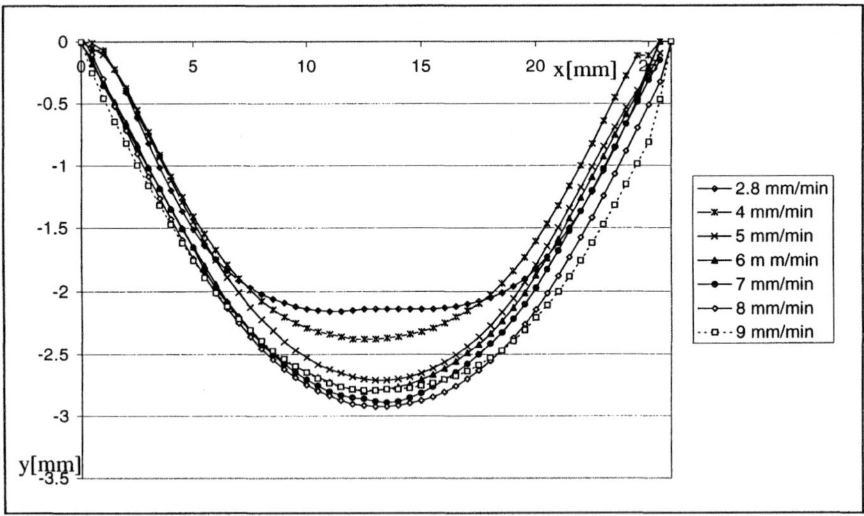

Fig 4 Phase boundaries of an FZ 25mm crystal for growth rates from 2.8 to 9 mm/min

The 100mm crystal shows an analog behavior. Fig. 5 depicts the striation pattern of a long sample of 140 mm in which the growth rate varies between 3.8 and 4.1 mm/min. Furthermore, the corresponding phase boundaries are drawn. The phase boundary at 4.1 mm/min is the phase boundary at structure loss. Fig 6 compares the phase boundaries for the various growth rates.

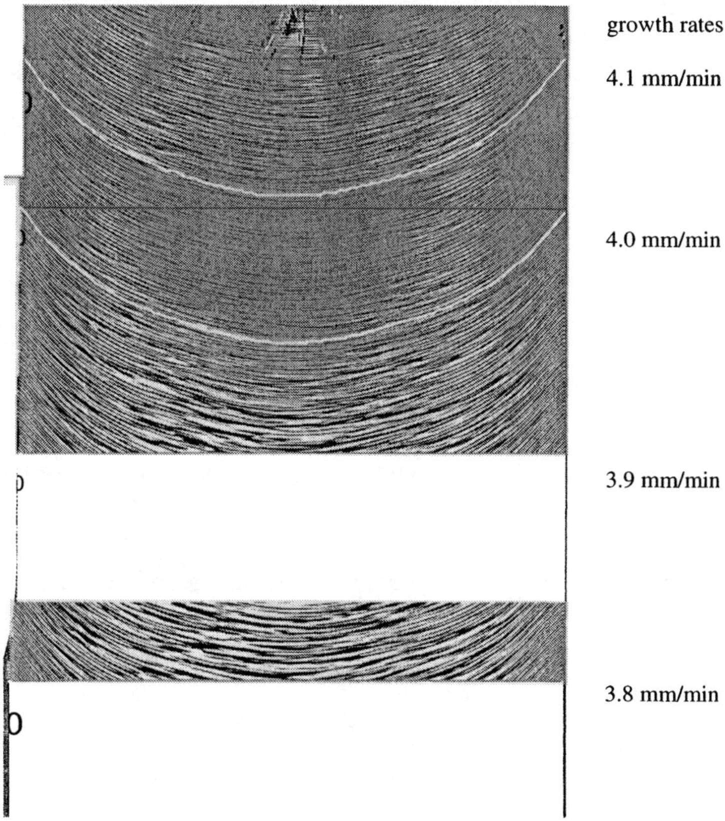

Fig. 5 Striation pattern of an axial cut (140mm) of a 100 mm crystal grown at growth rates of 3.8 mm/min to 4.1 mm/min including structure loss

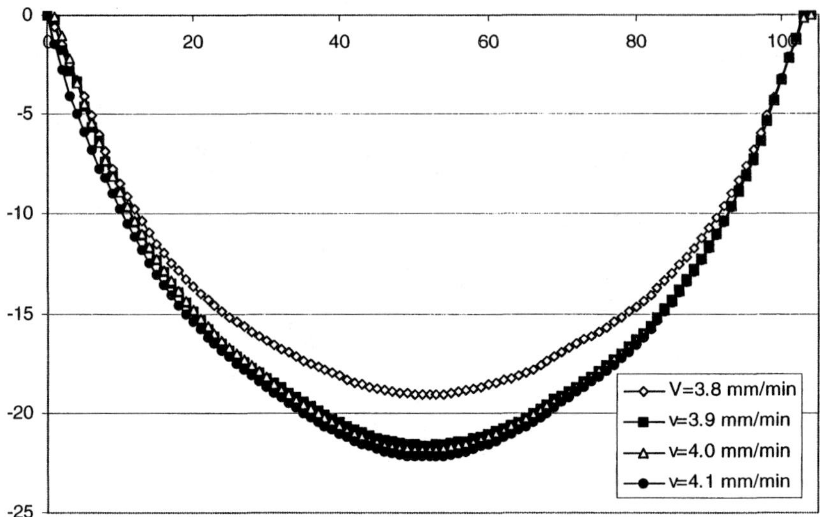

Fig 6 Phase boundaries of a 100mm FZ crystal grown at 3.8, 3.9, 4.0 and 4.1 mm/min

3.3 Minority carrier lifetime

The minority carrier lifetime τ as an important quality parameter for solar cell applications of silicon was measured for different growth rates with the μPCD method at our institute. For the 25mm crystal, the measured lifetimes τ raised from 450 μs to 500 μs with increasing pull rate according to the results of Ciszek [4], who also observed an increase of the lifetime for fast pull rates.

These μPCD τ values are comparatively low compared with the lifetimes measured at Topsil, where a common PCD method was used combined with a more detailed analysis of volume and surface recombination resulting in a volume recombination of 2-3 ms for the 150mm crystal. Additionally, this could be due to a better surface passivation.

Thus, the high growth rates did not degrade the measured lifetimes.

CONCLUSION

Dislocation-free FZ silicon can be grown with pull rates of 85% of the theoretical maximum value after Billig. This is considerably more than published values of about 60% for CZ crystals. These high growth rates do not degrade the minority lifetime, not even for large diameter crystals. Further development of special high-speed induction coils could improve maximum pull rates and the growth stability of large-diameter FZ crystals. Beside the economical benefits of the common FZ process like avoiding of costly crucibles and graphite parts, fast FZ growth has the advantage of considerably higher productivity. Therefor, it has a potential for high-efficiency solar cells.

REFERENCES

[1] E. Billig Proc. R. Soc. 229 346 (1955)
[2] H.J.Oh, et al.," High Purity SiliconVI ", Electrochem Soc.Proc. 2000-17, 44-53
[3] T.F. Ciszek, J. Applied Physics,49, 2 (1976)
[4] T.F. Ciszek, 14th European Photovoltaic Energy Conference Proc. 1997, 396-399

OBSERVATION OF RING-OSF NUCLEI IN AS-GROWN CZ SILICON CRYSTALS BY HIGHLY SELECTIVE REACTIVE ION ETCHING

Kenji Nakashima[1], Kozo Nakamura[2], Toshiaki Saishoji[2] YukihikoWatanabe[1], Yasuichi Mitsushima[1] and Naohisa Inoue[3]

E-mail:nakashima@mosk.tytlabs.co.jp

[1]Toyota Central R&D Labs., Inc.,
Nagakute, Aichi, 480-1192, JAPAN

[2]Komatsu Electronic Metals Co., Ltd., Technical Division
3-25-1, Shinomiya, Hiratsuka, Kanagawa, 254-0014, Japan

[3]RIAST, Osaka Prefecture Univ.,
1-2, Gakuencho, Sakai, Osaka, 599-8570, Japan

ABSTRACT

A highly selective reactive ion etching (RIE) method has been applied to the detection of the nuclei of ring-like distributed stacking faults (R-OSFs) in as-grown Czochralski (CZ) silicon wafers. The radial distribution of R-OSF nuclei in as-grown crystals was evaluated by the highly selective RIE method for the first time. This result directly shows that the OSF nuclei are very small oxygen precipitates existing in as-grown crystal. The size distribution of the R-OSF nuclei in the as-grown crystal was also evaluated for the first time. It was found that the size of the R-OSF nuclei distributes from less than 20 nm to about 200 nm for the wafer with oxygen concentration of 12.1×10^{17} cm^{-3}.

INTRODUCTION

It is well known that a ring-like distributed stacking fault region (R-OSF) can be observed after annealing in an oxidation ambient in Czochralski (CZ) silicon crystals[1)-4)]. Since the R-OSF region forms a boundary between regions of excellent and poor gate oxide integrity (GOI) yield of wafers, understanding the formation mechanism of the R-OSF region is very important for attaining excellent GOI yield. It has been believed that the formation of the OSFs is due to the agglomeration of excess silicon interstitials around oxygen precipitates which are already exist in the as-grown crystals[1)-5)]. This means that oxygen precipitates existing in as-grown crystal act as the nuclei of the OSFs. However, oxygen precipitates have not yet been observed in as-grown crystals with conventional techniques such as light scattering tomography (LST) or TEM, because the size is small or the density in not sufficient[2, 3), 5)].

Recently, we have proposed a highly sensitive method for detecting oxygen precipitates in silicon wafers which utilizes a highly selective reactive ion etching (RIE) [6)]. In this method, RIE is performed on a Si wafer containing oxygen precipitates under a condition that the etching rate ratio of Si to SiO_2 is very large, and consequently they are detected as Si cones having an oxygen precipitate at each top. The detection principle and a typical SEM image of such Si cones are shown in Fig.1. With this method, the size distribution of oxygen precipitates can be evaluated by observing the top of the Si cones[7)]. In addition, the spatial and depth distribution of oxygen precipitates can be estimated by evaluating the number and height of the cones. The detectable precipitate size limit of this method was about 15 nm under typical RIE conditions, as reported previously. After that, the detectable size limit has been improved to less than 10 nm by changing RIE conditions. LST is, on the other hand, capable of detecting an oxygen precipitate whose volume is equivalent to a sphere with 40 nm in diameter. For platelet oxygen precipitates, which cause the formation of OSFs, the detectable size limit of LST can be estimated to be about 200 nm in diagonal length because the thickness of the precipitates is about 2-3 nm. For platelet oxygen precipitates, therefore, the detectable size limit of the RIE method is two orders of magnitude smaller than that of LST.

In this study, we have attempted to detect oxygen precipitates, i.e. R-OSF nuclei, in an as-grown Si wafer using the RIE method. As a result, the distribution of R-OSF nuclei density has been successfully evaluated for the first time. In addition, the size distribution of the R-OSF nuclei has also been evaluated.

EXPERIMENTAL

Two kinds of boron doped CZ-grown Si(100) wafers including an R-OSF region were used in this study. The wafers were 150 mm and 200 mm in diameter. The oxygen concentrations of the wafers were $12.5x10^{17}cm^{-3}$ for 150 mm wafer and $12.1x10^{17}cm^{-3}$ for 200 mm wafers on the old ASTM scale, respectively; and the resistively of the wafer was about 10 ohm-cm for all wafers.

To reveal the R-OSF region of a 150 mm wafer, the wafer was annealed at 1000℃ for 16 h in dry O_2 after pre-annealing at 780℃ for 3 h. The R-OSF region of the wafer was revealed by X-ray topography and Wright etching.

The detection of the nuclei of R-OSFs was attempted using an LST method with a YAG laser of 1.06 um (Laser Scattering Tomography, MO - 401, Mitsui Mining & Smelting Co.) [8)] and a highly selective RIE method. With the LST method, the radial density distribution of light scattering tomography defects (LSTD) was measured. The highly selective RIE was performed in a conventional magnetic-field enhanced RIE system (Applied Materials Precision 5000 etch) using typical two conditions (A and B). Under the condition A, HBr, Cl_2, He and O_2 gases were used. Under the condition B, on the other hand, HBr, NF_3, He and O_2 gases were used. The detectable size limits for the conditions A and B were estimated to be 4 and 8 nm, respectively. For both conditions, selectivity of Si to SiO_2 was higher than 200. The etching depth was controlled about 6 um.

Following the etching, the density of OSF nuclei was evaluated by counting the number of Si cones per unit volume as a function of the distance from the wafer center. The size of R-OSF nuclei was measured by SEM observation of the top of Si cones.

RESULTS AND DISCUSSION

Detection of R-OSF nuclei in as-grown crystals

First, to reveal the R-OSF region of a 150 mm wafer, the wafer was annealed at 1,000℃ for 16 h in dry O_2 after pre-annealing at 780℃ for 3 h, and then observed by XRT. Also, the underlying OSFs were revealed by Wright etching and the radial distribution of the OSFs was then counting using an optical microscope. Figure 2 shows the XRT image and radial distribution of the OSFs. These results indicate that an R-OSF region is located between 40 mm and 55 mm in radius.

Next, detection of oxygen precipitates, i.e. the nuclei of R-OSFs, was attempted using the highly selective RIE method and LST for an annealed and as-grown 150 mm wafer. The annealing was done at 1100℃ for 2 h in O_2 to grow oxygen precipitates in the wafer. The RIE was performed under condition A. After the RIE, the density of Si cones formed on wafer surfaces was measured using an optical microscope. Figure 3 shows the radial distributions of Si cone density and LSTD density for the annealed wafer. The Si cone density is very high between 40 mm and 55 mm in radius. This radius range agrees with that of the R-OSF region shown in Fig.2, indicating that the oxygen precipitates detected as Si cones are the grown nuclei of the R-OSFs. Although the LSTD density is also high in the same range, its maximum is half of that of the Si cone density. The lower density of LSTD is probably due to the insufficient detectable size limit of LST and /or underestimation caused by the overlapping effect which occurs when the density of defects is higher than $10^8 cm^{-3}$. Figure 4 shows the radial distribution of Si cone density for the as-grown wafer. Although the oxygen precipitate sizes of the as-grown wafer are much smaller than those of the annealed wafer, the density distribution of Si cones is good agreement with the distribution for the annealed wafer. This fact indicates that the RIE method completely detected the R-OSF nuclei existing in the as-grown wafer, because the number of oxygen precipitates remains unchanged after the annealing. On the other hand, no LSTD was observed in the R-OSF range, though LSTDs attributed to void defects was observed with densities of about $10^6 cm^{-3}$ inside the R-OSF range. This indicates that the R-OSF nuclei sizes are smaller than the detectable size limit of LST as expected.

From the above results, it is concluded that the highly selective RIE method can detect R-OSF nuclei in an as-grown wafer, which cannot be detected by LST, the most useful conventional method. In addition, these results directly show that the nuclei of

R-OSF are very small oxygen precipitates existing in as-grown crystal.

Size of R-OSF nuclei

In the highly selective RIE method, oxygen precipitates in a wafer are detected as needle-shaped Si cones having an oxygen precipitate at each top. Since oxygen precipitates are hardly etched under conditions that the etching ratio of Si to SiO_2 is very large, their morphology remains almost unchanged after the RIE. Thus, the size of oxygen precipitates can be measured by observing the top of the Si cones using SEM or TEM.

In this experiment, the highly selective RIE was performed on an as-grown 200 mm wafer with an oxygen concentration of $12.1 \times 10^{17} cm^{-3}$ and the size distribution of oxygen precipitates, R-OSF nuclei, was measured by SEM observation of the top of the Si cones formed after the RIE. The RIE was performed under condition B. Figure 5 shows the radial distribution of Si cone density for the wafer. The Si cone density is high between 75 mm and 87 mm in radius. This range agrees with the radius range of the R-OSF region of this wafer. The size distribution of oxygen precipitates was measured by SEM observation of the top of Si cones formed at a radius of 81 mm. Figure 6 shows typical SEM images of the top of the Si cones. Single squares with various sizes, which reflect the platelet oxygen precipitates, were clearly observed as shown in Fig.6. Figure 7 shows the histogram of the size distribution of oxygen precipitates. The size of the R-OSF nuclei distributes from less than 20 nm to over 200 nm and more than half of the nuclei distribute less than 40 nm. For the size of the R-OSF nuclei, Sueoka et al. reported to be 10-40nm in radius as spherical precipitates by simulation[9)] and the size corresponds to about 50-200nm as platelet precipitates. Although further investigation is necessary to convince the size distribution of the nuclei, especially the disagreement of the distribution in the small-sized region, here we emphasize that the size distribution of oxygen precipitates smaller than the detectable size limit of LST was successfully evaluated by the highly selective RIE technique.

CONCLUSION

The highly selective RIE method has been applied to the detection of the R-OSF nuclei in as-grown crystals. The radial distributions of R-OSF nuclei in as-grown crystals are successfully evaluated by the RIE method for the first time. In addition, this result directly proves that the OSF nuclei are very small oxygen precipitates which exist in as-grown crystal. It was also found that the size of the R-OSF nuclei distributes from less than 20 nm to over 200 nm and more than half of the nuclei distribute less than 40 nm. Further investigation is necessary to convince the size distribution of the nuclei, especially the disagreement of the distribution in the small-sized region. We believe the highly selective RIE method is a powerful tool for understanding the behavior of oxygen precipitates in various crystals for advanced LSI.

ACKNOWLEDGEMENTS

The authors would like to thank Dr. Tomoyuki Yoshida for his useful comments and advice. The authors also wish to thank Shiho Ishihara for her technical assistance and Masakazu Kanechika for his helpful discussion.

REFERENCES

1) M. Hasebe, Y. Takeoka, S. Shinomiya and S. Naito, Jpn. J. Appl. Phys., 28, L1999 (1989).
2) K. Marsden, S. Sadamitsu, M. Hourai, S. Sumita and T. Shigematsu, Semiconductor Silicon, ed. by H.R. Huff, et al., (The Electrochem. Soc., Pennington N.J., 1994), p684.
3) K. Marsden, T. Kanda, M. Okui, M. Hourai and T. Shigematsu, Mater. Sci. Eng. B 36, 16 (1996).
4) K. Harada, H. Tanaka, T. Watanabe and H. Furuya, Jpn. J. Appl. Phys., 37, 3194 (1998).
5) K. Sueoka, M. Akatsuka, K. Nishihara, T. Yamamoto and S. Kobayashi, Mater. Sci. Forum 196-201, 1737 (1995).
6) K. Nakashima, Y. Watanabe, T. Yoshida and Y. Mitsushima, J. Electrochem. Soc., 147, 4294 (2000).
7) K. Nakashima, Y. Watanabe, T. Yoshida and Y. Mitsushima, High Purity Silicon VI, ed. by C. L. Claeys, P. Rai-Choudhury, M. Watanabe, P. Stalhofer and H. J. Dawson, (The Electrochem. Soc., Pennington N.J., 2000), p129.
8) K. Moriya and T. Ogawa, Jpn. J. Appl. Phys., 22, L207 (1983).
9) K. Sueoka, M. Akatsuka, M. Okui, and H. Katahama, Semiconductor Silicon, ed. by H.R. Huff et al., (The Electrochem. Soc., Pennington N.J., 2002), p540.

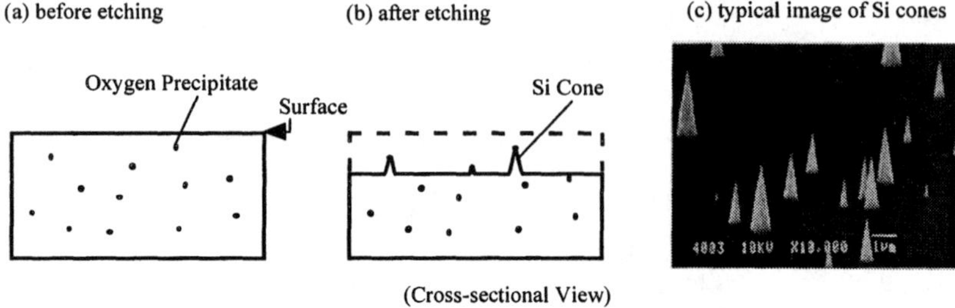

Fig.1 Schematic of the principle for detecting oxygen precipitates in silicon crystals by highly selective RIE method ((a), (b)) and typical image of Si cones after etching.

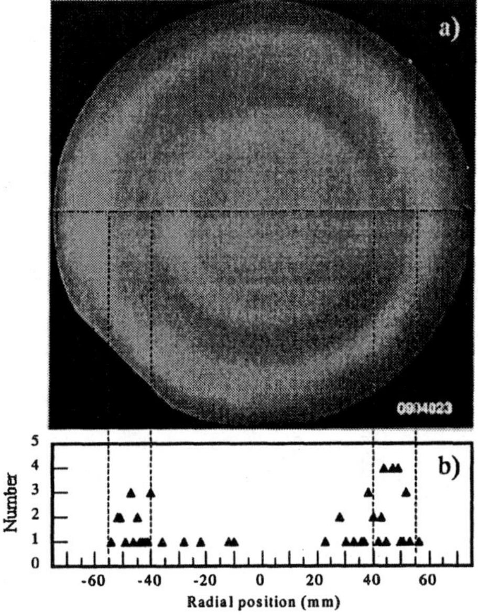

Fig.2 (a) X-ray topograph image and (b) radial distribution of OSFs for the 150 mm wafer after two-step annealing (780℃x3h + 1000℃x 16h).

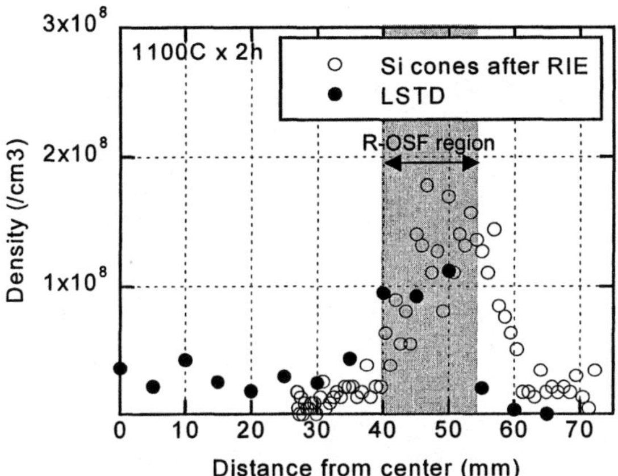

Fig.3 Radial distributions of LSTD density and Si cone density after highly selective RIE for the wafer annealed at 1100℃ for 2h in dry oxygen ambient.

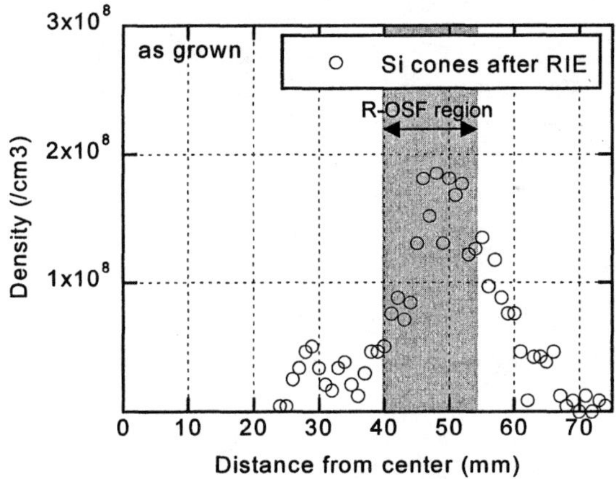

Fig.4 Radial distribution of Si cone density after highly selective RIE for the as-grown wafer.

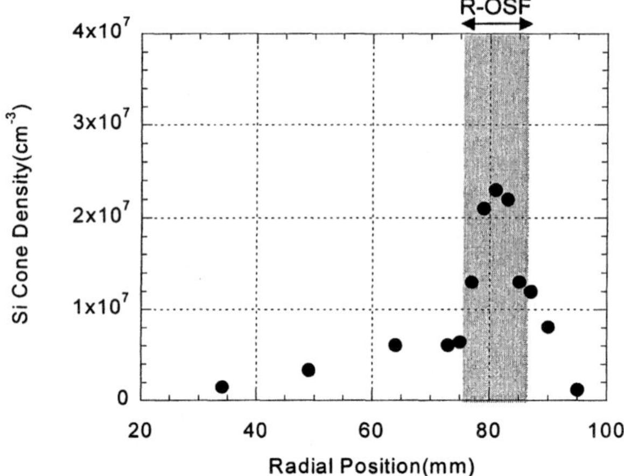

Fig.5 Radial distributions of Si cone density after RIE for as-grown wafer with oxygen concentration of $12.1x10^{17}$ cm^{-3}.

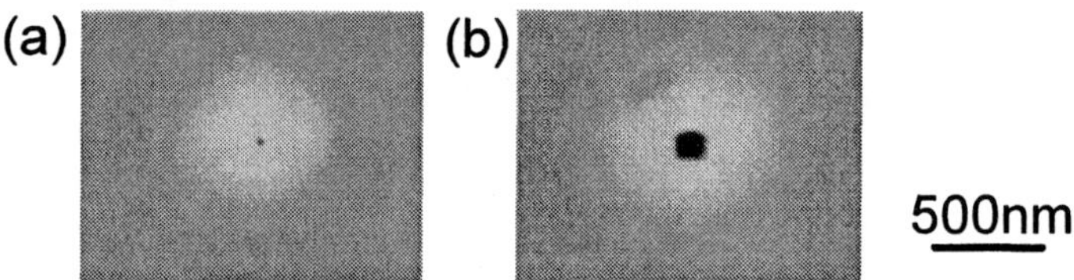

Fig.6 Typical SEM images of the top of Si cones for (a) small and (b) large oxygen precipitate..

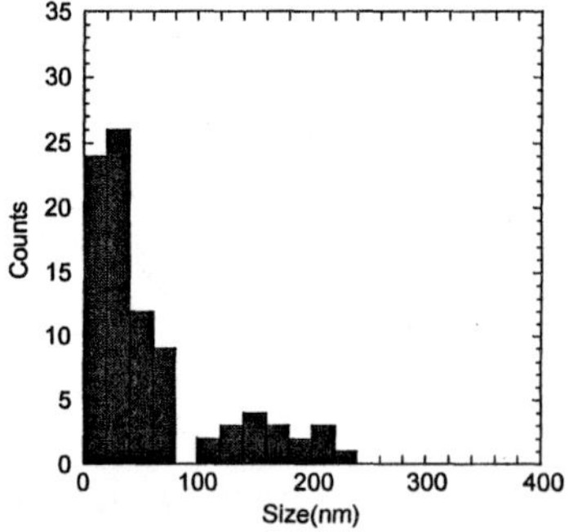

Fig.7 Size distribution of R-OSF nuclei evaluated by the top size of Si cones.

ON THE NUCLEATION OF BULK STACKING FAULTS IN CZ SILICON

Jan Vanhellemont, Olivier De Gryse and Paul Clauws

Department of Solid State Sciences, Ghent University,
Krijgslaan 281, B-9000 Ghent, Belgium

The nucleation of stacking faults (SF) at silicon oxide precipitates in silicon is discussed. A model is presented assuming that SF nucleation occurs by a sudden release of strain accumulated in the silicon oxide precipitate during its thermal history. This strain release occurs by emission of self-interstitials and leads to a very high super saturation of self-interstitials at the precipitate, resulting in a SF nucleus which grows by the Bardeen-Herring mechanism. In this view, bulk stacking fault nucleation is determined not only by the size of the grown-in oxide precipitates but also -and mainly- by the strain energy accumulated in them. This strain energy is controlled by intrinsic point defects during crystal pulling and thermal pretreatments. It is shown how the grown-in oxide precipitate size distribution can be determined from single step annealing experiments.

INTRODUCTION

The nucleation and growth of bulk stacking faults in Czochralski silicon has been studied for more than twenty five years now and the growth kinetics and the associated activation energy are well documented (1-4). Most of the analytical models for stacking fault growth describe the experimentally observed $(\text{time})^{3/4}$ dependence but fail to a large extent to give also a quantitative description of the saturation stacking fault length and surely are not able to predict the number of stacking faults formed. As the saturation length of the stacking faults depends strongly on their number -the total number of "available" self-interstitials is determined by the precipitation of the amount of interstitial oxygen above the solubility limit- one of the main reasons for the poor success of the modeling is the lack of understanding of the stacking fault nucleation process itself. During the last decade a lot of information on the formation of bulk stacking faults and more in particular its relation with the crystal pulling conditions has been obtained by studying the behavior of the so-called stacking fault ring (5,6). It is well-accepted now that, during cooling of the crystal, the central crystal part which is surrounded by the stacking fault ring is vacancy-rich while the crystal part outside the stacking fault area is self-interstitial-rich. The stacking fault ring itself occurs close to the transition between both regions and thus in an area where both types of intrinsic point defects are close to equilibrium. A lower density of stacking faults is formed inside the ring, while outside both SF formation and oxygen precipitation are strongly suppressed.

In the present paper a simple semi-quantitative model is presented explaining the nucleation of stacking faults on silicon oxide precipitates by a sudden release of self-interstitials when a critical level of grown-in strain energy is reached. It is shown that by considering conservation of self-interstitials and minimizing the total free energy, one can estimate the critical precipitate size needed for SF nucleation as well as the size of the SF nucleus itself. This critical precipitate size depends strongly on the thermal history of the material.

Few experimental observations are available on early stages of SF nucleation at oxide precipitates. Already more than twenty years ago, Takaoka et al (3) and Bender et al (7,8) observed the first stages of the formation of circular stacking faults by the Bardeen-Herring mechanism (Figure 1). All stacking faults nucleated during the first minutes of the high temperature treatment and reached rapidly a circular shape with a diameter of the order of 100 nm after which growth continued according to a $(\text{time})^{3/4}$ law. Wada et al (4,9) performed a detailed study of nucleation and growth of stacking faults over a wide temperature range. They also observed in all cases a rapid initial growth leading to stacking fault nuclei of about 100 nm diameter for the whole temperature range from 850 to 1200°C. Sadamitsu et al reported experimental observations of the initial stages of SF nucleation after low temperature pre-anneals to grow silicon oxide precipitates to the critical size needed for SF nucleation (10).

They explained qualitatively the SF nucleation as due to the strain field of the precipitates. Quantitative criteria for stacking fault nucleation were estimated by Marsden et al. (11, 12) and by Sueoka et al (13) suggesting that oxide precipitates with sizes larger than 70 nm and between 26 and 135 nm, respectively, can nucleate stacking faults. Kissinger et al (14) performed ramped anneals on wafers with a stacking fault ring and showed that the SF density increases with decreasing ramp rate and with decreasing start temperature of the ramp. In the present paper it is attempted to further clarify and quantify the criteria for stacking fault nucleation as well as the impact of pre-anneals on the stacking fault density.

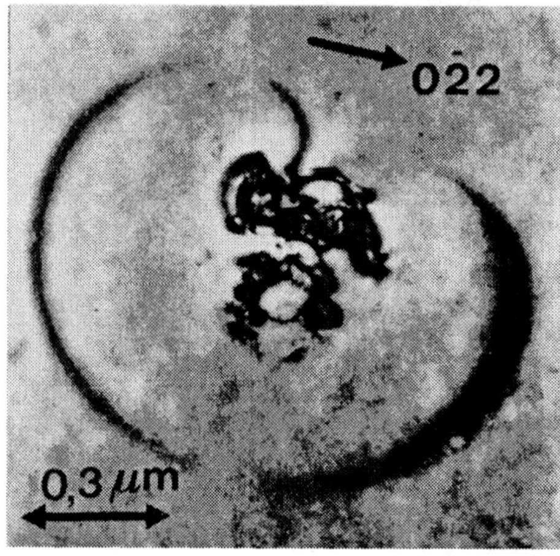

Figure 1: Stacking fault growing by the Bardeen-Herring mechanism observed after a 5 min nitrogen anneal at 1150°C [7]. (courtesy H. Bender)

A MODEL FOR STACKING FAULT NUCLEATION AT OXIDE PRECIPITATES

Following the idea that the rapid initial release of strain energy by self-interstitial emission is the key, the SF nucleation process at oxide precipitates can be understood as follows. In vacancy-rich material which is obtained for crystals grown with a pulling rate to axial temperature gradient ratio (6) larger than 2.2×10^{-5} $m^2 s^{-1} K^{-1}$, voids are formed by clustering of vacancies at temperatures above 1000°C (15). Below 1000°C the nucleation of silicon oxide precipitates in the cooling crystal increases with decreasing temperature as the super saturation of oxygen becomes larger than that of vacancies which are to a large extent already depleted by the high temperature void nucleation and growth. Grown-in silicon oxide precipitates are thus mainly formed during the cooling of the pulled crystal in the temperature range below 1000°C. As oxide precipitate growth is oxygen diffusion controlled, significant nucleus growth only occurs for temperatures above 850°C for typical crystal cooling rates which are of the order of 1°C/min. During silicon oxide precipitate growth elastic strain builds up due to the volume difference between the formed oxide and the consumed silicon. This elastic strain is partially released by the absorption of vacancies or, and mainly by the emission of self-interstitials. Both processes are only fully effective above 1000°C (16). At lower temperatures partial strain relief occurs and the strain energy of the precipitate increases during further cooling. Afterwards when heating the wafer to a higher temperature (e.g. above 1000°C), the strain energy accumulated at lower temperatures will be released rapidly by the emission of self-interstitials causing a very high local super saturation which can result in a stacking fault nucleus.

The critical precipitate radius r_p^c yielding enough self-interstitials to produce a stacking fault with critical radius R_{SF}^c can be estimated as follows. Let us assume first that partial strain release has occurred during the thermal history of the wafer by the interaction of the precipitate with on the average α intrinsic point defects per precipitated oxygen atom. For full strain relaxation one should have $\alpha \approx 0.6$ and $\alpha \approx 0.8$ for SiO_2 and SiO, respectively. In reality even at the high temperature (T) step whereby SF are formed, strain relaxation will not be complete and one will have $\alpha_{SiO_2}(T) < 0.6$ and $\alpha_{SiO}(T) < 0.8$. In the present model, strain release is assumed at the start of the high temperature treatment by the emission of [$\alpha_{SiO_x}(T)$- α] self-interstitials per oxygen atom in the precipitate. If all the released self-interstitials are incorporated in a circular stacking fault nucleus with radius R_{SF} one can write

$$\frac{4x[\alpha_{SiO_x}(T)-\alpha]}{3\Omega_{SiO_x}}\left(\frac{1+\delta}{1+\varepsilon}r_p^c\right)^3 \approx \frac{b(R_{SF})^2}{\Omega_{Si}} , \qquad [1]$$

with

$$\varepsilon = \frac{\delta}{1+\frac{4\mu}{3K}} .$$

b is the length of the Burgers vector $\mathbf{b} = \frac{a}{3}[111]$ of the stacking fault with a (≈ 0.54 nm) the lattice constant of silicon. Ω_{Si}, Ω_{SiO} and Ω_{SiO_2} are the molecular volumes of silicon and silicon oxide, respectively. μ is the shear modulus of the silicon matrix and K the compressibility of SiO_x. Both the linear misfit δ and the constrained strain ε are smaller than 0.1 and can therefore be neglected in first order approximation.

For all calculations in the present paper, the following values where used for the material parameters:

$\Omega_{Si} = 0.2x10^{-22} cm^3$, $\Omega_{SiO} = 0.345x10^{-22} cm^3$ and $\Omega_{SiO_2} = 0.45x10^{-22} cm^3$, $\mu \approx 6.41x10^{10}$ Pa and $K \approx 3.47x10^{10}$ Pa for SiO_2 (also used for SiO).

With $\alpha = 0$, assuming full strain relaxation and a stacking fault nucleus with radius of e.g. 100 nm, one obtains as lower limit of the corresponding SiO_2 precipitate radius of 16.4 nm. With $\alpha = 0.4$ the lower limit of the critical precipitate radius increases to 23.6 nm. Both critical precipitate sizes are still below the detection limit of typically 25 nm radius for the inspection tools based on infrared light scattering tomography. Figure 2 illustrates the dependence of r_P^c on α and $\alpha_{SiO_x}(T)$, both for SiO_2 and SiO precipitates and assuming a SF nucleus of 50 nm radius. Equation [1] is in good agreement with published TEM micrographs of SF nuclei and associated precipitate sizes.

A high α value as is expected in a vacancy-rich crystal leads to a large r_P^c value and thus to a low density of SF's. The nucleation rate of silicon oxide precipitates which lead to the required r_P^c radius decreases indeed exponentially with increasing radius. In the SF-ring area, α is smaller leading to a decrease of the required precipitate size for SF nucleation and thus to a higher SF density. In the interstitial-rich area outside the ring, the super saturation of interstitials hampers precipitate nucleation which therefore occurs at lower temperatures yielding precipitates which are too small to nucleate SF's.

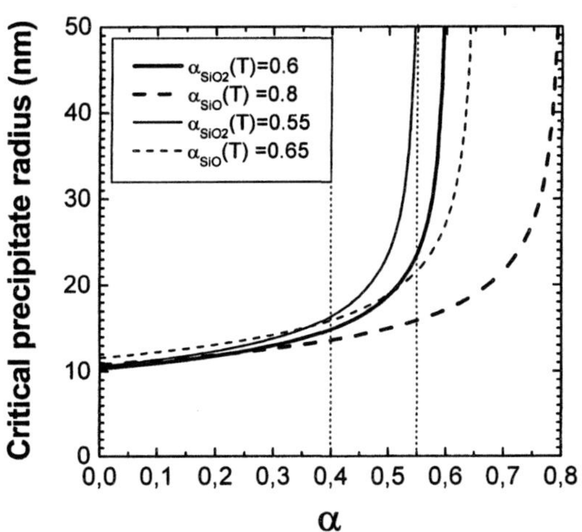

Figure 2: Critical silicon oxide precipitate size for SF nucleation according to [1], as a function of α and with $\alpha_{SiO_x}(T)$ as parameter. A critical stacking fault nucleus with radius of 50 nm is assumed and strain and misfit are neglected.

It is clear from the discussion above that α is one of the critical parameters governing stacking fault nucleation. In this stage of the discussion α is however still unknown. In the next paragraph it is shown how α can be estimated from calculation of the critical precipitate size for precipitate nucleation assuming a known self interstitial concentration and pre anneal temperature.

The critical precipitate radius for precipitate nucleation

During thermal treatment the critical radius r_c determines the stability of an oxide precipitate and thus also the nucleation rate of precipitates. This critical radius for precipitate nucleation is given by (17)

$$r_c = \frac{2\sigma_{SiO_x}}{\frac{xkT}{\Omega_{SiO_x}} \ln \frac{C_{OI}}{C_{OI}^*}\left(\frac{C_V}{C_V^*}\right)^{\varpi}\left(\frac{C_I}{C_I^*}\right)^{\iota} - E_s} \quad . \qquad [2]$$

σ_{SiO_x} is the interface energy between silicon oxide and silicon with $\sigma_{SiO} = 0.67 Jm^{-2}$ and $\sigma_{SiO_2} = 1.68 Jm^{-2}$ (19). C_{OI}, C_V and C_I are the interstitial oxygen, the vacancy and the self-interstitial concentration in the matrix, respectively and C_{OI}^*, C_V^* and C_I^* their thermal equilibrium values with $C_{OI}^* = 9x10^{22} cm^{-3} e^{-\frac{1.52eV}{kT}}$ (17). Ω_{SiO_x} is the molecular volume of SiO_x, k is the Boltzmann constant and T the temperature.

The strain energy E_S of a spherical precipitate can be estimated by

$$E_S = \frac{6\mu\delta^2}{1 + \frac{4\mu}{3K}} \quad . \qquad [3]$$

The linear misfit δ is given by

$$\delta = \left(\frac{\Omega_{SiO_x}}{(1+\alpha x)\Omega_{Si}}\right)^{\frac{1}{3}} - 1 \quad . \qquad [4]$$

The molecular volume Ω_{SiO_x} of all known silicon oxide phases is larger than Ω_{Si}, the molecular volume of silicon. In equation [2] it is assumed that this volume deficiency is partially relieved by the absorption of ϖ vacancies from and the emission of $-\iota$ self-interstitials into the matrix per precipitated oxygen atom, with $\alpha = \varpi - \iota$.

Neglecting the impact of strain and self interstitials [2] reduces to the well known classical expression for the critical radius

$$r_c = \frac{2\sigma_{SiO_x}}{\frac{xkT}{\Omega_{SiO_x}} \ln \frac{C_{OI}}{C_{OI}^*}} \quad . \qquad [5]$$

Due to the low concentration of vacancies after nucleation of oxide precipitates in the ring area, strain relaxation will be incomplete which is expressed by a decreasing α value. The lower the value of α is, the higher are the strain energy and the lower the stability temperature of the precipitate. A precipitate which grows at a temperature below that for SF nucleation can thus accumulate a significant amount of strain energy. This makes the precipitate unstable when the temperature is increased so that the precipitate either has to dissolve or to release part of this strain energy by self interstitial release. As in the ring area also the self-interstitial concentration is lower than outside the ring, the second process is more likely due to the high super saturation of oxygen. Inside the ring a significant amount of vacancies is still present which allow to reduce the build-in strain and to increase the stability temperature. For that reason only a small number of SF's are formed there. Outside the ring a super saturation of self-interstitials exists which suppresses oxide precipitate nucleation leading to a very low SF density. It is possible however to go beyond this qualitative discussion as is shown in the following.

Precipitate nucleation in vacancy-poor silicon: impact of strain

Let us first assume that the oxide precipitates leading to SF nucleation are formed in a vacancy poor crystal so that $\varpi = 0$ and $\alpha = -\iota$, thus reducing [2] to

$$r_c = \frac{2\sigma_{SiO_x}}{\frac{xkT}{\Omega_{SiO_x}} \ln \frac{C_{OI}}{C^*_{OI}} \left(\frac{C_I}{C^*_I}\right)^{-a} - E_s} \quad . \qquad [6]$$

Although [6] clarifies the impact of various parameters on the critical radius of an oxide precipitate, it is difficult to make quantitative predictions as a number of important parameters such as C_I and α are not a priori known, nor is the intimate relation between these parameters taken into account. This makes the practical use of the equations rather limited. The reason that the relation between the parameters is not expressed quantitatively is that [6] was obtained by minimizing the total free energy G of the system only with respect to the radius or $\frac{\partial G}{\partial r} = 0$. If one minimizes G for a self interstitial rich material also with respect to strain energy or which is equivalent, with respect to δ, one can write $\frac{\partial G}{\partial \delta} = 0$, resulting in (18)

$$\ln \frac{C_I}{C^*_I} = \frac{4\mu\Omega_{Si}\left\{\frac{3K\delta[3K(1+\delta)+4\mu]^4 \Omega_{SiO_x}}{kT(3K+4\mu)^4} - x(1+\delta)^2 \ln \frac{C_{OI}}{C^*_{OI}}\right\}}{4(1+\delta)^2 \mu\Omega_{Si} + 3K\Omega_{SiO_x}} \quad . \qquad [7]$$

If C_I is known, δ can be calculated from [7], α from [4] and finally r_c from [6]. At a given temperature r_c thus depends only on the oxide phase and the super saturation of oxygen and self interstitials. The application of [7] is illustrated in Figure 3 showing δ and α as a function of temperature with the self interstitial super saturation as parameter.

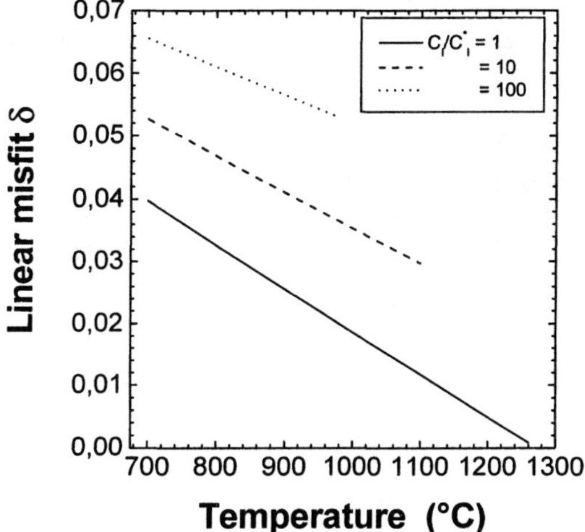

Figure 3: δ (top) and α (bottom) calculated as a function of temperature using [7] and [4], respectively, with $\frac{C_I}{C_I^*}$ = 1, 10 and 100. The calculations were performed assuming SiO_2 precipitates and an initial interstitial oxygen concentration C_{OI} of 10^{18} cm^{-3}.

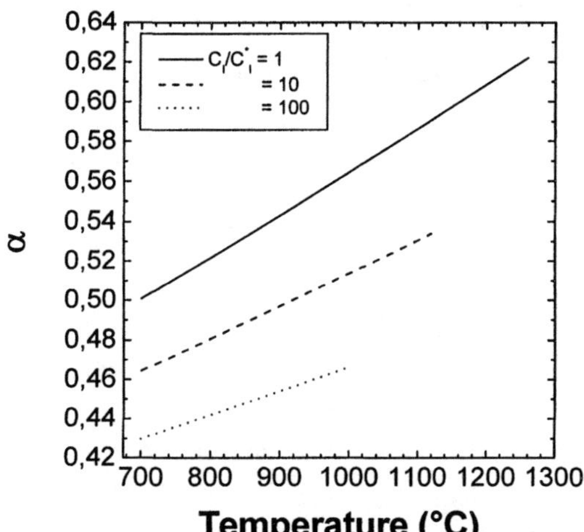

The corresponding critical radius according to [6] is shown in Figure 4. For the calculations it is assumed that $C_{OI} = 10^{18}$ cm^{-3} and that SiO_2 precipitates are formed. The results suggest that for temperatures below 1000°C (typical for pre anneals), α ranges between 0.4 and 0.55. These limits are schematically indicated by the vertical lines in Figure 2 suggesting that the critical precipitate radius for stacking fault nucleation lies somewhere between 15 and 25 nm for SiO_2 and between 10 and 15 nm for SiO precipitates if one assumes full strain relaxation. As expected, for lower $\alpha_{SiO_x}(T)$ values the precipitate radius for stacking fault nucleation increases.

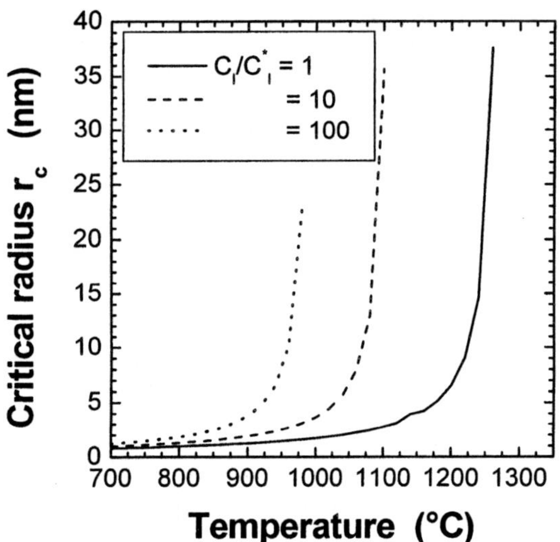

Figure 4: The critical precipitate radius for SiO_2 precipitate nucleation according to [6], taking into account relation [7] and assuming $\frac{C_I}{C_I^*}$ = 1, 10 and 100.

DETERMINATION OF THE GROWN-IN PRECIPITATE SIZE DISTRIBUTION FROM SINGLE STEP ANNEALS

It is well known since the early 1980's that thermal pretreatments at temperatures below 1000°C lead to an increase of the stacking fault density. This is illustrated in Figures 5 and 6 summarizing the experimental results of Takaoka et al (3) and Bender et al (7,8). In view of the model presented above the explanation is that during the pretreatment preexisting oxide precipitates grow and at the same time their strain energy increases. Due to this more precipitates will exceed the critical radius for stacking fault nucleation. In the next paragraphs it is shown how this kind of experiments can yield information on the grown-in precipitate size distribution, on the critical precipitate size for SF nucleation and how this approach can be used to predict SF densities as a function of pretreatment.

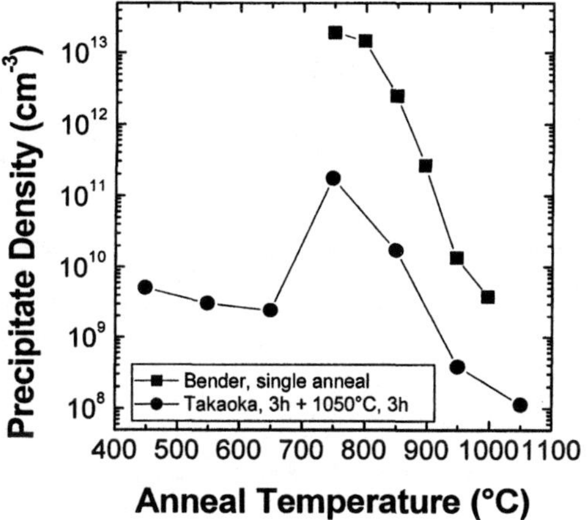

Figure 5: Experimental results of Takaoka et al (3) and Bender et al (7,8) showing the impact of thermal pretreatments on oxide precipitate densities.

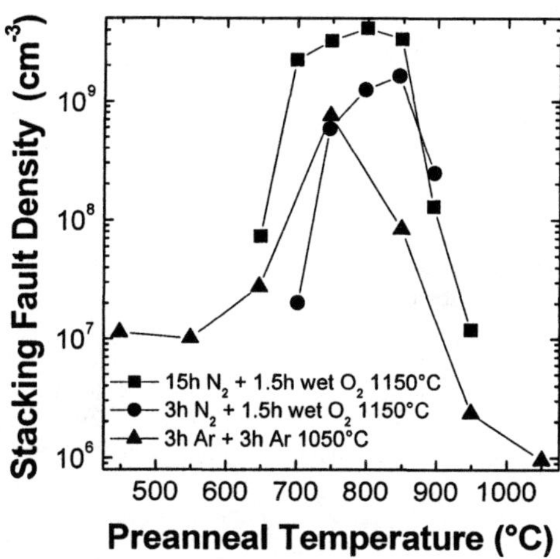

Figure 6: Experimental results of Takaoka et al (3) and Bender et al (7,8) showing the impact of thermal pretreatments on SF densities.

Determination of oxide precipitate size distribution from single step annealing experiments

Assuming the number of precipitates $N_P(r)$ decreasing exponentially with increasing precipitate radius, i.e.

$$N_P(r) = N_0 e^{-Ar} \quad , \qquad [8]$$

the number of precipitates with radius larger than r, $N_t(r)$, is given by

$$N_t(r) = \int_r^{\infty} N_0 e^{-Ar} dr = \frac{N_0}{A} e^{-Ar} \quad , \qquad [9]$$

and similarly, the number of stacking faults N_{SF}, can be estimated by

$$N_{SF} = \frac{N_0}{A} e^{-Ar_P^c} \quad . \qquad [10]$$

Assuming an initial precipitate size distribution whereby the precipitate density decreases exponentially with radius, the initial precipitate size distribution can be determined as follows. For a single step anneal at temperature T_1 all precipitates larger than $r_c(T_1)$ will grow. Assuming that no additional nucleation occurs and anneals long enough to observe precipitates by TEM, LST or etching, the total number of precipitates observed after a single anneal at temperature T is given by

$$N_t(r_c(T)) = \frac{N_0}{A} e^{-Ar_c(T)} \quad , \qquad [11]$$

and thus

$$\ln N_t(T) = \ln\left(\frac{N_0}{A}\right) - Ar_c(T) \quad . \qquad [12]$$

N_0 and A summarize the thermal history of the wafer and depend thus on the crystal pulling process and thermal pre-treatments (e.g. RTP step).

As an example, a fit of [12] to the data of Bender et al. in Figure 5 is shown in Figure 7 using [5] and thus neglecting impact of strain and self interstitials. A very good fit is obtained yielding A = 24.4 and $N_0 = 1.81\text{x}10^{22}$ for SiO_2 and A = 30.6 and $N_0 = 2.26\text{x}10^{22}$ for SiO, respectively. The data points at 650°C and 1000°C were not used in the fit.

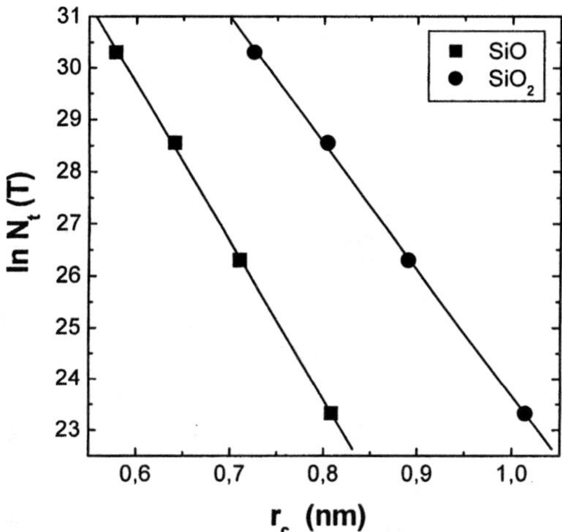

Figure 7: Best fit to the precipitate data of Bender et al (7,8) using [12] and thus neglecting the impact of strain and self interstitials.

CONCLUSIONS

In the present paper it was shown that bulk stacking fault nucleation in Cz silicon can be understood assuming transformation of the strain energy of grown-in silicon oxide precipitates with sufficient size into released self-interstitials which collapse into a SF nucleus. The grown-in strain depends on the intrinsic point defect concentration during oxide nucleus formation and growth and thus on the thermal history of the crystal. Both the crystal pulling conditions, especially the crystal cooling after solidification and also the interstitial oxygen content are dominant process and material parameters determining the bulk stacking fault density distribution in annealed oxygen-rich Cz wafers.

The number of stacking faults formed during a subsequent high temperature treatment is determined by the number of oxide precipitates exceeding the critical size/strain for SF formation after the pre-treatment. It is shown how the grown-in oxide precipitate size distribution in the starting wafer can be determined from single step anneals allowing stacking fault density and precipitate size distribution prediction after a subsequent high temperature anneal.

Work is going on to extend the present model to spheroidal precipitates and to the formation of prismatic loops (prismatic punching) as an alternative to stacking fault nucleation for strain release (18). This will allow predicting also critical plate like precipitate sizes for stacking fault or prismatic loop nucleation.

ACKNOWLEDGEMENT

Part of this work was performed with the financial support of the Science Foundation of Flanders (FWO) (project number: FWO-G.0051.97.N).

REFERENCES

1. J.R. Patel, K.A. Jackson and H. Reiss, J. Appl. Phys., **48**, 5279 (1977).
2. S. Kishino, S. Isomae, M. Tamura and M. Maki, J. Appl. Phys., **49**, 3255 (1978).
3. H. Takaoka, J. Osaka and N. Inoue, Jpn. J. Appl. Phys., **18**, 179 (1979).
4. K. Wada, N. Inoue and J. Osaka, Mat. Res. Soc. Symp. Vol. **14**, 125 (1983).
5. M. Hasebe, Y. Takeoka, S. Shinoyama and S. Naito, Jpn. J. Appl. Phys., **28**, L1999 (1989).
6. E. Dornberger and W. von Ammon, J. Electrochem. Soc., **143**, 1648 (1996).
7. H. Bender, J. Van Landuyt, S. Amelinckx, C. Claeys, G. Declerck and R. Van Overstraeten, Inst. Phys. Conf. Ser., **60**, 313 (1981).
8. H. Bender, PhD thesis, University of Antwerp (1984).
9. K. Wada and N. Inoue, in "Defects and Properties of Semiconductors: Defect Engineering", eds. J. Chikawa, K. Sumino and K. Wada, KTK Scientific Publishers, Tokyo, 169 (1987).
10. S. Sadamitsu, M. Okui, K. Sueoka, K. Marsden and T. Shigematsu, Jpn. J. Appl. Phys., **34**, L597 (1995).
11. K. Marsden, T. Kanda, M. Okui, M. Hourai and T. Shigematsu, Materials Science and Engineering B,**36**, 16 (1996).
12. K. Marsden, S. Sadamitsu, T. Yamamoto and T. Shigematsu, Jpn. J. Appl. Phys., **34**, 2974 (1995).
13. K. Sueoka, M. Akatsuka, T. Yamamoto and S. Kobayashi, Materials Science Forum Vols., **196-201**, 1737 (1995).
14. G. Kissinger, J. Vanhellemont, U. Lambert, D. Gräf, T. Grabolla and H. Richter, Jpn. J. Appl. Phys., **37**, L306 (1998).
15. J. Vanhellemont, Appl. Phys. Lett., **69**, 4008 (1996).
16. W.J. Taylor, Materials Chemistry and Physics, **34**, 166 (1993).
17. J. Vanhellemont and C. Claeys, J. Appl. Phys., **62**, 3960 (1987); ibid., **71**, 1073 (1992).
18. To be published.
19. J. Vanhellemont, J. Appl. Phys., **78**, 4297 (1995).
20. S. Senkader, G. Hobler and C. Schmeiser, Appl. Phys: Lett., **69** (15), 2202 (1996).

Impurities in Silicon

GROWN-IN DEFECTS IN NITROGEN-DOPED CZOCHRALSKI SILICON

DEREN YANG*, XUEGONG YU

STATE KEY LAB OF SILICON MATERIALS, ZHEJIANG UNIVERSITY,

HANGZHOU 310027, PEOPLE'S REPUBLIC OF CHINA

ABSTRACT

The effect of nitrogen doping on the grown-in defects in Czochralski (CZ) silicon has been investigated. It is demonstrated that nitrogen can enhance grown-in oxygen precipitation at high temperature in nitrogen-doped CZ (NCZ) silicon, based on the formation of nitrogen-related complexes. Furthermore, the oxygen precipitation behavior in NCZ silicon, which contains both vacancy-type and interstitial-type defects distinguished by an OSF-ring in the oxidized wafer, is in sharp contrast to that in the CZ silicon. The density of crystal originated particles (COP) in as-grown NCZ silicon is higher, while the size is smaller. The COPs can be annihilated at relatively lower temperatures and therefore a wider denuded COP-free zone (DZ) can be obtained, compared to those in CZ silicon samples. Finally, the impact of nitrogen on the voids, the oxygen precipitation and the OSF ring is discussed.

* Correspondent author: mseyang@dial.zju.edu.cn

INTRODUCTION

Voids have been identified to be a result of vacancy aggregation in the temperature range 1100°C~1070°C during crystal growth (1,2), which can degrade gate oxide integrity (GOI) and enhance leakage current, if existing at the surface of the wafer (3,4) and has been referred to with different names based on measurement method used, such as crystal originated particles (COPs), (5), flow pattern defects (FPDs)(6), and light scattering tomograph defects (LSTD)(7). Recently, it has been clarified that nitrogen doping can increase the density of grown-in voids but decreases their sizes during crystal growth (8,9). Furthermore, by first principle calculation, it was derived that nitrogen could bind vacancies to form N_2–V_2 complexes, thus altering the diffusion and concentration of free vacancies and the void formation (10,11). On the other hand, grown-in oxygen precipitation can be enhanced at high temperature by nitrogen doping, which can improve the thermal stability of grown-in oxygen precipitates and is benefit for the IG capability (12,13) Meanwhile it has been reported that nitrogen can causes N_mO_n ($m,n>1$) complexes at low temperatures (14-18). In addition, it is reported that the nitrogen can affect the behavior of the OSF-ring in mixed-type wafers. However, the exact mechanism of the impact of nitrogen on oxygen precipitation, voids and the OSF ring is not clear yet.

In this paper, the grown-in defects including oxygen precipitates, voids and the OSF ring in NCZ silicon were investigated. Based on the experimental results, the conceptual model to elucidate the formation of oxygen precipitates, voids and the OSF ring in NCZ crystal has been developed.

EXPERIMENTAL

To investigate the effect of nitrogen doping on the grown-in defects in silicon crystals, two 150 mm <100> *p*-type silicon crystals were grown under almost the same conditions, besides the protective gases. One crystal was CZ silicon grown in an Ar ambient; the other was NCZ silicon grown in a nitrogen ambient. The crystals were grown under vacancy-rich conditions with an average pulling rate of about 1 mm/min. The interstitial oxygen (Oi) concentrations determined by Fourier transform infrared spectroscopy (FTIR) with a calibration coefficient of 3.14×10^{17} cm^{-2} were 5.5~9.6 $\times10^{17}$ atoms/cm^3. The substitutional carbon concentrations were below the detection limit of FTIR. The nitrogen concentrations in the samples used for the experiments were also below the detection limit of FTIR. However, they could be roughly estimated to be in the range of $2\text{~}4\times10^{14}$ atoms/cm^3 from the concentration of nitrogen–oxygen complexes which are electrically active and positively correlated with the nitrogen concentration (14,15).

First, the samples of CZ and NCZ silicon were annealed in a N_2 ambient under two different conditions: 1) After annealing at 1270°C for 1 h, the samples were cooled down to 1070°C with a controlled rate of 1°C/min. 2) The samples were annealed at 1150°C for 0~32 h. After heat treatments, the Oi of all the samples were measured by FTIR. On the other hand, the distribution of COPs in silicon wafers sampled from different portions of two crystals was determined by means of a laser particle counter after SC1 cleaning for 1h. Hydrogen annealing at 1050~1200°C for 1~2 h was employed to investigate the annihilation behavior of COPs in the NCZ and CZ silicon wafers. It should be mentioned that the sampling wafers for both the crystal were adjacent. Therefore, it can be believed that the amounts of grown-in COPs are nearly the same on the sampling wafers.

In order to investigate the effects of nitrogen doping on oxygen precipitation in mixed-type NCZ crystal which contains an OSF-ring zone after oxidation, the samples were annealed in a N_2 ambient under different conditions: 1) 1050°C, 32 h; 2) 800°C, 6 h + 1050°C, 16 h. After annealing, the bulk micro-defects (BMDs) on the cross-section of samples were observed by an optical micrograph.

EXPERIMENTAL RESULTS AND DISCUSS

The effect of nitrogen on oxygen precipitation and voids at high temperature during crystal growth

It is well known that during crystal growth, grown-in oxygen precipitates could be generated due to the post-annealing in crystal pullers. Accordingly, we simulated the oxygen precipitation during the crystal growth, and the results are shown in Fig. 1 (12). Obviously, after 1270°C/1 h annealing, the increase of [Oi] in the NCZ crystal is higher than that in the CZ crystal, directly indicating that there are more grown-in oxygen precipitates in the NCZ crystal than that in the conventional CZ crystal. During the subsequent cooling from 1270 to 1070°C, the reduction of [Oi] in the NCZ crystal is also more evident than that in the CZ crystal. These two facts suggest that the nitrogen doping effectively enhances grown-in oxygen precipitation at high temperatures during the crystal growth.

In order to investigate the effect of nitrogen on oxygen precipitation at high temperature, the samples were subjected to 1150°C annealing for different times in the range of 1–32 h. The [Oi] in both crystals as a function of the annealing time is shown in Fig. 2a. It can be observed that the oxygen precipitation of the NCZ silicon increased more rapidly than that of the CZ silicon, indicating that the nitrogen doping could enhance oxygen precipitation even at 1150°C. Figure 2b shows the density of BMDs in the samples versus the annealing time. It can be seen that the BMD density of the NCZ samples was much higher than that of the CZ samples. Therefore, it is reasonably considered that the nitrogen doping offers numerous heterogeneous nucleation sites for oxygen precipitation, and therefore oxygen precipitates could be generated due to the fast diffusion of oxygen atoms towards the nucleation sites at the high temperature, even if the oxygen super-saturation is very small.

The density distribution of COPs in CZ and NCZ silicon along the as-grown crystal axis is shown in Fig.3. It can be seen that the density of COPs in the NCZ silicon is higher than that in the CZ silicon. It shows that nitrogen in silicon crystals resulted in denser COPs. More interesting, unlike the density of COPs in the CZ silicon remaining almost constant along the axis of crystal, the density of COPs in the tail part is higher than that in the head part in the NCZ crystal, which can be considered to be a result of the segregation of nitrogen. However, after reaching a critical value, the density of COPs decreases until zero and the size also decreases along the axis. Therefore, we consider that when the nitrogen concentration reaches a critical value, the size of COPs is too small to be observed by the laser particle counter with a detection limit of 0.11μm. Accordingly, nitrogen doping is believed to be responsible for generating denser grown-in COPs with smaller sizes. Our previous paper has clarified that the COPs on the surface of NCZ wafers can be removed more easily than those on the surface of CZ wafers by high temperature annealing in a H_2 atmosphere (12). After annealing in hydrogen at 1200℃, the depth profiles of the COPs of the CZ and NCZ silicon wafers are shown in Fig. 4a. It is obvious that the denuded zone (DZ) of the NCZ silicon with lower COP density was about 10μm, wider than that of the CZ silicon wafer. Furthermore, the COP depth profiles of the NCZ silicon wafers subjected to annealing in hydrogen at different temperatures are shown in Fig. 4b. It can be also seen that the number of the COPs in the near-surface region of the NCZ silicon wafers reduced significantly with the increase of the annealing temperature.

In CZ silicon, the residual vacancies after void formation are essential for the nucleation of oxygen precipitates during the further cooling process below 900°C (19). The concentration variation of vacancies with cooling temperature is shown by the dotted line in Fig. 5. However, for NCZ silicon the density of grown-in oxygen precipitates is much larger than that in CZ silicon, which is enhanced not only by vacancies, but also by nitrogen. By first-principle calculation, a N–N pair is stable and can bind V to form N_2–V_2 complexes, which can stably exist above the formation temperature of voids (20,21). In fact, we have measured N–N pairs existing in the tail of NCZ crystals by means of FTIR (15), which remained due to rapid cooling of the crystal. We consider that N–N pairs exist in silicon near the melting temperature. With the decrease of temperature, N_2–V_2 complexes will form at high temperatures, resulting in the decrease of N–N pair concentration. There is an equilibrium between N–N and N_2–V_2 by the reaction around 1150°C as follows:

$$N_2 + 2V = N_2\text{–}V_2 \tag{1}$$

The equilibrium constant k is smaller than 1. At high temperature before void formation, the concentration of free vacancies is higher, so the above reaction favors the N_2–V_2 generation. After the vacancies are consumed by void formation, the N_2–V_2 cannot produce any more. Furthermore, it is considered that both N–N and N_2–V_2 can bind oxygen to form complexes as N_mO_n (m,n.1) and N_2–V_2–O_X (X>1), respectively, but their formation temperatures are different. The N–N can directly bind oxygen to form N_mO_n in the temperature range of 750~450°C via reaction as below:

$$m\text{N–N} + 2n\text{O} = 2N_mO_n \tag{2}$$

N*m*O*n* is electrically active, which can become the nuclei of small oxygen precipitates generated in subsequent heat treatments. At high temperatures, N_2–V_2 can bind oxygen to form N_2–V_2–O_X complexes, which also become the nuclei of oxygen precipitates by the reaction

$$N_2\text{–}V_2 + X\text{O} = N_2\text{–}V_2\text{–}O_X \tag{3}$$

Therefore, prior to the formation of voids, the nuclei of oxygen precipitates can grow up by the rapid diffusion of oxygen and the absorption of a considerable amount of vacancies at high temperature. Accordingly, the surviving vacancies contributing to the formation of voids during the subsequent cooling are reduced. As a result, the characteristic formation temperature of voids controlled by the super-saturation of vacancies definitely decreases to some extent. When nitrogen concentration is above a critical value, the voids will not be formed, and the concentration of vacancies as a function of the cooling temperature in NCZ silicon is shown in Fig. 5 by the solid line. According to Voronkov's theory, it is obvious that the size of the voids decreases and the density of voids increases in NCZ silicon due to the reduction of the vacancy concentration as a result of nitrogen-doping enhanced oxygen precipitation prior to void formation. When subjected to annealing, the voids with smaller sizes in NCZ silicon can be annihilated at lower temperatures. As the concentration of doped nitrogen increases, the formation temperature of voids will become even lower and the size of voids will be greatly reduced.

The effect of nitrogen on the OSF-ring

Here, the mixed-type crystals refers to the crystals containing a "P band", which gives rise to a well-known narrow ring of oxidation induced stacking faults (OSF- ring) on oxidized wafers. It has been reported that the OSF-ring will become wider and move toward the inner of wafer by nitrogen, furthermore, the density of OSF in ring zone of NCZ silicon is much higher than that of CZ silicon. Therefore in NCZ silicon, the density and size of grown-in oxygen precipitates were determined by the concentration of nitrogen, while they were normally determined by the concentration of vacancies in CZ silicon.

Fig. 6a shows the BMD distribution in the void and ring zones of the NCZ silicon after two-step anneal (800°C,6h + 1050°C,16h). It is obviously that the density of the precipitates in the OSF-ring was much higher, compared to that in the void region. Generally, the first-step low temperature anneal is used to form the nuclei of oxygen precipitates and determines the density of

BMD. So it can be considered that in NCZ silicon, the precipitates nuclei in the OSF-ring region are more than those in the void region, which is different from in CZ silicon.

When both kinds of the samples were annealed at 1050°C for different time, the oxygen precipitates is shown in Fig. 6b. It can be observed that the BMD density in the void zone of the NCZ silicon was much higher than that in the ring zone. It indicates that the density of the grown-in precipitate nuclei in the void zone of the NCZ silicon, which size is larger than the critical size at 1050℃, is much higher than that in the ring zone.

During crystal growth, the value of V/G decreases from the center to the edge across the radius; therefore, the incorporated vacancy concentration in the grown crystal also decreases in the radial direction. Thus, during NCZ silicon crystal growth, at high temperature, the N_2–V_2 complexes and then the N_2–V_2–O_X complexes generate more significantly in the void zone than t in the ring zone due to the vacancy distribution as mentioned above. Consequently, the void zone has more numerous large grown-in oxygen precipitates formed at high temperature in comparison with the ring zone. And the vacancy concentration will decrease and OSF ring will move toward inner of the NCZ wafer, as shown in Fig. 7a. Based on those as-grown oxygen precipitates, in subsequent high temperature annealing, the oxygen precipitation in the void zone is thus more significant than that in the OSF-ring zone. Actually, in the above discussion, we have neglected the formation of oxide particles in the ring zone, because their density is much lower in comparison with that of grown-in precipitates enhanced by nitrogen doping at high temperatures. It is inferred that the concentration of N–N in the ring zone is higher than that in the void zone, because the N–N loss contributing to the formation of N_2–V_2 complexes at high temperatures in the ring zone is weaker than that in the void zone. Therefore, at low temperatures the formation of N_mO_n complexes, which offer the heterogeneous nuclei for small grown-in oxygen precipitates, is more remarkable in the ring zone than that in the void zone. Those smaller oxygen precipitates can grow up during subsequent medial temperature annealing, consequently, with two-step annealing of 800°C, 6 h + 1050 °C, 16 h, the oxygen precipitation in the ring zone is much heavier than that in the void zone. The density distribution of oxygen precipitates along radius in NCZ silicon is shown by the solid line in Fig. 7b, which is different from the CZ silicon shown by the dot line in Fig. 7b (19).

CONCLUSIONS

The effect of nitrogen doping on grown-in defects including oxygen precipitates, voids and the OSF ring in NCZ silicon crystals has been investigated. It was found that nitrogen doping can enhance the formation of grown-in oxygen precipitates at high temperatures, thus leading to significant loss of vacancies prior to the generation of voids during NCZ crystal growth. Furthermore, oxygen precipitate nucleation in the mixed type NCZ crystal was in sharp contrast to that in the mixed type CZ crystal, as a result of different concentrations of N_2–V_2–O_X and N_mO_n which play critical roles in the enhancement of oxygen precipitation at high and low temperatures during crystal growth. After the oxygen precipitation prior to the void formation, which was enhanced by nitrogen doping, the surviving vacancies in relatively lower concentration resulted in denser voids with smaller sizes in the NCZ crystal. Thus, the voids in NCZ silicon could be annihilated at lower temperatures.

ACKNOWLEDGMENTS

The authors would like to thank the Natural Science Foundation of China (No. 50032010) and 863 programs (No.2002AA31) for finance support.

REFERENCES

1. H. Yamagishi, I. Fusegawa, N. Fujimaki, and M. Katayama, *Semicond. Sci. Technol.* **7,** A135 (1992).
2. M. Itsumi, H. Akiya, T. Ueki, M. Tomita, and M. Yamawaki, *J. Appl. Phys.* **78,** 5984 (1995).
3. S. Huth, O. Breitenstein, A. Huber, and U. Lambert, *J. Appl. Phys.* **88,** 4000 (2000).
4. J. G. Park, G. S. Lee, K. D. Kwack, and J. M. Park, *Jpn. J. Appl. Phys.* **39,** 197 (2000).
5. J. Ryuta, E. Morita, T. Tanaka, and Y. Shimanuki, *Jpn. J. Appl. Phys.* **29,** L1947 (1990).
6. H.Bender, J.Vanhellemont and R.Schmolke, *Jpn. J. Appl. Phys.* **36,** 1217 (1996).
7. K.Moriya and T.Ogawa, *Jpn. J. Appl. Phys.* **22,** 307 (1983).
8. B. M. Park, G. H. Seo, and G. Kim, *J. Cryst. Growth* **222,** 74 (2001).
9. W. von Ammon, R. Holzl, J. Virbulis, E. Dornberger, R. Schmolke, and D. Graf, *J. Cryst. Growth* **226,** 19 (2001).
10. H. Sawada and K. Kawakami, *Phys. Rev. B* **62,** 1851 (2000).
11. K. Tanahashia, H. Haradaa, A. Koukitsub, and N. Inoue, *J. Cryst. Growth* **225,** 294 (2001).
12. Xuegong Yu, Deren Yang, Xiangyang Ma, Jiangsong Yang and Duanlin Que, *J. Appl. Phys.* **92**, 188 (2002).
13. Xiangyang Ma, Xuegong Yu, Ruixin Fan and Deren Yang, *Appl. Phys. Lett.* **81**, 496 (2002).
14. D. Yang, D. Que, and K. Sumino, *J. Appl. Phys.* **77,** 943 (1995).
15. D. Yang, X. Ma, R. Fan, J. Zhang, L. Li, and D. Que, *Physica B* **273–274,** 308 (1999).
16. Q. Sun, K. H. Yao, and H. C. Gatos, *J. Appl. Phys.* **71,** 3760 (1992).
17. K. Aihara, H. Takeno, Y. Hayamizu, M. Tamatsuka, and T. Masui, *J. Appl. Phys.* **88,** 3705 (2000).
18. K. Nakai, Y. Inoue, H. Yokota, A. Ikari, J. Takahashi, A. Tachikawa, K. Kitahara, Y. Ohta, and W. Ohashi, *J. Appl. Phys.* **89,** 4301 (2001).
19. V. V. Voronkov and R. Falster, *J. Cryst. Growth* **204,** 462 (1999).
20. K. Sumino, I. Yonenaga, M. Imai, and T. Abe, *J. Appl. Phys.* **54,** 5016 (1983).
21. H. Kageshima, A. Taguchi, and K. Wada, *Appl. Phys. Lett.* **76,** 3718 (2000).

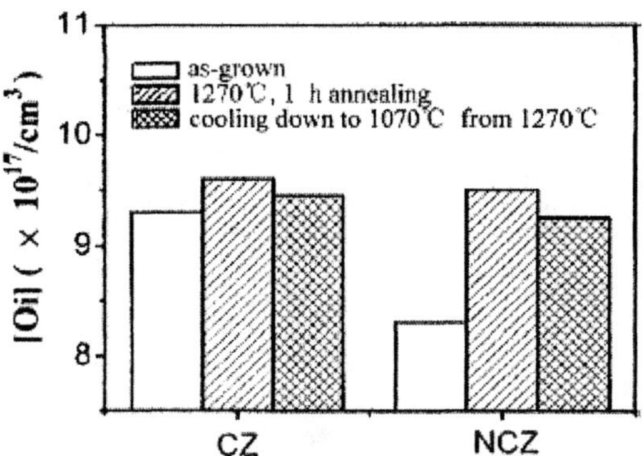

Fig.1 Variation of [Oi] in the CZ and NCZ crystals before and after heat treatments.[12]

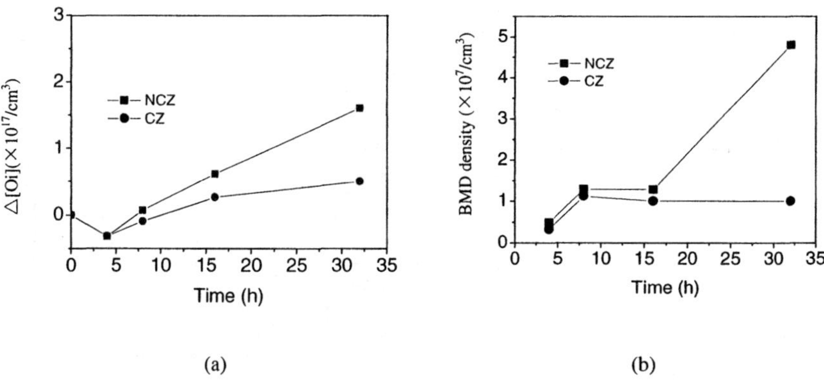

Fig. 2 Variation of [Oi] and BMD density in the NCZ and CZ silicon as a function of the annealing time at 1150°C.

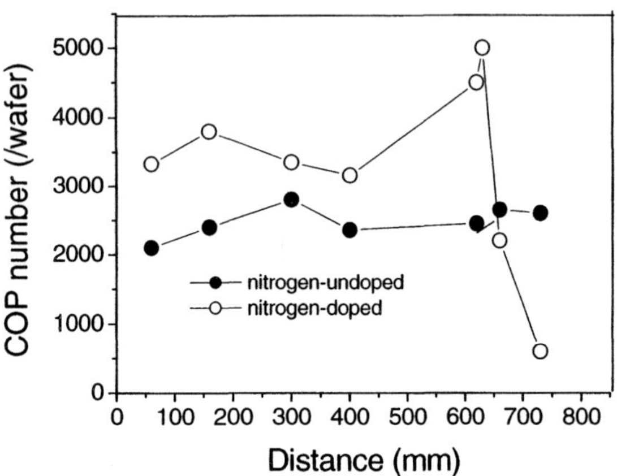

Fig. 3 Density of COPs at different parts in CZ and NCZ silicon crystals.

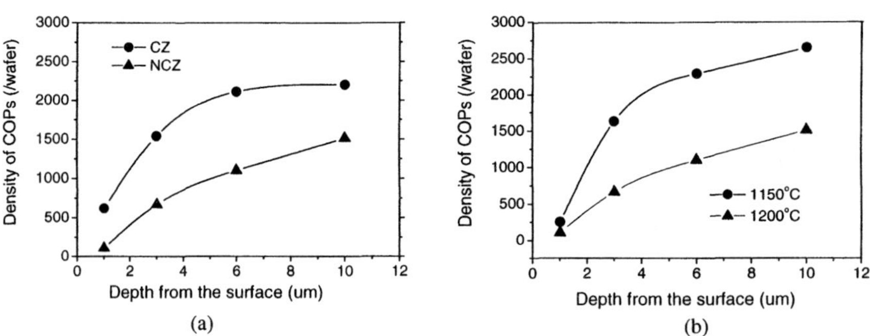

Fig. 4 Density variation of COPs along the depth in the CZ and NCZ silicon crystals after annealing in H_2 ambient.

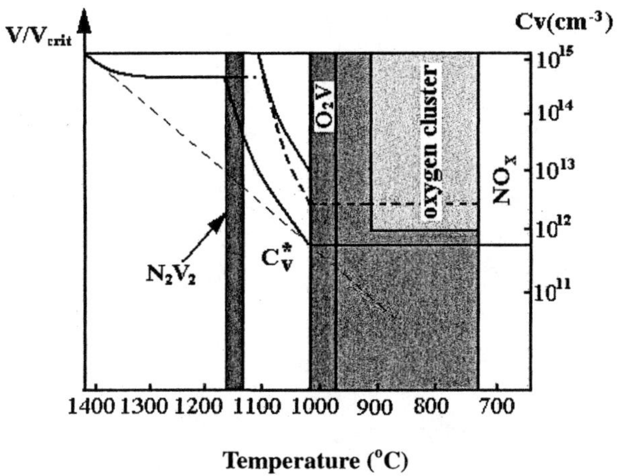

Fig. 5 Illustration of vacancies with temperature in CZ and NCZ silicon.

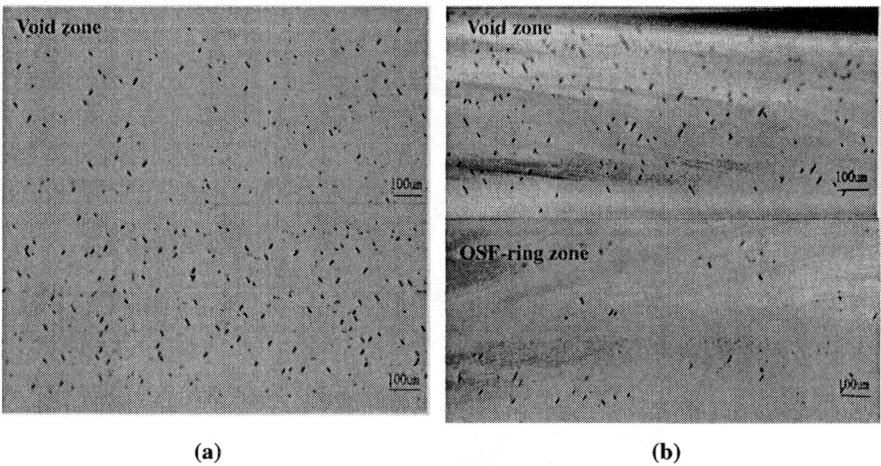

Fig. 6 BMD distribution in different zone of the mixed-type NCZ silicon after different annealing (a) 800 °C/6 h + 1050 °C/16 h; (b) 1050 °C/32 h.

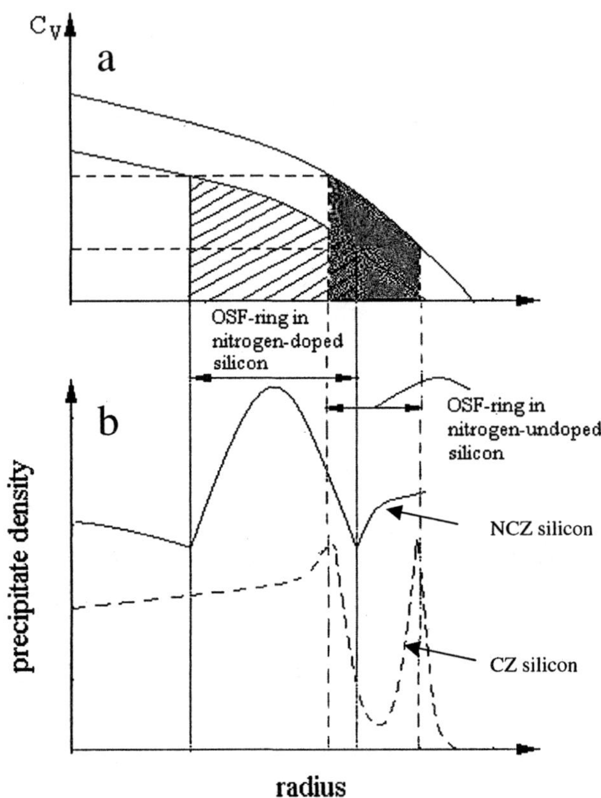

Fig. 7 Vacancy (a) and grown-in oxygen precipitate density (b) distribution along the radius in as-grown CZ and NCZ silicon.

Infrared Measurement of nitrogen concentration in CZ-Si

A. Hashimoto*[1], T. Matsumoto[1], D. Funao[1], N. Inoue[1,2]

RIAST, Osaka Prefgecture University[1]
1-2, Gakuencho, Sakai, Osaka, 599-8570, Japan
JEITA nitrogen measurement WG[2]
3-11, Kanda-Surugadai, Chiyoda, Tokyo 101-0062, Japan

ABSTRACT

We have improved the measurement accuracy of nitrogen concentration in Czochralski (CZ) silicon by infrared absorption spectroscopy and extended the adaptation for samples. The elimination method of interfering absorption is established. On the other hand, we enable to measure the low resistivity samples that are used widely for manufacturing devices. The measurement accuracy is evaluated by comparing the results with SIMS results and those of FZ samples.

INTRODUCTION

Nitrogen doping technology has attracted much attention as a defect control technology [1], and an accurate metrology of nitrogen concentration of the CZ crystal is needed. We have previously established the basic measurement procedure of nitrogen concentration in Czochralski (CZ) silicon crystal by infrared absorption spectroscopy [2, 3]. The samples used for these studies were high resistivity samples, because the background absorption was small. The measurement method established initially was simple so that many people could use easily. Therefore, in this paper we examine more accurate procedures. In addition, we extend the method to apply low resistivity samples, because the wafers that are used widely for manufacturing devices have low resistivity. The method that eliminates interfering absorption near three peaks of nitrogen is established. We confirm the measurement accuracy by comparing the results with SIMS results and those of FZ samples.

EXPERIMENTAL

Samples are nitrogen doped CZ silicon crystals and FZ silicon crystals, which were grown by the most major wafer venders in the world. As reference samples, non N-doped crystals were prepared. The range of estimated nitrogen concentration is from 1.9E14 to 3.6E15 $/cm^3$. The sample resistivity is from 1.72 to 8.5 Ωcm. Samples whose thickness is 10 mm are used in this study. The both surfaces of the samples were mirror polished.

For infrared absorption measurement, a Fourier transform infrared (FT-IR) spectrometer was used. There are three nitrogen absorption peaks to be measured to obtain total nitrogen concentration in CZ-Si. They are peaks at 963 cm^{-1} (NN), 996 cm^{-1} (NNO) and 1018 cm^{-1} (NNOO), respectively. Basic measurement conditions are summarized below. Temperature: room temperature, beam diameter: about 7 mm, wave number resolution: 2 cm^{-1}, detector: MCT, number of scans: typically 4000 (noise level less than 0.0001 in absorbance).

To obtain the absorption coefficient, basically the following data processing procedures are performed. First, difference spectrum is obtained by subtracting reference spectrum from sample spectrum. Thickness correction is done by subtracting the reference spectrum after multiplying sample thickness / reference thickness, so that lattice vibration absorption is fully cancelled. In case that nitrogen peak is detected clearly and there are neither interfering subsidary peak nor slowly changing background absorption, straight baseline is drawn between the end point of nitrogen - related absorption peak as reported previously.

But usually, some unwanted absorption is left. The nitrogen related peaks are located on the strong and wide lattice vibration absorption band at about 960 cm^{-1} as an example shown in Fig. 1. Therefore only a slight thickness difference between the sample and reference results in the big absorption band. Moreover, there are many absorption peaks, mainly due to accompanying impurities. It is necessary to separate these unwanted peaks. Finally, nitrogen-related peaks are delineated and absorbance is obtained.

The relation between the nitrogen concentration C and absorption coefficient alpha is given by using the conversion factor k as follows

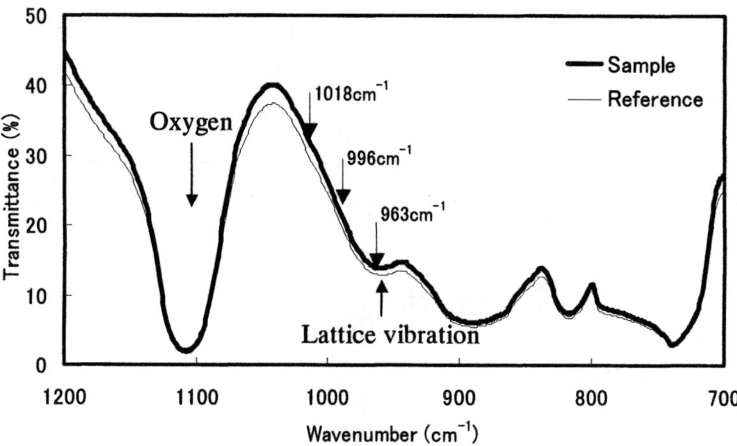

Fig. 1 Location of nitrogen-related absorption peaks.

$$C = k\alpha. \tag{1}$$

For FZ crystal, the absorption coefficient α is that of the peak at 963 cm^{-1}. On the other hand, the absorption coefficient of CZ crystal is the sum of α_{963}, α_{996} and α_{1018}. Moreover, the absorption coefficient is obtained by weighting with the oscillator strength as follows,

$$\alpha = \alpha_{963} + \frac{D_{963}^2}{D_{996}^2}\alpha_{996} + \frac{D_{963}^2}{D_{1018}^2}\alpha_{1018} \tag{2}$$

where C is concentration, α is absorption coefficient, k is conversion coefficient and D is dipole moment. Here, D_{963}, D_{996}, D_{1018} are calculated theoretically and their values are 11.8, 9.13 and 17.4 debyes respectively.

RESULTS AND DISCUSSION

Analytic procedure

The background of the peak at 963 cm^{-1} is generally determined by the lattice absorption peak as an example shown in Fig. 2. As this is smooth, we can remove it by fitting the smooth curved line to the spectrum as examples shown by the dotted lines in Fig. 2(a) and (b). The resulting spectrum includes sub-peaks on both sides of the peak at 963 cm^{-1}. When we use CZ reference, it has holes as marked by arrows in Fig. 2(a). They are due to the absorbance of CZ reference as marked in Fig. 2(c). When we use FZ reference, most of these sub-peaks are well separated as shown in Fig. 2(b). As a result, nitrogen peak is clearly delineated which is similar to the peak of FZ samples shown in Fig. 2(d).

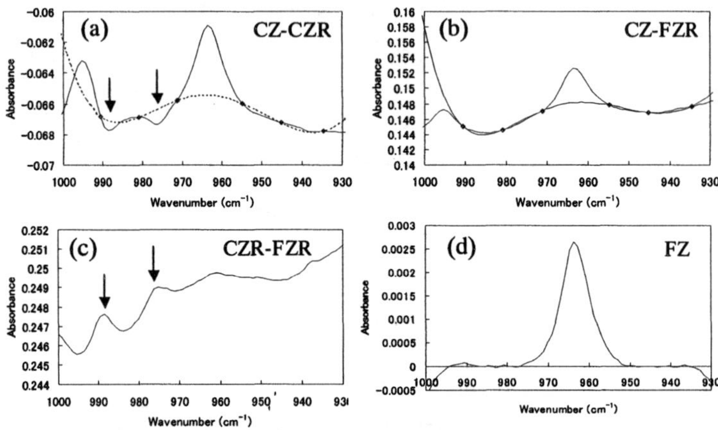

Fig. 2 Analytical procedure of the peak at 963 cm^{-1}. (a) CZ – CZ reference leaves dips marked by arrows, (b) FZ reference give smooth background, (c) Dips are due to absorption in CZ reference, (d) Nitrogen peak in FZ sample for comparison to the peak shown in (b).

On both sides of the peak at 996 cm^{-1}, interfering absorptions appear in the difference spectrum between sample and reference, like the case of peak at 963 cm^{-1}. In contrast to the case of 963 cm^{-1}, sub-peaks shown in the difference spectrum using FZ reference Fig. 3(a) (marked by arrows in) are overlapping the object. As a result, it is difficult to eliminate the interference. It is necessary to choose adequate reference sample. Difference spectrum using a CZ reference has dips on both sides in some cases. This was found to be due to unwanted absorption peak in the CZ reference as shown in Fig.3(b) (shown in (b')). By canceling these sub-peaks by weighted subtraction of these peaks as shsown in Fig. 3(c), we resolve the peak at 996 cm^{-1} successfully as shown in Fig.3(d).

As for the peak at 1018 cm^{-1}, we have already examined to eliminate the oxygen related peak on both sides as an example shown in Fig.4(a). This time we improve the accuracy by Gauss fitting to these peaks. First, the smooth curved baseline is fitted as shown by the dotted line in Fig.4(a). After the subtraction, the peak at 1018 cm^{-1} is accompanied by the peaks at 1008 cm^{-1} and 1026 cm^{-1}

respectively as shown in Fig.4(b). They are related to oxygen and the spectrum is reproduced by the subtraction of N-free FZ sample spectrum from N-free CZ sample spectrum. By subtracting this spectrum with a proper weight from measurement sample, these sub peaks are cancelled. As a result, the peak at 1018 cm^{-1} is successfully resolved by Gauss fitting as shown in Fig.4(d).

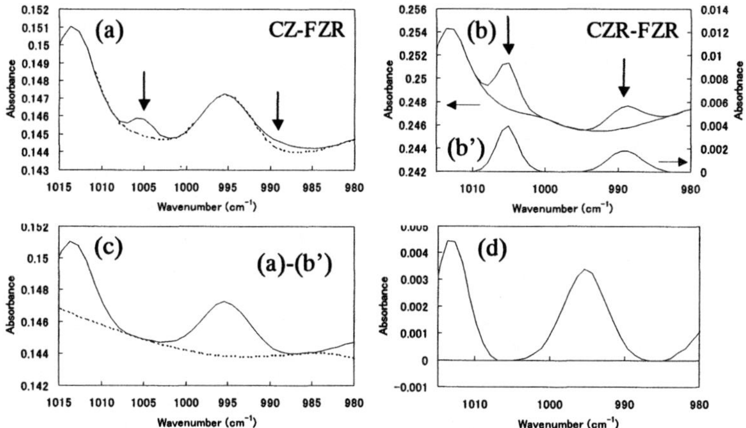

Fig. 3 (a) Sub-peaks overlap on peak at 996 cm-1. (b) They are due to absorption in CZ samples. (c) By removing them by using the synthesized spectrum (b'), smooth background is obtained. (d) The peak is resolved.

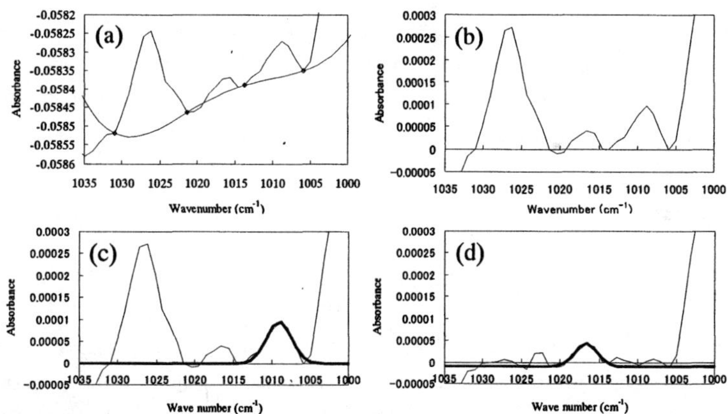

Fig. 4 (a) Peak at 1108 is accompanied by oxygen related peaks on both sides. (b) Elimination of background absorption. (c) Cancellation of oxygen peaks by Gauss fitting. (d) Nitrogen peak is also resolved by Gauss fitting.

Evaluation of the accuracy

It was easy to measure the CZ samples with high resistivity and FZ samples. Measurement of low resisitivity samples become possible now. The results of low resisitivity samples are compared to those of high resisitivity samples as shown in Fig.5. Both results agree well within the experimental accuracy. Thus, the measurement procedures for low resistivity samples are established.by the analytical procedures described above.

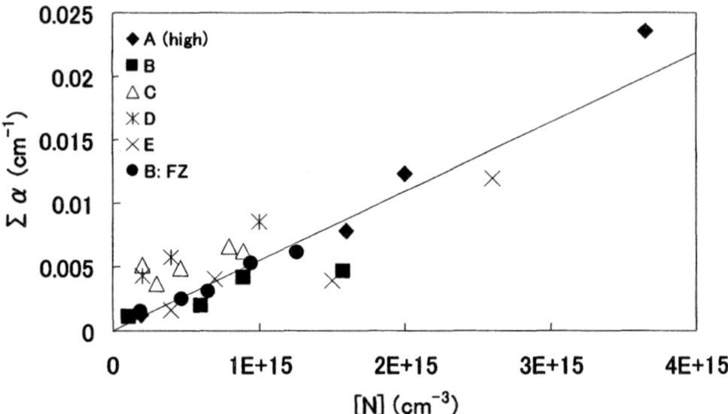

Fig. 5 Relation between the weighted sum of absorption coefficients of three peaks and the estimated concentration by the segregation theory.

Further, comparison of the results of CZ crystal to FZ crystal is done on samples with SIMS data as summarized in Fig. 6. As the measurement method of nitrogen concentration in FZ crystal has been established, this comparison enables us to examine the accuracy of measurement method for CZ crystal. Both data lie nearly on one line which is the calibration for IR and CPAA previously reported. Therefore, it is confirmed that nitrogen concentration of CZ crystal samples can be measured correctly.

SUMMARY

To measure nitrogen concentration more accuracy, we examine the procedures to analytically eliminate the interfering absorptions. For the peak at 963 cm^{-1}, the lattice vibration absorption is eliminated by slowly changing curved baseline. As the sub-peaks on both sides are separate from the objective peak, it is not difficult to cancel them. For the peak at 996 cm^{-1}, there are also sub-peaks on both sides. As they are overlapping to the objective peak, it is difficult to eliminate them. Ideal

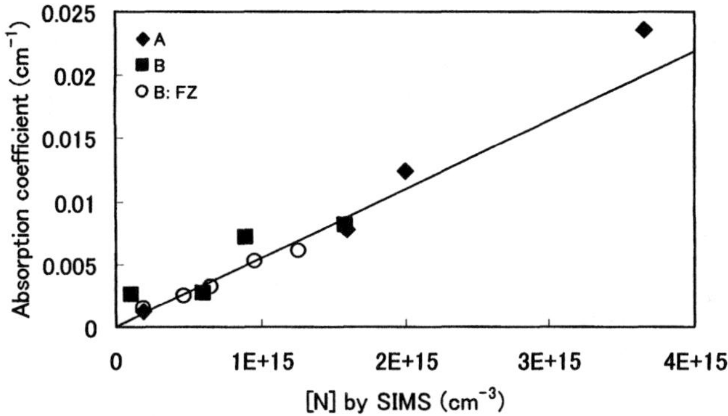

Fig. 6 Relation between the sum of absorption coefficients and SIMS concentration of CZ and FZ samples. Straight line represents the previous calibration on FZ samples by IR and CPAA.

sub-peak spectrum is created by the differential spectrum of CZ reference and FZ reference, and weighted subtraction of the peak is employed. As a result, more accurate absorption coefficient is obtained as evidenced by the better linear relationship to estimated [N]. For the peak at 1018 cm^{-1}, the sub-peaks on both sides are oxygen related and well reproduced by the differential spectrum of CZ reference and FZ reference. Gaussian fitting is employed to cancel thse sub-peaks accurately.

As a result, absorption coefficient in low resistivity samples and high resisitivity samples agree well. Moreover relation between absorption coefficients and SIMS concentration of CZ and FZ samples agree well with the previous calibration between IR and CPAA of FZ samples. These results confirm that the present methods provides the accurate measurement method of nitrogen concentration in CZ silicon.

ACKNOWLEDGEMENT

The authors thank Japan Electronic and Information Technology Industries Association WG members who accommodate us with samples. And we thak Dr Shirai froom Toshiba Ceramics and Dr Tanahashi and Dr Kaneda from Fujitu Laboratorieswho for valuable discussions.

REFERENCES

[1] Y. Itoh, T. Nozaki, T. Masui and T. Abe: Appl. Phys. Lett., 47, (1985) 488.

[2] T. Matsumoto, Y. Yamanaka, H. Harada and N. Inoue: Materials Sci. Eng. B91-92 (2002) 164.

[3] N. Inoue, K. Shingu and K. Masumoto, Semiconductor Silicon 2002 (Electrochem. Soc.) Vol. 2, p. 875.

[4] R. Jones, C. Ewels, J. Goss, J. Miro, P. Deak, S. Oeberg and F.B. Rasmussen: Semicond. Sci. Technol., 9, (1994) 2145.

Segregation of Nitrogen in CZ Silicon

N. Inoue*, **
*RIAST, Osaka Prefecture University
1-2, Gakuencho, Sakai, Osaka, 599-8570, Japan
**Japan Electron. Information Technology Association (JEITA)
3-11, Kanda-Surugadai, Chiyoda, Tokyo 101-0062 Japan

ABSTRACT

Accuracy of nitrogen concentration estimation by the segregation theory is evaluated by accurate measurement by infrared absorption spectroscopy and secondary ion mass spectroscopy. The ratio of measured concentration/ estimated concentration scatters from 1/3 to 4, showing that the estimation includes a serious error in some cases. In about half of examined crystals, measured concentration tends to increase with the solidified fraction, more rapidly than the estimated concentration. Empirical relation between the distribution coefficient and solid solubility at melting point is obtained.

INTRODUCTION

Nitrogen (N) doping has attracted much attention because of defect control ability. Generally nitrogen concentration ([N]) is estimated by the segregation theory, because the measurement method has not been established. We have developed the accurate measurement procedures for infrared absorption spectroscopy (IR) and secondary iron mass spectroscopy (SIMS) [1]. In this paper we measure the N concentration by IR and compare with the estimated [N] by the segregation theory. Accuracy of estimation is evaluated. In addition we draw the empirical relation between the segregation coefficient and solid solubility of impurities in silicon.

METHOD OF ESTIMATION

Details of the method of estimation are not provided by the wafer vendors. Therefore we calculate the concentration by the well known formula for the case of normal freezing and compare to the reported value with solidified fraction. In the case of normal freezing, solid concentration Cs of dopant with distribution coefficient k at the solidified fraction x from melt with initial concentration Co is given by the following formula:

$$Cs = Coko(1-x)^{(ko-1)}. \qquad (1)$$

Equilibrium distribution coefficient ko of N is 0.0007 [2]. Change in this case is illustrated in Fig. 1. It was found that the calculated concentration can be fitted with the reported value by choosing adequate initial melt concentration Co. Therefore it is thought that all wafer vendors use eq. (1) to estimate the solid concentration.

EXPERIMENT

Ten sample crystals with estimated Cs and corresponding x were supplied from five wafer vendors to JEITA. Estimated [N] ranges from 2x10^14 to 4x10^15 atoms/cm^3. Nitrogen concentration was measured by IR on 28 samples (10 mm thick) by the standard procedures established and improved by JEITA [3]. Both N (963 cm^-1 line) and N-O complexes (996 cm^-1 and 1018 cm^-1 lines) were measured. Correction for lattice vibration absorption and coexisting oxygen peaks was performed. Twelve samples including FZ samples were measured by SIMS to examine the accuracy of measurement. Most of the results of IR measurement agreed with the results of SIMS for most samples within several ten % as shown in Fig. 2. Thus the IR measurement accuracy is estimated to be within several ten % except low concentration samples. Thus estimated [N] is evaluated by the IR measured [N] below.

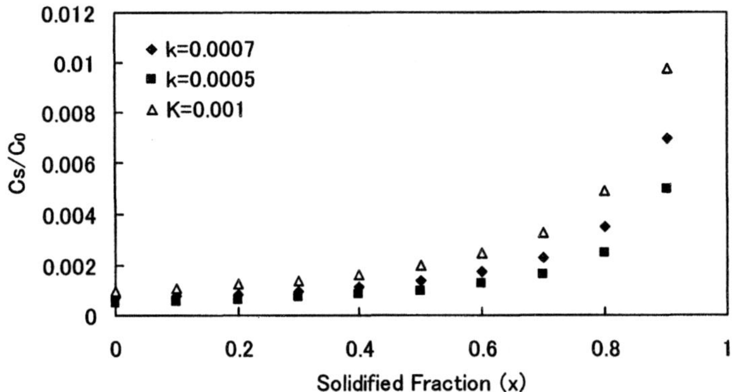

Fig. 1 Calculated concentration change in the case of normal freezing.

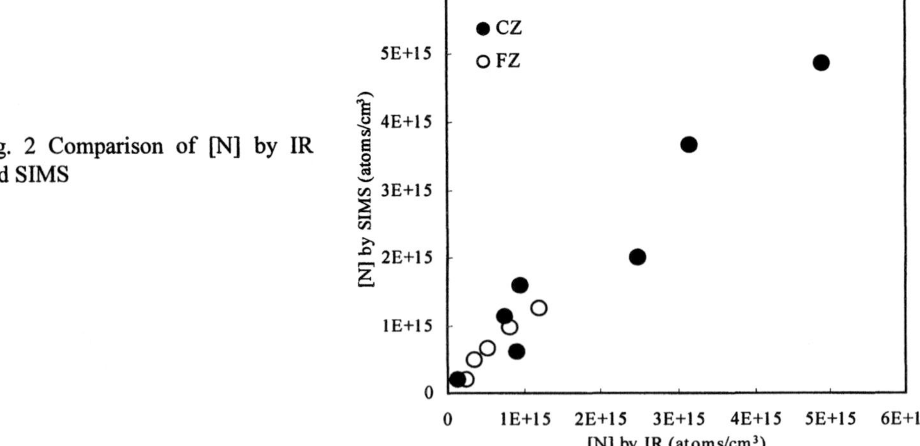

Fig. 2 Comparison of [N] by IR and SIMS

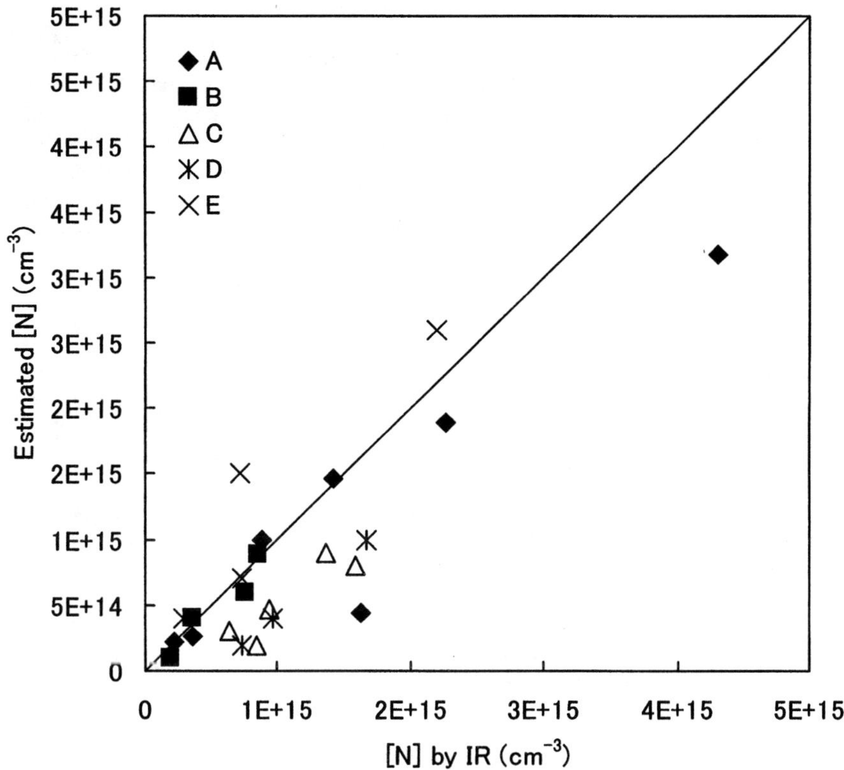

Fig. 3 Comparison of [N] by segregation theory and N by IR.

RESULTS AND DISCUSSION

Estimated concentration is compared to measured concentration by infrared absorption as shown in Fig. 3. Close correlation is observed, suggesting the successful estimation as a whole. It further supports that the IR measurement is accurate. Estimated values of some samples deviate from the measured one much, remarkable in the low concentration range. The deviation changes from vendor to vendor. The magnitude of deviation is characterized by the ratio of IR measured [N]/ Segregation estimated [N] as summarized in Fig. 4 in terms of solidified fraction. In many samples, the ratio is in the range of +/- 100%. This deviation is a few times as large as the difference between IR and SIMS results. Estimated [N] scatters from 1/4 to 3 of the measured [N], depending on the sample. Therefore some of estimated [N] include big error. Half of the specimens has the ratio less than 1, but others have the ratios more than 1. This suggests some loss of N from the melt. To clarify the sources of deviation, the ratio is plotted against solidified fraction. Data of some crystals having the distinct change are connected by lines in Fig. 4 to clarify the tendency. In about a half of crystals the ratio tends to increase with the 00%. This deviation is a few times as large as the difference between IR and SIMS results. Estimated [N] scatters from 1/4 to 3 of the measured [N], depending on the sample. Therefore some of estimated [N] include big error. Half of the specimens has the ratio less than 1, but others have the ratios more than 1. This suggests some loss of N from the melt. To clarify the sources of deviation, the ratio is plotted against solidified fraction. Data of some crystals having the distinct change are connected by lines in Fig. 4 to clarify the tendency. In about a half of crystals the ratio tends to increase with the increase of solidified fraction. There are few crystals showing the reverse tendency. To compare with the theory, estimated, measured and normalized concentration of a crystal is plotted against solidified fraction in Fig. 5. The normalized concentration increases steeper than the theoretical prediction. This can not be explained by changing the distribution coefficient, because the solidified fraction dependence in eq. (1) does not change much by changing ko because ko is included as (ko-1). Therefore the assumptions themselves must be examined, such as complete mixing, no diffusion boundary layer and so on. Moreover, evaporation of nitrogen must be examined.

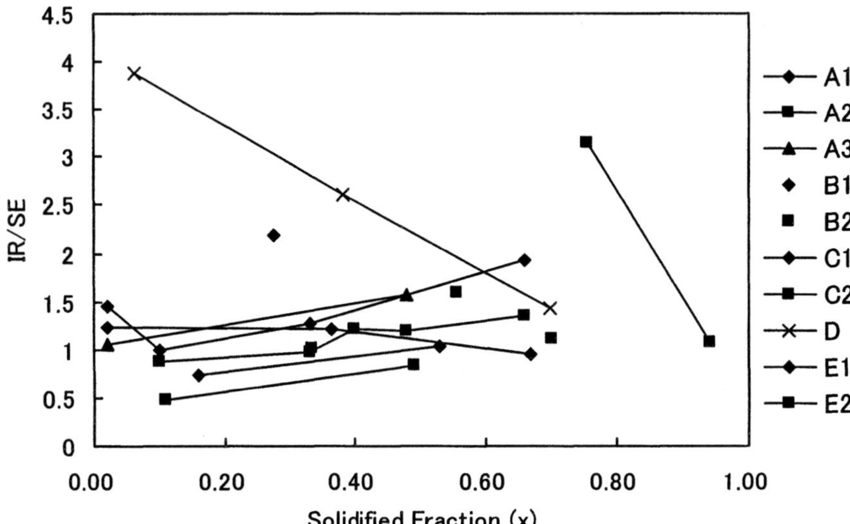

Fig. 4 Ratio of [N] by IR/ [N] by estimation plotted for solidified fraction. Solidified fraction dependence is indicated by lines.

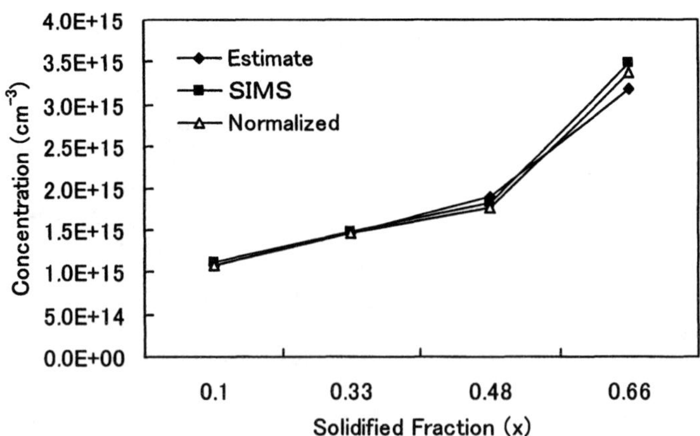

Fig. 5 Comparison of calculated concentration and measured concentration.

There have been few theoretical examination on the distribution coefficient of nitrogen in silicon. Empirical relationship of distribution coefficient to atomic radius had been reported as shown in Fig. 6 [4]. However, N deviates 4 orders of magnitude from the expected value as shown in Fig. 5. We make another approach: Relationship between the distribution coefficient and solid solubility at the melting

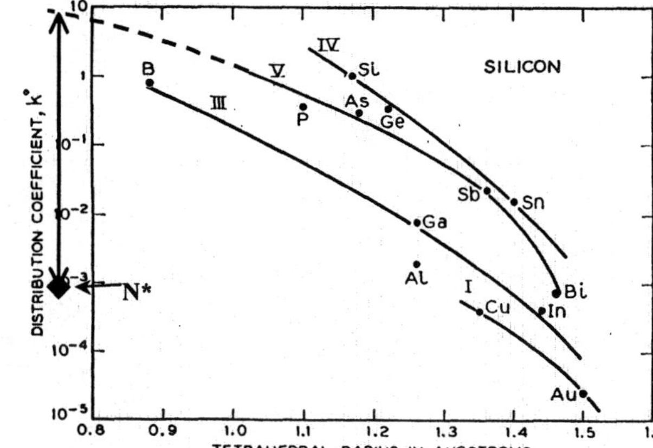

Fig. 6 Empirical relation between the segregation coefficient and atomic radius reported by Trumbore [4].

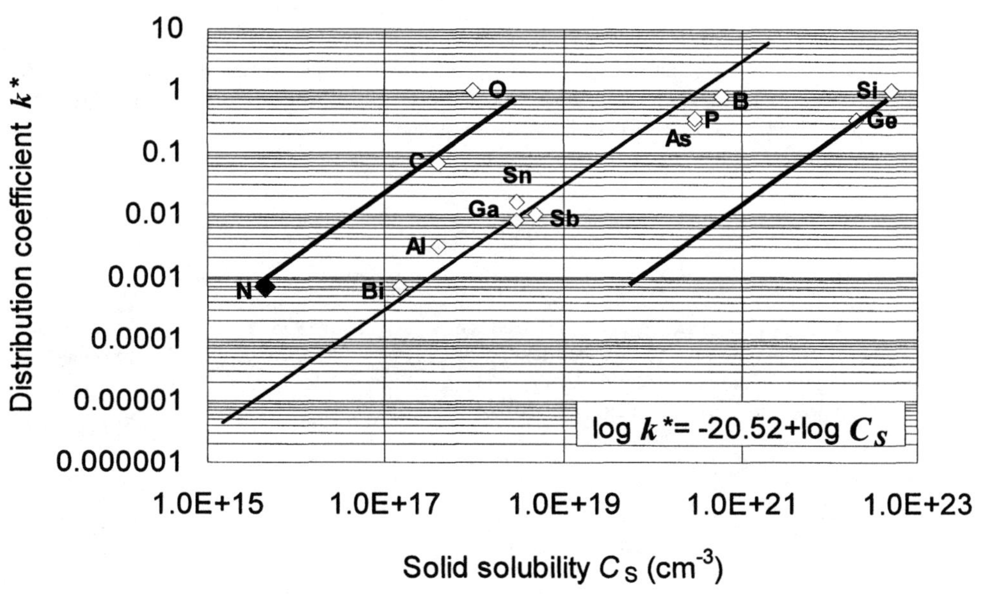

Fig. 7 Empirical relation between the segregation coefficient and solid solubility at melting point

point is examined. The result is shown in Fig. 7. Distribution coefficient and solid solubility has linear relationship in logarithm. It is to be noted that most dopants in third and fifth columns of periodic table lie on the same line with slope 1. On the other hand light elements of O, C and N lie on another line parallel to the dopant line and forth column species of Si and Ge form the other group. Therefore, it is shown that N behaves like the other light elements, in contrast to the previous study. The reason of linear relationship and the origin of grouping is discussed in terms of thermodynamics elsewhere.

SUMMARY

In summary, accuracy of nitrogen concentration [N] estimation by the segregation theory was evaluated by infrared absorption measurement. Samples with estimated concentration and known solidified fraction are provided by wafer vendors. Infrared measurement was done on all samples and SIMS measurement was done on some of the samples with the established procedures by JEITA. IR and SIMS data agree well with each other, within a few ten %, assuring the good accuracy of both measurement methods. Majority of estimated [N] ranges within +/- 100% of the IR measured [N], but some scatter from 1/4 to 3. Thus, it is deduced that estimation by segregation theory sometimes includes serious error. In about half of crystals, the ratio of IR [N]/ estimated [N] increased with the increase of solidified fraction. This tendency can not be reproduced by changing the distribution coefficient. Therefore, there may be problems in the normal freezing approximation, or there may be loss or increase of N in the melt such as evaporation. Empirical linear relationship between the segregation coefficient and solid solubility of dopants is found. Nitrogen forms groups with other light elements of oxygen and carbon.

ACKNOWLEDGEMENT

The author is grateful for help in the theoretical analysis and members of JEITA nitrogen measurement working group for providing samples and data. Discussion with S. Uda from Mitsubishi Material is very fruitful. He thanks to T. Matsumoto and M. Hamada for help in analysis and manuscript preparation.

REFERENCES

[1] N. Inoue, K. Shingu and K. Masumoto, Semiconductor Silicon 2002, p. 875.
[2] Y. Yatsurugi, N. Akiyama and Y. Endo, J. Electrochem. Soc. 120 (1973) 975.
[3] N. Inoue, N. Fujiyama and Yagi, Electrochem. Soc. 2002 Fall Meeting, Proc. Symp. Diagnostic Techniques of Semiconductor (to be published).
[4] F. A. Trumbore, Bell Syst. Technical J. (1960)205.

THEORETICAL ANALYSIS OF NITROGEN COMPLEXES IN CZ Si

D. Funao[1] I. Ohkubo[1] N. Inoue[1]
A. Karoui[2] F. S. Karoui[2] G. A. Rozgonyi[2]
1: RIAST, Osaka Prefecture University
1-2 Gakuen-cho, Sakai city, Osaka 599-8570, Japan
2: Materials Science and Engineering Dept.
North Carolina State University, Raleigh, NC 27695-7916

Nitrogen doping has attracted much attention because it affects the formation of defects in silicon. It is important for analysis of defect formation and measurement of nitrogen concentration to clarify the configurations in silicon or the origins of infrared absorptions. Therefore, we analyze the configurations and its vibrations theoretically by using molecular orbital method.

INTRODUCTION

Nitrogen doping has attracted much attention because it affects the formation of defects in silicon and enhances the precipitation of oxygen [1]. To clarify these mechanisms, it is important to clarify the configuration and behavior of nitrogen related complexes. The infrared absorption by the local vibration is widely used for such purposes in case of oxygen and carbon. Therefore it is necessary to examine the infrared absorption measurement of nitrogen related complexes in silicon.

Nitrogen forms interstitial NN pair in FZ crystal. This NN pair has two infrared absorption peaks at $963cm^{-1}$ and $764cm^{-1}$ and nitrogen concentration is measured by using either peak [2]. In the CZ crystal, NN pair further makes complexes with oxygen (NNO and NNOO). And they also show infrared absorption peaks at $996cm^{-1}$ and $1018cm^{-1}$ to which $963cm^{-1}$ peak of NN pair shifts [3], [4]. We call these configurations nitrogen-oxygen complexes. We theoretically study infrared absorption of these complexes to improve measurement accuracy of nitrogen concentration in CZ silicon. To sum absorption coefficient of these peaks, it is necessary to know their oscillator strength and make weighted sum. We determined the structure by molecular orbital method and made vibration analysis classically. We determined oscillator strength from translation dipole of each configuration. On the other hand, it is theoretically suggested that NN makes complex with vacancy. They are NNV and NNVV [5]. We call these configurations nitrogen-vacancy complexes. It is thought that nitrogen-vacancy complexes NNV and/or NNVV play some role to suppress formation of the defects. But, there is no method to research these configurations by experiment. We are expecting that nitrogen-vacancy complexes can be detected by infrared absorption, and for this reason, we are theoretically studying them to predict the infrared absorption.

Previously we have found four types of vibrations of each NNV and NNVV from 500cm^{-1} to 2000cm^{-1} in the infrared region. But, two of each are infrared inactive, so there are only two vibrations to appear. As the result, there are four infrared absorption peaks which appear between the two NN peaks at 764cm^{-1} and 963cm^{-1}. In the previous study, the numbers of silicon atoms of silicon clusters are different from each other among NN, NNO, NNOO, NNV and NNVV. In this point, the effects from silicon clusters to the core structure are not equal, and there are some possibilities of ambiguity in the relative infrared peak position or oscillator strength of these configurations.

This time, we aim at more accuracy of comparison, we use same silicon clusters among these configuration for vibration analysis.

ANALYSIS

In this study, molecular orbital method is adopted for optimization of structure to analyze vibration. The core structures of NNV and NNVV are same to what Kageshima proposed. First, sample structures are made to simulate reported structure. Then, their structures are optimized by molecular orbital method. We use semi-empirical method that includes approximation from experiments, because it is necessary to efficiently calculate large number of vibration modes of structures that include large amount of atoms. Hamiltonian that is used for the optimization is 'MNDO-d'. After the optimization of the structure, vibration mode, vibration frequency and translation dipole are calculated by the valence force theory.

In the previous analysis, numbers of surrounding silicon atoms for various structures has been different as described above, 40 atoms for NN, NNO and NNOO, 34 atoms for NNV and 42 atoms NNVV. This time, we improve these sample structures by two points. First, number of silicon atoms surrounding core structures of NN, NNO, NNOO, NNV and NNVV are equalized to compare the location of absorption peaks and oscillator strengths more correctly. Next, numbers of silicon atoms are increased to 62 to make the structure more actual. These sample structures are shown in Fig.1.

RESULTS AND DISCUSSION

Structure optimization

The core structures after optimization of nitrogen-oxygen complexes and nitrogen-vacancy complexes are shown in Figs. 2 and 3. Bond lengths of NNO structure become shorter than that of NN by the effect that the oxygen compresses the core diamond structure strongly. This agrees with the experimental result that the peak wavenumbers of NNO and NNOO are higher than that of NN.

Comparing with the previous result, the bond length shrinkage becomes bigger, because the bigger silicon cluster helps compression by the oxygen more efficiently. In the case of NNOO, because there is one more oxygen atom than NNO, it is expected that the bond lengths become shorter than those of NN or NNO. But, looking at the optimization result of NNOO, not all length the bond lengths are shorter than that of NN. This must be due to that the silicon shell compresses the two oxygen atoms effectively in the horizontal direction, but not effectively in the vertical direction. It is found that to achieve the good result, it is necessary to use bigger silicon shell.

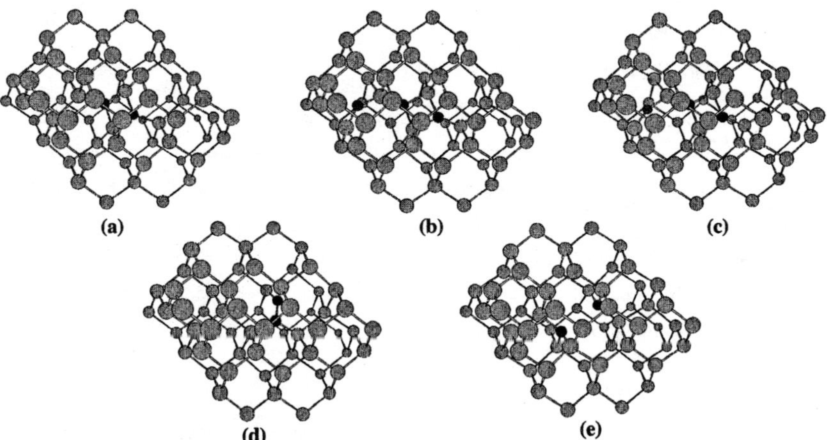

Fig. 1. Structures used for the vibration analysis. (a) NN: $Si_{62}N_2H_{54}$ (b) NNO: $Si_{62}N_2OH_{54}$ (c) NNOO: $Si_{62}N_2O_2H_{54}$ (a) NN: $Si_{61}N_2H_{54}$ (a) NN: $Si_{60}N_2H_{54}$ Terminal hydrogen is not shown.

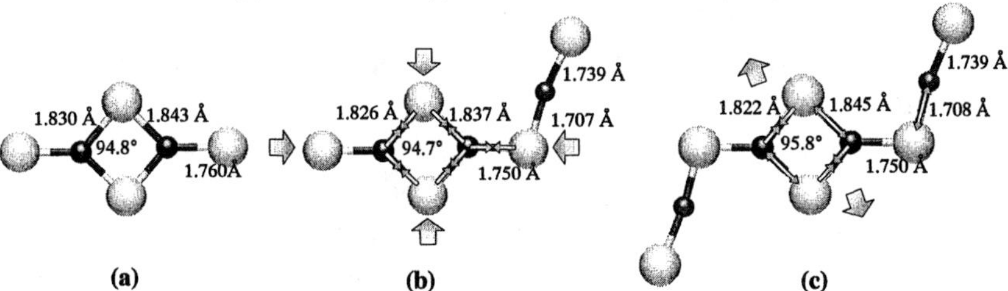

Fig. 2. (a) NN, (b) NNO and (c) NNOO core structures optimized by molecular orbital method.

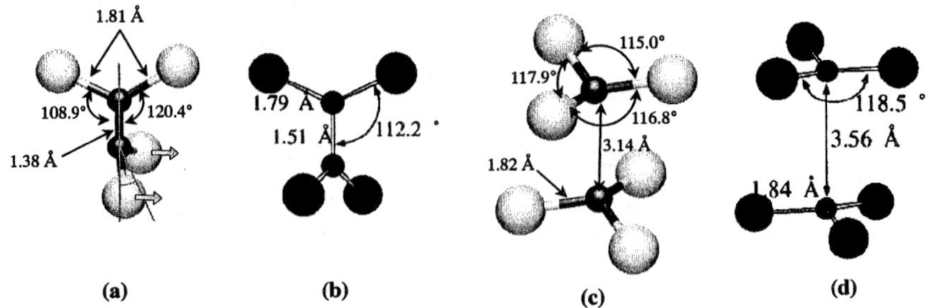

Fig. 3. (a) NNV and (c) NNVV core structures optimized by molecular orbital method.
(b) NNV (d) NNVV structures proposed by Kageshima.

Next, the core structures of nitrogen-vacancy complexes are compared to that of Kageshima's. About NNVV, its optimized structure becomes almost same as reported, and bond lengths become shorter than our previous structure, because of the effect from bigger silicon clusters. But about NNV, N-N-Si_2 core structure is not in plane but bended, though these atoms were in plane in the previous study. The reason is that are NNV can not be located at the center of the silicon shell, while NN, NNO, NNOO, NNV and NNVV are located at the center, and the compression from silicon shell is not isotropic. We must use bigger shell.

Vibration analysis

Next, we discuss the results of vibration analysis on these structures. The results are summarized in Table.1. Vibration modes of nitrogen-oxygen complexes are shown in Fig.4. About NN vibration, there are H_2O-type asymmetric stretching mode and BF_3-type asymmetric stretching mode. The wavenumber of the H_2O-type is 840cm^{-1}, and that of the BF_3-type is 1022cm^{-1}. These wavenumbers correspond to the observed 764cm^{-1} and 963cm^{-1} respectively. Now, we consider relative location of vibrations of each configuration. About NNO vibration, there are H_2O-type asymmetric stretching mode and BF_3-type asymmetric stretching mode. The wavenumber of the H_2O-type is 847cm^{-1}, and that of the BF_3-type is 1024cm^{-1}. They correspond to 801cm^{-1} and 996cm^{-1} respectively in the experiment. In the calculation, NNO's wavenumbers shift to higher side from NN's which agree with the experiment. About NNOO vibration, there is BF_3-type asymmetric stretching mode at 1022cm^{-1}, which appears at 1018cm^{-1} in the experiment. But, in this calculation result, NNOO wavenumber does not shift higher from NNO wavenumber. This result comes from compression of oxygen to core described above. About translation dipole of NN, NNO, NNOO, there is not big change compared with the previous data, and there is a tendency that the translation dipole of BF_3-type group becomes larger in order of NN (963cm^{-1}), NNO (996cm^{-1}) and NNOO

($1018cm^{-1}$).

Structure	Vibration mode	Exp. (cm^{-1})	Calc (cm^{-1})
NNV	H_2O-type (bending)	-	735 737
NN	H_2O-type	764	840
NNO	H_2O-type	801	847
NNVV	H_2O-type	-	896
NNVV	BF_3-type	-	898
NNVV	BF_3-type	-	907
NNVV	H_2O-type	-	914
NNV	H_2O-type (stretching)	-	927 972
NN	BF_3-type	963	1022
NNO	BF_3-type	996	1024
NNOO	BF_3-type	1018	1022
NNOO	O-self	1026	1104

Table 1. Normal vibration modes, measured frequencies and calculated frequencies of NN, NNO, NNOO, NNV and NNVV.

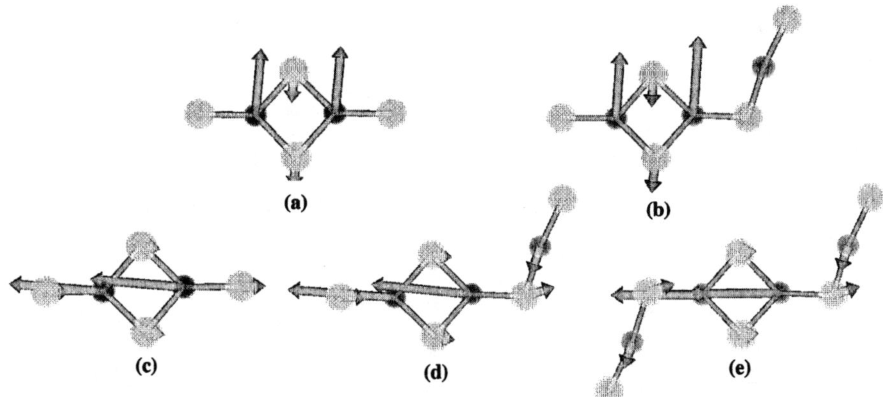

Fig. 4. Normal vibration mode by H_2O-type asymmetric stretch of (a) NN (b) NNO Normal vibration mode by BF_3-type asymmetric stretch of (c) NN (d) NNO (e) NNOO

Next, we compare the result of NNV and NNVV vibration to nitrogen-oxygen complexes (Fig.5 and Fig.6), About the NNVV, there are four vibration modes, two H_2O-types and two BF_3-types. Two modes (their wavenumbers are 896cm^{-1} and 898cm^{-1}) are infrared inactive, because translation dipoles are zero, and two modes (their wavenumbers are 907cm^{-1} and 914cm^{-1}) are expected to be detected by infrared absorption. It can be predicted that the absorption peaks of NNVV appear between two vibration peaks of NN. And these translation dipoles are enough large compared to that of NN at 764cm^{-1} or 963cm^{-1} , therefore it can be detected by infrared absorption if concentration is enough large. About NNV vibration, there are three types of vibrations, H_2O-type bending mode, H_2O-type asymmetric stretching mode and stretching of N-N bond. These modes are same as that previously obtained. But in more detail about H_2O-type bending mode, there was one wavenumber previously, but the modes break into three wavenumbers here. This comes from that N-N-Si_2 is not in plane but bended. The translation dipoles of these become much smaller than those previously, and the wavenumbers become much lower than previously. The reason of it must be this vibration mode is much affected by the bended core structure. Accordingly, these wavenumbers of 735cm^{-1} and 737cm^{-1} are dubious, and analysis of NNV must be improved. As the results of NNO and NNVV agree well with the previous ones, the previous results of NNOO and NNV are reliable.

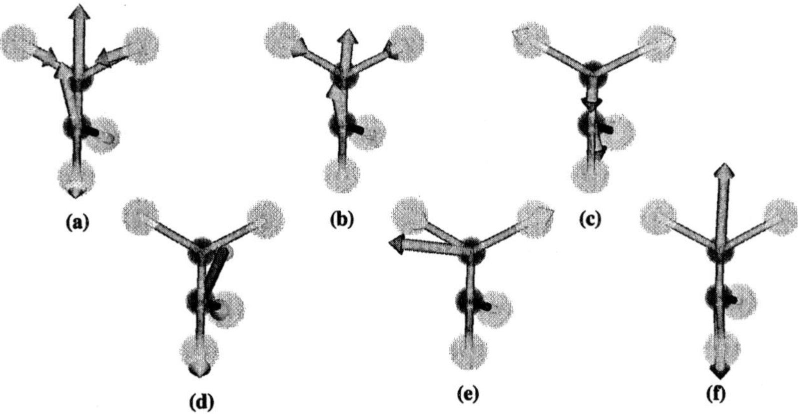

Fig. 5. Normal vibration mode of NNV by (a), (b) and (c) H_2O-type bend, (d) and (e) H_2O-type asymmetric stretch and (f) N-N bond's stretch

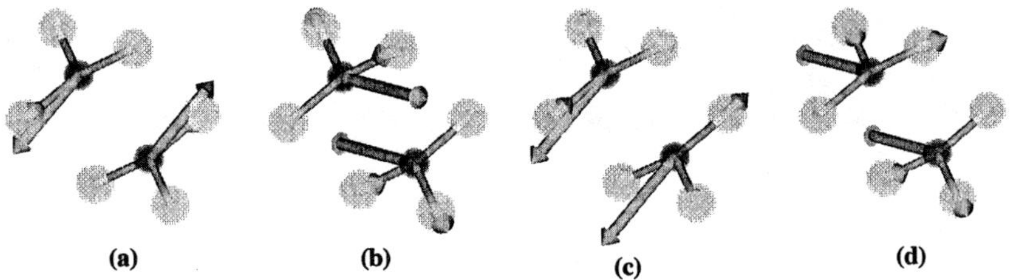

Fig. 6. Normal vibration mode of NNVV by (a) and (b) H_2O-type asymmetric stretch. (b) and (c) BF_3-type asymmetric stretch.

SUMMARY

We analyze vibration of NN, NNO, NNOO, NNV and NNVV under the same silicon clusters in theoretically, at the same time, these silicon clusters are made lager, and try to predict of these infrared absorption wavenumber and these oscillator strength more accurately.

As the result, wavenumbers of NNO absorption shifts higher compared to that of NN by repetitive, and the accuracy is improved. In addition, absorption peak of NNVV can be detected between two absorption peaks of NN. The translation dipoles of NNVV are enough large compared to that of NN, therefore it can be detected by infrared absorption if concentration is enough large. On the other hand, the bond length of NNOO cannot reproduce as previous length. The atoms in core structure of NNV are not in one plane. As NNOO, the compression of silicon cluster located roof or ground position is not enough to core structure, and for improving it, the number of silicon atoms in this position should make bigger. And as NNV, compression from silicon cluster is not isotropic from the problem in the located position about core structure, and for improving it, the number of surrounding silicon atoms should make bigger to reduce the effect of asymmetry structure. On the other words, we improve the accuracy of vibration analysis to put the core structure in isotropic surroundings by making silicon cluster large after this.

ACKNOWLEDGEMENTS

This work was partially supported by the JSPS for the future program.

REFERENCES

[1] M. Iida et al., DEFECTS IN SILICON III(The Electrochem. Soc., 1999) 499.

[2] Y. Itoh, T. Nozaki, T. Masui and T. Abe, Appl. Phys. Lett. **47** (1985) 488.

[3] R. Jones, S. Oberg, F. B. Rasmussen and B. B. Nielsen: Phys. Rev. Lett. **72** (1994) 1882

[4] R. Jones, C. Ewels, J. Gosss, J. Mori, P. Deak, S. Oeberg and F. B. Rasmussen: Semicon. Sci. Technol. **9** (1994) 2145.

[5] H. Kageshima et al., Appl. Phys. Lett., 76 25 (2000) 3718.

LOW TEMPERATURE DOPING OF SILICON BY HYDROGEN PLASMA TREATMENTS

R. Job, A. G. Ulyashin, Y. Ma, W. R. Fahrner
University of Hagen, Department of Electrical Engineering (LGBE),
P.O. Box 940, D-58084 Hagen, Germany

E. Simoen[1)], J. M. Rafi[1)], C. Claeys[1, 2)]
[1)] IMEC, Kapeldreef 75, B-3001 Leuven, Belgium
[2)] KU Leuven, Department of Electrical Engineering, Kasteelpark Arenberg 10, B-3001 Leuven, Belgium

F. J. Niedernostheide, H. J. Schulze
Infineon AG, Balanstr. 59, D-81541 München, Germany

ABSTRACT

In this paper we review our recent studies on hydrogen enhanced thermal donor (TD) formation in Czochralski (Cz) silicon. Applying hydrogen plasma treatments under appropriate conditions, TD formation is strongly enhanced and therefore a rapid n-type doping by TDs can occur even in deep wafer regions due to the high diffusivity of hydrogen. Doping by TDs can result in a counter doping of p type Cz Si, and therefore deep p-n junctions can be observed either just after a plasma hydrogenation or after a post-hydrogenation annealing, i.e. after applying a "1-step-process" or a "2-step-process", respectively. Graded p-n junctions with a depth of 100 μm or more can be easily realized at temperatures below 500 °C and process times in the order of 1 hour. The controlled TD formation can be used as a low thermal budget technology for the fabrication of devices with deep p-n junctions, e.g. high voltage diodes.

INTRODUCTION

Since several decades the properties of hydrogen in silicon faced a strong research interest. Numerous investigations were carried out to study the effects of hydrogen in the silicon host lattice and its interactions with various impurities and defects. For instance the role of hydrogen in connection with the passivation of various defects, impurities, deep levels, etc. [1, 2] was extensively investigated and up to now is still of great interest. For the item of this article some catalytic effects of hydrogen in silicon are essential and will be briefly discussed, i.e. the enhancement of oxygen diffusion and thermal donor (TD) formation by hydrogen.

In the present article our recent studies on hydrogen enhanced thermal donor formation will be reviewed and some outlines for future applications will be discussed. The presented studies were carried out with regard to the development of a new doping method with low thermal budget based on TD formation. Hydrogen operates as a kind of catalyst for these processes. Since hydrogen exhibits a high diffusivity, TDs can be created also in rather deep wafer regions. The controlled hydrogen enhanced TD formation can be used in order to fabricate devices with deep p-n junctions at low process temperatures (< 500 °C) and rather short process times. Deep p-n junctions are useful for certain semiconductor devices, e.g. high voltage diodes.

We developed two process routes for hydrogen enhanced thermal donor formation, which will be described in more detail in the next paragraph. During these process routes TDs cause counter doping of initial p-type Cz Si, and therefore deep p-n junctions can be observed either just after an plasma hydrogenation or after a post-hydrogenation annealing. By such methods graded p-n junctions with a depth of for instance 100 μm or more can be easily realized at temperatures below 500 °C and processing times in the order of 1 – 2 hours. Some test diodes fabricated with p-type Cz Si substrates on the base of the hydrogen enhanced TD formation were investigated. Also three layer transistor like test structures, i.e. p-n-p structures, were fabricated by hydrogen enhanced thermal donor formation. The properties of these test devices are analyzed and future prospects will be discussed throughout the text.

HYDROGEN ENHANCED THERMAL DONOR FORMATION

When we started our investigations, it was known that atomic hydrogen, which was introduced into n-type Cz Si by plasma treatments at temperatures between about 300 and 500 °C, causes an enhanced rate of TD formation [3-6]. This process finally results in n-type doping by TDs at temperatures below 500 °C, which is added to the initial n-type doping of the wafers. The created TDs belong to the family of the well-known thermal oxygen donors (TODs) or so called 'old' thermal donors, which are in fact thermal double donors (TDDs) [7-12]. The maximum concentration of TDs, which could be obtained in n-type Cz Si, depends on the concentration $[O_i]$ of interstitial oxygen in the substrates, the process temperatures and the dose of the incorporated atomic hydrogen. TD concentrations up to $[TD] \sim 10^{16}$ cm^{-3} could be observed after a few hours processing at 400 °C, where a rate of about $3.5 \cdot 10^{15}$ donor states/hour was observed for the TD formation close to the wafer surface [6]. It was also known that hydrogen causes an enhanced diffusion of O_i [13, 14]. A hydrogen atom, which diffuses through the Si lattice and approaches an interstitial oxygen atom, reduces the activation energy for the O_i migration through the Si lattice. Therefore, the probability that O_i atoms can reach appropriate defect lattice sites for the TD formation is strongly enhanced.

Some experimental results concerning hydrogen enhanced TD formation in p-type material were also published [15] before we started our investigations. But for these plasma hydrogenated p-type Cz Si samples an initial period of about 3 hour plasma exposure time at 400 °C was observed without an enhanced formation of TDs, i.e. the TD formation rate was similar to the one, which could be observed for furnace annealed samples [15]. So, some discrepancies occured for the hydrogen enhanced TD formation in p- and n-type Cz Si, since such kind of an incubation period was not seen for the plasma hydrogenated n-type material [6].

Due to our point of view, this difference between hydrogen plasma treated p- and n-type Cz Si might arise from the fact that the investigations of p-type Cz Si in [15] were carried out with bulk techniques (IR absorption spectroscopy), which are incapable to

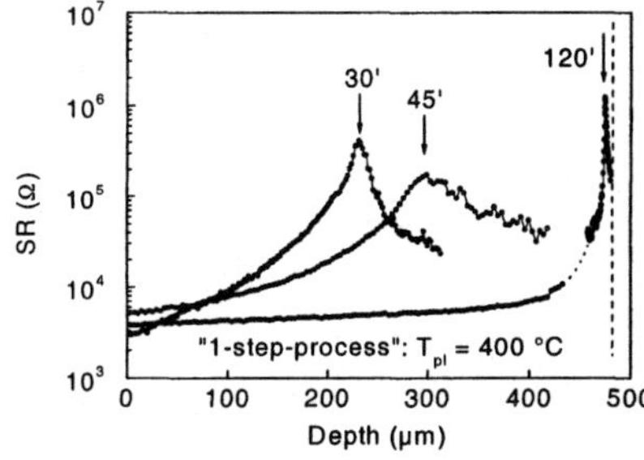

Fig. 1: SRP profiles of H-plasma treated Cz Si (wafer thickness: 480 μm), treatment time: 30, 45 120 min, substrate temperature [21]: 400 °C. The p-n junctions are marked by arrows.

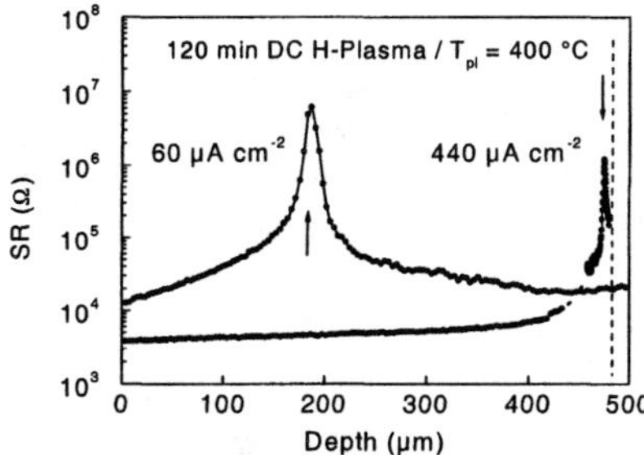

Fig. 2: SRP profiles of plasma treated Cz Si: 60 μAcm^{-2} = low hydrogen dose, 440 μAcm^{-2} = high hydrogen dose. The p-n junctions are marked by arrows (wafer thickness: 480 μm).

discriminate bulk and surface properties [16]. A depth resolving spreading resistance probe (SRP) analysis [17] is a more appropriate method for the analysis of TD formation in plasma hydrogenated Cz Si. This holds especially for early stages of TD formation, where the n-type region due to TDs is still located rather close to the wafer surface. Indeed, having applied the SRP method, we could show that the hydrogen enhanced TD formation can also be observed after short time hydrogen plasma treatments of p-type Cz Si [18-21].

We developed two process routes for a controlled counter doping of p-type Cz Si by TDs: (i) a "1-step-process", where the p-n junction appears just after a plasma hydrogenation at elevated temperatures (400 - 450 °C) [18, 21], (ii) a "2-step-process", where after a hydrogen plasma treatment at lower temperatures (~ 250 °C) an annealing at 400 – 450°C is required for the p-n junction formation [20].

Fig. 1 shows the resistance profiles obtained by SRP measurements on Cz Si samples treated by the "1-step-process" [18]. The substrates used for these experiments were (100)-oriented boron doped (12 Ωcm) wafers, which were polished on one side. The thickness of the wafers was about 500 μm. The concentration of interstitial oxygen was $8 \cdot 10^{17}$ cm^{-3}, as could be seen by IR-absorption measurements. The concentration of substitutional carbon was less than $10^{16} cm^{-3}$ (specified). SRP measurements were done with a 2/4-point probe instrument with tungsten carbide tips. To carry out the SRP measurements the samples were mechanically beveled under low angle (up to 10°) on a rotating quartz plate. On the beveled surface of the wafer the thermoelectric microprobe method was also applied to prove the type of conductivity as function of the wafer depth [17]. The curves in Fig. 1 were obtained from samples, which were treated by a DC H-plasma for various process times [18]. The plasma hydrogenations were done at 400 °C under a pressure of $1 \cdot 10^{-2}$ mbar, a plate voltage of 500 V, and a current density of 440 $\mu A/cm^2$. No subsequent annealing was carried out. The peaks in the spreading resistance curves (Fig. 1) indicate the p-n junctions. The n-type doping due to TDs occurred between the wafer surface and the resistance peaks. A p-n junction depth of ~ 200 μm can be observed after 30 min hydrogenation time under the applied conditions [18]. A prolongation of the plasma treatment yields a penetration of the p-n junctions to deeper wafer regions. After a 2 hour hydrogenation time the sample was almost completely

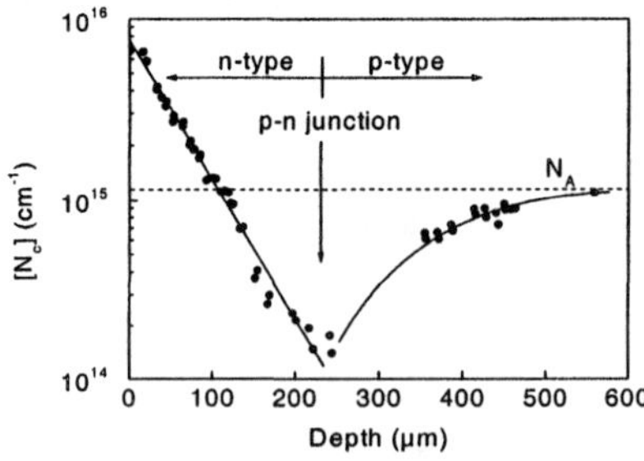

Fig. 3: "1-step-process": free carrier concentrations due to doping by TDs. The data points were obtained by C(V) measurements [21].

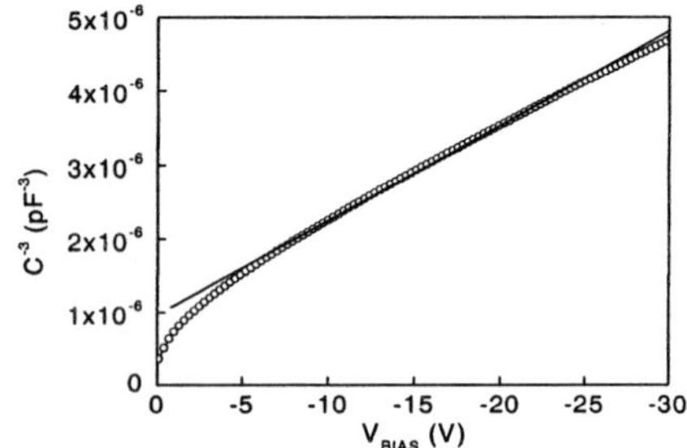

Fig. 4: "1-step-process": linear behavior of C^{-3} vs. bias voltage hints at a linear graded p-n junction [21].

counter doped by TDs except for a narrow region close to the samples backside. If the DC plasma exposure is done at lower current density the p-n junction location is less deep than for the case shown in Fig. 1 (e.g. ~ 190 μm after 2 hour exposure at 60 μA/cm^2 [18]). A comparison of the SRP profile of the 2 hours plasma exposure for two different current densities is shown in Fig. 2. This shows that the hydrogen dose plays a major role in the TD formation process.

One well-known problem of SRP measurements is the conversion of the resistivity to a doping or carrier concentration. Although some algorithms for solving this problem are developed [22-24], the results of such transformations can be sometimes questionable [25]. Therefore C(V) measurements for the determination of the carrier concentrations are often more appropriate despite of some limitations of this method [17, 26]. The C(V) measurements were performed at a frequency of 1 MHz using a 'HP4284A LCR'-meter. Two mercury probes in the arrangement of a guard ring structure were used to provide a Schottky contact to the samples surface. For depth profiling of the carrier concentrations chemical etching and/or mechanical polishing were applied to overcome the limitations of C(V) measurements caused by the avalanche breakdown of the metal-semiconductor barrier at high bias voltages. As can be seen from such C(V) measurements, the electron concentration originating from the TDs shows an exponential decrease between the surface and the depletion zone of the p-n junction (see Fig. 3) [18, 21]. One always should keep in mind that one TD complex provides two carriers, since we are always dealing with TDDs. The capacitance to the minus third power, C^{-3} exhibits an almost linear dependence on the reverse bias voltage, if $|V_{bias}|$ exceeds 5 V (Fig. 4). This hints at a p-n junction, which can be approximated roughly by a nearly linear graded junction [27].

Similar results can be obtained from p-type Cz Si wafers, which were treated by a "2-step-process" for hydrogen enhanced TD formation. Fig. 5 exhibits the SRP profiles of a 5-10 Ωcm p-type Cz Si wafer obtained after a RF H-plasma hydrogenation at 250 °C for 1 hour and subsequent post-hydrogenation annealing at 450 °C in air for different annealing times between 30 min and 3 hours [20]. The plasma was applied at a frequency of 110 MHz , a power of 30 W, a pressure of 0.333 mbar and a hydrogen flux of 200 sccm. After 30 min annealing the SRP profile exhibits a p-n junction which is located still rather close to the surface of the wafer. Longer annealing times result in a rather fast shift

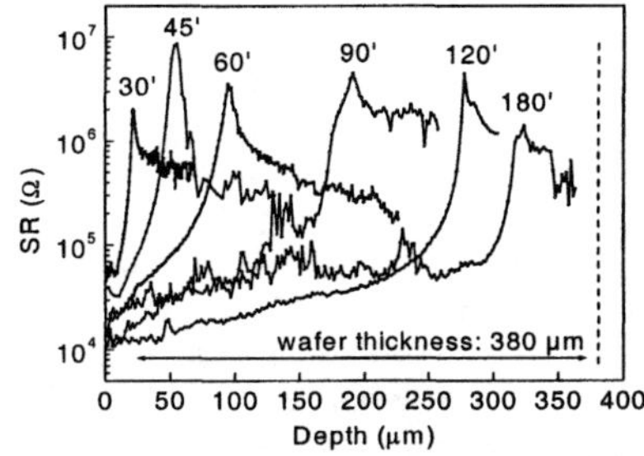

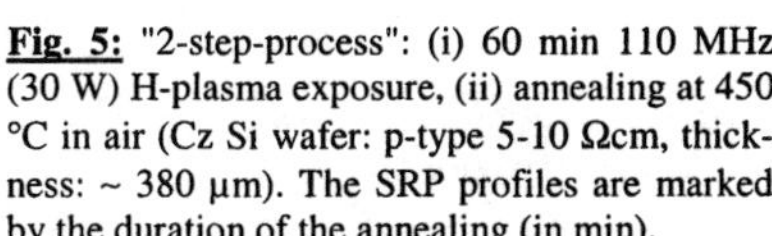
Fig. 5: "2-step-process": (i) 60 min 110 MHz (30 W) H-plasma exposure, (ii) annealing at 450 °C in air (Cz Si wafer: p-type 5-10 Ωcm, thickness: ~ 380 µm). The SRP profiles are marked by the duration of the annealing (in min).

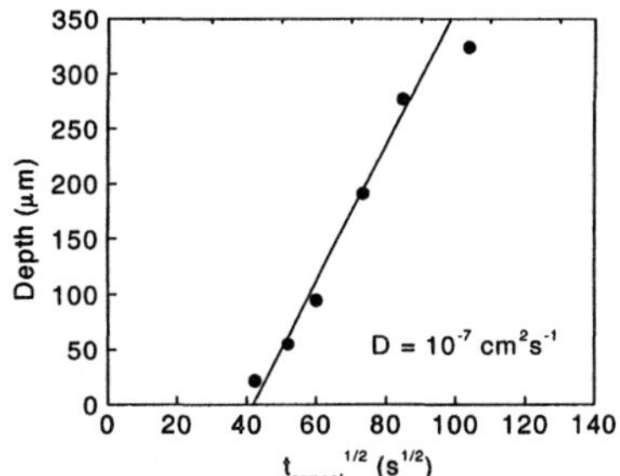

Fig. 6: p-n junction depth in dependence on the annealing time. From the slope of the linear part of the curve a diffusion D constant can be estimated [20].

of the p-n junction into deeper regions of the wafer. After 3 hours annealing the p-n junction is already located rather close to the backside of the wafer. Applying the annealing for more than 4 hours at 450 °C a complete conversion from initial p-type Cz Si to n-type can be achieved by TDs with a rather constant SRP profile [20].

We have shown in [20] that the hydrogen enhanced TD formation reflects the diffusion of atomic hydrogen and that the p-n junction depths (peak location in Fig. 5) are proportional to a mean distance L of the H diffusion profile. The TD formation by the "2-step-process" hints at a decay of molecular hydrogen species during the post-hydrogenation annealing step at around 400 - 450 °C [20]. It is known that immobile molecular H species can be formed in p-type Si during a plasma hydrogenation at 200 - 250 °C (e.g. [28]) close to the wafer surface. They decay again during a subsequent annealing at 400 - 450 °C. Therefore at 450 °C a large amount of mobile atomic hydrogen is present in the Cz Si wafer and can support the diffusion of interstitial O_i. As a consequence the TD formation can be found even in very deep wafer regions. The created donor states dissolve at 600 °C annealing for 15 min (as expected for 'old' thermal donors [11, 12, 18]). In the "1-step-process" the mobile atomic hydrogen is directly inserted from the plasma (not by the indirect way via the hydrogen molecular decay in the "2-step-process"). Therefore, the p-n junctions generated in the samples by the "1-step-process" are deeper compared with samples subjected to the "2-step-process" under otherwise similar conditions [18].

It is an accepted model that neutral atomic H^0 is the species, which supports an enhanced diffusion of interstitial oxygen atoms and therefore enhances the TD formation in n-type Si [13, 14]. However, as we have discussed in [20], most probably H^0 is also responsible for the enhancement of the TD formation in p-type Cz Si. We have shown that the penetration of the counter doped n-type region (due to TDs) into the p-type wafer bulk can be described by Fick's diffusion law. In Fig. 6 the depths of the p-n junctions shown in Fig. 5 are plotted in dependence on the corresponding annealing times. Up to annealing times of about 1.5 hours the p-n junction depths are proportional to the square root of the annealing time. According to the relation that $L = (4 \cdot D \cdot t)^{1/2}$, where L is the mean diffusion length, from the slope of the linear part of the curve in Fig. 6 one can calculate the diffusion constant. At longer annealing (t > 1.5 hours) this simple relation due to the diffusion mechanism of Fick does not hold anymore. The reason for this fact might originate in an out-diffusion of hydrogen from the wafer, if the annealing times are too large. The estimated diffusion constants D can be attributed to H^0. At 450 °C the experimentally found value of $D \approx 1 \cdot 10^{-7}$ cm^2s^{-1} comes close to the extrapolated one from the "Van Wieringen-Warmholtz" equation ($D_{VWW} = 4.36 \cdot 10^{-6}$ cm^2/s) [20, 29]. Therefore, we conclude that the diffusion of H^0 plays the major role for the fast TD formation. This is supported by the results from [30], where a much faster diffusion of H^0 than of charged H^+ ions was predicted theoretically. From our recent experimental results presented in [20] we could estimate that $D(H^0)/D(H^+) \approx 10^5$ in agreement with [30].

It is well known (e.g. [1, 2]) that atomic hydrogen should exist in p-type silicon mainly as positively charged H^+ ions and only a small amount of neutral atomic H^0 atoms might exist in such material; on the other hand H^0 and perhaps negatively charged H^- should be found in n-type Si. Due to our observation of a similar hydrogen enhanced TD formation in n-type and p-type Cz Si, one can conclude that a significant concentration of neutral atomic H^0 should also exist in p-type Cz Si. H^0 acts as a catalyst for the TD formation in both p- and n-type Cz Si due to the mechanisms described for instance in [31, 32].

PREPARATION AND CHARACTERIZATION OF TD-DIODES

Several test diodes have been prepared by hydrogen enhanced thermal donor formation. These diodes were prepared either by the "1-step-process" or by the "2-step-process". For this purpose different kinds of plasma hydrogenation setups were employed. For the preparation of the TD-diodes p-type (100)-oriented Cz Si wafers were used. The wafers had a diameter of 3 inch and a thickness of about 400 μm. The p-type conductivity was specified with 12 – 20 Ωcm, i.e. the initial boron concentration in the wafers was about $1 \cdot 10^{15}$ cm^{-3}. The impurity concentrations of interstitial oxygen and substituted carbon were specified with $7 - 8 \cdot 10^{17}$ cm^{-3} and $< 5 \cdot 10^{16}$ cm^{-3}, respectively.

In the following section we want to discuss the characteristics of some TD-diodes prepared under different conditions. The preparation of TD-diode No. 1 was done by a "1-step-process". The plasma hydrogenation was carried out in a μ-wave plasma assisted chemical vapor deposition (MPCVD) reactor. The hydrogen plasma was excited at a frequency of 2.45 GHz and a power of 800 W. The hydrogen flux and the pressure in the MPCVD-reactor were set to 200 sccm and 8 mbar, respectively. The plasma hydrogenation was applied for 30 min at a process temperature of 450 °C. After the plasma exposure no subsequent annealing was carried out. Finally, contact metallization was simply done by a photolithography process and deposition of aluminum. The contact size at the frontside of TD-diode No. 1 was 1 mm^2, while the whole wafer backside was covered by aluminum. After the metallization single diodes were cut from the wafer. The edges of the diodes were mechanically polished on a rotating quartz plate.

No other subsequent preparation procedures were carried out to further improve the diode characteristics. Especially it is to point out that neither a surface passivation nor a proper edge termination have been carried out to reduce leakage currents via the surface and the edges of the diodes. It is also important to note that no additional n^+ and p^+

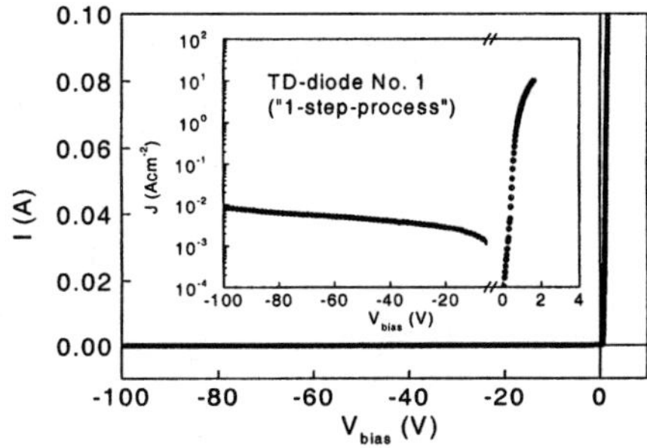

Fig. 7: I(V) curve of TD-diode No. 1 prepared by a 30 min plasma hydrogenation at 450 °C ("1-step-process"). Contact area: 1 mm^2. The insert shows the logarithmic scale.

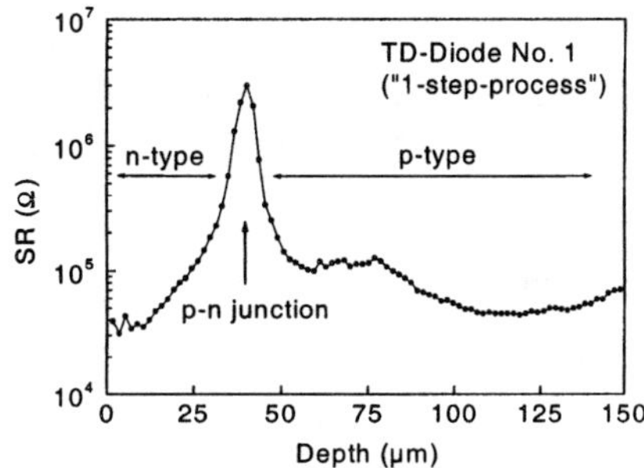

Fig. 8: SRP profile of TD-diode No. 1, p-n junction depth: 40 μm.

doping were applied (e.g. with [P] ≈ 10^{20} cm^{-3} and [B] ≈ 10^{19} cm^{-3} at the front- and backside of the wafer) to improve the properties of the metallic contacts. Therefore it is to underline that under these simplified process conditions the prepared TD-diodes can be only regarded as test devices. For industrial prototypes or real devices some technological steps have to be applied for a sufficient strong reduction of the leakage current. Anyway, for the purposes of our study to obtain information about general properties of TD-diodes and related technological trends, the simple test TD-diodes are sufficient.

In Fig. 7 the current-voltage (I(V)-) curve of TD-diode No. 1, which was prepared by the 30 min MPCVD plasma hydrogenation at 450 °C is shown. All I(V) curves presented in this article were measured at room temperature. The p-n junction depth of this diode was about 40 μm, as could be observed by SRP measurements (Fig. 8). As mentioned above, the p-n junction depth is strongly related to the hydrogen concentrations, which are incorporated into the Cz Si wafers. The p-n junction of TD-diode No.1 was not as deep as could be expected from the p-n junction locations shown in Fig. 1 and 2, which were obtained by a "1-step-process" at 400 °C using a DC hydrogen plasma setup. For this latter hydrogenation setup the hydrogen incorporation and therefore the TD formation was more effective than in case of the MPCVD plasma hydrogenation. It is to point out that one always has to take into account that the degree of the hydrogen enhanced TD formation can be significantly changed if the processes are carried out in different hydrogenation reactors. If various setups are used for the preparation of TD-diodes, each of them has to be calibrated with regard to the amount of hydrogen, which is incorporated into the wafer during the TD formation process. Otherwise the depth of the p-n junctions and therefore the properties of the TD-diodes can vary significantly, even when a substrate material with the same interstitial oxygen concentration is used.

Nevertheless, the I(V) characteristic of TD-diode No. 1 is quite promising, up to a reverse bias voltage of –100 V no breakdown occurs (the –100 V limit was given by the 'HP4156A' semiconductor parameter analyzer). However, the leakage current is rather high. Normalizing the current to the contact size (1 mm^2) TD-diode No. 1 exhibits a leakage current density of about 10^{-2} Acm^{-2} at a reverse bias of –100 V.

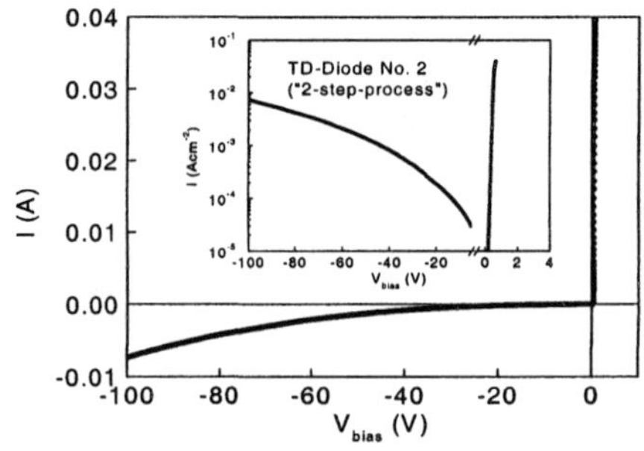

Fig. 9: I(V) curve of TD-diode No. 2 prepared by a 60 min 110 MHz H plasma treatment at 250 °C and subsequent annealing at 450 °C/air for 30 min ("2-step-process"). Contact area: 1 cm^2. The insert shows the logarithmic scaling.

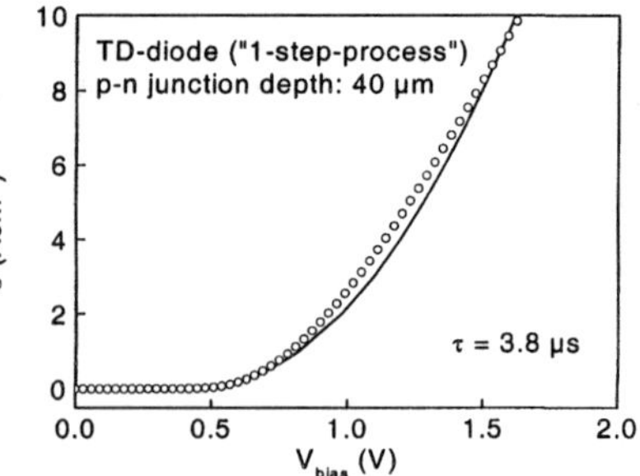

Fig. 10: The $J(V_{bias})$ characteristic of TD-diode No. 1 in comparison with the simulated curve of the device simulator MEDICI.

The second example, TD-diode No. 2, was prepared by a "2-step-process", i.e. by a 110 MHz RF plasma hydrogenation at 250 °C for 1 hour (50 W power, 200 sccm hydrogen flux, 0.333 mbar pressure) and a subsequent post-hydrogenation annealing at 450 °C for 30 min on a hotplate in air. After annealing the sample was rapidly cooled down to room temperature. Aluminum contacts have been fabricated directly after the annealing procedure. The contact size was 1 cm^2. Further processing was similar to diode No. 1. The I(V) curve of TD-diode No. 2 is shown in Fig. 9. The current density at a reverse voltage of –100 V is again approximately 10^{-2} Acm^{-2}.

Although the characteristics of the presented test diodes under reverse bias are rather poor, the results are promising for future applications, since the leakage currents should decrease strongly if the diodes are provided with proper edge termination and a surface passivation. This technical problem should cause no principle limitations to the preparation of TD-diodes, which can be used as real devices, because such a reduction of leakage is done by standard procedures in the industrial processes. Some future prospects and technical trends of TD-diodes can be obtained by simulation and are discussed in the next chapter.

NUMERICAL SIMULATION OF TD-DIODES

The numerical simulations were carried out using the commercial software package MEDICI. Fig. 10 exhibits the forward current-voltage behavior of a simulated TD-diode. In this simulation the characteristic geometrical and electrical parameters are chosen as observed experimentally for the test diode No. 1 with a total thickness of 380 µm. The p-type doping concentration of the substrate was set to $1 \cdot 10^{15}$ cm^{-3} corresponding to the mean boron concentration in 12 – 20 Ωcm p-type Cz Si. According to the C(V) measurements (Fig. 3) an exponential donor concentration profile with a surface concentration of $1 \cdot 10^{16}$ cm^{-3} leading to a p-n junction depth of 40 µm was chosen. The overall doping profile with a graded p-n junction, which was used for the simulation of the TD-diode No.1, is shown in Fig. 11.

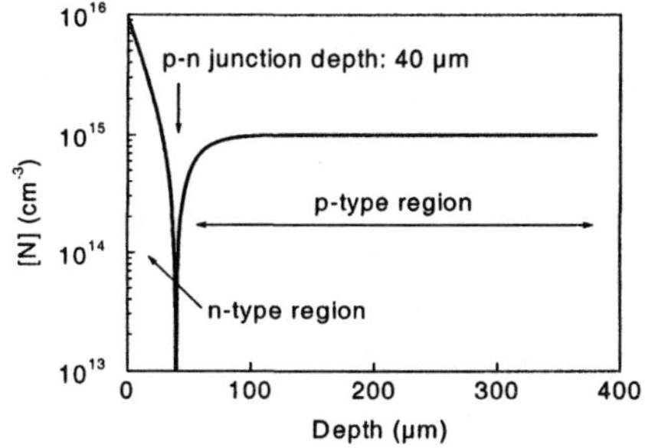

Fig. 11: Carrier profile used for the simulations in Figs. 10, 12, 13. P-n junction depth: 40 µm.

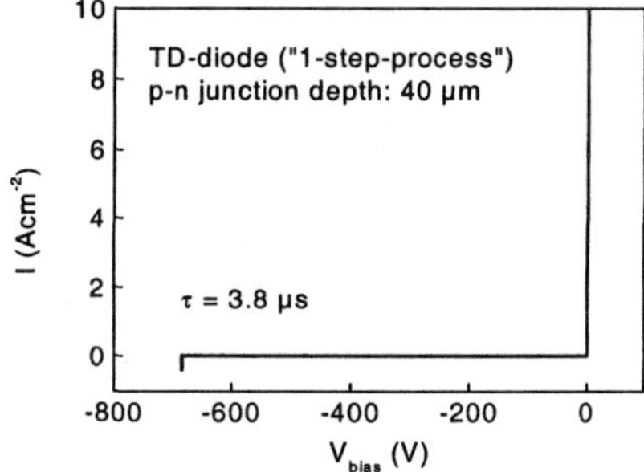

Fig. 12: $J(V_{bias})$ simulation of TD-diode No. 1: avalanche breakdown occurs at about –700 V.

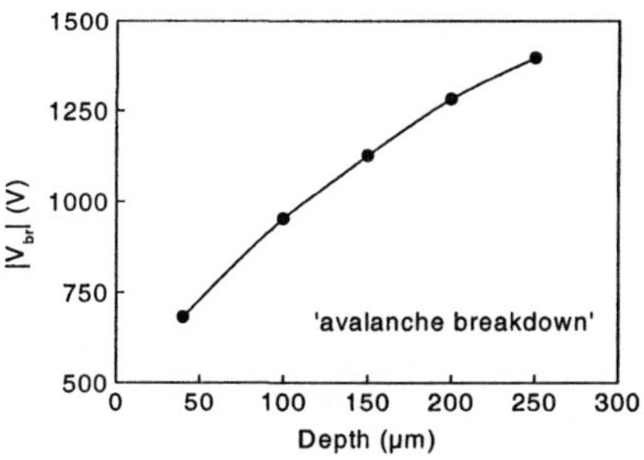

Fig. 13: Calculated avalanche breakdown voltages under reverse bias in dependence on the p-n junction depths d (d = 40 µm).

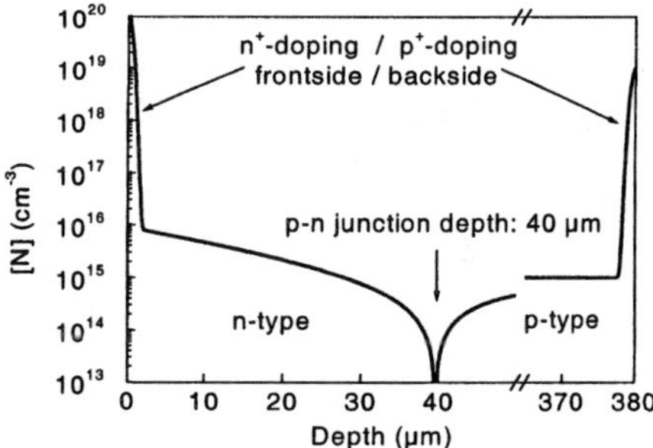

Fig. 14: Doping profiles used for simulations taking into account n^+ and p^+ doping at the front- and backside of the wafer.

Except for the simulated curve of the current density J at forward bias, the experimentally observed J(V) curve is also shown in Fig. 10. The relatively good agreement of the measured J(V) characteristic with the simulated one was achieved by fitting the carrier lifetime τ to a value of 3.8 µs. This value of τ was used for all numerical calculations presented below, though a part of the measured voltage drop is attributed to the bad contact properties. Further, one can expect that the carrier lifetime in the processed diodes is vertically inhomogenous, i.e. it is somewhat lower in the thermal donor region than in the non-disturbed region of the bulk material. This, however, was not taken into account for the simulations. Nevertheless, the calculations are very useful to illuminate general trends for the investigated diode types.

The high reverse leakage current density which was observed experimentally for TD-diode No. 1 (about 10^{-2} Acm^{-2} at V_{bias} = –100 V, Fig. 7) is attributed to surface currents from the wafer edge. Assuming that an optimal edge termination is provided to suppress such leakage the simulations show that diodes of such type should be able to sustain reverse voltages up to –700 V before carrier multiplication due to avalanche breakdown leads to a steep increase of the current density.

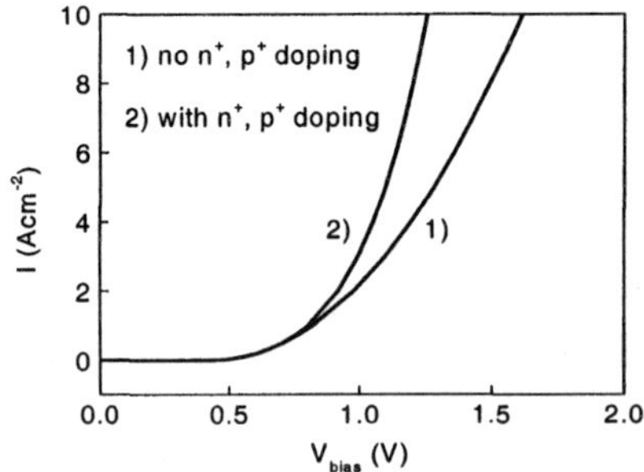

Fig. 15: Comparison of forward diode characteristics with and without n$^+$ and p$^+$ doping at the front- and backside of the wafer.

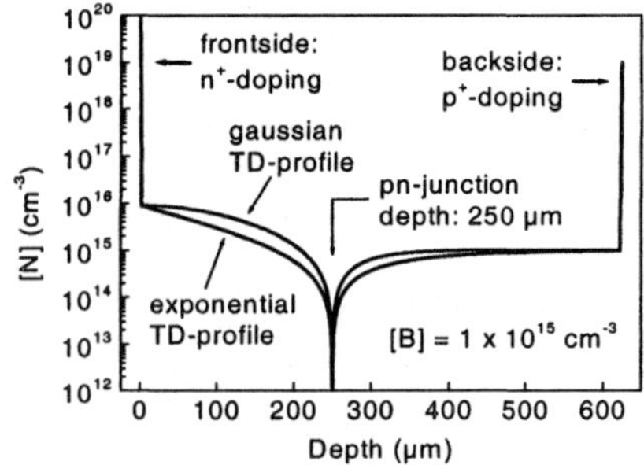

Fig. 16: Simulated doping profiles with exponential and Gaussian TD-profile.

The breakdown voltage can be significantly enhanced by increasing the depth of the p-n junction, keeping the surface concentration fixed. This leads to a lower donor concentration gradient which in turn reduces the maximum field at the p-n junctions at a fixed voltage. As it is obvious from Fig. 13 the breakdown voltage can be approximately doubled by increasing the p-n junction depth from 40 μm to 250 μm.

The effect of an additional n$^+$ and p$^+$ doping at the front- and backside of the wafer, as shown in Fig. 14, is shown in Fig. 15. In comparison to a TD-diode without n$^+$ and p$^+$ doping at the wafer surface, the forward J(V) characteristic is clearly improved.

To approach even higher breakdown voltages it is necessary to reduce the p-type doping concentration of the starting material significantly. In Fig. 17 the dependencies of the breakdown voltage on the p-n junction depth are compared for three different concentrations of the p-doped substrate. Evidently, breakdown voltages up to 6 kV can be achieved for the lowest considered substrate doping concentration ($2 \cdot 10^{13}$ cm^{-3}). In this case, the breakdown voltage for higher p-n junction depths (≥ 150 μm) is limited by the fact that the space charge region extends through the whole p-region (punch-through effect). Concerning the breakdown voltage the exponential shape of the TD profile is superior to that of conventional diffused donor profiles of Gaussian type, as demonstrated in Fig. 18 for profiles with the same peak concentration and the same depth of the p-n junction. Thus we may conclude that apart from the advantages arising from the low thermal budget necessary for the generation of hydrogen enhanced TD profiles, the higher breakdown voltage gives an additional plus in comparison to 'classical' diodes with in-diffused p-n junctions.

FORMATION OF 3-LAYER P-N-P TEST DEVICES BY HYDROGEN ENHANCED TD-FORMATION

As we already outlined before, TD (or more correctly TDD) formation is essentially related to the interstitial oxygen concentration in the Cz Si substrates. The maximum concentration of TDs is dependent on the third power of [O_i] (e.g. [10, 33]). Near the

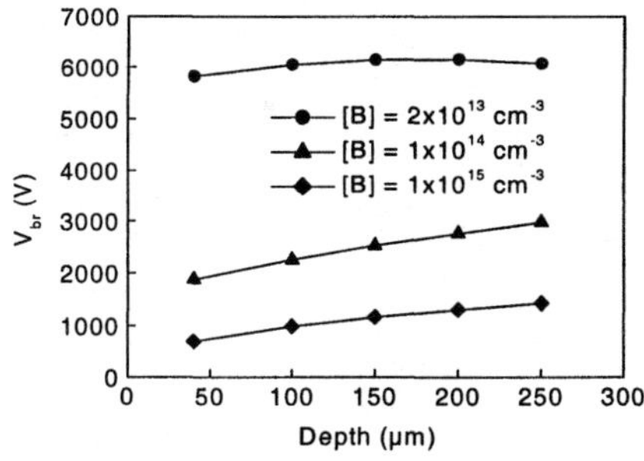

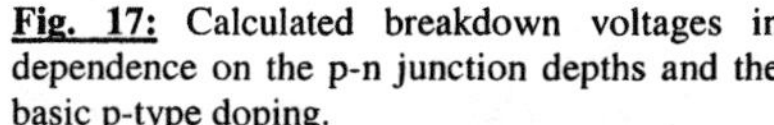

Fig. 17: Calculated breakdown voltages in dependence on the p-n junction depths and the basic p-type doping.

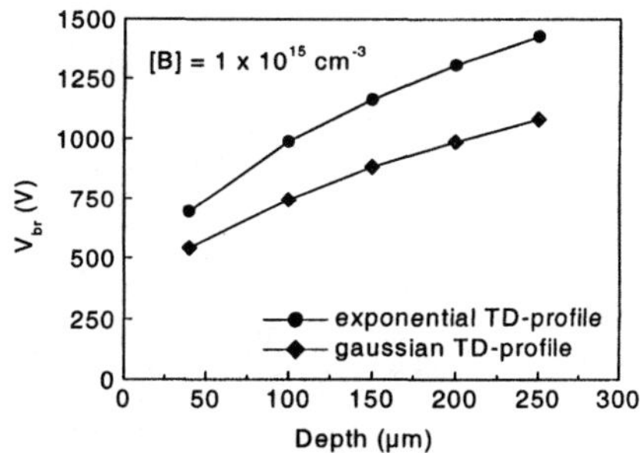

Fig. 18: Calculated avalanche breakdown voltages; comparison of the results for exponential and Gaussian TD-profiles.

surface of Cz Si wafers an oxygen denuded zone can be formed by oxygen out-diffusion, where $[O_i]$ is not high enough to generate oxygen precipitates under any heat treatment conditions [34]. Oxygen out-diffusion occurs in Cz Si wafers at sufficiently high temperatures in any ambient. The denuded zone formation, oxygen precipitation and subsequent intrinsic gettering of impurities at precipitates in deeper wafer regions can be obtained by so called "high-low-high" annealing series [35-37].

Oxygen lean regions in Cz Si have a strong impact on the TD formation. In comparison to the wafer bulk, in the denuded zone the TD formation rates are significantly reduced. Therefore one can expect that in the denuded zone counter doping of p-type Cz Si does not occur. Only in deeper wafer regions with sufficiently high $[O_i]$ n-type counter doping can be observed again. Therefore, in the framework of the "2-step-process" after appropriate post-hydrogenation annealing, one can observe two p-n junctions in initial p-type wafers, a deep one due to the hydrogen enhanced TD formation in deep wafer regions (like for the TD-diodes) and a second one close to the wafer surface due to the denuded zone and suppressed TD formation [38]. So, finally such a substrate with denuded zone exhibits a vertical p-n-p doping sequence (i.e. a vertical bipolar transistor structure): p-type at and near the surface (denuded zone), n-type up to a controllable depth due to n-type counter doping by TDs, and again p-type in very deep wafer regions, where the TD concentration is not high enough to achieve counter doping. If the post-hydrogenation annealing is applied for sufficiently long times, a complete counter doping from p-type to n-type occurs throughout the entire wafer, with exception of the denuded zone regions at the front- and backside, i.e. again a p-n-p structure is obtained.

Test devices with a p-n-p structure were fabricated on pre-annealed Cz Si substrates. For this first investigation we used a (100)-oriented p-type Cz Si wafer with a resistivity specified with 10 – 20 Ωcm. The interstitial oxygen and substitutional carbon concentrations were in the same order as for the 12 – 20 Ωcm material used for the preparation of the TD-diodes. First of all, i.e. before plasma hydrogenation, the wafer was annealed for 4 hours at 1100 °C in flowing O_2 ambient followed by a 30 min annealing at 900 °C in flowing N_2 atmosphere. After the pre-annealing the oxide layers were removed by HF etching. During these pre-annealing steps oxygen out-diffusion and formation of oxygen denuded zones occur at the front- and backside of the wafer. Further processing of this pre-treated wafer was done in accordance with the "2-step-process", i.e. by an 1

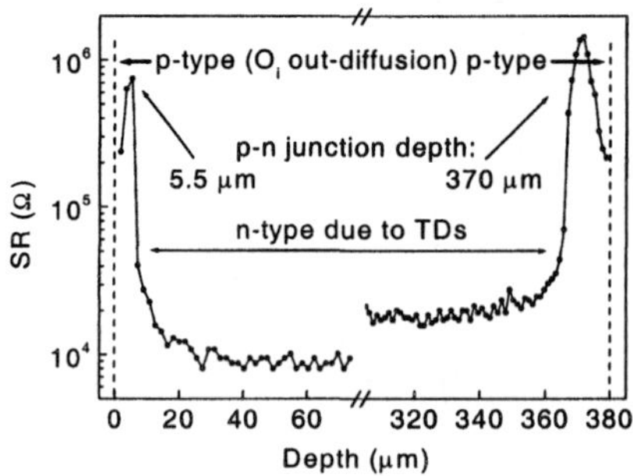

Fig. 19: SRP profiles showing two p-n junctions at the front- and backside of the wafer, due to hydrogen enhanced TD formation in a p-type wafer with oxygen denuded zones.

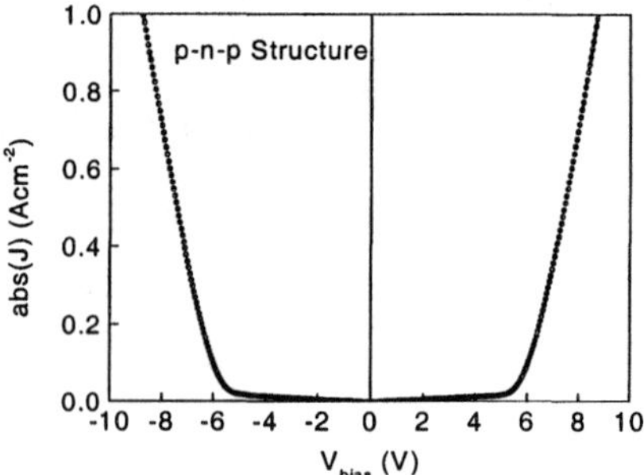

Fig. 20: $J(V_{bias})$ characteristics of a p-n-p test device structure. Mention that the absolute values of J are plotted in this figure.

hour plasma hydrogenation at 250 °C (110 MHz, 50 W) and subsequent annealing at 450 °C. But before annealing the Al contacts were deposited on the wafers frontside (1 mm^2) and backside (entirely). The annealing was applied for 2 hours at 450 °C in air. During the annealing (i) the Al contacts were burned in to achieve good ohmic behavior and (ii) the whole wafer was counter doped to n-type by TD formation, apart from the denuded zones at the front- and backside. As in all described cases before, neither a proper edge termination nor surface passivation were applied.

Fig. 19 shows the SRP profile of the prepared p-n-p test structure. Two p-n junctions near the front- and backside of the wafer can be observed at a depth of about 5.5 μm and 370 μm, respectively (the wafer thickness was about 380 μm). Between these p-n junctions the wafer exhibits n-type behavior, the surface and subsurface regions are p-type. The p-n-p structure is nearly symmetric, the p-type region at the backside of the wafer is only somewhat broader than at the frontside. The reason for this slight asymmetry might be that n-type doping by TDs did not convert the entire wafer as excepted, but only comes very close to the backside, or oxygen out-diffusion was somewhat less effective at the wafers frontside. In agreement with its nearly symmetric structure the p-n-p test device also exhibits a symmetric current density vs. bias voltage characteristic (Fig. 20). The leakage current density is of the same order as in the TD-diodes. The reason is of course again the missing edge termination and surface passivation.

SUMMARY

Under appropriate conditions the incorporation of hydrogen into Cz Si by a plasma treatment results in an enhanced formation rate of TDs, i.e. oxygen related 'old' TDDs. As a consequence n-type doping can be achieved at rather low temperatures and in deep wafer regions. TD concentrations up to 10^{16} cm^{-3}, or somewhat higher, can be obtained. Therefore, rapid counter doping and a p-n junction formation can be observed in p-type Cz Si, if the initial boron concentration does not exceed 10^{16} cm^{-3}. At temperatures below 500 °C and process times in the order of 1 – 2 hours, such doping methods by TDs can be employed for the preparation of diodes with deep graded p-n junctions (with typical p-n junction depth in the order of 100 μm or more). We developed two process routes for hydrogen enhanced thermal donor formation: i.e. a "1-step-process", where the p-n junction appears just after a plasma hydrogenation at appropriate temperatures (400 - 450 °C), and a "2-step-process", where after the plasma hydrogenation at lower temperatures (~ 250 °C) a subsequent annealing at 400 - 450°C is required for a p-n junction formation. On pre-annealed wafers with denuded zones also p-n-p test devices (vertical bipolar transistor structure) were processed. From MEDICI simulations we could conclude that TD-diodes exhibit promising properties for future high voltage applications.

REFERENCES

1. J.I. Pankove, N.M. Johnson (Eds.), "Hydrogen in Semiconductors", Academic Press, San Diego (1991).
2. S.J. Pearton, J.W. Corbett, M. Stavola, "Hydrogen in Crystalline Semiconductors", Springer-Verlag, Berlin, Heidelberg, New York (1992).
3. H.J. Stein, S.K. Hahn, Appl. Phys. Lett. **56**, 63 (1990).
4. H.J. Stein, S.K. Hahn, J. Appl. Phys. **75**, 3477 (1994).
5. H.J. Stein, S.K. Hahn, in: K. Sumino (Ed.), "Defect Control in Semiconductors", Vol. 1, Elsevier Science Publishers B.V., North Holland, Amsterdam, (1990), p. 211.
6. H.J. Stein, S.K. Hahn, J. Electrochem. Soc. **142**, 1242 (1995).
7. C.S. Fuller, J.A. Ditzenberger, N.B. Hannay, E. Buehler, Phys. Rev. **96**, 833 (1954).
8. W. Kaiser, Phys. Rev. **105**, 1751 (1957).
9. C.S. Fuller, R.A. Logan, J. Appl. Phys. **28**, 1427 (1957).
10. W. Kaiser, H.L. Frisch, H. Reiss, Phys. Rev. **112**, 1546 (1958).
11. J. Michel, L.C. Kimerling, "Electrical Properties of Oxygen in Silicon", in: F. Shimura (Ed.), "Oxygen in Silicon", Academic Press, San Diego (1994), p. 251.
12. P. Wagner, J. Hage, Appl. Phys. A **69**, 123 (1989).
13. A.R. Brown, M. Claybourn, R. Murray, P.S. Nandhra, R.C. Newman, J.H. Tucker, Semicond. Sci. Technol. **3**, 591 (1988).
14. R. Murray, A.R. Brown, R.C. Newman, Mater. Sci. Engineer. B **4**, 299 (1990).
15. R.C. Newman, J.H. Tucker, A.R. Brown, S.A. McQuaid, J. Appl. Phys. **70**, 3061 (1991).
16. C.D. Lamp, D.J. James II, Appl. Phys. Lett. **62**, 2081 (1993).
17. D.K. Schroder, "Semiconductor Material and Device Characterization" (2nd Edition), John Wiley & Sons, New York (1998).
18. R. Job, D. Borchert, Y.A. Bumay, W.R. Fahrner, G. Grabosch, I.A. Khorunzhii, A.G. Ulyashin, MRS Symp. Proc. Series **469**, 101 (1997).
19. A.G. Ulyashin, Y.A. Bumay, R. Job, G. Grabosch, D. Borchert, W.R. Fahrner, A.Y. Diduk, Solid State Phenomena **57-58**, 189 (1997).
20. R. Job, W.R. Fahrner, N.N. Kazuchits, A.G. Ulyashin, MRS Symp. Proc. Series **513**, 337 (1998).
21. A.G. Ulyashin, Y.A. Bumay, R. Job, W.R. Fahrner, Appl. Phys. A **66**, 399 (1998).
22. W. Vandervorst, T. Clarysse, J. Electrochem. Soc. **137**, 679 (1990).
23. S.C. Choo, M.S. Leong, Chandra B.T. Liem, K.C. Kong, Sol. Stat. Electron. **33**, 783 (1990).
24. H.L. Berkowitz, R.A. Lux, J. Electrochem. Soc. **128**, 1137 (1981).

25. E.C. Andre, Jap. J. Appl. Phys. **30**, 1511 (1991).
26. N. Sieber, H.E. Wulf, Phys. Stat. Sol. (a), **126**, 213 (1991).
27. S.M. Sze, "Physics of Semiconductor Devices", John Wiley & Sons, Inc., New York (1981).
28. S.M. Myers, M.I. Baskes, H.K. Birnbaum, J.W. Corbett, G.G. DeLeo, S.K. Estreicher, E.E. Haller, P. Jena, N.M. Johnson, R. Kirchheim, S.J. Pearton, M.J. Stavola, Rev. Mod. Phys. **64**, 559 (1992).
29. A. Van Wieringen, N. Warmholtz, Physica **22**, 849 (1956).
30. D. Mathiot, Phys. Rev. B **40**, 5867 (1989).
31. R.C. Newman, R. Jones, in: F. Shimura (Ed.), "Oxygen in Silicon", Academic Press, San Diego (1994), p. 289.
32. S.K. Estreicher, Mat. Sci. Eng. R **14**, 319 (1995).
33. J, Michel, L.C. Kimerling, in F. Shimura (Ed.): "Oxygen in Silicon", Academic Press, San Diego (1994), p. 251.
34. F. Shimura, in F. Shimura (Ed.): "Oxygen in Silicon", Academic Press, San Diego (1994), p. 577.
35. K. Nagasawa, Y. Matsushita, S. Kishino, Appl. Phys. Lett. **37**, 622 (1980).
36. E.M. Murray, J. Appl. Phys. **55**, 536 (1984).
37. S. Isomae, S. Aoki, K. Watanabe, J. Appl. Phys. **55**, 817 (1984).
38. A.G. Ulyashin, A.N. Petlitskii, R. Job, W.R. Fahrner, Electrochemical Society Proceedings, Vol. **98-13**, 425 (1998).

Surface Structure of Hydrogen Annealed Silicon Wafer using Ozonized Water and Dilute HF Cleaning

Hisatsugu Kurita, Koji Izunome, Hiromi Nagahama,
Takao Ino[*)], Jyunsei Yamabe[*)], Naoya Hayamizu[*)] and Naoaki Sakurai[*)]
Wafer Processing Technology, Silicon Division, Toshiba Ceramics Co, Ltd.
6-861-5, Higashikou, Sheirou-machi, Kitakanbara-Gun, Niigata, 957-0197, Japan
[*)] Manufacturing Technology Center, Toshiba Corporation
33 Shin-Isogo, Isogo-ku, Yokohama-City, Kanagawa, 235-0017, Japan

ABSTRACT

The mechanism, of which the surface step and terrace structure of a Hydrogen annealed silicon (Si) wafer can be preserved after ozonized (O_3) water and dilute HF (DHF) water cleaning, was investigated using a XPS (X-ray Photoelectron Spectroscope), an AFM (Atomic Force Microscope) and a TEM (Transmission Electron Microscope).

The oxide thickness after 30 sec in O_3 water is 0.75nm on both Si(100) and Si(111) surface by XPS. It is well known that SiO_2 films are removed isotropic in DHF water. It is suggested that the surface structure of the hydrogen annealed Si wafer is preserved in oxidation of Si wafer with O_3 water.

The surface structure and micro-roughness (Rms) of these wafers are preserved after O_3 water and DHF water treatment by AFM. The Rms of these wafers without O_3 water treatment increases, however, due to the generation of a lot of protrusions on Si(100) surface. The protrusion, consisting of crystalline Si, has the height of 0.6nm and the width of 15nm by TEM.

Thus, a formation model for the protrusions is proposed, the generation of depressions with preference native oxide growth at the site of the type-C defects on Si(100) surface and the remain of the protrusions with etching in DHF water.

1. INTRODUCTION

As the design rules of MOS ULSI shrink to quarter micron gate lengths, the surface micro-roughness of silicon (Si) is becoming more crucial for developing device performance and reliability. [1)] This is because the wafer surface affects the dielectric breakdown, which is directly related to the reliability of thin oxide films grown on Si wafers.

It is widely known that polished Czochralski-grown Si (CZ-Si) wafer annealed in a hydrogen atmosphere at high temperature, which is so-called hydrogen annealed Si wafer, have a periodic structure consisting of terraces and single-layer steps. [2] The terrace width increases with decreasing mis-orientation angle of Si(100) wafer. [3] Hydrogen annealed Si wafer can improve the gate oxide breakdown properties of SiO_2 films, because defects in a device active layer shrink and annihilate during hydrogen annealing in terms of oxygen out-diffusion. [4] M. Murata et al. reported that the leakage current of SiO_2 films with a thickness of 1.4 nm is generated at the step edge of triple-layer steps. [5] Thus, it is expected that hydrogen annealed Si wafer with a very small mis-orientation angle have extremely high reliability for thin gate oxides on integrated circuits.

The micro-roughness of the hydrogen annealed Si wafer surface increases when it is cleaned with a SC-1 solution ($NH_4OH/H_2O_2/H_2O$) because of this etches the Si surface anisotoropically. [6] On the other hand, S. Verhaverbeke et al. reported that the periodic surface structure of a hydrogen annealed Si wafer was preserved when it was cleaned with ozonized (O_3) water and dilute HF (DHF) water. [7] It is suggested that isotropic oxidation proceeds on the hydrogen annealed Si wafer surface during O_3 water and DHF water cleaning. T. Hattori et al. presented results that the micro-roughness after cleaning repetitively with O_3 water and DHF water had the same value as a polished Si wafer before cleaning. [8] However, the micro-roughness (Rms) in the area of 2um x 2um of polished Si wafer is very large with a value of 0.25 nm, as that of a hydrogen annealed Si wafer, which is below 0.1 nm.

Thus, no paper has been presented that describes the variation of the micro-roughness of a hydrogen annealed Si wafer after repetitive cleaning with O_3 water and DHF water.

The purpose of the present study is to clarify the mechanism by which the step and terrace structure of a hydrogen annealed Si wafer surface are preserved after cleaning with O_3 water and DHF water. To determine the oxidation behavior in O_3 water, we first investigated the oxide growth rate on Si wafers with various crystal growth directions. Next, we investigated the effect of SiO_2 films etching in DHF water and the effect of oxidation in O_3 water during the cleaning process on the surface structure and micro-roughness of the hydrogen annealed Si wafer.

2. Oxidation of Si(100) and Si(111) surface in O_3 water

2-1 Experimental

The samples used in this experiment were p-type 200mm diameter CZ-Si wafers. The first of these had a Si(100) surface and the other had a Si(111) surface. Both had a resistivity of 17-22 Ω cm and an oxygen atom concentration of $1\text{-}1.5\text{x}10^{18}$atoms cm^{-3} (The conversion factor of old ASTM for oxygen is $4.81\text{x}10^{17}$atoms cm^{-3}). The wafers were cut into 20mm x 20mm test specimens, and were then treated in HF water to remove the native oxide layer. Then the samples were treated in 20ppm O_3 water.

The atomic ratio of Si atoms was measured by X-ray photoelectron spectroscopy (XPS, Surface Science Instruments, SSX-100). The photoelectron escape angle (θ) between the detector axis and the normal direction to the Si surface was kept at 35° in this experiments. The measurement area was 10 mm x 10 mm. The oxide thickness, d, was evaluated using the intensity ratio of the shifted peak from the oxide to the unshifted peak from the Si substrate in an Si_{2p} XPS spectrum, and can be expressed as follows,

$$d=\lambda_A \sin\theta \ln[(n_B M_A \lambda_B)/(n_A M_B \lambda_1) I_A/I_B+1] \quad (1)$$

where subscript A is native oxide (SiO_2), subscript B Si, λ the electron escape depth of Si 2p photoelectrons in Si and SiO_2, (which are 2.8 nm for Si and 3.6 nm for SiO_2 respectively,), n the atomic density of Si in the Si substrate and the SiO_2, (which are 5.0 x 10^{28} m^{-3} for Si and 2.5 x 10^{28} m^{-3} for SiO_2 respectively,), M the molecular weight, and I the intensity of the Si_{2p} spectrum including Si(IV) (SiO_2) and the intermediate oxidation states (suboxides; Si(I), Si(II) and Si(III)).

2-2 Results and discussion

The atomic ratio of Si(IV) and suboxides evaluated from the Si_{2p} spectrum as a function of O_3 water treatment time on Si(100) surface and Si(111) surface are shown in Fig. 1. The atomic ratio of Si(IV) increases with increasing O_3 water treatment time, and become constant about 15% after 30 sec for Si(100) surface and Si(111) surface. The variation of the atomic ratio of Si(IV) is therefore independent on the crystal directions. The atomic ratio of suboxides, however, shows a small variation between Si(100) surface and Si(111) surface at an early stage in the oxidation. The ratio of suboxides on Si(111) surface is 9% at an early stage in the oxidation, and then decreases to 7% with increasing the O_3 water treatment time. On the other hand, the ratio of suboxides on Si(100) surface is 6% at an early stage in the oxidation, and then keeps a constant value of 6%. The difference in the ratio of suboxides on Si(100) surface and Si(111) surface in the early stage oxidation is considered that the surface density of Si(111) surface is higher than that of Si(100) surface. Oxygen atoms in O_3 water can react more easily with Si atoms on Si(111) surface in the initial stages.

The relationship between the O_3 water treatment time and the oxide thickness is indicated in Fig. 2. The oxide thickness on both Si(100) surface and Si(111) surface after 30 sec is 0.75 nm. M. Hirose et al. found that the native oxide thickness in deionised water (DIW) saturates at about 0.76 nm in various type and doped CZ-Si(100) wafers. [9)] The oxidation rate on Si surface in DIW varies with the crystal direction, and the native oxide initially generates as islands on Si surface. [9)] The oxide thickness coincides with that of a structural model of the SiO_2/Si(100) interface proposed by Herman et al. [10)] The model shows that layer-by-layer oxidation occurs on a hydrogen-terminated Si (Si-H) surface, and that the oxidation proceeds from the Si-Si back bond of the Si-H bonds. [11)]

The oxide thickness in our study is in good agreement with that of Herman's structural model. It is suggested that the chemical oxidation in O_3 water occurs from the Si-Si back bonds in near-surface region, and is independent on the crystal directions. It is, therefore assumed that the surface structure of a hydrogen annealed Si wafer is preserved by the isotropic oxidation of Si in O_3 water.

3. Effect of DHF water and O_3 water treatment on the structural variation of hydrogen annealed Si wafer surface

3-1 Experimental

The wafers used in this experiment were p-type CZ-Si(100) 200mm diameter with mis-orientation angles of 0.04° or less toward [011] direction. They had a resistivity of 1.0 Ω cm and an oxygen concentration of 9×10^{17} atoms cm^{-3}. The samples were annealed in hydrogen atmosphere at 1200 degree C for 1 hour, using a commercially available vertical furnace.

In order to evaluate the variation in the surface structure of hydrogen annealed Si wafer with respect to different DHF water and O_3 water treatments, the samples were treated in a repetitive single-wafer spin cleaning system. A wafer was held horizontally and rotated in a spin cup made from Teflon. The cleaning solutions (20ppm O_3 water, 1% DHF water and DIW) were applied to the rotating wafer successively through nozzles both wafer sides, and then finally the wafer was dried at a high-speed rotation. All the cleaning and drying procedures were performed at room temperature. To evaluate the effect of SiO_2 films etching in DHF water (STEP2) and the effect of oxidation in O_3 water and DIW after DHF water treatment (STEP3 and STEP4), the samples were successively treated with STEP1 to STEP5, as shown in Table 1. In STEP 2, in order to remove the SiO_2 film on the Si wafer completely, the samples were treated with DHF water for 20sec.

Table 1. Cleaning sequence for each sample

No.	Cleaning sequence				
	STEP1	STEP2	STEP3	STEP4	STEP5
	O_3 water	DHF water	O_3 water	DIW	Dry
1	No cleaning	←	←	←	←
2	30sec	3sec	0sec	20sec	40sec
3	30sec	20sec	0sec	20sec	40sec
4	30sec	3sec	30sec	20sec	40sec
5	30sec	20sec	30sec	20sec	40sec

In order to observe the step and terrace structure of the hydrogen annealed Si wafers in this experiment, the surface structure of each sample was observed at the area of 3 um x 3 um by atomic force microscope (AFM, Veeco Instruments, NanoScope Ⅲ) with a tapping mode. The AFM was equipped with an active damper and a sound insulation to eliminate the influence of vibration during measurements in a standard room environment. The noise floor (detection limit value) was 0.03 nm. The surface micro-roughness (Rms) was evaluated in AFM images at the area of 3 um x 3 um.

3-2 Results and discussion

The AFM images and Rms values of the samples are shown in Fig. 3 and Table 2, respectively. The hydrogen annealed Si wafer has the single atomic-step and terrace structure, [6, 7] and the terrace width is 1.7 um, corresponding to mis-orientation angle of 0.01° or less. [3] The step and terrace structure of all of the samples after cleaning are preserved, as shown in Fig. 3 (b), (c), (d) and (e), respectively. These results are in good agreement with those in the Verhaverbeke's report. [7]

Table 2. Surface micro-roughness (Rms) of samples by AFM images (3um x 3um)

Sample No.	1	2	3	4	5
Rms (nm)	0.073	0.135	0.131	0.092	0.080

The variation of the Rms as a function of DHF treatment time is shown in Fig. 4. If we consider the samples treated with O_3 water in STEP 3 after DHF water shown in

Fig. 4, there is only a little variation in the Rms of the samples. In particular, the Rms of sample 5, (treated successively with DHF water for 20sec and then with O_3 water), has the same value as that of the hydrogen annealed Si wafer (sample 1). The Rms values of samples without O_3 water treatment in STEP 3 increase after cleaning shown in Fig. 4. There are, however, a lot of protrusions on the terraces after cleaning, as in Figs. 3 (b), (c), (d) and (e). The density of the protrusions was calculated by a graphical analysis of Fig. 3.

The variation of the calculated protrusion density as a function of DHF water treatment time in STEP 2 is indicated in Fig. 5. The density of the protrusions on the samples is between 1.3 x 10^{10} cm^{-3} and 3.0 x 10^{10} cm^{-3}. The density of the protrusions on the samples with O_3 water treatment in STEP3 decreases with increasing DHF treatment time, as shown in Fig. 5. The density of the protrusions on the samples without O_3 water treatment in STEP3 has a constant value, is independent on DHF treatment time, as shown in Fig. 5. To clearly show the cause of the Rms increment in Table 2, the relationship between the Rms and the protrusion densities are plotted in Fig. 6. As shown in Fig. 6, the Rms increases with increasing the density of the protrusions exponentially. It can be, thus, speculated that the generation of protrusions on Si(100) surface may cause the Rms increment of the sample after DHF water and O_3 water cleaning.

Next, we will analyze the morphology of the protrusions on the surface, and discuss a formation model for the protrusions on Si(100) surface, on the basis of Hamer's model of defects into Si(100)-2x1 surface. [12)]

4. Formation model of protrusions on Si(100) surface

To clarify the structure of the protrusions on Si(100) surface, we can analyze AFM images and also observe the protrusions at the SiO_2/Si(100) interface by transmission electron microscope (TEM).

4-1 Experimental

The distance between protrusions, the coverage ratio of the protrusions to the Si surface and the width of the protrusions were obtained by a graphical analysis in Fig. 3. The height of the protrusions was also calculated by the cross-sectional profile in Fig. 3.

Cross-sectional TEM image samples were made at the position of the protrusions. The TEM samples were prepared by mechanical grinding, dimple grinding, and finally argon-ion milling. After ion milling, the morphology of the specimens was observed by TEM. TEM observations were carried out using a H-90000UHR (HITACHI) at 300kV with a bright-field image from the [011] direction.

4-2 Results and discussion

The distance between protrusions and the coverage ratio of the protrusions as a function of DHF water treatment time are shown in Figs. 7 and 8, respectively. The protrusion heights and widths are also shown in Figs. 9(a) and 9(b).

The distance between protrusions increases with DHF treatment time, as shown in Fig. 7, both with and without O_3 water treatment in STEP3. As shown in Fig. 8, the coverage ratio of the protrusions is between 2% and 10%, and becomes smaller with the increasing DHF treatment time, both with and without O_3 water treatment in STEP3.

The height of the protrusions is between 0.2 nm and 0.9 nm, and becomes smaller with increasing DHF water time for samples 4 and 5 with O_3 water treatment in STEP3, as shown in Fig. 9 (a). On the other hand, for samples 2 and 3 without O_3 water treatment in STEP3, the height of the protrusions are larger when compared with samples 4 and 5 in Fig. 9 (a). The width of the protrusions is between 10 nm and 130 nm, and there is little variation within the measurement errors shown in Fig. 9 (b).

The oxidation rate on the surface of Si in DIW and in air is extremely slow compared with that in O_3 water and varies with the crystal direction. [9] It is supposed that the protrusions grow heterogeneously in DIW, [9] which coincides with the native oxide growth mechanism so-called island-shaped structure.

Typical TEM images of the protrusions at the SiO_2/Si(100) interface are shown in Figs. 10(a) and 10(b). Two protrusions can be seen at the SiO_2/Si(100) interface in Fig. 10(a). As shown in Fig. 10(b), the protrusions consist of crystalline Si, which edges are surrounded by Si(111) surfaces without a contamination. The height of the protrusion is about 0.6 nm, corresponding to fourth-atomic Si(100) layers and the width of the protrusion is about 15nm in Fig. 10(b). The distance between protrusions is about 68nm. As mentioned above, the protrusion observed in Figs. 10(a) and 10(b) is in good agreement with that in Fig. 3.

These results suggest that the protrusion may be formed during etching process of SiO_2 films in DHF water. Generally, SiO_2 films are etched isotropy in DHF water. Therefore the oxidation on Si(100) surface before etching has no uniformity. In other words, there are a lot of defects on the Si(100) surface, which induce the irregular oxidation.

It has been reported that type-C defects cover 2.9% of the area on Si(100)-2x1 surface under an ultrahigh vacuum by the scanning tunneling microscopy (STM) analysis. [12] Type-C defects is two half-dimmers, [11] and have a strong preference for oxidation. Oxygen molecules near Si wafer easily dissociate and absorb atomically onto the type-C defects, and then oxygen atoms can easily be inserted into the Si-Si back bonds. [12] The effective size of the type-C defect, estimated from the topographic line and the measured tunneling barrier height, is about 0.1nm in height and about 2nm in width.[11] From the point of view at the coverage ratio of the protrusions and preferential oxidation properties, we consider the type-C defects may exist on the terraces of hydrogen annealed Si wafer.

The following models are proposed about the formation of the protrusions. A schematic diagram of the formation of protrusions is shown in Fig. 11. The native oxide is grown on Si surface in air after hydrogen annealing. Oxygen molecules dissociate easily and absorb preferentially on the type-C defects in the initial stages of native oxide growth. (Fig. 11 (a)) Oxygen atoms are then easily inserted into the back bonds near the type-C defects. The native oxide growth rate on Si(111) is slower than that on Si(100). Then the depressions which edges surrounded by Si(111) are induced in the SiO_2/Si interface. (Fig. 11 (b)) The SiO_2 films are removed by DHF water treatment. The protrusions, which consist of crystalline Si, with edges surrounded by Si(111) surfaces may appear on the terraces. From a morphological aspect, the protrusions become smaller with a little etching of the Si. (Fig. 11 (c)) The oxidation behavior with O_3 water treatment has no dependence on whether it is Si(100) surface or Si(111) surface. The SiO_2 films are formed isotropy on Si surface. The protrusion remains as it is. (Fig. 11 (d))

5. CONCLUSION

The mechanism, of which the surface step and terrace structure of a Hydrogen annealed silicon (Si) wafer can be preserved after ozonized (O_3) water and dilute HF (DHF) water cleaning, was investigated using a XPS (X-ray Photoelectron Spectroscope), an AFM (Atomic Force Microscope) and a TEM (Transmission Electron Microscope). The oxide thickness after 30 sec is 0.75 nm on both Si(100) and Si(111) surfaces, suggesting that the surface structure of a hydrogen annealed Si wafer is preserved in the oxidation of Si wafer by O_3 water.

We investigated the effect of SiO_2 films etching in DHF water and the effect of oxidation in O_3 water, on the surface structure and micro-roughness of the hydrogen annealed Si wafer by AFM and TEM analysis. The step and terrace structure of the hydrogen annealed Si wafer is preserved after O_3 water and DHF water treatment. The Rms of the samples treated with O_3 water after the DHF water is significantly same as that of the hydrogen annealed Si wafer. The Rms of the samples without O_3 water treatment increases, due to the generation of a lot of protrusions exist on the terraces after cleaning, and this effects on the Rms of the samples. The protrusion consists of crystalline Si by TEM. It is proposed that a formation model for the protrusions, the generation of depressions at the site of the type-C defects with native oxide growth and the formation of the protrusions with etching in DHF water.

REFERENCES

1. M. Miyashita, T. Tusga, K. Makihara, and T. Ohmi, *J. Electrochem. soc.*, **139**, 2133 (1992).
2. L. Zhong, A. Hojo, Y. Matsushita, Y. Aiba, K. Hayashi, R. Takeda, H. Shirai, H. Saito, J. Matsushita and J. Yoshikawa, *Phys. Rev.* **B, 54,** R2304 (1996).
3. K. Izunome, Y. Saito and H. Kubota, *Jpn. J. Appl. Phys.*, **31**, L1277 (1992).
4. H. Kubota, M. Numao, T. Amai, M. Miyashita, S. Samata and Y. Matsushita, in *SEMICONDUCTOR SILICON/1994*, H. R. Huff, W. Bergholz and K. Sumino, Editors, PV 94-10, p. 225, The Electrochemical Society Proceedings Series, Pennington, NJ (1994).
5. M. Murata, N. Tokuda, D. Hojo and K. Yamabe, in *HIGH PURITY SILICON IV*, C. L. Claeys, P. Rai-Choudhury, M. Watanabe, P. Stallhofer, and H. J. Dawson, Editors, PV 2000-17, p. 425, The Electrochemical Society Proceedings Series, Pennington, NJ (2000).
6. Y. Yanase, H. Horie, Y. Oka, M. Sano, S. Sumita and T. Shigematsu, *J. Electrochem. soc.*, **141,** 3259 (1994).
7. S. Verhaverbeke, T. Futatsuki, R. Messoussi and T. Ohmi, in *SEMICONDUCTOR SILICON/1994*, H. R. Huff, W. Bergholz and K. Sumino, Editors, PV 94-10, p. 1170, The Electrochemical Society Proceedings Series, Pennington, NJ (1994).
8. T. Hattori, T. Osaka, A. Okamoto, K. Saga, and H. Kuniyasu, *J. Electrochem. soc.*, **145,** 3278 (1998).
9. M. Hirose, T. Yasaka, M. Takakura, S. Miyazaki, *Solid State Technology,* 43 (1991).
10. F. Herman, J. P. Batra, V. Kasowski, in *The Physics of SiO_2 and Its Interface,* S. Pantelieds, Editor, p. 333, Pergamon Press, New York, (1978).
11. T. Takahagi, A. Ishitani, H. Kuroda, Y. Nagasawa, H. Ito and S. Wakao, *J. Appl. Phys.,* **68,** 217 (1990).
12. R. J. Hamers and U. K. Kohler, *J. Vac. Sci.,* **A7 (4),** 2854 (1989)
13. M. Udagawa, Y. Umetani, H. Tanaka, M.Itoh, T. Uchiyama, Y. Watanabe, T. Yokotsuka and J. Sumita, *Ultramicroscopy,* **42-44,** 946 (1992)
14. J. M. Gibson, X. Tong, R. D. Twesten and F. M. Ross, in *25th International Conference on Solid State Devices and Materials,* S-III-1, p. 77, Extended Abstracts, Chiba, (1993).
15. Y. Sugita, S. Watanabe, *Jpn. J. Appl. Phys.,* **37,** 217 (1998).
16. Y. Sugita, S. Watanabe, and N. Awaji, *Jpn. J. Appl. Phys.,* **35,** 5437 (1996).

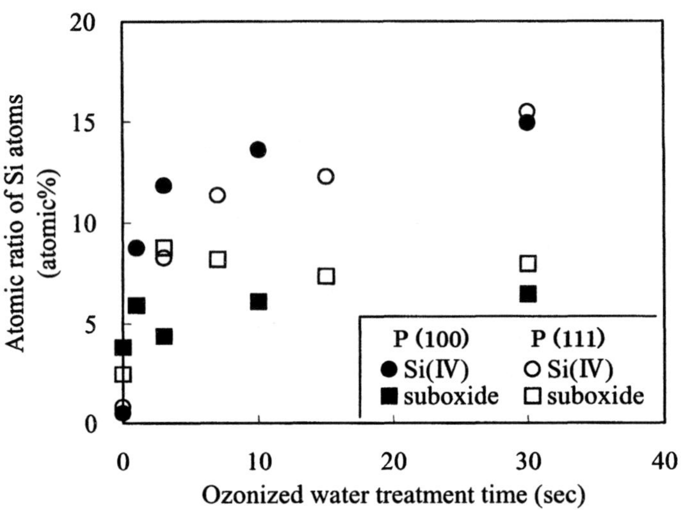

Fig.1 : Atomic ratio of Si atoms on Si (100) wafer surface and Si (111) wafer surface measured by XPS as a function of ozonized water treatment time.

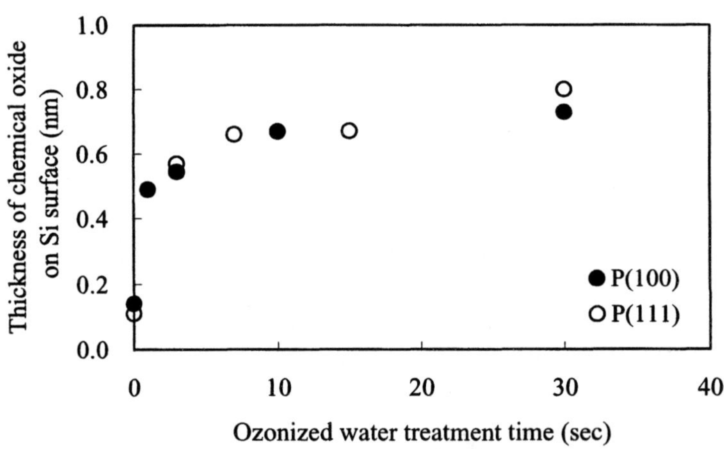

Fig.2 : Thickness of chemical oxide on Si (100) wafer and Si(111) wafer surface measured by XPS as a function of ozonized water treatment time.

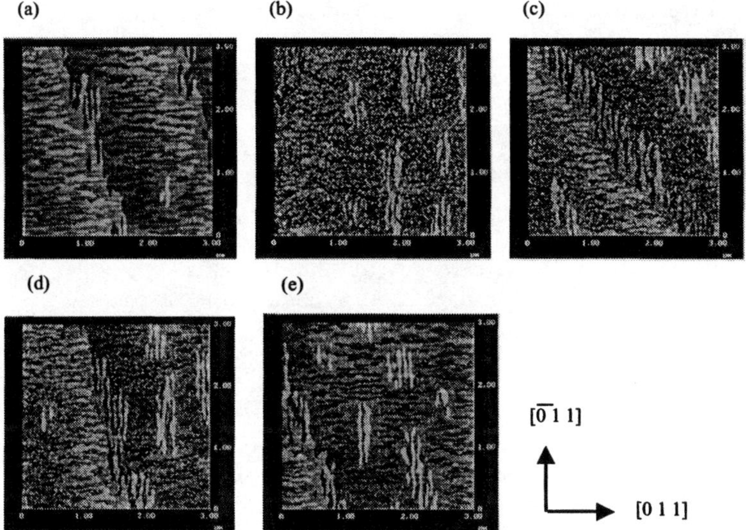

Fig.3 : AFM images ($3 \times 3 \mu m^2$) of H_2 annealed Si(100) surface of samples with misorientation angle of 0.02 degrees.
(a) sample 1, (b) sample 2, (c) sample 3, (d) sample 4 and (e) sample 5

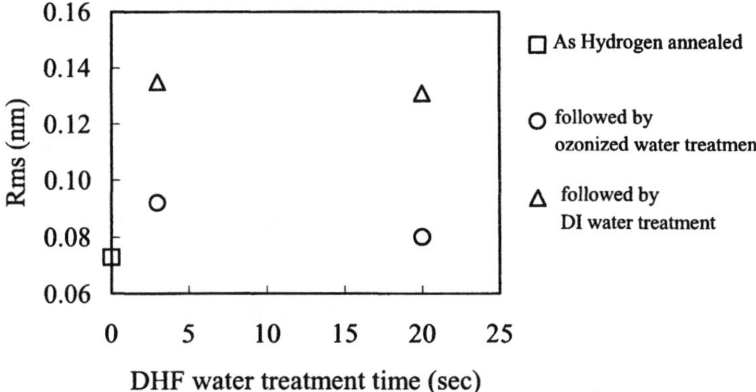

Fig.4 : Surface micro-roughness (Rms) of H_2 annealed Si(100) samples as a function of DHF water treatment time.

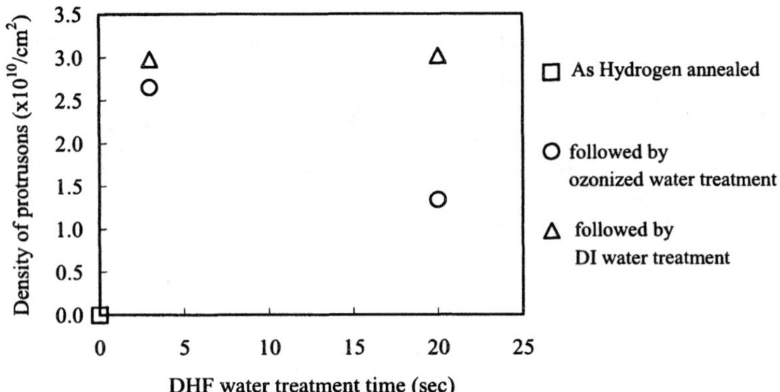

Fig.5 : Surface density of protrusions on H_2 annealed Si (100) samples as a function of DHF water treatment time.

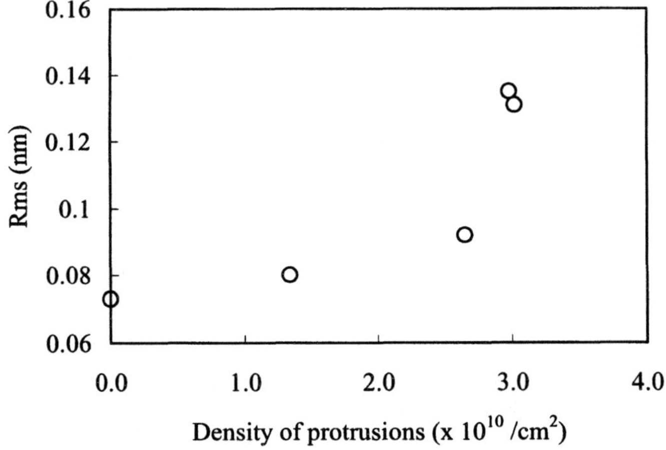

Fig.6 : The dependence of surface micro-roughness (Rms) of H_2 annealed Si (100) samples on density of protrusions.

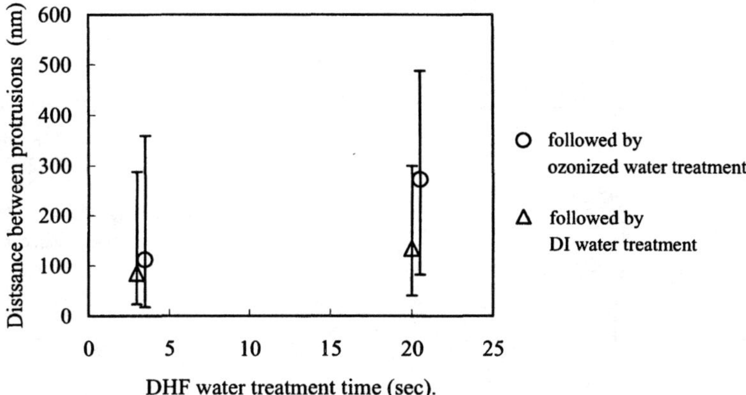

Fig.7 : Distance between protrusions on H_2 annealed Si(100) samples as a function of DHF water treatment time.

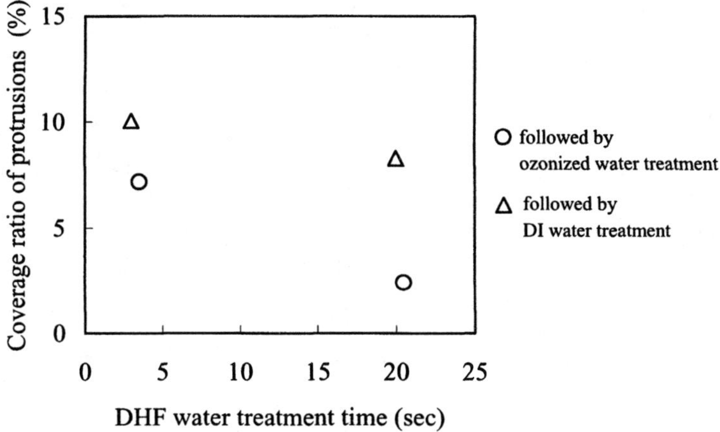

Fig.8 : Coverage ratio of protrusions on H_2 annealed Si(100) samples to the surface as a function of DHF water treatment time.

(a)

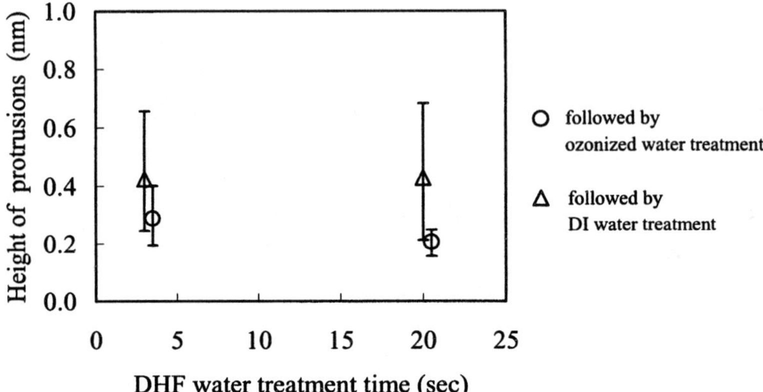

(b)

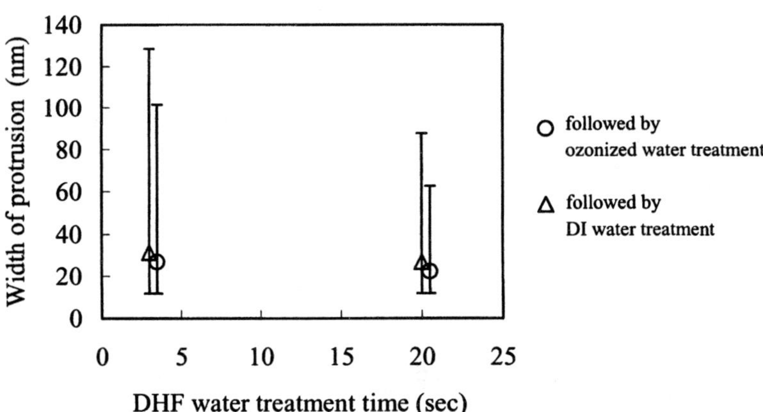

Fig.9 : The morphology of protrusions on H_2 annealed Si(100) samples as a function of DHF water treatment time.
(a) Height of protrusion, (b) Width of protrusion

(a)

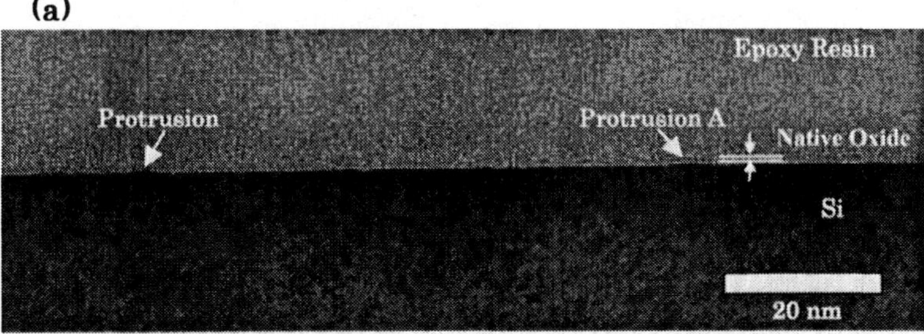

(b)

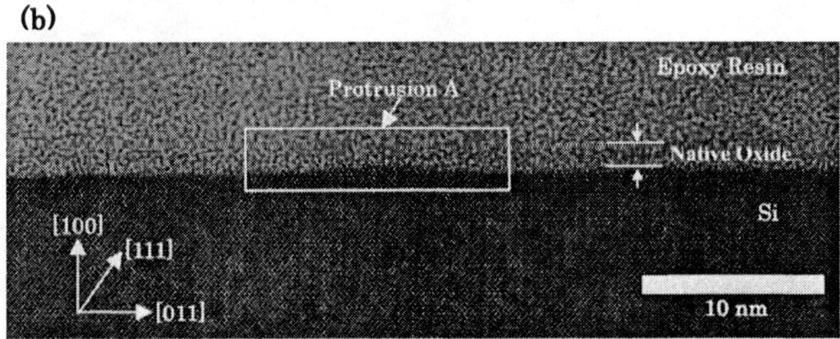

Fig.10 : TEM images of SiO_2/Si (001) interface for H_2 annealed Si (100) (sample 3) with a misorientation angle of 0.04 degree.
(a) an overall view (b) an enlargement of protrusion A

(a) Absorption of oxygen in the initial stages of Native oxide growth

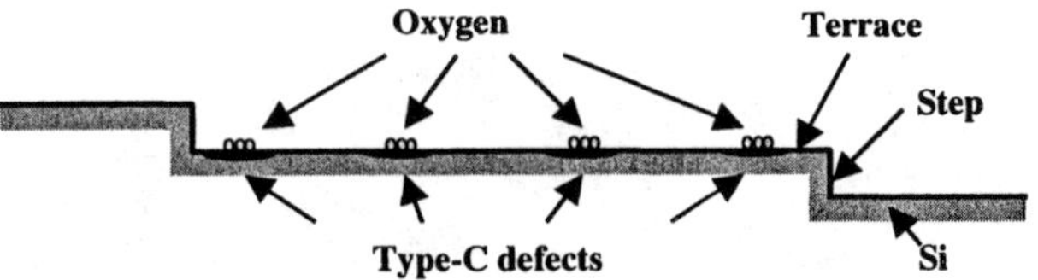

(b) Generation of depressions by the back bond oxidation

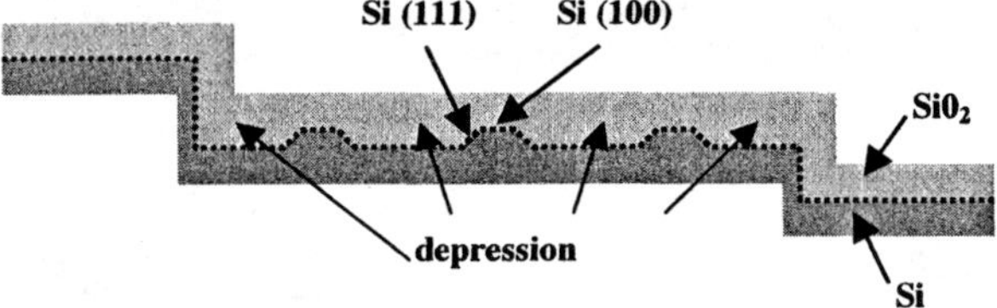

(c) SiO_2 films removal in DHF treatment

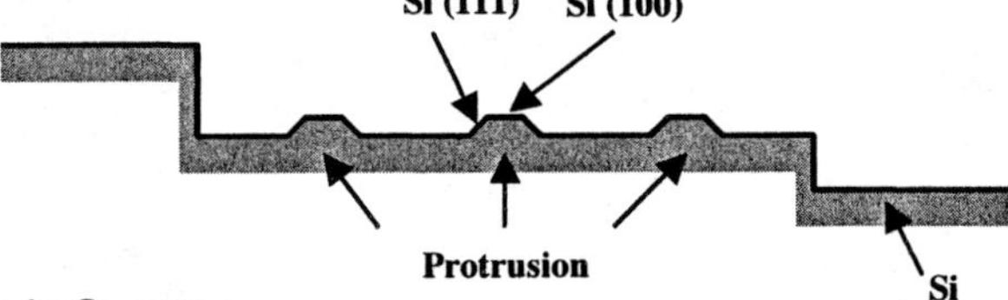

(d) Oxidation in O_3 water

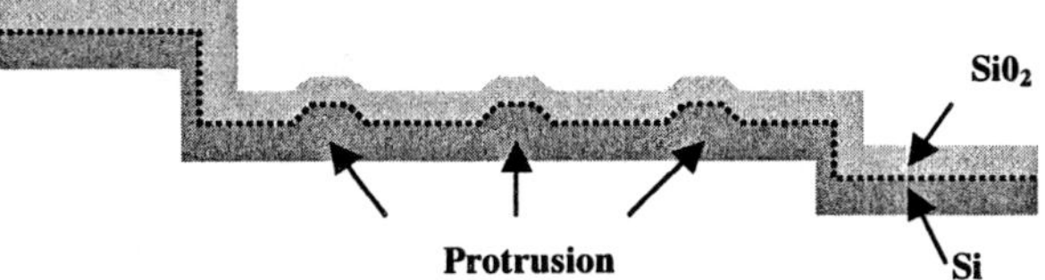

Fig.11 : Schematic diagram of the formation of protrusion.

DISLOCATION LOCKING BY OXYGEN IN SILICON: NEW INSIGHTS TO OXYGEN DIFFUSION AT LOW TEMPERATURES

S. Senkader, A. Giannattasio, R. J. Falster*, and P. R. Wilshaw
Department of Materials, University of Oxford, UK
*MEMC Electronic Materials SpA, Novara, Italy

ABSTRACT

We investigated locking of dislocations by oxygen atoms in Czochralski silicon. Experiments were carried out in the temperature range 350-850°C for different annealing times (10 s to 10000 h) and three different oxygen concentrations. It has been observed that the locking of dislocations as a function of annealing time has five well-defined regimes. From the temperature dependence of the unlocking stress of dislocations the binding energy of oxygen to dislocations has been deduced. Experimental results have indicated that at temperatures below 700°C an enhanced transport of oxygen to dislocations takes place. Numerical simulations of oxygen transport to dislocations showed that the effective diffusivity of oxygen at lower temperatures is different than "normal" diffusivity and can be several orders of magnitude larger. At lower temperatures the value of effective diffusivity becomes dependent on oxygen concentration and has an activation energy of about 1.5 eV. Possible mechanisms leading to "enhanced" oxygen transport are discussed.

INTRODUCTION

It is very important that silicon wafers resist plastic deformation during processing; not only because of the dislocations and the device issues associated with them but also because of the enormous demands placed on wafer flatness by the requirements of the lithography. Since mobile dislocations determine the mechanical strength of a crystal, a retardation or suppression of dislocation movement would lead to an improved mechanical stability of the crystal. Resistance to dislocation motion in silicon can be strongly influenced by various impurities, most notably oxygen (1, 2).

Silicon grown using the Czochralski (Cz) technique contains interstitial oxygen in concentrations of 10^{17}-10^{18} cm^{-3}. Since oxygen is incorporated into a growing crystal at the melting temperature it is supersaturated at temperatures typical to device processing and forms oxide precipitates during subsequent annealing treatments. While oxide precipitates and other defects associated with them (mainly dislocations and stacking faults) are useful for gettering metallic impurities excessive precipitation can lead to dislocation generation resulting in plastic deformation. Thus, the presence of oxygen can be both beneficial and detrimental from the viewpoint of device production (3).

The mechanical strength of a Si wafer depends on the concentration of oxygen left after precipitation. At elevated temperatures oxygen atoms diffuse to the dislocation core regions resulting in immobilization of dislocations, an effect termed dislocation locking (4). The stress needed to start a locked dislocation moving is called the unlocking stress. It depends on the amount of oxygen diffused to the dislocation core which is a function of oxygen content of the Si wafer and its thermal history. Since the interaction between oxygen atoms and dislocations is one of the main factors influencing the mechanical strength of a Si wafer the occurrence of warpage can be reduced or avoided if an understanding the nature of the oxygen-dislocation interaction is attainable. Then one can predict their effect on the material properties to improve the yield of device production by finding suitable material and processing conditions.

One of the key parameters, which determines the behaviour of oxygen in silicon, is the diffusivity of oxygen. Both oxygen precipitation and oxygen transport to dislocations are controlled, to a large extend, by the magnitude of diffusivity. For temperatures larger than 700°C the diffusivity is well characterized by using different experimental techniques and it is found that oxygen mass transport is by single atomic hops with an activation energy of about 2.5 eV. At lower temperatures, however, several observations indicate that the rate of oxygen transport is anomalous and deviates from expected behaviour of single atomic jumps. One of the earliest indications of this behaviour was the thermal donor formation. Thermal donors are oxygen containing electrically active clusters formed in the temperature range of about 350-550°C. The rates of thermal donor formation, corresponding to an activation energy of about 1.7-1.8 eV, are much larger than those expected from diffusion by single atomic jumps (5). Therefore it is necessary to invoke mechanisms assuming fast diffusing oxygen containing species to explain the kinetics of thermal donors. Furthermore there is evidence from oxygen precipitation and outdiffusion studies for a possible long range anomalous transport of oxygen. It has been observed that the high nucleation rate of oxygen precipitates and unexpected long range outdiffusion of oxygen from wafer surfaces at lower temperatures cannot be explained with normal oxygen diffusivity data extrapolated from high temperature measurements. Unfortunately the techniques used for the determination of oxygen diffusivity at high temperatures are not suitable for low temperature measurements and it is difficult to obtain reliable data of diffusivity. It is however possible to use well characterized dislocations and locking effect of oxygen atoms to obtain quantitative information on oxygen transport in a broad range of temperature (well below 400°C). In this article we present our work on understanding oxygen-dislocation interactions in silicon and show how one can obtain quantitative information on oxygen transport at low temperatures.

We have investigated the role of oxygen in locking of dislocations systematically in an extensive temperature range (350-850°C) for different time scales (from 10 s to 10000 h) and for different oxygen concentrations. We have also developed a model to simulate the dislocation locking effect of oxygen. Some of the results were published elsewhere (6, 7). Here we present results obtained mainly using samples with low oxygen content by which we can investigate the transport behaviour of oxygen at lower temperatures as well as their binding characteristics to dislocations.

EXPERIMENTAL METHOD

Rectangular samples with dimensions 0.65 mm x 3 mm x 25 mm were cleaved from dislocation free (001) Cz-Si wafers (p-type, 10 Ωcm). Samples with three different oxygen concentrations were used: $2.6x10^{17}cm^{-3}$ (labelled as L-C_O), $6.3x10^{17}cm^{-3}$ (M-C_O), and $10.4x10^{17}cm^{-3}$ (H-C_O) (DIN 50438-I). After chemomechanical polishing of cleaved surfaces well defined dislocation arrays were produced in a two stage procedure: First the sample surface was damaged in a controlled manner to introduce dislocation sources. Then samples were stressed by employing a four-point bending technique at high temperatures to produce dislocation half loops extending from damaged areas. This produced dislocation half loops up to 200 μm in diameter extending from damaged areas. After the generation of dislocations surface damage was removed to prevent further dislocation generation during later stages of experiments.

Samples were annealed at temperatures between 350 and 850°C for oxygen to diffuse to dislocations to generate a locking effect. Anneals of different durations (10 s to 10000 h) were carried out depending on the temperature of heat treatment. After this high temperature treatment a surface layer of about 50 μm was removed to prevent any influence of oxygen outdiffusion. Afterwards samples were stressed using a three-point bending technique to determine the unlocking stress. Since during the three-point bend test each dislocation array experiences a different stress, only those exposed to a stress larger than the unlocking stress grow while other dislocation loops remain the same size as after the four-point bending. A preferential etch was then used to determine the position of dislocation loops in the specimen from which the unlocking stress was deduced. The temperature of the three-point-bend test was always the same for all specimens (550°C).

For control purposes some specimens from nitrogen-free float-zone Si wafers were prepared to check whether there is any locking effect in the absence of oxygen. Furthermore some samples from wafers with higher vacancy content (8) were used to test the possible influence of intrinsic point defects. All these samples were tested under the same procedure as those used for the standard ones. All high temperature treatments were carried out in flowing argon atmosphere.

EXPERIMENTAL RESULTS AND DISCUSSION

In figures 1(a)-1(d) the unlocking stress is shown for four different temperatures and three different oxygen concentrations. We found that for all temperatures investigated (350-850°C) the unlocking stress increases almost linearly with increasing annealing time. As figure 1 shows that with both increasing temperature and oxygen concentration

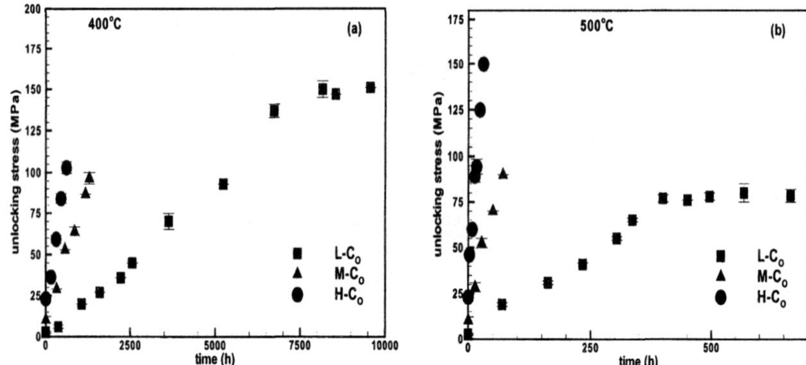

Figure 1: Experimental results on the unlocking stress as a function of annealing time for three different oxygen concentrations ($2.6 \times 10^{17} cm^{-3}$, L-$C_O$; $6.3 \times 10^{17} cm^{-3}$, M-$C_O$; $10.4 \times 10^{17} cm^{-3}$, H-$C_O$) and annealing temperatures (a) 400°C, (b) 500°C, (c) 600°C, (d) 650°C.

the rate of the unlocking stress increases indicating that more oxygen atoms have diffused to the dislocations.

We noted that in control samples prepared from float-zone Si no locking was observable. In these samples all dislocations were mobile irrespective of the annealing temperature and applied stress values studied. Moreover, samples from wafers with high vacancy content (estimated to be ~100 times greater than in normal wafers) had similar unlocking stress values suggesting that for the investigated temperature range points defects have no or negligible influence on oxygen transport and dislocation locking.

From measurements performed at many different temperatures and times, which is not presented in this paper, the dependence of the unlocking stress on temperature and time can be found. This is shown schematically in figure 2. We observed that the transport of oxygen to the dislocation core and its segregation to the dislocation show well-defined regimes. At the early stages of annealing the concentration of oxygen at the core is small and far from equilibrium. Therefore the oxygen concentration at the core, and so the unlocking stress, increases with time. At higher temperatures and oxygen concentrations the transport to the core is quicker so that the increase in the unlocking stress is higher for faster temperatures and oxygen concentrations, as can be seen from figure 1. Following this initial rise a steady-state is achieved when a local-equilibrium with the background oxygen concentration is established. The steady-state manifests itself as a saturation of

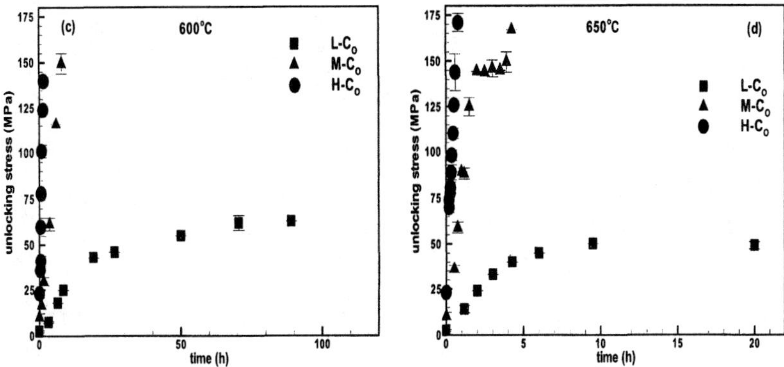

Figure 1: (continued)

the unlocking stress. The oxygen concentration at the core is then time independent as the absorption of oxygen atoms by the core is balanced by the emission of oxygen atoms from the core. After some time at the saturation unlocking stress oxygen precipitation begins to occur at the dislocation core. This is associated with a sudden rise in the unlocking stress as annealing is continued. With the increasing annealing time the unlocking stress saturates again. The value of this second saturation stress is also dependent on the annealing temperature and oxygen concentration. It decreases with increasing temperature and increases with increasing oxygen content. After long times spent in the second saturation regime the unlocking stress starts to decline rapidly with increasing annealing time.

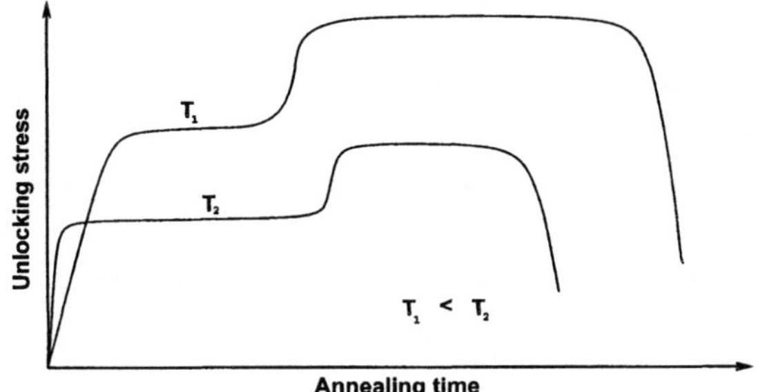

Figure 2: The unlocking stress as a function of annealing temperature depicted schematically.

The behaviour of first two regimes is controlled by the transport of oxygen to the dislocation and the thermodynamics that control the core oxygen concentration. The transport of oxygen to the dislocation is controlled by the diffusion process. The interaction between an oxygen atom and a dislocation is characterized by the binding energy of oxygen atoms to the dislocation. Since in the equilibrium state the oxygen atoms at the core will establish a Maxwell-Boltzmann concentration distribution the first saturation regime is determined by the concentration of oxygen at the far field of diffusion C_O, and the binding energy between an oxygen atom and a dislocation, ΔG. Assuming that the unlocking stress is proportional to the number of oxygen atoms at the dislocation core the unlocking stress τ_u can be written as

$$\tau_u = KC_O \exp(-\frac{\Delta S}{k})\exp(\frac{\Delta H}{kT}) \quad [1]$$

where for the oxygen-dislocation binding energy the thermodynamic relationship ΔG=ΔH-$T\Delta S$ is used. In equation [1] K is the proportionality factor, ΔS is the entropy change, ΔH is the enthalpy change, k is Boltzmann's constant, and T is the absolute temperature.

An Arrhenius plot of the saturation values of unlocking stress can be used to determine the value of ΔH. In figure 3, values of the unlocking stress normalised by the oxygen concentration are shown for the temperatures between 400°C and 850°C. Also shown are two straight lines representing equation [1] with different values of binding energy. The striking observation from figure 3 is that not only one but two different oxygen-dislocation binding enthalpies can be deduced. For temperatures larger

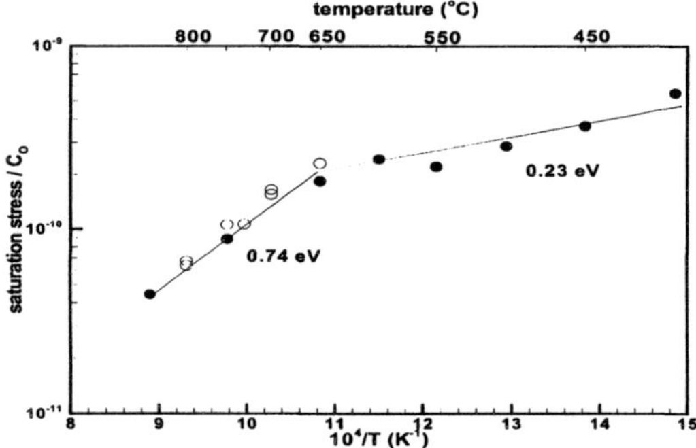

Figure 3: The saturation stress normalised by the oxygen concentration as a function of reciprocal temperature. Closed circles indicate the results obtained using low oxygen samples. Previous data obtained using samples with high and medium oxygen concentration (open circles) (6) are also shown.

than 650°C ΔH is 0.74 eV. The binding enthalpy measurements presented here from the low oxygen specimens are in very good agreement with our previous data obtained from specimens with higher oxygen content in the temperature range 650-850°C. However, the low oxygen specimens have allowed us to measure the saturation behaviour at much lower temperatures than previously possible. Our new data has shown a markedly different behaviour in the low temperature range of 450-650°C. In this regime a binding enthalpy of 0.23 eV has been found.

MODELLING AND DISCUSSION

Since one of our aims was to obtain information on oxygen transport at temperatures less than 700°C we have used a model (6) to calculate the number of oxygen atoms at the core by simulating the oxygen transport to the dislocation. The unlocking stress was assumed to be proportional to the number of oxygen atoms at the dislocation core and the oxygen diffusivity was used as a fit parameter to obtain agreement between the calculated results and experiments. We described the oxygen diffusion to the dislocation core by

$$\frac{\partial C_O}{\partial t} = D_O \nabla\left(\nabla C_O + \frac{C_O}{kT} \nabla(\Delta G) \right) \qquad [2]$$

where D_O is the oxygen diffusivity. We solved equation [2] numerically using a cylindrical domain where the dislocation core is in the centre. The outer calculation boundary is chosen to be sufficiently large so that a zero flux boundary condition could be used. The dislocation core boundary has been modelled such that both capture and reemission of atoms to/from the core were considered. We assumed that the capture is controlled by the oxygen concentration next to the core, C_1, the concentration of available sites at the core, C_a, and the concentration of occupied sites in the core, C_c, that is

$$capture \propto C_1 \frac{C_a - C_c}{C_a} \qquad [3]$$

Emission of oxygen atoms from the core is assumed to be a function of the oxygen concentration in the core and the escape probability of oxygen atoms from the core:

$$emission \propto C_c \exp\left(-\frac{\Delta G}{kT} \right) \qquad [4]$$

Figure 4 shows the calculation results together with the experimental data for two different temperatures. The agreement is very good between simulation results and experimental ones. Simulations could quantitatively describe the behaviour of unlocking stress. The slope of the initial rise, saturation time and saturation stress could be calculated correctly for different oxygen concentrations. During calculations it was observed that the rate of initial rise was mainly determined by the value of diffusivity. Since at the early stages of a heat treatment reemission of oxygen atoms from the core is less noticeable because the dislocation core is still effectively "empty" of oxygen and so the whole process is controlled mainly by oxygen diffusion and capture at the core. As the concentration of oxygen at the dislocation increases, it gradually approaches the concentration which is in equilibrium with the background oxygen concentration. The rate of capture of oxygen atoms to the core and their thermal reemission become equal so that saturation develops.

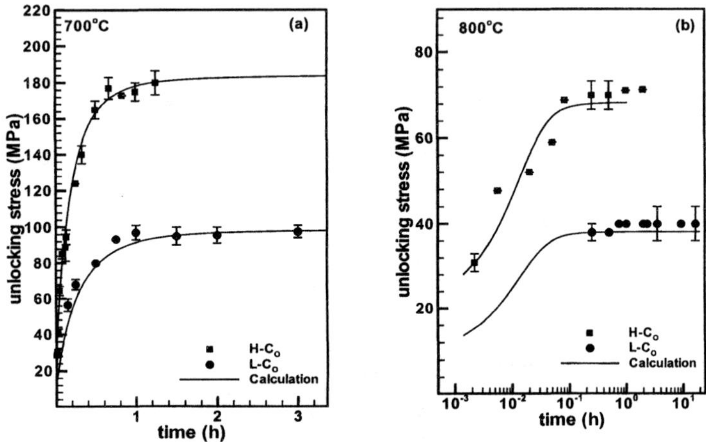

Figure 4: Calculated (lines) and experimental (symbols) results on the unlocking stress as a function of annealing time for temperatures (a) 700°C and (b) 800°C.

Figure 5 shows the results of oxygen diffusivity together with data from the literature. Our results for temperatures larger than 700°C agree well with the results from the literature. Below this temperature, however, the effective diffusivity is controlled by another activation energy of about 1.5 eV while the magnitude of the diffusivity becomes dependent on oxygen concentration. We, therefore, obtained three different diffusivity values depending on the concentration of oxygen. In low oxygen content samples

$$D_O^{L-C_O} = 2.04 \times 10^{-7} \exp\left(-\frac{1.51\,eV}{kT}\right) \quad \text{cm}^2\,\text{s}^{-1}\,. \qquad [5]$$

In samples with medium oxygen content

$$D_O^{M-C_O} = 7.33 \times 10^{-7} \exp\left(-\frac{1.52\, eV}{kT}\right) \qquad \text{cm}^2\ \text{s}^{-1}\ . \qquad [6]$$

And samples with high oxygen content

$$D_O^{H-C_O} = 2.16 \times 10^{-6} \exp\left(-\frac{1.55\, eV}{kT}\right) \qquad \text{cm}^2\ \text{s}^{-1}\ . \qquad [7]$$

Comparison of the binding enthalpy data from figure 3 with the diffusivity data suggests that below 650-700°C the movement of oxygen in Si is by a species other than the single oxygen atoms and that this also results in dislocation locking by a species different from that at high temperatures.

Our results are in agreement with the aforementioned impression given in the literature. At temperatures greater than 700°C there is an excellent agreement between our data and literature data. At lower temperatures, below to 350°C the lowest temperature we could perform our experiments, transport of oxygen seems to be controlled by a new mechanism different than single oxygen jump. At this temperature, where data sensitive only to single oxygen atom jumps is available in the literature, our results indicate a three orders of magnitude "enhancement" in oxygen transport. Thus, the possibility that the transport might be dominated by fast diffusing single oxygen atoms

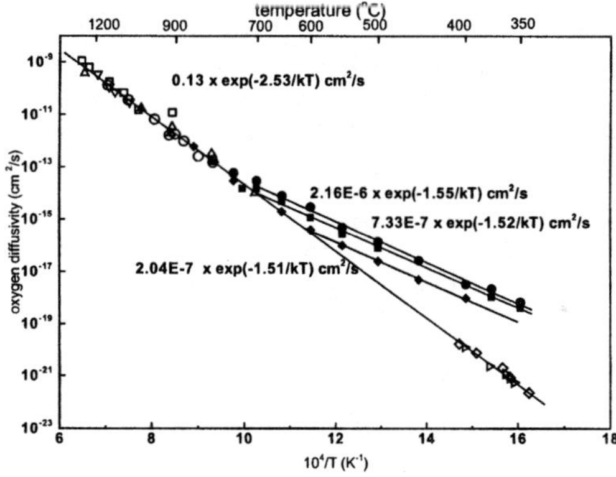

Figure 5: Oxygen diffusivity as a function of reciprocal temperature. Open symbols are data from the literature (9). Closed symbols are values obtained in this work for different oxygen concentrations (2.6×10^{17}cm^{-3} (diamonds); 6.3×10^{17}cm^{-3} (squares); 10.4×10^{17} cm^{-3} (circles)). Best fit to each data group is also given.

is eliminated. We also demonstrated experimentally that the transport of oxygen in samples with different point defect concentrations is indistinguishable. So any hypothesis connecting fast oxygen transport with any kind of oxygen-point defect complex would be unsubstantiated. Of the oxygen only complexes oxygen dimers are major candidates for fast diffusing species which we observe. The existence of stable oxygen dimer in silicon has been indeed demonstrated experimentally (10). It has been shown that analysis of generation kinetics of thermal donors would require a dimer diffusion with an activation energy of about 1.3 eV (11). Furthermore several theoretical studies concluded that the oxygen dimer in Si is stable and can diffuse with an activation energy of about 1.5 eV [e.g., Ref. (12)].

On the assumption that for temperatures below 700°C the oxygen dimer is responsible for the transport of oxygen one can write the "effective" diffusivity, D_O^{eff}, as (7)

$$D_O^{eff} = D_O + 2\frac{C_{O2}}{C_O} D_{O2} \qquad for \quad C_O >> C_{O2} \qquad [8]$$

where D_{O2} is the diffusivity of an oxygen dimer and C_{O2} is the concentration of oxygen dimers. If oxygen dimers are in equilibrium with single oxygen atoms their concentration will be

$$C_{O2} \propto C_O^2 \exp\left(\frac{\Delta H_{b2}}{kT}\right) \qquad [9]$$

where ΔH_{b2} is the binding enthalpy of the oxygen dimer. From equations [8] and [9] one can see that the effective diffusivity at lower temperatures has to be proportional to oxygen concentration if oxygen dimers are the species responsible for "enhanced" oxygen transport. Our experimental results have shown that at lower temperatures diffusivity is in fact concentration dependent. Using equations [8] and [9] it is straightforward to show that the effective diffusivity would become proportional if dimers contribute to much of the transport of oxygen. A simple analysis of our experimental data of the absolute values of diffusivity at temperatures between 400 and 650°C shows the diffusivity to increase on average by a factor of 1.61, 5, and 3.3 as the oxygen concentration increases by a factor 10.4/6.3=1.65, 10.4/2.6=4, and 6.3/2.6=2.42, respectively. This is clearly consistent with the above analysis and shows that oxygen dimers are capable of explaining the observed increase in effective diffusivity observed at low temperatures.

CONCLUSIONS

The experimental investigation of dislocation locking by oxygen atoms in the temperature range 350-850°C for different annealing times and different oxygen concentrations demonstrated that the transport of oxygen to a dislocation core and its segregation at the dislocation show well-defined regimes. It has been observed that with increasing annealing time the number of oxygen atoms at the dislocation core increases until a steady-state (saturation) regime is achieved. Subsequently precipitation of oxygen at the core takes place determining the locking effect according to the precipitate distribution.

Analysis of the saturation regime as a function of temperature produced information on oxygen-dislocation binding energy. It has been found that for temperature larger than 650°C the binding is characterized by an enthalpy of about 0.74 eV. For lower temperatures a markedly different binding enthalpy of about 0.23 eV has been obtained. Simulations of experimental results revealed that to account the oxygen transport to dislocation at lower temperatures the effective diffusivity of oxygen has to be enhanced. By comparing the binding enthalpy data with the diffusivity data of oxygen it has been suggested that below 650°C the transport and binding of oxygen in silicon are by a different species other than single oxygen atoms. Interpretation of our results showed that the effective diffusivity deviates from normal diffusivity at around 700°C and follows an Arrhenius curve down to lower temperatures with an activation energy of about 1.5 eV. The value of the effective diffusivity becomes dependent on the oxygen concentration but is unchanged with different point defect content. With the assumption that the fast diffusing species could be oxygen dimers it has been shown it is possible to account all experimental results.

REFERENCES

1. K. Sumino, in *Handbook on Semiconductors,* S. Mahayan, Editor, p. 73, Elsevier, Amsterdam(1994).

2. K. Sumino and I. Yonenaga, *Semicond. Semimetals,* **42**, 449 (1994).

3. A. Borghesi, B. Pivac, A. Sassella, and A. Stella, *J. Appl. Phys.*, **77**, 4169 (1995).

4. S. M. Hu and W. Patrick, *J. Appl. Phys.*, **46**, 1869 (1975).

5. *Early Stages of Oxygen Precipitation in Silicon,* R. Jones, Editor, Nato Advanced Study Series 3: High Technology, Kluwer Academic, Dordrecht (1996).

6. S. Senkader, K. Jurkschat, D. Gambaro, R. J. Falster, and P. R. Wilshaw, *Philos. Mag. A,* **81**, 759 (2001).

7. S. Senkader, P. R. Wilshaw, and R. J. Falster, *J. Appl. Phys.*, **89**, 4803 (2001).

8. R. Falster, D. Gambaro, M. Olmo, M. Cornara, and H. Korb, *Mat. Res. Soc. Symp. Proc.*, **510**, 27 (1998).

9. J. C. Mikkelsen, Jr., *Mat. Res. Soc. Symp. Proc.*, **59**, 19 (1986).

10. L. I. Murin, T. Hallberg, V. P. Markevich, and J. L. Lindström, *Phys. Rev. Lett.*, **80**, 93 (1998).

11. D. Åberg, B. G. Svenson, T. Hallberg, and J. L. Lindström, *Phys. Rev. B*, **58**, 12944 (1998).

12. M. Ramamoorthy and S. T. Pantelides, *Solid State Commun.*, **106**, 243 (1998).

CHEMICAL AND STRUCTURAL CHARACTERIZATION OF OXIDE PRECIPITATES IN HEAVILY BORON DOPED SILICON BY INFRARED SPECTROSCOPY AND TRANSMISSION ELECTRON MICROSCOPY

O. De Gryse, P. Clauws, and J. Vanhellemont
Department of Solid State Sciences, Ghent University, Krijgslaan 281, B-9000 Gent, Belgium

O. Lebedev and J. Van Landuyt
University of Antwerp-EMAT, Groenenborgerlaan 171,B-2020 Antwerpen, Belgium

E. Simoen and C. Claeys*
IMEC, Kapeldreef 75, B-3001 Leuven, Belgium
*Also at E.E. Dept, KU Leuven, Kasteelpark Arenberg 10, B-3001 Leuven, Belgium

Infrared absorption spectra of oxygen precipitates in boron doped silicon with a boron concentration between 10^{17} and 10^{19} cm^{-3} are analyzed, applying the spectral function theory of the composite precipitates. The aspect ratio of the platelet precipitates has been determined by transmission electron microscopy measurements. Our analysis shows that in samples with moderate doping levels (<10^{18} B cm^{-3}) SiO_γ precipitates are formed with stoichiometry as in the lightly doped case. In the heavily boron doped (>10^{18} cm^{-3}) samples, however, the measured spectra of the precipitates are consistent with a mixture of SiO_2 and B_2O_3, with a volume fraction of B_2O_3 as high as 0.41 in the most heavily doped case.

INTRODUCTION

Interstitial oxygen is the main grown-in impurity in Czochralski-grown silicon. Being highly supersaturated it agglomerates into oxide precipitates of various morphologies during thermal treatments. The precipitation kinetics and the resulting precipitate morphology depend on the experimental conditions and material parameters, the doping level being one of them. Although relatively well understood in moderately doped silicon, oxygen precipitation kinetics and precipitate morphology are strongly different in low-resistivity silicon (1,2,3). To use the internal gettering properties of oxide precipitates in a controlled way, a quantitative understanding of oxygen precipitation and of the chemical composition of the oxide precipitates in heavily doped silicon is needed.

As shown previously for moderately doped silicon (4) absorption spectra related with oxide precipitates can be analyzed quantitatively using a modified Day-Thorpe approach (5), provided the precipitate matrix consists of 2 phases. When the aspect ratio of the platelet precipitates is determined from transmission electron microscopy analyses and the dielectric functions of the constituents of the precipitate are known, the algorithm proposed in Ref. (4) can compute the reduced spectral function (5,6,7), together with the volume fractions of the two phases. In case of lightly doped silicon the precipitates were shown to consist of SiO_γ with $\gamma=1.17\pm0.14$ (4).

In this contribution the precipitate absorption spectra of boron doped samples with boron concentrations $[B_s]$ ranging between 10^{17} and 10^{19} cm^{-3} are analyzed.

EXPERIMENTAL

Two sets of samples with increasing boron concentration are used in this study as shown in Table I and Table II. The samples in Table I received a two-step heat treatment (32 h at 700°C + 16 h at 900°C under Ar flow) and are labeled set I. The samples in Table II received a three-step heat treatment (500°C/ 16h+ 800°C/16h + 1050°C/8h) and are labeled set II. The latter set was already used for a different study (8), but the chemical composition of the precipitates was not analyzed at that time.

For the determination of the interstitial oxygen concentration $[O_i]$, specimens with a thickness between 10-30 μm were prepared and measured at low temperature (8, 9). In the case of the samples B3A2 and B6A2 the oxygen concentration after anneal was determined from the absorption bands due to interstitial oxygen and A–center in the low temperature absorption spectra of irradiated samples. The latter approach was followed because the remaining interstitial oxygen concentration ($[O_i]$) after anneal was expected to be very low, making the ultra-thin sample method difficult to apply as the detection limit is around 2 10^{17} cm^{-3}. Also, because of the small $[O_i]$ after anneal, the absolute error on $\Delta[O_i]$ (= difference between initial and final $[O_i]$) due to irradiation was expected to be rather low. The initial oxygen concentration and the precipitated oxygen concentration $\Delta[O_i]$ are shown in the third and the fourth column of Table I and Table II.

The absorption spectra of the oxide precipitates are obtained from equivalent pieces of sample that are subsequently irradiated with a high dose of 2 MeV electrons (8) to obtain transparency. The remaining $[O_i]$ is determined from 6 K infrared measurements and a scaled 1107 cm^{-1} absorption peak is then subtracted from the absorption spectrum at room temperature, resulting in the oxide precipitate absorption spectrum. As was shown in reference 8 the absorption spectrum of the oxide precipitates is not altered by the irradiation itself.

Transmission electron microscopy (TEM) observations were performed at 200 keV along the [0 0 1] direction (plan view) and along the [011] direction (cross-section) on the four heat treated samples of set I, and the aspect ratio, defined as the ratio of the thickness to the edge length, was measured (Table I).

Table I: Initial ($[O_i]$) and precipitated ($\Delta[O_i]$) interstitial oxygen concentration for samples of set I, together with the resonance wave numbers due to polyhedral (ν_s) and platelet (ν_d) precipitates and the corresponding aspect ratio y as obtained from a TEM analysis are given.

Sample name	$[B_s]$ (10^{17} cm^{-3})	$[O_i]$ (initial) (10^{17} cm^{-3})	$\Delta[O_i]$ (10^{17} cm^{-3})	ν_s (cm^{-1})	ν_d (cm^{-1})	y (TEM) (10^{-2})
B14A	1.0	8.24	7.30	1123	1226	4.3 ± 2
B12A	4.8	8.00	6.93	1123	1226	2.7
B3A2	39	6.8	5.2	-	1220	3.5 ± 2
B6A2	91	9.2	9.1	-	1224	5.5 ± 2

Table II: Initial ($[O_i]$) and precipitated ($\Delta[O_i]$) interstitial oxygen concentration for samples of set II, together with the resonance wave numbers due to polyhedral (ν_s) precipitates.

Sample name	$[B_s]$ (10^{17} cm^{-3})	$[O_i]$ (initial) (10^{17} cm^{-3})	$\Delta[O_i]$ (10^{17} cm^{-3})	ν_s (cm^{-1})
B1	39	7.1	5.1	1083
B2	38	5.5	4.2	1085
B3	41	5.2	3.8	1086
B4	110	6.7	5.4	1086

THEORY, MODEL AND SIMULATIONS

In reference 4 an algorithm was developed which permits to extract the volume fraction of one phase in a two-component material from the room temperature absorption spectrum of the oxide precipitates. The key is to use the Bergman-Milton spectral representation of the dielectric function of a two-component material (5, 6), or the numerically more accessible reduced spectral function $g(x)$ as introduced by Day and Thorpe (5). This permits to separate the optical parameters of the two components from the geometrical information of the composite, the latter represented by the spectral function (6,7). Although the determination of the reduced spectral function from a reflectance or absorption spectrum is an ill conditioned problem, it is shown to be reliable if a priori knowledge is introduced as additional constraints in the minimization procedure (4,5,10).

Isolation of the precipitate absorption bands in irradiated silicon is not so straightforward, as broad bands due to irradiation damage occur. Therefore, the algorithm is extended with a baseline, i.e. the experimental absorption spectrum is now fitted with the absorption due to precipitates superimposed on a not a priori defined straight baseline. So, it is assumed that the absorption due to irradiation damage can be treated as a straight baseline in a limited wave number region. This turns out to be only the case for set I. For set II a curved baseline is necessary, but tests with the latter give no convincing results.

In the Bergman-Milton (BM) spectral representation (6,7) the geometrical information is summarized in the not *a priori* defined (reduced) spectral function. For a two-component material the effective dielectric function is given by (5):

$$\epsilon_p = \epsilon_{pm} \cdot \left(1 - \frac{\sigma_{pp}}{s_{pm}} - \int_0^1 \frac{g(x)}{x.(s_{pm} - x)} .dx \right), \qquad [1]$$

with ϵ_p, ϵ_{pm}, ϵ_{pp} the effective dielectric function of the composite, of the host and of the inclusion in the composite, respectively. σ_{pp} is the weight of the delta function at the origin and $g(x)$ is the reduced spectral function (5), while the variable s_{pm} is given by $s_{pm}=(1-\epsilon_{pp}/\epsilon_{pm})^{-1}$. The reduced spectral function $g(x)$ does not have singularities, goes to zero at x=0 and x=1, and is positive for x between 0 and 1 (5). Furthermore the zeroth moment gives the volume fraction of the inclusion p_{pp}, and for an isotropic material the first moment satisfies:

$$\int_0^1 g(x).dx = \frac{1}{3}.p_{pp}.(1-p_{pp}). \qquad [2]$$

The spectral functions are extracted by minimizing the chi-squared function $\chi^2 = \chi_\alpha^2 + \chi_{constraints}^2$, with $\chi_\alpha^{\ 2} = \lambda_0 \sum_i \left\| \alpha(\nu_i) - \left[\alpha(\nu_i, \sigma_{pp}, g(x), f_s, f_d, y) + a\nu_i + b \right] \right\|^2$.

$\alpha(\nu_i)$ is the experimental absorption spectrum with i running over all the data points, $\lambda_0 = 1/max(\alpha)$ a normalization factor to ensure that all the absorption spectra are treated alike; a and b define the straight baseline. The simulated absorption $\alpha(\nu_i, \sigma_{pp}, g(x), f_s, f_d, y)$ due to spherical and oblate spheroidal particles is calculated with the continuum approach (11, 4) and depends on the volume fraction of the spheres (f_s) and oblate spheroidal particles (f_d), on the corresponding aspect ratio (y) and on the dielectric function of the silicon matrix and the effective dielectric function of the composite material of the precipitates ($\in_p$) given by Eq. [1].

Since the core of the model is equation [1], which is an inhomogeneous Fredholm equation of the first kind, the problem is ill conditioned (5). The standard strategy to tackle this is to insert *a priori* knowledge into the chi-squared equations by means of a set of constraints ($\chi^2_{constraints}$):

$$\begin{aligned} \chi^2_{constraints} = {} & \lambda_1.(\delta m_1)^2 + \lambda_2.\left(\left(\Delta[O_i] - [O]_{pp} \right) / \Delta[O_i] \right)^2 + \\ & \lambda_3.\left(\nu_s^e - \nu_s^f\right)^2 + \lambda_4.\left(\nu_d^e - \nu_d^f\right)^2 + \lambda_5.\left(\nabla g(x)\right)^2 \end{aligned}, \qquad [3]$$

with $\delta m_1 = \int_0^1 g(x).dx - \frac{1}{3}.p_{pp}.(1-p_{pp}) \Big/ \int_0^1 g(x).dx$ the first-moment constraint, given as a relative deviation from the first moment of the spectral function, ensuring internal consistency of the reduced spectral function, with $[O]_{pp}$ the concentration of oxygen in the precipitates, given by

$$[O]_{pp} = (f_s + f_d).p_{pp}.\frac{2}{v_{SiO_2}} \qquad [4]$$

in the case the precipitates consist of a mixture of SiO_2 and Si, and by

$$[O]_{pp} = (f_s + f_d) \cdot \left(p_{pp} \cdot \frac{2}{v_{SiO_2}} + (1 - p_{pp}) \frac{3}{v_{B_2O_3}} \right) \quad [5]$$

in the case of a mixture of SiO_2 and B_2O_3. v_{SiO2} (=4.3 × 10^{-23} cm^3) and v_{B2O3} (=4.2 ×10^{-23} cm^3) are the volumes of a SiO_2 and B_2O_3 molecule, respectively. The second term in Eq. [3] states that all the disappeared interstitial oxygen ($\Delta[O_i]$) went to the precipitates ($[O]_{pp}$). The third and the fourth term stress the importance of the resonance wave numbers for spheres and platelets (subscript *s* and *d* for sphere and disc, and superscript *e* and *f* for experiment and fit), since those are closely related with the transverse (ν_{TO}) and longitudinal (ν_{LO}) optical modes, which in turn determine the dielectric function (12). The last term is a smoothing term, to prevent wild oscillations in $g(x)$ (4, 5).

The constrained fit to the absorption spectrum contains two large steps. In the first step, the reduced spectral function is given by an analytic expression that automatically satisfies the first moment relation:

$$g(x) = \begin{cases} \dfrac{C.(x - x_L)^{1-\alpha}.(x_U - x)^{\beta}}{x^{-1+\gamma}} & 1 > \alpha > 0, 3 > \beta > 0, \\ & 3 > \gamma > 0, 0 \le x_L \le x \le x_U \le 1, x_L < x_U \; . \\ 0 & otherwise \end{cases} \quad [6]$$

This is an extension of the analytic function of Ghosh and Fuchs (13) that can produce a more skewed form. This function allows accurate computations of integrals using Jacobi-Gauss quadrature rules (14). σ_{pp} can be found from Eqs. [2] and [6]. The goal of this first fit is 1/ find an accurate baseline, 2/ find a solution that reproduces the resonance frequencies of the polyhedral and platelet precipitates in the absorption spectrum and 3/ satisfies the first-moment relation. The last requirement is automatically fulfilled. In the second fit this solution is used as the initial value. In this case the reduced spectral function is approximated by a histogram with 64 bins. This two-fit approach is rather robust and always leads to a solution.

The dielectric function of SiO_2 was re-extracted (see Table III) from experimental data of Grosse *et al.* (15), using the analytic solution of a Gaussian broadened Lorentzian oscillator model (16). For vitreous B_2O_3 the dielectric function of Galeener *et al.* (17) is used, and for Si ϵ=11.7, as at 900°C the silicon phase is crystalline in composite suboxides close to SiO (18).

Table III: Parameters for the dielectric function of SiO_2 describing the anti-symmetric stretching (AS) modes with notations as in Reference 16 and with ϵ_∞=2.23.

Mode	ν_{TO} (cm^{-1})	ν_p (cm^{-1})	ν_τ (cm^{-1})	σ (cm^{-1})
AS1	1070	788	5	21
AS2	1178	232	20	36

The algorithm was first tested extensively. It is robust against uncertainties of 15% in $\Delta[O_i]$ and against approximations in the shape (sphere instead of a truncated polyhedron, oblate spheroid instead of a square platelet) for both SiO_2/Si and SiO_2/B_2O_3 mixtures. It is always capable of finding the volume fraction p_{pp} to within 7% or better, except for a SiO_2/Si mixture with platelets with a large aspect ratio (y=0.06), where the absence of a clearly distinct resonance ν_d is responsible for a 12% deviation in p_{pp}.

RESULTS AND DISCUSSION

Morphology of the oxide precipitates

The morphology of the oxide precipitates in moderately doped and in heavily boron doped silicon are different for the medium (900°C) as well as for the high (1050°C) temperature heat treatments. In the moderately doped samples of set I both polyhedral and platelet precipitates are found in the absorption spectra. In the heavily boron doped case the platelet morphology is dominant in the absorption spectra after a medium temperature heat treatment (set I), while after a high temperature heat treatment only polyhedral precipitates are observed (set II).

Results for the samples of set I

Analogous to the lowly doped case, the oxide precipitates were assumed to consist of a mixture of SiO_2 and Si. In the moderately doped materials (B14A and B12) this turns out to be the case. However, in the heavily doped samples ($\geq 10^{18}$ B/cm^3) the concentration constraint (Eq. [4]) cannot be met: the oxygen concentration in the precipitates underestimates ΔO_i by a factor of 2.5-3. This indicates that the mode strength of ϵ_{pp} is too strong and that not all the oxygen vibrates with a mode in the Reststrahlen region of the AS-mode of SiO_2. The above suggests that the oxide precipitates consist of SiO_2 and another material with a low dielectric constant and that not all the oxygen in the precipitates contributes to the absorption spectrum between 1050 and 1250 cm^{-1}. This means that if the oxide precipitates consist of two components, then the second component probably contains oxygen and its dielectric constant should be rather low. As the samples are heavily doped with boron, B_2O_3 should be considered the prime candidate. First, boron is the only dominant impurity available, next to oxygen. Second, B_2O_3 has a low dielectric constant ($\epsilon_{B2O3} \approx 3.1$ around 1000 cm^{-1}) [*Galeener80*]. Third, the *Reststrahlen* region of B_2O_3 lies between 1260 and 1550 cm^{-1}, which is outside the *Reststrahlen* region of SiO_2 (1070-1245 cm^{-1}).

The results of the fits are summarized in Table IV. The results on the moderately doped samples B14A and B12A fall in the uncertainty region of volume fractions in the case of moderately doped material: $p_{SiO2} \approx 0.71 \pm 0.04$ (4). The precipitated oxygen concentration is retrieved to within 15%. And the deviations from the first moment of the spectral function (δm_1) are within 5%. In the algorithm we gave this restriction a large weight in order to pursue internal consistency of the solution. It is also this restriction that is, at least partly, responsible for the rather narrow confidence intervals of the results in Table IV. Comparing the aspect ratio with the TEM values in Table I, we see the values deviate by less than 0.02, i.e. they fall in the confidence region of the TEM results. Moreover, the trends in the aspect ratio are consistent with the TEM results. In Fig. 1 we give the fit to the absorption spectrum of B12A, together with the corresponding reduced spectral function.

The accuracy of the results on the heavily boron doped samples is comparable to the moderately doped case. The oxygen concentration found in the precipitates agrees to within 15% with the precipitated oxygen concentration, the deviations from the first moment of the spectral function (δm_1) are less than 5%, and the aspect ratios agree even better with the results as obtained from TEM. The absorption spectrum and corresponding reduced spectral function of B6A2 are shown in Fig. 2.

Table IV: Set I: Results for the experimental spectra. The fitted absorption spectra were calculated, assuming spherical and oblate spheroidal particles.

		Solution						
Name	$\Delta[O_i]$ (10^{17} cm^{-3})	Mix	p (10^{-2})	f_s (10^{-5})	f_d (10^{-6})	y (10^{-2})	$[O]_{pp}$ (10^{17} cm^{-3}) (%)	δm_1 (%)
B14A	7.3	SiO_2+ Si	71 [-0,+2.2]	1.3 [-0.06,+0.04]	5.8 [-0.3,+0.06]	3.1 [-0.01,+0.4]	6.3 (14)	5.0
B12A	6.9	SiO_2+ Si	70 [-0.04,+1.9]	1.3 [-0.1,+0.03]	5.4 [-0.01,+0.03]	1.5 [-0,0.01]	5.9 (15)	5.2
B3A2	5.2	SiO_2+ B_2O_3	68 [-0.1,+6]	0.003 [-0,+0.2]	8.3 [-0,+0.7]	3.4 [-0,+0.03]	4.5 (14)	1.3
B6A2	9.1	SiO_2+ B_2O_3	59 [-0.01,+0.5]	0.03 [-0.01,+0.03]	17 [-0.03,+0.2]	4.6 [-0,+0.2]	9.8 (7.6)	5.2

B6A2 was constrained fitted between 1030 and 1300 cm^{-1}. Using the resulting particle parameters (f_s, f_d) and the reduced spectral function (σ_p, $g(x)$) of B6A2 in Table IV and Fig. 2 we calculated the precipitate absorption spectrum between 1000 cm^{-1} and 1600 cm^{-1} and compared this to the experimental absorption spectrum in Fig. 3. The experimental absorption spectrum is corrected with one straight baseline between 1000-1600 cm^{-1}, in order to avoid any distortion of the line shape of the various contributions. The two particle surface modes due to B_2O_3 are clearly resolved in the experimental absorption spectrum and the simulated spectrum agrees qualitatively over the Reststrahlen regions of the two components SiO_2 and B_2O_3. The strong absorption band at 1400 cm^{-1} also occurs in the B3A2 absorption spectrum, although less pronounced. The latter is according to expectations since there is less B_2O_3 incorporation (see Table IV). On the other hand it did not occur in any of the other room temperature absorption spectra of other irradiated samples (as grown or heat treated), indicating it is not an irradiation induced absorption band. The additional absorption peaks at the high wave number side in Fig. 3 are probably due to irradiation induced defects. The reasonable agreement between the experimental and simulated absorption spectrum between 1030 cm^{-1} and 1450 cm^{-1} lends additional support to the SiO_2/B_2O_3 composite precipitate hypothesis

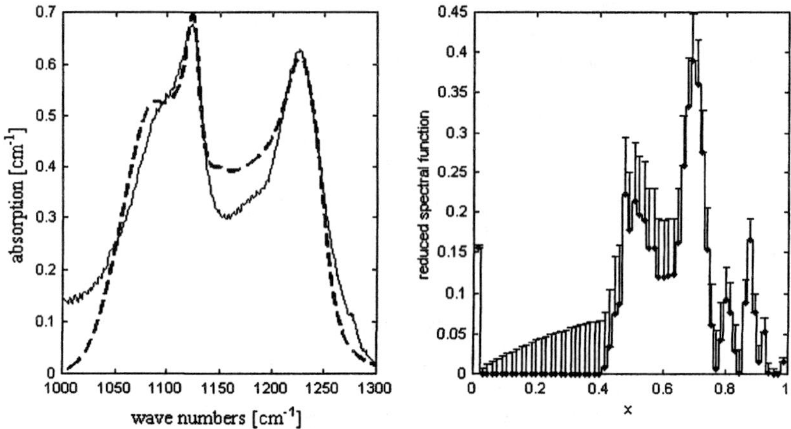

Fig. 1: Left: Experimental (full line) and fitted (thick dashed line) SiO_2/Si precipitate absorption spectrum at room temperature of B12A. The experimental spectrum is corrected with a straight baseline by the algorithm. Right: The reduced spectral function.

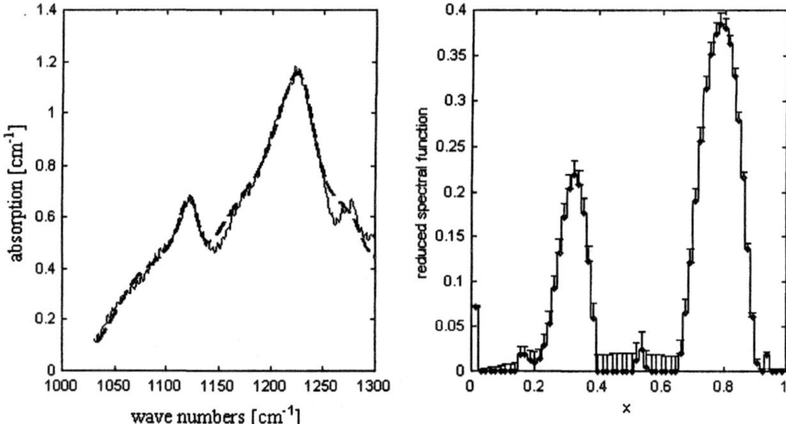

Fig. 2: Left: Experimental (full line) and fitted (thick dashed line) SiO_2/B_2O_3 precipitate absorption spectrum at room temperature of B6A2. The experimental spectrum is corrected with a straight baseline by the algorithm. Right: The reduced spectral function.

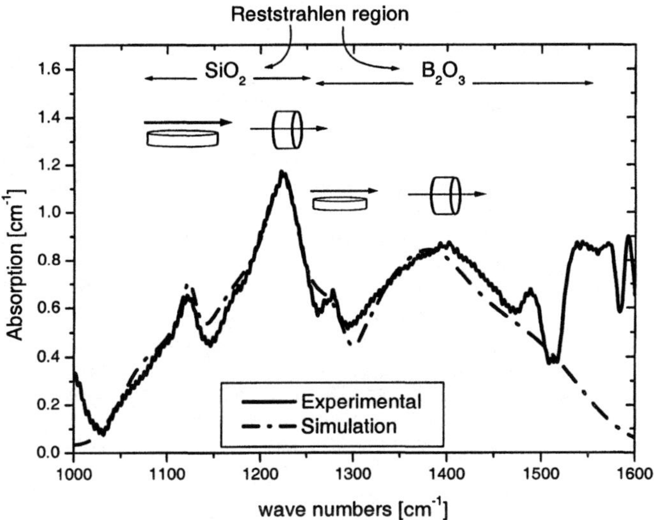

Fig. 3: Experimental and simulated absorption spectrum of sample B6A2. The Reststrahlen region of both constituents of the oxide precipitates and the orientation of the light vectors with regard to the principal axes of the precipitate particles are shown.

Influence of the uncertainty with regard to the dielectric function of SiO_2 and B_2O_3 on the results of set I

The fit to the absorption spectrum is also dependent on the dielectric functions. To estimate the influence of the choice of the dielectric function of SiO_2 we repeated the fits with the dielectric function for thermal oxidized silicon, grown at 900°C, as determined by Ishikawa *et al.* (19). The results are shown in the second to the fourth column of Table V. The obtained volume fractions of SiO_2 in the precipitates deviate by less than 5% as compared to the first fit, except for the most heavily doped sample, where the deviation amounts to 8%.

Table V: Influence of the choice of dielectric functions.

	Influence of $\in(SiO_2)$			Influence of $\in(B_2O_3)$		
Name	p (10^{-2})	δ[O] (%)	δm_l (%)	p (10^{-2})	δ[O] (%)	δm_l (%)
B14A	73	4.8	3			
B12A	68	-1.4	0.1			
B3A2	70	-4	1.5	71	9.7	0.7
B6A2	64	-11.4	0	64	3	4.6

In a last test the influence of the choice of the dielectric function of B_2O_3 is estimated. For this test we used the dielectric function that we extracted from an absorbance spectrum, measured by Wong (20). This B_2O_3 film was first grown on a silicon substrate at 350°C and subsequently heated at 700°C for 10 minutes (20). At this temperature the fraction of building units that consists of boroxol rings is already reduced as compared to B_2O_3 grown at low temperature (21). Again, from the last three columns in Table V we see that the deviation to the original solution is comparable to the analogous deviation for a different dielectric function of SiO_2.

We can conclude that the algorithm is capable of performing good fits to the experimental absorption spectra. Compared to the old algorithm in reference 4, the replacement of the absolute constraint by the relative constraint on the first moment rule of the spectral function gives solutions with narrower uncertainty regions. We estimated the sensitivity of the solution to the choice of the dielectric function of SiO_2 and, in the case of the heavily boron doped silicon, also to the choice of the dielectric function of B_2O_3. In the case of the lowly doped material and the sample with a boron concentration of 4 10^{18} cm^{-3} the deviation on the volume fraction of SiO_2 in the oxide precipitates is less than 5%. In the case of the most heavily doped sample the deviation is still less than 10%.

Estimate of the SiO_2 volume fraction in the precipitates of the samples of set II

A SiO_2 spherical particle would give a Fröhlich frequency around 1090 cm^{-1}, while in the case of SiO_γ precipitates in lightly doped silicon the Fröhlich frequency lies between 1100-1125 cm^{-1} (4). However, in the present samples of set II Fröhlich frequencies are as low as 1083-1086 cm^{-1} (see Table II). This indicates a difference in the chemical and/or structural composition, compared to the oxide precipitates in lowly doped material. In order to find an explanation we tried to find the minimum of the Fröhlich frequency for SiO_2/Si and SiO_2/B_2O_3 composites in silicon. However, the use of an effective medium approximation implies an a priori choice regarding the geometry of the composition. As the number of possible geometries is infinite, one can never perform an exhaustive search. In order to avoid an a priori choice with regard to the geometry of the composite we used the analytical model (Eq. [6]). Based on the analytical spectral function we wrote an algorithm that searches the minimum Fröhlich frequency for a spherical particle in a silicon host. The minimum Fröhlich frequencies are shown in Fig. 4. Both the SiO_2 / Si and the SiO_2/B_2O_3 system can have a geometry with decreasing Fröhlich frequency as the volume fraction p_{SiO2} decreases. The absolute minimum of ν_F for the SiO_2/Si system lies at 1085.4 cm^{-1} for a volume fraction of p_{SiO2} =0.74. The lower limit of the experimental resonance frequency can only be reproduced assuming a SiO_2/B_2O_3 mixture, but based on Fig. 4 it is not possible to exclude either of the two possible mixtures. However, the low values of the Fröhlich frequency, as compared to the lightly doped case (4), and the analogous resistivities as compared to the samples of set I supports the assumption of SiO_2/B_2O_3 precipitates in the samples of set II. As the position of the Fröhlich frequency in the samples of set II is found between 1083 and 1086 cm^{-1} (see Table II), it follows from Fig. 4 that the upper limit of p_{SiO2} is at 0.85, in the assumption that the oxide precipitates consist of a SiO_2/B_2O_3 composite. There are not enough data to confirm a strong dependence of p_{SiO2} on the substitutional boron concentration for the high temperature anneals (1050°C).

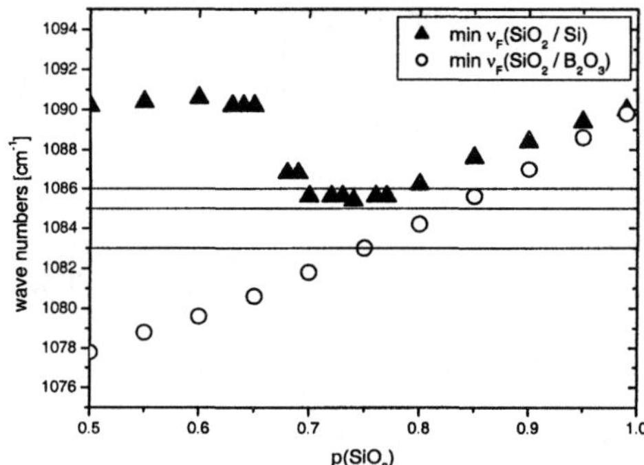

Fig. 4: Minimum of the Fröhlich frequency in silicon of a two-component mixture as a function of the volume fraction of SiO_2 (p(SiO2)). The horizontal lines indicate the position of the Fröhlich frequency found in the experimental absorption spectra of the heavily doped silicon samples after a three-step anneal (see Table II).

SUMMARY AND CONCLUSIONS

The analysis of the absorption spectra of annealed and subsequently irradiated heavily boron doped silicon, following the removal of the contribution of interstitial oxygen in the absorption spectra, combined with the measurement of the interstitial oxygen concentration in ultra-thin samples, has allowed for the first time a quantitative analysis of the chemical (SiO_2/Si or SiO_2/B_2O_3) and structural ($g(x)$, p_{PP}) composition of the oxide precipitates in heavily boron doped silicon.

The morphology of the oxide precipitates in moderately doped and in heavily boron doped silicon are different for the medium (900°C) as well as for the high (1050°C) temperature heat treatments. In the moderately doped samples both polyhedral and platelet precipitates are found in the absorption spectra. In the heavily boron doped case the platelet morphology is dominant in the absorption spectra after a medium temperature heat treatment, while after a high temperature heat treatment only polyhedral precipitates are observed.

Using the extended algorithm, we have obtained the spectral function and the corresponding volume fraction of SiO_2 in the SiO_2/Si mixture in the moderately doped samples of set I as well as in the SiO_2/B_2O_3 mixture in the heavily boron doped samples of set I. The assumption of oxide precipitates consisting of a SiO_2/B_2O_3 mixture is shown to be a plausible one. The reasonably good agreement of the experimental and simulated absorption spectrum of sample B6A2 over the Reststrahlen regions (1000-1600 cm^{-1}) of both components and the absence of the broad, strong 1400 cm^{-1} absorption band in all the other irradiated room temperature absorption spectra, also contributes to the SiO_2/B_2O_3 precipitate hypothesis.

The value of the Fröhlich frequency of the oxide precipitates in heavily boron doped silicon after a three step heat treatment is low when compared to the moderately doped counterpart (4) and it is shown that a SiO_2/B_2O_3 mixture is possible, although an analysis based solely on the position of the Fröhlich frequency cannot exclude the possibility of SiO_γ precipitates. Based on the position of the Fröhlich frequency of the polyhedral precipitates in the heavily boron doped samples of set II we were able to estimate un upper limit of the volume fraction of SiO_2 in the oxide precipitates, assuming it consists of SiO_2/B_2O_3.

We may conclude that the presented detailed analysis of infrared absorption spectra strongly supports the hypothesis of oxide precipitates composed of a SiO_2/B_2O_3 mixture with a substantial volume fraction of B_2O_3.

ACKNOWLEDGEMENTS

This work was performed with financial support from the National Science Foundation (FWO-Vlaanderen).

REFERENCES

1. S. Hahn, F.A. Ponce, W.A. Tiller, V. Stojanoff, D.A.P. Bulla and W.E. Castro, *J. Appl. Phys.*, **64**, 4454 (1988)
2. T. Ono, E. Asayama, H. Horie, M. Hourai, K. Sueoka, H. Tsuya and G.A. Rozgonyi, *J. Electrochem. Soc.*, **146**, 2239 (1999)
3. K. Sueoka, M. Akatsuka, M. Yonemura, T. Ono, E. Asayama, and H. Katahama, *J. Electrochem. Soc.*, **147**, 756 (2000)
4. O. De Gryse and P. Clauws, J. Van Landuyt, O. Lebedev, C. Claeys, E. Simoen, and J. Vanhellemont, *J. Appl. Phys.*, **91**, 2493 (2002)
5. A.R. Day and M.F. Thorpe, *J. Phys.: Condens. Matter,* **11**, 2551 (1999)
6. D.J. Bergman, *Phys. Rep.*, **43**, 377 (1978)
7. G.W. Milton, *J. Appl. Phys.*, **52**, 5286 (1981)
8. O. De Gryse, P. Clauws, L. Rossou, J. Van Landuyt, J. Vanhellemont, and W. Mondelaers, in *High Purity Silicon V*, C.L. Claeys, P. Rai-Choudhury, M. Watanabe, P. Stallhofer, H.J. Dawson, Editors, PV **98-13**, p. 398, The Electrochemical Society Proceedings Series, Pennington, NJ (1998).
9. O. De Gryse, P. Clauws, L. Rossou, J. Van Landuyt, and J. Vanhellemont, *Rev. Sci. Instrum.*, **70**, 3661 (1999)
10. A.R. Day, A.R. Grant, A.J. Sievers, and M.F. Thorpe, *Phys. Rev. Letters,* **84**, 1979 (2000)
11. L. Genzel and T.P. Martin, *Surface Science,* **34**, 33 (1973)
12. C.T. Kirk, *Phys. Rev. B,* **38** (2), 1255 (1988)
13. K. Ghosh and R. Fuchs, *Phys. Rev. B,* **38**, 5222 (1988)
14. W. Gautschi, *Mathematics and Computers in Simulation,* **54**, 403 (2000)
15. P. Grosse, B. Harbecke, R. Meyer and M. Offenberg, *Appl. Phys. A,* **39**, 257 (1986)
16. R. Brendel and D. Bormann, *J. Appl. Phys.*, **71**, 1 (1992)
17. F.L. Galeener, G. Lucovsky, and J.C. Mikkelsen Jr., *Phys. Rev B,* **22**, 3983 (1980)
18. A.E. Widmer, G. Harbeke, L. Krausbauer, and E.F. Steigmeier, *J. Electrochem. Soc.*, **133**, 1881 (1986)
19. K. Ishikawa, K. Suzuki, and S. Okamura, *J. Appl. Phys.*, **88**, 7150 (2000)
20. J. Wong, *J. Electrochem. Soc.,* **127**,62 (1980)
21. A.K. Hassan, L.M. Torell, L. Börjesson, and H. Doweidar, *Phys. Rev. B,* **45**, 12797 (1992)

Process- and Irradiation-Induced Defects in Silicon

Backside Grinding Induced Stress Measurement by Raman Spectroscopy

Masaharu Watanabe, Yasushi Kozuki, Yasuyo Tomitsuka,
and
Tomohisa Nakamura*, Atsuyoshi Inoue*

SEZ Japan Inc
1-24-1 Hongo, Bunkyou-ku, Tokyo 113-0033 Japan
and
*Center for Analytical Chemistry and Science, Inc.
*8-3-1 Chu-o, Ami, Inashi, Ibaraki 300-0032, Japan

1. Introduction

In addition to high-speed devices trend, LSI packaging trend like SIP (system in package) drives the final silicon wafer thickness thinner and thinner. Target of device chip thickness in near future is 50μm[1]. One of the issues is the stress induced in the back ground silicon. The conventional thinning process is back grinding that is a high speed process with good control. Abrasive machining of silicon is reviewed elsewhere[2] and it leaves the mechanical damage layer and underneath stressed region that is called a subsurface stress or damage. Back grinding of silicon is a kind of brittle materials machining similar to the ceramics grinding[3]. Depth profile of the subsurface stress is important for the fabrication of very thin chips. Subsurface damages and their characterization is reviewed[4,5]. Cracks induced by the mechanical stress is more dominant than plastic deformation induced by the thermal stress. Damage profile or stress profile models are proposed. One model close to this study is shown in Fug.1[6].

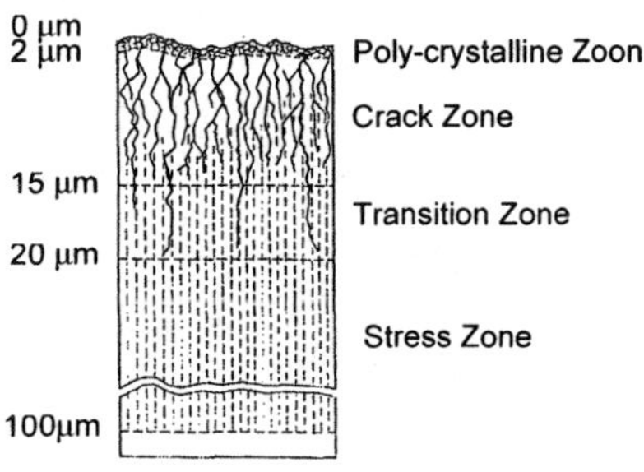

Figure 1. One of the residual damage model of ground silicon.

Micro-Raman spectroscopy is widely used to detect local stress in materials science and technology. It has a high special resolution down to 1μm without special instrument. It has been applied to measure local stress in LSI processes, especially; measurement of the cleaved surface shows local subsurface stress distribution near the surface[7,8]. It is used to local stress profiling or characterizing the grinding damage[9,10,11]. In this study, we use it to evaluate the back grinding induced stress in thin silicon wafers. Other interesting characterization of damage are reported elsewhere[12,13,14,15,16].

2. Experiments

2.1 Sample preparation

Stress in the back grinding after device fabrication process was interested in here. For the simplicity of the experiment, plain polished wafers were used instead of actual device wafers. Silicon wafers used were (100) P type CZ 725 μm thick conventional single side polished wafers. First, back side of wafers were ground to 275 μm using #360 grinding. Then another 25 μm was ground by #2000

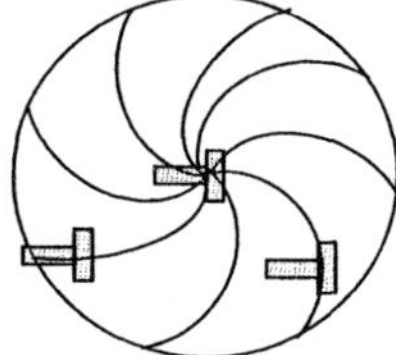

Figure 2. Sampling position of Raman measurement chips. Center, R/2 and edge. Sample edge and parallel or perpendicular to grinding mark.

or #1200 grinding. In case of #360 grinding finish, wafers were continuously ground down to 250μm by #360 grinding. Grit sizes were 40-60μm, 8-16μm, 4-6μm for #360, #1200 and #2000 grinding respectively. Sample chips were made by cleaving along or perpendicular to the wafer grinding marks at three positions that were center, R/2 and near edge of the wafer, as shown in Fig. 2. Raman was measured on (110) cleaved surface. Cracks were observed by SEM after etching the chips for 30sec by diluted Secco etchant.

2.2 Raman measurements

Raman shift due to the stress in a crystal depends upon optical configuration, such as angle and polarization of incident light and scattered light. It also depends upon stress tenser. This complicated situation makes the quantitative analysis of Raman data difficult. However, once the conversion coefficient is established, Raman shift is always a good tool of the stress analysis. Incident light was perpendicular to the (011) cleaved surface of (100) wafer. Usually Raman shift is measured on (100) surface. Kanbayashi et al[7] give a conversion coefficient $2.49 \times 10^2 \times \nu(cm^{-1})$ MPa to (011) surface for the optical configuration similar to our experiment. The light was not intentionally polarized but, practically, it was somewhat polarized because it came through mirrors. This conversion coefficient is same to that of (100) surface.

To measure stress distribution around the grinding damage and to detect how deep it extends, cleaved (011) surface was probed using the high-resolution ($0.2cm^{-1}$) triple grating JASCO NR-1800 micro-Raman spectrometer. Incident light was unpolarized 514.5 nm Ar laser beam perpendicular to the (011) surface. Raman scattering light was detected, without polarizer, by liquid nitrogen cooled CCD. Beam diameter was 1 μm and laser power at sample surface was 1mW to avoid Raman shift due to the sample heating. Temperature coefficient silicon $520cm^{-1}$ peak wavelength[17,18] is $0.025cm^{-1}deg^{-1}$, that is 6.2 MPa deg^{-1}. The peak wave number could be read to 0.01 cm^{-1} by a curve fitting. Standard deviation of repeated measurements shown in the Figure 3 to 5. Room temperature was controlled 1 to 2°C This may change Raman peak position, sample by sample. Sample chamber temperature was not intentionally controlled but it would be constant during one sample measurement.

3. Surface and subsurface stress measured by Raman spectroscopy

3.1 Depth profile of grinding stress

Raman data of #2000, #1200 and #360 ground wafers are shown in Figure 3, 4 and 5. Raman peak wave number of all samples are plotted from surface to 100 μm depth in the same scale factor in the figures. There are some distinct features in these depth profiles. One is a strong compressive stress, that is a higher wave number shift of the Raman peak, is observed at the ground surface (depth = 0) of all samples. Coarse grit size, smaller mesh number grinding, induces more Raman shift. That means stronger compressive stress like the transition zone in Fig.1. Especially, #360 ground samples show very strong surface stress.

If there is additional stress region beneath the surface, that subsurface stress would be observed as an extension of the surface stress region or some swell in the depth profile of the stresses. Figure 2(c) shows a swell clearly. This will be an indication of subsurface stress region. Closely looking at Fig.3, 4 and 5, some data show the swells.

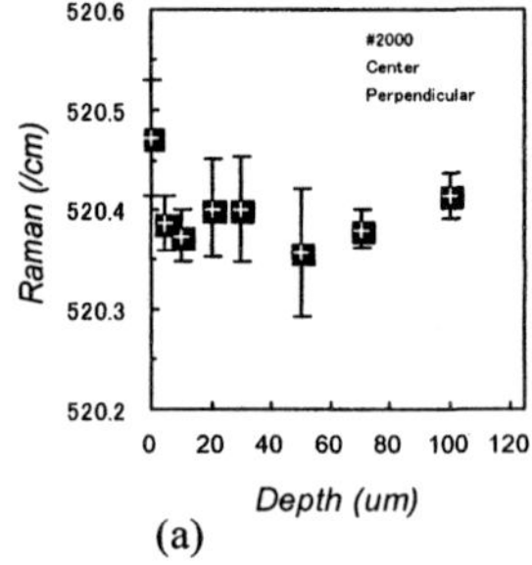

(a)

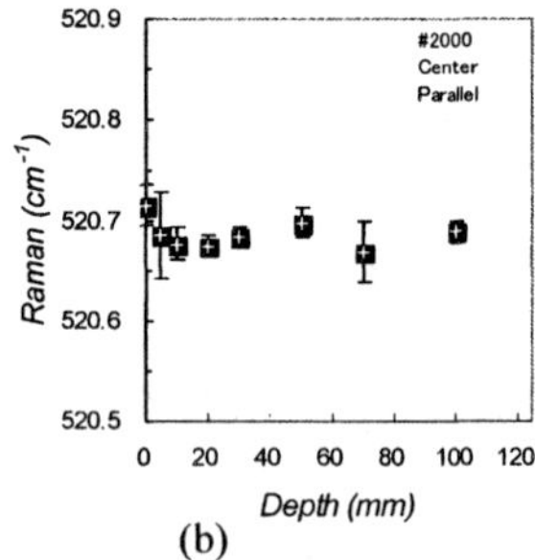

(b)

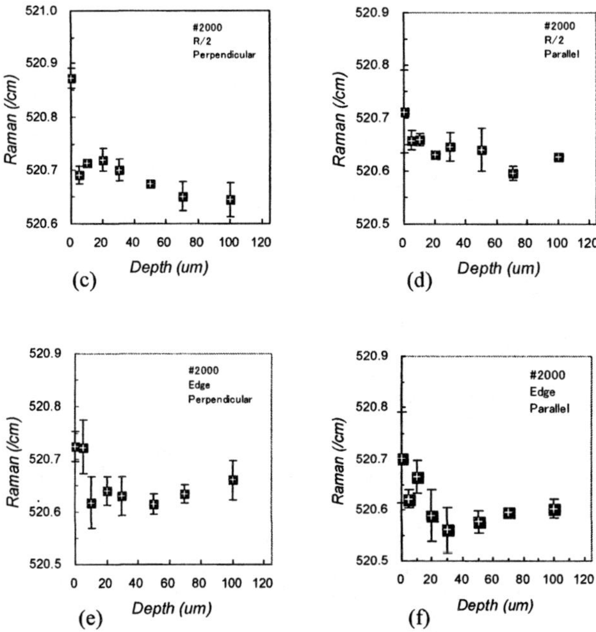

Figure 3. Depth profile of Raman shift of #2000 ground wafer. Sampling positions are shown in Fig.2. Perpendicular means measured (011) surface is perpendicular to the grinding tracks. Parallel means measured (011) surface is parallel to the grinding tracks

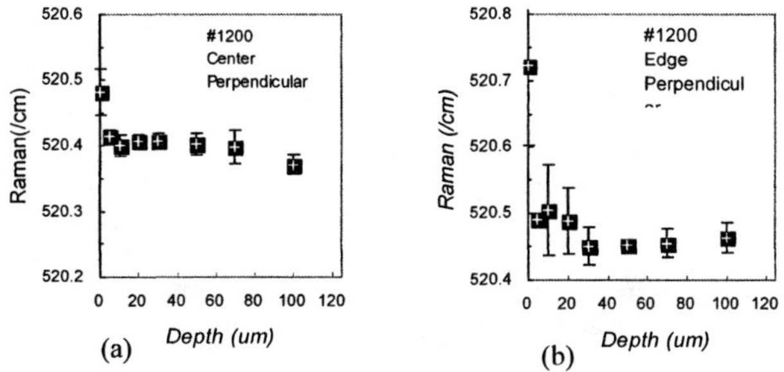

Figure 4 Depth profile of Raman shift of #1200 ground wafer. Sampling positions are shown in Fig.1

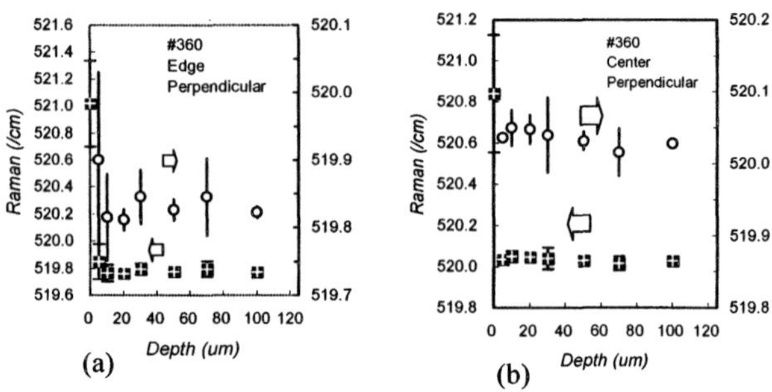

Figure 5 Depth profile of Raman shift of #360 ground wafer. Sampling positions are shown in Fig.1

These faint swells are with in the standard deviation of measurement fluctuation so that it will not be recognized without subsurface stress assumption. Of cause, some data do not show such an additional stress zone. Although the swell is not so large but it is clear that some samples show the swell of which peak wave number is at about 10 to 20μm depth.

3.2 Surface stress

As discussed below, median plane is close to the stress-free plane if the stressed layer thickness is thin compared to wafer thickness. Back grinding stress is expected to be localized in near-surface layer. So that, stress in

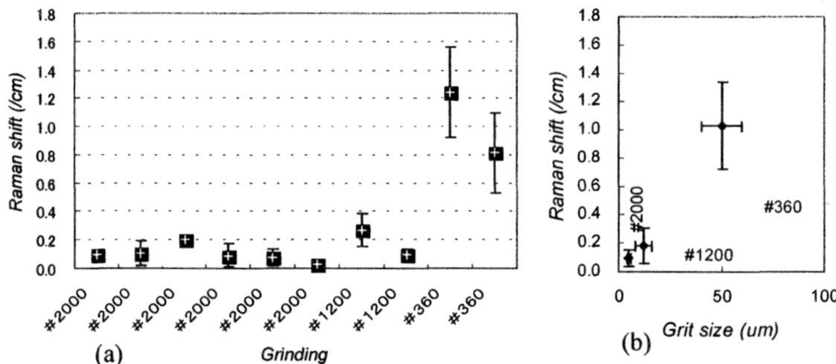

Figure 6. Raman shift due to surface stress. Fig. 6(a) is a measured stress of the samples. Fig. 6(b) is an average stress vs grit size.

the most of the wafers will be linearly changing due to warp. Raman peak wave number around median plane will be equal to that of stress-free silicon. Weighted mean of data at 50μm, 70μm, 100μm depth are used as Raman peak wave number of the stress-free silicon for each samples. Raman peak wave number of surface (depth = 0 μm) is always higher than that of the stress-free silicon. Raman shift due to the surface stress is the difference of these two peak wave numbers. Figure 6(a) shows the Raman shift depending on the grinding. Raman shift increases as the grit size increases, that is, from #2000 to #360 grinding. Average Raman shift is plotted against the grit size as shown in Figure 6(b). The Raman shift seems to depend linearly on the grit size. Raman shifts shown in Fig 6(b) are converted to the stress using the conversion coefficient discussed above. They are 24±14, 45±31, 256±76 MPa for #2000, #1200 and #360 grinding. Wu et al detected Raman shift, 23.0 MPa compressive stress for the ductile mode cutting surface from machined silicon[19] and 172.5 MPa stress for the brittle mode cutting surface. These indicate that the damage of #2000 ground surface of wafers investigated here is similar to the ductile mode cutting. Damage of #360 ground surface corresponds to the brittle mode cutting.

3.3 Cracks

Grinding induces two types of stresses that induces defects. One is thermal stress due to high temperature at the abrasive tips. Temperature and stress are high enough to cause plastic deformation. The other stress is mechanical stress. This causes cracks around grinding tracks. Grinding of brittle materials such as ceramics, introduces both of cracks and plastic deformation. While plastic deformation region is limited to near surface, crack, especially, median crack penetrates deeper than plastic deformation defects. Fig.7 shows a typical crack system induced by the grinding. SEM observation of etched samples reveals typical median cracks as shown in Fig.8 and variety of cracks, as shown in Fig.9. Some of them are same as reported earlier by Pei et al[20]. An interesting one, Fig. 7(c), is a crack of which roof of lateral crack was detached during Secco etching. Floor looks an accumulation of radial cracks. Depth of cracks is 0.5 μm in #2000 ground sample and 5 μm in #360 ground sample. Although observations are limited, these depths well correspond to the surface stress detected by Raman. Grinding using #360 grit introduces about 10 times deeper cracks. It is interesting that surface stress is also about 10 times higher in #360 ground sample, that is 24±14 vs 256±76 MPa.

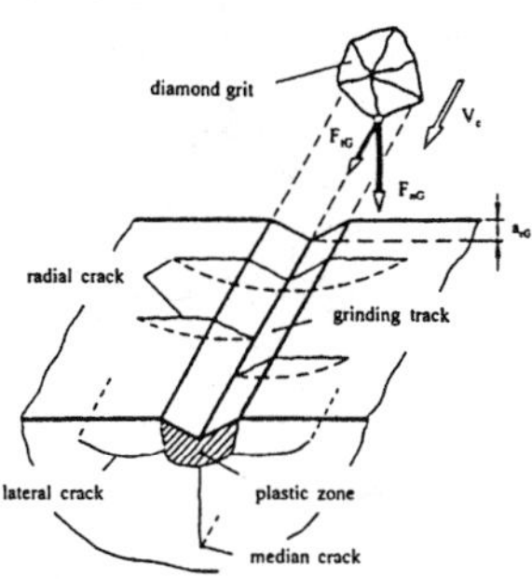

Figure 7. Typical crack system of brittle material grinding[3].

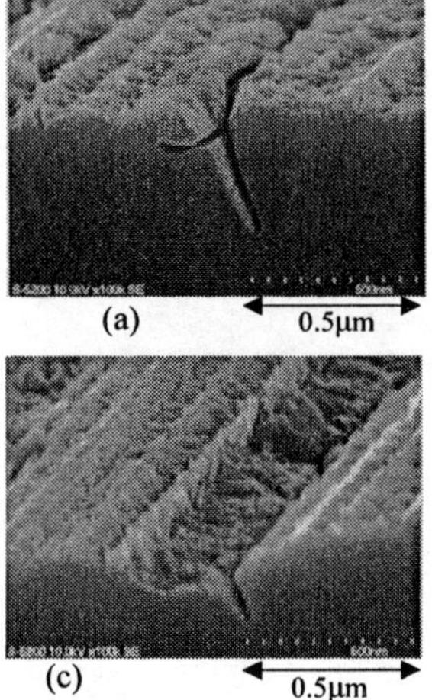

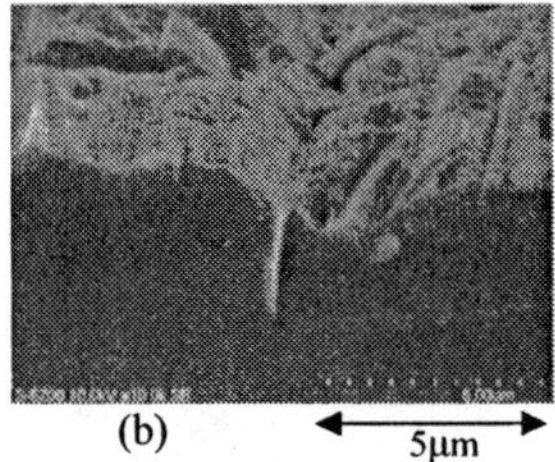

Figure 8 Median cracks after Secco etching of #2000 ground (a) and #360 ground (b) samples. Fig. 7(c): Roof of the median crack was detached during Secco etching

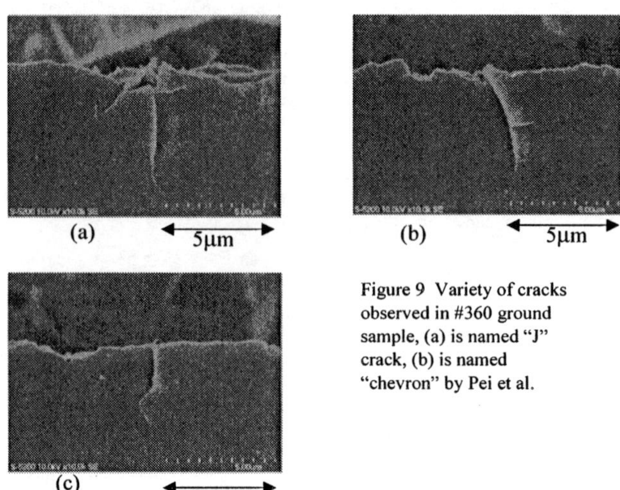

Figure 9 Variety of cracks observed in #360 ground sample, (a) is named "J" crack, (b) is named "chevron" by Pei et al.

Takano et al[21] studied damage in #800 (grit size = 20 to 40 μm) ground CZ (111) wafers by step etching and x-ray topography. Cracks were observed down to 9 μm deep from the ground surface. However, Pei et al[20] reported much deeper cracks, as shown in Fig.10, up to 20μm deep. These deep cracks will induce a stress in that deep subsurface region. Although cracks are a common surface damage of machined brittle materials, more SEM observation is required to find that deep cracks in our samples

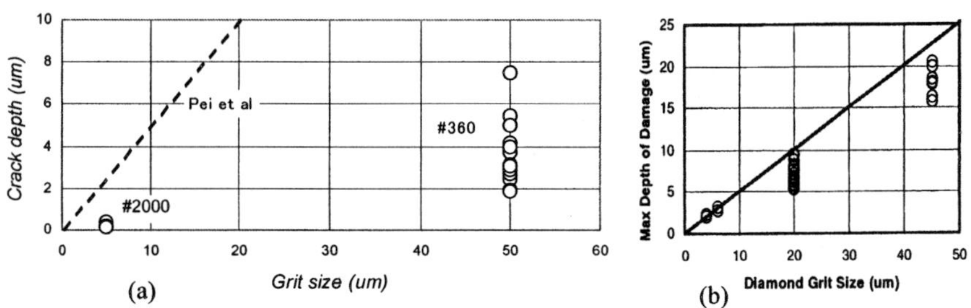

Figure 10. Crack depth in #360 and #2000 ground samples observed by SEM (a) and reported by Pei[20] et al (b).

4. Discussions and stress model

Machining like grinding induces the damage just beneath the surface and subsurface damage. In general, The damage consists of three layers,

First layer: Skin surface layer which is not a single crystal silicon, like poly crystalline silicon, amorphous silicon or silicon other than diamond structure.

Second layer: Plastics deformation zone where are dislocations, twins and micro-cracks.

Third layer: Cracks such as median crack, lateral crack, radial crack and deep crack as shown in Fig.1. There would be an additional region where elastic deformation beneath the surface damage region extends. Many damage models like Fig.1 have been proposed but data are still needed for understand the back grinding process. It is still ambiguous what causes the deep stress.

As shown in Fig. 3 to 5, the stresses are detected by micro-Raman spectroscopy. One is the distinct surface damage stress and the other is the subtle deep crack stress. The surface stress is compressive of which Raman shift is higher than 50 to 100 μm deep region almost free of stress. The other is the some swell in the stress depth profile of some samples. The swell is detected beyond the surface stress region in some samples when the extra stress zone is intentionally looked for in the depth profile. The swell is not so large compared to standard deviation of measurement fluctuation but it is still clear that some sample has the additional stress at 10 to 20μm deep. These two types of the stress regions are shown in Fig. 11.

The surface stress region in Fig. 3 to 5 diminishes quickly in about 5 μm. Depths of the cracks observed by SEM after the etching are almost equal to the depths of the surface stress observed by Raman. These are noted as *Surface Damage* in Fig.11. Various cracks in the surface damage are typical cracks of the ground brittle materials. The compressive stress is linearly depending on the grit size. Crack depths depend on the grit size. These data can be analyzed within a framework of the brittle materials machining theory or models. There may be plastic deformation at the bottom of the grinding tracks as shown in Fig. 7 or poly crystalline region as shown in Fig.1. None of them is observed. It may be due to the detection limit of the micro-Raman spectrometer.

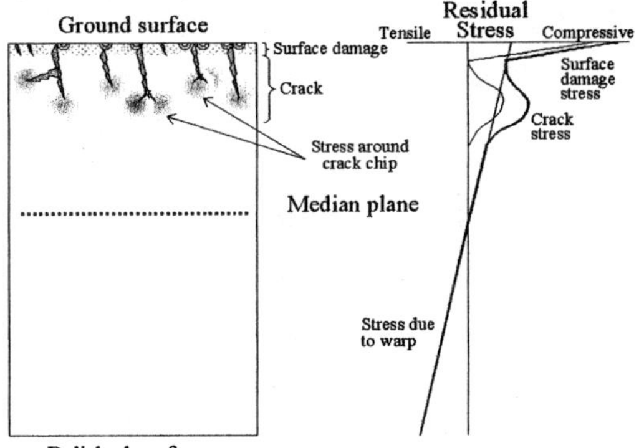

Figure. 11 Grinding damage and stress model. Compressive stress near the surface caused by caused by surface damage such as cracks. Deep cracks will induce stresses around its chips.

On the other hand, stress in the transition zone shown in Fig.1 is not so clearly shown by the Raman data. When the swell appears in the depth profile, it indicates that some extra stress is located in that depth. The swell peak is about 10 to 20 μm deep. This will be the transition zone in Fig.1 where some deep cracks extend. Deep cracks of about 20 μm depth are reported by Pei et al[20] and Tonshoff et al[22]. These cracks would correspond to the transition zone stress of Fig.1. They are notes as *Crack stress* in Fig.11. Stress around a crack is localized at crack chips so that the stress due to the deep cracks will be localized around some depth to where the deep cracks extend. Density of the deep crack will not be so high that it will not always be observed by micro-Raman spectroscopy. Its stress will not be so large because of the low density. So that, their Raman shift is close to the detection limit. Statistical distribution of the crack length or stress around crack tip will induce a complicated stress profiles. However, this depth of stress zone is also reported by others. Lundt et al[16] detect subsurface damage down to about 15 μm by SIRD (Scanning Infrared Depolarization) method. Magic mirror and micro PCD lifetime measurement show a damage model that elastic strained layer is laterally localized but extends more than 10 μm deep[23]. These are consistent with the deep stress observed as a swell, in Fig. 3 to 5, by micro-Raman depth profiling of the grinding damage. The two types of stresses observed are schematically shown in Fig.11. It looks similar to the model in Fig.1. One difference is shallow surface layer of poly crystalline silicon, shown in Fig.1, is not detected in this Raman. Other difference is the depth of crack zone. It extends 10 to 15 μm in Fig.1 but cracks in Fig.10(a) extend about 7.5 μm. Transition zone consisting of 10 to 15 μm deep cracks in Fig.1 will correspond to the swell shape stress profile observed in Fig. 3 to 5. For the quantitative discussions, the observed deep stresses are too subtle and obviously more data are needed.

Stress profile in bi-layer plate is linearly changing from one surface to the opposite surface. This linear stress profile is caused by warp, which balances stress in the surface layer to the stress in the substrate. Situation is the same in case of thermal stress, in a plate, caused by arbitrary near surface temperature distribution. Expansion or shrink is localized to the temperature distribution. Stress in the rest of the plate is caused by warp to balance the stress in the whole plate. Thermal stress analysis is more similar to the residual stress analysis because both analyze local expansion/shrink of material caused by temperature or damage profile. Thermal stress in thick plate with arbitrary temperature profile[24] is

$$\sigma_x(z) = \frac{\alpha E}{1-\nu}\left\{-T(z) + \frac{1}{2t}\int_{-t}^{t} T(z)dz + \frac{3z}{2t^3}\int_{-t}^{t} T(z)zdz\right\}$$
$$= \frac{E}{1-\nu}\left\{-\alpha T(z) + \varepsilon_0 + \frac{z}{\rho}\right\}$$

where σ_x is thermal stress parallel to the surface, α is thermal expansion coefficient, E is Young's modulus, ν is Poison's ratio, z is depth, T(z) is temperature distribution, 2t is thickness of plate, ε_0 is an average strain and ρ is radius of curvature. This means that thermal stress consist of three stress components. First is a constant that is related to the average strain, second is linearly changing stress due to curvature and third is local stress induced by the temperature distribution. Similar argument can be applied to the stress profile caused by the grinding. Besides a constant stress, local stress component and linearly changing warp stress are shown in Fig.11. The local stress component consists of two parts in the stress profile, surface damage stress and crack stress. Linearly changing stress profile is due to warp. Warp is a result of the stress balancing. All of these stress components are additive as shown in Fig.11.

5. Summary

Two residual stresses of the ground Si (100) surface are detected by micro-Raman spectroscopy. One is the residual stress from surface down to a few μm deep depends on grit size of grinding. It is 24±14, 45±31, 256±76 MPa for #2000, #1200 and #360 grinding. These indicate that the stress of #2000 grounding is similar to the ductile mode cutting. Stress of #360 grinding is similar to the brittle mode cutting. Subtle residual stress is observed in some samples at some depth around 10 to 20 μm. Various cracks, typical brittle material grinding crack, are observed by SEM after Secco etching. These cracks extend several μm deep from the ground surface and cause the surface stress. On the other hand, deep subsurface stress may be caused by the deep cracks down to 20μm. That deep cracks are reported elsewhere[20] but are not observed by SEM yet so that more SEM observation needs to be done to confirm this model.

Acknowledgement
Authors thank to Ms. S. Itoh for her helpful discussions.

References

[1] ASET 2nd annual meeting, 2001, Tokyo
[2] H. K. Tonshoff, W. v. Scmieden, I. Inasaki, W. Konig, G. Spur: Annal. CIRP 39, 621, 1990
[3] Kun Li, T. Warren Liao: J. Mater. Process. Technol. 57, 207, 1996
[4] D. A. Lucca, E. Brinksmeir and G. Goch: Annal. CIRP 47, 669, 1998
[5] U. Bismayer, E. Brinksmeier, B. Gutter, H. Seibt, and C. Menz: Precisin Eng. 16, 139, 1994
[6] H. F. Hadamovsky et al: Werstoffe der I-albeitertechnik., Dt. Verl. fur Grunsdtoffindudtie, 2 Aufl. 1990 cited in PhD Theses Danierla Trenklerb
[7] S. Kanbayashi, T. Hamasaki, T. Nakakubo, and M. Watanabe, H. Tango, 18th Conf. Solid State Device and Materials, 1986, p415, Publ. by Japan Soc. of Applied Physics (1986)
[8] S. Nadahara, S. Kanbayashi, M. Watanabe, and T. Nakakubo, 19th Conf. Solid State Device and Materials, 1987, p327, Publ. by Japan Soc. of Applied Physics (1989)
[9] H. Kobayashi, K. Mizuhara, N. Morita and Y. Yoshida: J. Japan. Soc. Abrasive Technol. 38, 37, 1994
[10] R. G. Sparks and M. A. Paesler: Precision Eng. 10, 191, 1988
[11] Y. Gogotsi, C. Baek, and F. Kirscht: Semicon. Sci. Technol. 14, 936, 1999

[12] Z. Zhong and V. C. Venkatesh: J. Mater. Process. Technol. 44, 179, 1994
[13] H. S. Oh, J.R. Kim, T. H. Kim, J. J. Yu, H. L. Lee, J.H. Lee, D.Rice: ECS Proc. 99-1, 100, 1999
[14] C. Y. Choi, J. H. Lee, and S. H. Cho: J. Appl. Phys. 84, 168, 1998
[15] C. Y. Choi, J. H. Lee, S. H. Cho: Solid State Electronics 43, 2011, 1999
[16] H. Lundt, M. Kerstan, A. Huber, and P. O. Hahn: Semiconductor Silicon 1994, ECS Proc. 94-10, p218, 1994
[17] M. A. Lourenco, D. J. Gardiner, V. Gouvernayre, M. Bowden, J. Hedley, D. Wood: J. Mater. Sci. Lett. 19, 771, 2000
[18] K. Mizuhara, S. Takahashi, J. Kurokawa, N. Morita, Y. Yoshida: Mar. Res. Soc. Proc. 505, 501, 1998
[19] T-C. Wu, N. Morita and Y. Yoshida, Trans. Japan Soc. Mech. Eng. **59**, 283, (1993) in Japanese
[20] Z. J. Pei, S. R. Billingsley, S. Miura: Int. J. Mach. Tools Manuf. 39, 1103, 1999
[21] Y. Takano, M. Ogirima, M. Maki: Japan. Sci. Prom. #145 Committee 17th meeting document p25, 1982 July. in Japanese
[22] H. K. Tonshoff, B. Karpuschewski, M. Hartmann, C. Spenbler: Machining Sci. Technol. 1, 33, 1997 cited in ref.5.
[23] H. S. Oh, J. R. Kim, T. H. Kim, J. J. Yu, H. L. Lee, J. H. Lee and D. Rice, "Defects in Silicon III" Electrochem. Soc. Proc. **99-1**, 100, (1999)
[24] Y. Takeuchi: "Analysis of Thermal Stresses" p13, Nisshin Publishing, in Japanese

DETECTION OF METAL CONTAMINATION IN INTERNALLY GETTERED WAFERS

M. L. Polignano, P. Bacciaglia, D. Caputo, C. Clementi, B. Padovani
ST Microelectronics, Via Olivetti 2, 20041 Agrate Brianza (MI), Italy

F. Priolo, INFM and Department of Physics,
University of Catania, Corso Italia 57, 95129 Catania, Italy

T. Simpson, Shin-Etsu Handotai Europe,
Livingston, West Lothian, EH54 7DA, United Kingdom

Abstract

In this paper we address the problem of detecting metal contamination in the active regions of internally gettered wafers. Iron and nickel were considered as the examples of a contaminant in the solid solution and of a surface-precipitated contaminant, respectively.
SPV measurements were found unsuitable to estimate the iron contamination of the denuded zone, as the region probed by SPV measurements is usually deeper than the denuded zone. Vice versa both DLTS and generation lifetime measurements are limited to the region of interest, i.e. the denuded zone. However generation lifetime provides some sensitivity to interstitial iron only, so dissociation of the FeB pairs is required to detect iron by this technique.
Generation lifetime was found to be very sensitive to nickel contamination in MCZ material. This technique is also suitable for internally gettered wafers, however in internally gettered CZ and epitaxial wafers no generation lifetime degradation was observed, indicating efficient nickel gettering.

INTRODUCTION

The detrimental effect of transition metal contamination over device performances is known since a very long time [1], and more recently it was shown [2] that even a very low level contamination ($\approx 10^{11}\text{cm}^{-3}$) can be harmful for present generation devices. For this reason, several techniques have been developed for the control of metal contamination in device processing, and among these techniques those relying on measurements of carrier lifetime are most often used for monitoring contamination. Examples of this sort are the Surface PhotoVoltage (SPV) method [3, 4], measuring the variation of surface potential upon illumination, the detection of PhotoConductivity Decay by microwave reflection (μ-PCD) [5, 6], measurements of the photocurrent collected by a silicon-electrolyte contact (Elymat) [7]. Some of these techniques were also extended to detect surface-segregated metals [8].
However, metal contamination can usually be reduced but not completely suppressed, and for this reason in device processing residual metal impurities are removed from the active areas of devices and confined in suitably created regions of the wafers,

where they are electrically inactive. This result can be obtained by specific process steps, which are referred to as metal gettering processes, and require the creation of getter sites, suitable for the capture of metal impurities and the segregation of dissolved metal impurities into getter sites. Gettering processes can be roughly distinguished into intrinsic gettering processes, using oxygen precipitates and related bulk defects as the gettering sites [9], and extrinsic ones, which require the formation of specific heavily doped or damaged gettering region far enough from device regions. Gettering techniques are also classified according to the gettering mechanism. Gettering by segregation exploits the increased solid solubility of metals in gettering region [10, 11, 12]. Gettering by heterogeneous precipitation uses extended defects (whether in the bulk or at wafer backside) as the nuclei for metal precipitation, thus preventing metal precipitation at wafer frontside.

In the presence of intrinsic gettering, recombination lifetime techniques are generally unsuitable for the detection of metal contamination in the active regions, as the denuded zones are typically thinner than the regions probed by recombination lifetime techniques. Indeed, the region probed by these techniques is usually very thick, from a few hundred microns to the whole wafer thickness, depending on the specific technique and on the sample under inspection [13]. A specific application of the SPV technique was developed to determine the thickness of the denuded zone [14], however this method provides no information about the residual contamination in the denuded zone.

In this paper we address the problem of detecting both contaminants in the solid solution and surface-precipitated contaminants in internally-gettered wafers. Iron and nickel were chosen as examples of a contaminant in the solid solution and of a surface-precipitated contaminant, respectively.

EXPERIMENTAL DETAILS

Experimental techniques

Various method were tested for their ability to detect metal contamination in internally gettered wafers. Recombination lifetime was measured by SPV, and iron in the solid solution in the bulk was obtained by dissociation of iron-boron pairs. Deep Level Transient Spectroscopy (DLTS) was used for measuring the residual iron concentration in the denuded zone. Interface state density was measured on non-patterned wafers by a non-contact system using corona charge deposition, and by C–V measurements of Al:Si gate capacitors. Generation lifetime was extracted from capacitance-time (C–t) measurements obtained by a Hg-probe.

Surface Photovoltage. Surface photovoltage (SPV) measurements are carried out by illuminating the sample with light of various wavelengths. Generated minority carriers are collected in a depletion region at the wafer surface and produce a variation in surface potential, which is recorded as a function of light wavelength. In the standard SPV technique [3], diffusion length is extracted from these data by assuming that sample thickness is much larger than diffusion length. This hypothesis often fails in present wafers, and the actual wafer thickness must be taken into account. An "enhanced SPV" [15] technique is available for this purpose. The enhanced SPV equation is obtained by imposing the correct boundary condition at the wafer backside, i.e. by taking into account backside surface recombination. Both standard and enhanced SPV are only sensitive to bulk recombination, so SPV cannot detect surface-segregated contaminants. SPV measurements are always obtained under very low injection conditions. In this work the SPV tool in the FAaST SDI system was used. This instrument is equipped with a set of filters producing monocromatic light at seven different wavelengths, ranging from 800nm to 1μm.

SPV measurements allow iron to be identified in p-type silicon by splitting of the iron-boron (FeB) pair [16]. Iron-boron pairs can be dissociated into interstitial iron Fe_i and substitutional boron either by baking [17, 18] or by illumination with a light of a suitable intensity [16]. Under low injection conditions, interstitial iron is more effective as a recombination center than the FeB pair, so that lifetime decreases upon dissociation [18]. Iron concentration is calculated from diffusion length measurements before and after dissociation. It is important to remark that iron under the form of FeB pairs is detected only, so precipitated iron cannot be revealed by SPV measurements.

In SPV measurements the probed region is of the order of light penetration depth plus carrier diffusion length [13], so it is typically larger than the denuded zone depth. Threfore, in the presence of a denuded zone SPV measurements are affected by iron concentration both in the denuded zone and in the defective bulk region. SPV measurements of iron concentration can be easily interpreted only if the probed region is uniform for what concerns both bulk defects and iron distribution. For this reason, iron concentration in the defective bulk region was measured after removing the denuded zone at wafer backside.

Deep Level Transient Spectrocopy. DLTS measurements were carried out to measure iron concentration in the near-surface region. 0.785 mm^2 area Schottky diodes were used for these measurements.

A DLS-83D instrument by Semilab was used. In this instrument lock-in integration is used for averaging capacitance transients, and temperature can be scanned from abot 30 K to 300 K. Alternatively, constant temperature spectra can be obtained as a function of the frequency of excitation pulses [19] in the range 0.5 Hz–2 KHz. Both methods were used in this work. The differential DLTS method [20, 21] was used in order to reduce the edge effect. In this method two filling pulses are applied, a first pulse in the beginning of the integrating period and a second pulse a half period later. In the lock-in integration, the integral of the capacitance transient caused by the first pulse is taken with positive sign, and the the integral of the capacitance transient caused by the second pulse is taken with negative sign. The edge effect contribution is approximately independent of voltage, so this contribution is cancelled out by this procedure, resulting in a more precise concentration measurement.

In our measurements, samples were reverse biased at -5 V, and filling pulses with 4.5 V and 0.5 V amplitudes were applied during each integrating period. Pulse width was 20μs, however the signal was verified to be independent of this parameter in the range 5–100μs.

As most iron is under the form of iron-boron pairs in p-type silicon, iron concentration was measured from the intensity of the FeB related peak, which shows up at temperatures of 50–60 K (depending on frequency) and corresponds to an energy level located at about 0.1 eV above valence band. Like in the case of SPV measurements, precipitated iron is not revealed by these measurements.

Vice versa DLTS is very different from SPV from the point of view of the extension of the probed region. Indeeed DLTS measurements explore the space charge region of the junction only, which is smaller than the denuded zone, so DLTS data are a measure of iron concentration in the denuded zone.

Generation lifetime measurements. Generation lifetime data were extracted from measurements of capacitance vs. time (C–t) according to the pulsed capacitor method [22]. Measurements were obtained by a Solid-State Measurements 5100 HG Probe system equipped with a mercury probe that acts as a gate electrode. In the pulsed capacitor method a deep depletion region (i.e. a non-steady state depletion region) is created by applying a suitable voltage pulse to the gate electrode and the return of the system to steady state is monitored by measuring the system capacitance as a function of time. In order to control the extension of the lateral depletion region the silicon surface was initiallly put into accumulation condition by a suitable corona charge deposition. Generation lifetime was extracted from C–t data by the well-known Zerbst method [23, 24].

The sensitivity of generation lefetime measurements depends on the position in the energy gap of the deep level responsible for carrier generation [22]. Indeed generation lifetime is very sensitive to levels located close to midgap, but this sensitivity exponentially decreases with $\mid E_i - E_t \mid / k_B T$, where E_t is the energy level responsible for generation, E_i is the intrinsic Fermi level, k_B is the Boltzmann constant and T is measurement temperature. The iron-boron pair has two levels, located at 0.1 eV above valence band and 0.29 eV below conduction band. Both these levels are far from midgap, so the FeB pair is expected to have little efficiency for generation. Interstitial iron has a level located 0.385 eV from valence band, so closer to midgap than the FeB levels. Therefore interstitial iron is expected to be more effective for generation than the FeB pair, and for this reason generation lifetime measurements were repeated after optical dissociation of the FeB pair. The optical dissociation tool in the SPV instrument was used to this purpose.

Non-contact measurement of interface state density. Non-contact measurements of interface state density were obtained by a sequence of corona charge deposition at the oxide surface and measurements of the contact potential difference between a suitable non-contact metal electrode and the silicon substrate [25]. This procedure was automatically carried out by a FAaST (Film Analysis and Substrate Testing) tool from SDI. In this tool, the contact potential difference between the silicon wafer and the electrode is measured in the dark and under illumination with a light intensity high enough to make the surface band bending vanish under illumination. So, the surface band bending can be obtained as the difference between the measurements of the contact potential difference in the dark and under illumination. Measurements are repeated after subsequent depositions of a known charge amount, so the surface band bending V_s and hence the charge Q_{sc} in the space charge region at the silicon surface are obtained as a function of the charge deposited at the oxide surface. Because of charge neutrality, the sum of the variations of the space charge region charge ΔQ_{sc} and of the interface state charge ΔQ_{it} must equal the corona charge ΔQ_c deposited on the oxide surface at each step. The incremental interface state charge can, therefore, be obtained as $\Delta Q_{it} = \Delta Q_c - \Delta Q_{sc}$ and the distribution of interface states is obtained as a function of surface band bending as $D_{it} = \partial Q_{it} / \partial V_s$. The obtained distribution can be expressed in terms of the energy E in the energy gap by considering that $V_s = 0$ (flat band conditions) corresponds to $E = E_F$ (Fermi level of the silicon substrate).

Elymat. In this technique excess carriers are generated by a laser beam, and the diffusion length is extracted from photocurrent measurements. The contribution of surface recombination is suppressed by immersion in a HF solution [26] which also provides Schottky contacts for carrier collection. Carrier diffusion length data are obtained by injecting carriers at wafers frontside and measuring the photocurrent at wafer backside (backside photocurrent, BPC) or at wafer frontside (frontside photocurrent, FPC [27]).

Under our experimental conditions BPC measurements are possible provided $L_{\text{diff}} \geq 100\mu\text{m}$, otherwise FPC measurements must be used.

As previously stated, bulk lifetime is measured by passivating surface states and hence suppressing surface recombination velocity by immersion in a HF solution. In this case, the backside phocurrent I_{BPC} depends on bulk recombination only. However, the Elymat technique can be easily extended to the measurement of a surface-related parameter. Indeed, if surface recombination is not zero because of some by some interface property (e.g. the presence on an oxide-silicon interface), the collected photocurrent I^*_{BPC} depends both on bulk lifetime and on surface recombination velocity s. Subsequent measurements of $I^*_{\text{BPC}}(L_{\text{diff}}, s)$ and $I_{\text{BPC}}(L_{\text{diff}})$ allow both bulk lifetime and surface recombination velocity to be obtained [28]. In the case of an oxide-silicon interface, the oxide must be etched off and the surface passivated by HF for measuring I_{BPC}.

Surface-precipitated metals too increase carrier recombination at wafer surface, so surface recombination velocity can also be used for the detection of metal precipitates at wafer surface [29]. In that case, a short silicon etching is required in order to remove the precipitate-rich layer and so measure bulk lifetime. The surface recombination velocity related to metal precipitates is then evaluated from BPC measurements carried out before and after removing precipitates.

Sample preparation

p-type, 10Ωcm resistivity wafers with oxygen concentration in the range $6\text{–}7.5 \cdot 10^{17}\text{cm}^{-3}$ were used in this work. Iron-contamined samples were prepared by dipping in an contamined solution and high temperature (1100°C, 40 min) diffusion. The so-obtained iron concentration in the solid solution in silicon was measured by SPV and is reported in fig.1 as a function of iron concentration in the contaminating solution. The very good linear correlation down to low contamination levels ensures a well-controlled contamination process. A denuded zone was then created by a HI–LO treatment, consisting of a 1100°C 4 hour out-diffusion treatment followed by a 780°C 3 h 10 min nucleation treatment. The denuded zone created by this process was previously characterized [14] and was shown to be of the order of 30μm. The oxide grown in these thermal treatments was then stripped and the wafers were oxidized at 1000°C to a thickness of 1115 Å. This thermal treatment also enhances precipitate growth. A moderate temperature (800°C, 1 h) furnace annealing was used to enhance iron segregation into getter sites. The segregation temperature was chosen on the basis of previous experiments about gettering efficiency [10]. In addition, some samples received a final Rapid Thermal Treatment at 1100°C to check whether this treatment can redistribute previously gettered iron through the wafer. Finally wafers received a N_2:H_2 annealing to reduce interface states, In order to obtain the best sensitivity in our tests, Si-SiO_2 interface states should be as low as possible in the absence of metal contamination, otherwise such interface states would represent a background when studying the effect of the metal contaminant on interface states and generation lifetime. For this reason, the wafers were annealed in a forming gas environment (N_2:H_2).

The nickel experiment included internally gettered polished Czochralski (CZ) materials from two different suppliers, a p/p$^+$ epitaxial substrate and a Magnetic field CZ (MCZ) substrate, which is assumed to provide no gettering and is therefore a reference for evaluating gettering efficiency. For this reason, unnecessary thermal treatments were avoided in MCZ wafers to prevent any bulk defect growth. On the other hand, p/p$^+$ epitaxial material is often reported to provide optimum gettering, and this test was designed to check the gettering ability of this substrate even with moderate thermal treatments. Vice versa, in polished CZ wafers some thermal treatments were specifically included in order to ensure bulk defect formation and hence gettering ability.

Polished CZ wafers received the same HI-LO treatment as used in the iron experiment (1100°C, 4 h denuding treatment followed by a 780°C 3 h 10 min nucleation treatment). The oxide was then stripped and all the wafers were oxidized at 1000°C to a thickness of 1000 Å.

Nickel contamination was obtained by nickel implantation. 6 inch wafers were implanted in a 75 mm diameter area, in order to obtain a non-contaminated reference region on each wafer. Low doses (10^{11}cm^{-2} to 10^{12}cm^{-2}) were considered, in order to limit ourselves to realistic contamination levels in non-optimum process conditions. After oxidation the wafers were implanted with 1 MeV energy. After the implantation, the wafers were thermally treated in an inert environment at 780°C for 1 h 50 min to recover the implantation damage. A thermal treatment at 1000°C for 230 min was added in CZ polished wafers to enhance precipitate growth by partially simulating the thermal cycle of a real process. MCZ and epitaxial wafers did not receive this treatment. The wafers were subsequently treated at 800°C for 1 h 40 min to allow nickel segregation in getter sites or at wafer surfaces during wafer cool-down. Complete segregation is made easier by a slow cool down which is typical of furnace treatments. Finally, wafers received a forming gas annealing for interface state reduction. A few nickel-implanted wafers were also used to build capacitor structures by depositing, masking and etching an Al:Si layer.

EXPERIMENTAL RESULTS

Iron contamination

Iron concentration in the defective bulk region was measured by SPV at wafer backside after removing the denuded zone. Fig.2 reports these data vs. initial concentration, and shows that most iron in the solid solution in the bulk after processing, irrespective of the final thermal treatment (whether furnace or RTP). At high concentration levels only some iron inactivation (precipitation?) is observed in furnace treated wafers.

Fig. 3 compares SPV data of iron concentration measured at the wafer frontside with the denuded zone to Deep Level Transient Spectroscopy (DLTS) data. DLTS data refer to a space charge region close to the surface, so within the denuded zone. Vice versa the SPV probed region is partially located in the defective bulk region, which is expected to have a higher iron concentration than the denuded zone because of gettering, and for this reason SPV data are sistematically higher than DLTS data.

Figure 1: Iron concentration measured by SPV after drive-in vs. iron content in the solution

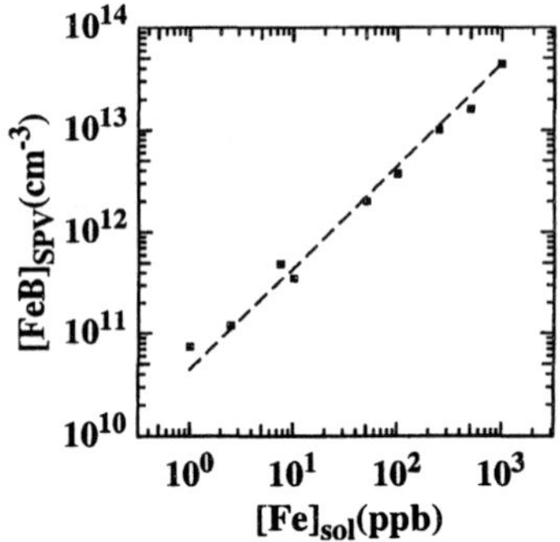

Figure 2: SPV iron concentration data in the bulk of internally gettered wafers vs. initial iron concentration

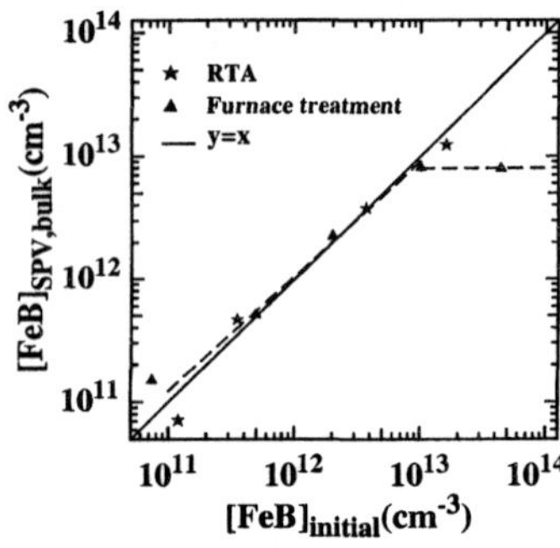

It is worth recalling that quantification of SPV iron concentration data was obtained by comparison with DLTS data [18] in wafers with a uniform iron distributon, so no difference between these techniques is expected in the absence of in-depth variations of iron concentration. Fig. 3 shows that SPV cannot be used to evaluate the gettering efficiency of bulk defects. However, DLTS is not very suitable for monitoring purposes, because it is time consuming and allows a very limited area only to be explored. So, data in fig. 3 could be used as a calibration curve to estimate the iron concentration in the active region from SPV measurements.

Gettering efficiency is estimated by comparing DLTS data of iron concentration to the initial iron concentration,as shown in fig. 4. These data show that gettering efficiency decreases with increasing iron concentration from about 90% at $10^{11}cm^{-3}$ iron contamination to 50 % at $10^{13}cm^{-3}$ iron contamination, irrespective of teh final thermal treatment (whether furnace or RTP treatment). From data in figs. 2 and 4 it is observed that even the iron gettered in the defective bulk region is under the form of iron-boron pairs (hence in the solid solution) up to a concentration of about $10^{13}cm^{-3}$.

Generation lifetime measurements were obtained both before and after optical dissociation of the FeB pair and are reported in fig. 5 vs. the iron concentration as obtained by DLTS. Before dissociation of the FeB pairs generation lifetime is independent of iron concentration. Vice versa after dissociation a generation lifetime decrease is observed starting from $10^{12}cm^{-3}$ iron concentration. So, as expected, interstitial iron is more effective than the FeB pair in carrier generation, however a $10^{12}cm^{-3}$ sensitivity is not enough for present needs.

Non-contact measurements of interface state density were also carried out, but no relation of this parameter to iron contamination could be observed.

Nickel contamination

Generation lifetime was found to be very sensitive to nickel contamination in MCZ material, and at relatively high nickel doses, the measured data are limited by the instrument velocity (fig. 6). A $10^{11}cm^{-2}$ nickel dose induces a generation lifetime drop of about two orders of magnitude with respect to the non-contamined region,

Figure 3: DLTS vs SPV iron concentration data in internally gettered wafers

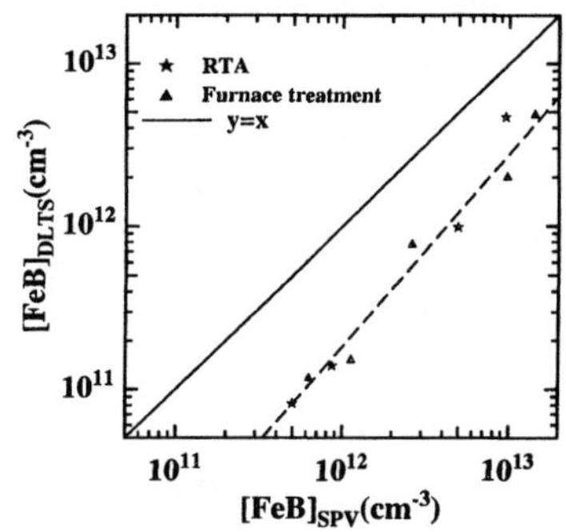

Figure 4: Iron concentration in the denuded zone (from DLTS measurements) vs. initial iron concentration.

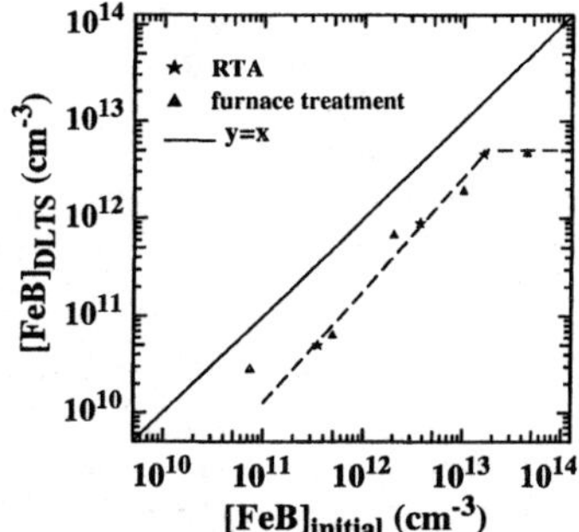

suggesting that a $10^{10}cm^{-2}$ nickel segregation could be easily detected. Vice versa in internally gettered CZ and epitaxial wafers no or very little generation lifetime degradation was observed, indicating efficient nickel gettering. In polished CZ material a weak generation lifetime reduction is observed at the highest nickel dose ($10^{12}cm^{-2}$), while no such effect is detected in epitaxial material, thus confirming that this material provides the most effective gettering [10]. It is worth noting that, though generation lifetime is in principle a bulk property, the effective datum measured by the pulsed capacitor method may well be affected by surface-segregated metals through the surface generation in the lateral depletion region.

Figure 5: Generation lifetime vs. iron concentration as obtained from DLTS measurements.

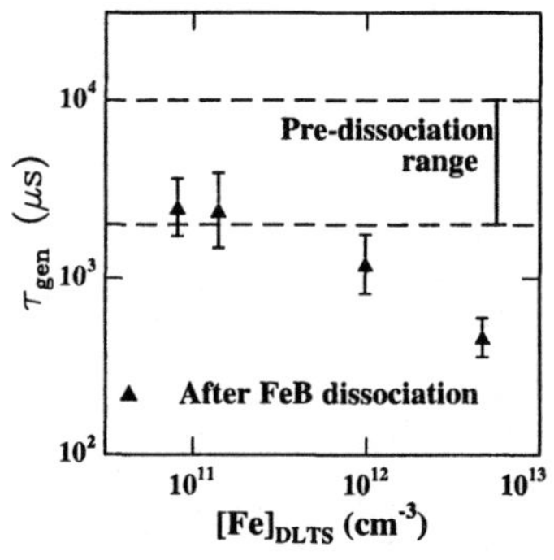

Figure 6: Generation lifetime in MCZ, CZ and epitaxial wafers vs. nickel dose.

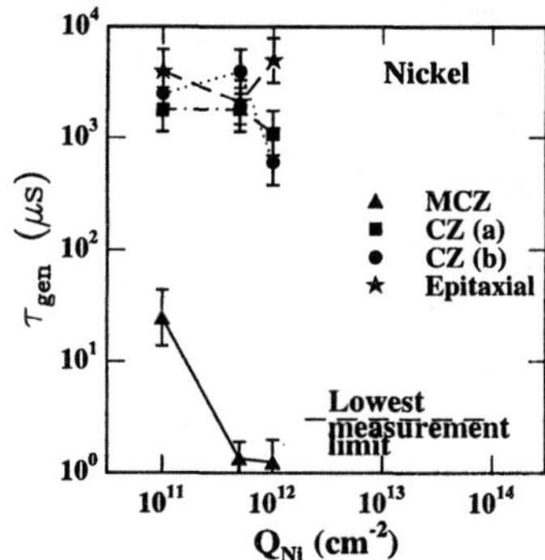

In MCZ wafers nickel completely segregates at wafer surface, so it cannot be detected by SPV measurements. Indeed, in these wafers the SPV diffusion length was larger than 1 mm irrespective of nickel dose (enhanced SPV measurements).

In internally gettered CZ material nickel decorates bulk defects and in some cases can be detected by SPV measurements in the bulk region as an increase in the recombination activity of such defects. This effect is better revealed by measuring the diffusion length in the precicitate-rich region after removing the denuded zone. Fig. 7 shows the so-obtained diffusion length map of a nickel-implanted wafer. It is immediately recognized that the central simmetric precipitation pattern and the 4 inch nickel-implanted region are superimposed in this map. However, bulk defect density can be significantly different from wafer to wafer, and so is the precipitation-related carrier diffusion length (from about 80 μm to 4μm in this test). Therefore, no quantification of metal contamination can be obtained by this sort of measurements. In addition, no information can be obtained by SPV about any residual metal in the denuded zone.

Nickel contamination is efficiently revealed in MCZ wafers by surface recombination velocity data in the Elymat (Electrolytic Metal Tracer) technique. Fig. 8 reports the surface recombination velocity related to nickel segregation as a function of dose. The results of the present test are compared to previous results obtained by a shorter thermal budget (800°, 30 min, [29]). It is shown that by increasing the thermal budget, the sensitivity at low doses is reduced, and the dependence of surface recombination velocity on dose is made stronger. Unfortunately the Elymat measurement of surface recombination velocity is based upon backside photocurrent measurements, and so it cannot be used when bulk lifetime is too low for measuring backside photocurrent, as it often happens in internally gettered wafers.

Figure 7: SPV diffusion length map in the precipitate-rich region of an internally gettered nickel-implanted CZ wafer.

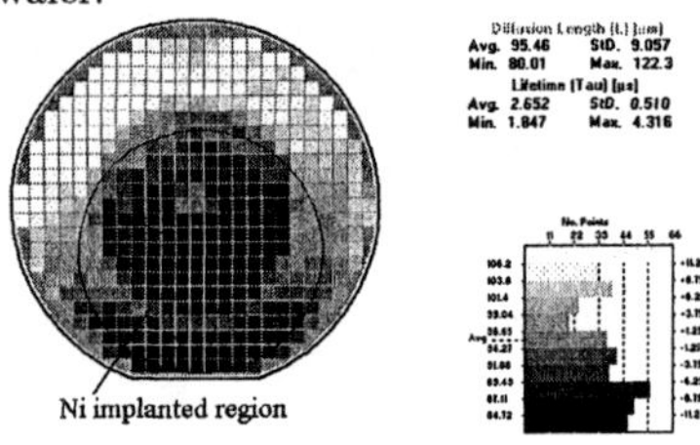

Figure 8: Surface recombination velocity vs. nickel dose in MCZ wafers for two different thermal treatments.

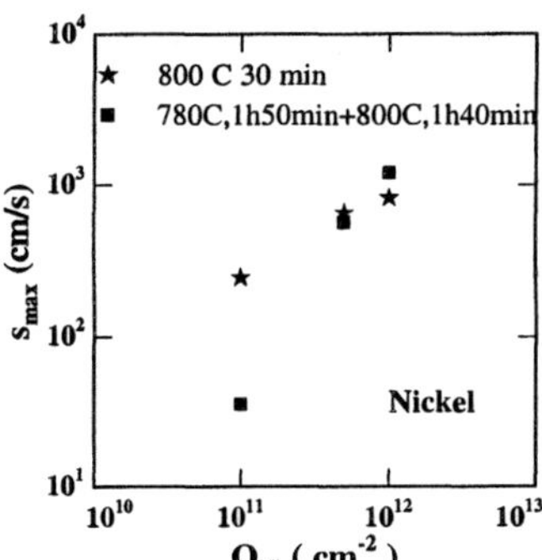

Fig.9 shows the interface state density spectrum as a function of energy in the nickel-contaminated and in the non-contaminated region of the $10^{12}cm^{-2}$ implanted sample. These data were obtained by the non-contact technique described above. Nickel contamination does not affect the midgap value of interface state density, but it does affect the spectrum in the upper part of the band-gap. It is worth noting however that the midgap value, or the minimum of the spectrum, are often used for process monitoring purposes. As the midgap datum seems not to be significant for what

concerns metal contamination, we report an averaged interface state density over the explored energy interval as a function of nickel dose (fig.10). In this plot, the increment in the interface state density with respect to the non-contamined region is reported, in order to account for fluctuation in the background value. Figs. 9 and 10 show that nickel contamination does affect interface state density, but this effect is not dramatic.
Measurements of capacitor structures were also carried out, but as previously discussed only the lower half of the energy gap could be explored. Therefore, no effect could be detected, in agreement with data in fig.9.

Figure 9: Interface state density spectrum in the nickel-contamined region and in the non-contamined region.

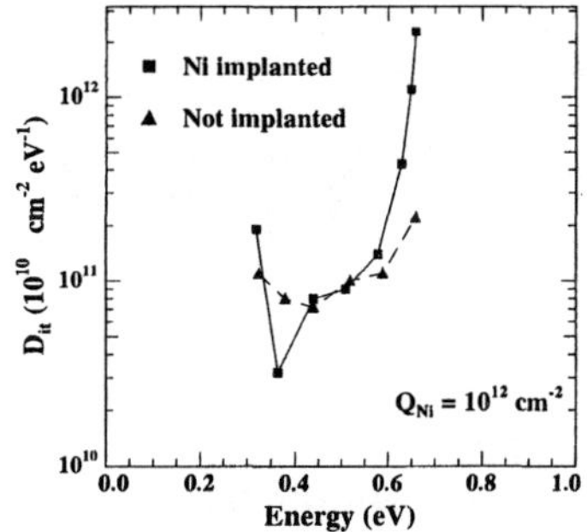

Figure 10: Average increment of the interface state density with respect to the non-contamined region vs. nickel dose.

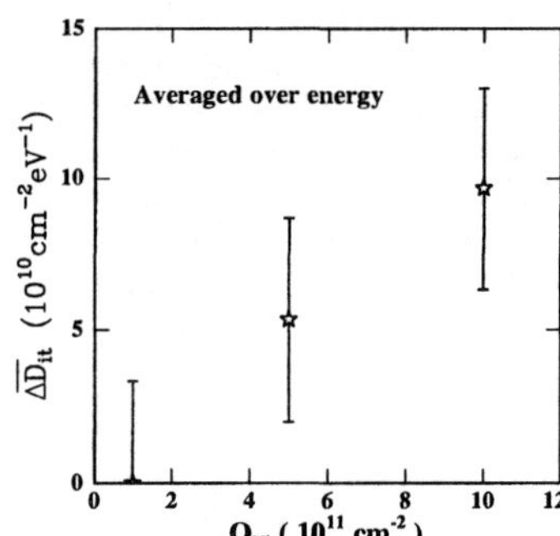

Data in figs. 9 and 10 show that the COCOS method provides a limited sensitivity to metal segregation. As a consequence, no contamination effect could be detected in internally gettered wafers.

CONCLUSIONS

The problem of detecting metal contamination in the active regions of internally gettered wafers was discussed. Iron and nickel were considered as the examples of a contaminant in the solid solution and of a surface-precipitated contaminant, respectively.
By comparing DLTS measurements of iron concentration at the end of the process to initial SPV data the gettering efficiency of bulk defects was estimated in the range 90% to 50 %, depending on the initial iron concentration. The gettering efficiency of bulk defects is independent of the final thermal treatment (whether furnace or RTP treatment).
SPV measurements in the defective bulk region showed that even the iron gettered in this region is under the form of FeB pairs, hence in the solid solution. This fact suggests that the mechanism of iron gettering in the defective bulk region is not precipitation at bulk defects, but rather a a segregation mechanism [10]. Indeed in this

mechanism contaminants are gettered because of an increased solid solubity of the impurity in the gettering region, so this mechanism allows gettered impurities to be in the solid solution.
SPV measurements were found unsuitable to estimate the iron contamination of the denuded zone, as the region probed by SPV measurements is usually deeper than the denuded zone. Vice versa both DLTS and generation lifetime measurements are limited to the region of interest, i.e. the denuded zone. However generation lifetime provides some sensitivity to interstitial iron only, so dissociation of the FeB pairs is required to detect iron by this technique. In addition, also in the presence of interstitial iron generation lifetime measurements are less sensitive than DLTS.
Vice versa, generation lifetime was found to be very sensitive to nickel contamination in MCZ material. This technique is also suitable for internally gettered wafers, however in internally gettered CZ and epitaxial wafers no generation lifetime degradation was observed, indicating efficient nickel gettering. In MCZ wafers nickel completely segregates at wafer surface, so it cannot be detected by SPV measurements, whereas it is efficiently revealed by surface recombination velocity data in the Elymat technique. Unfortunately this technique is not suitable for internally gettered and epitaxial wafers. Vice versa the COCOS interface state density measurement is suitable for internally gettered wafers, but it provides a very limited sensitivity to metal segregation. So, generation lifetime is the most suitable technique to detect surface-segregated metals in internally gettered wafers.

References

[1] A. Goetzberger and W. Shockley, J. Appl. Phys. **31**, 1821, (1960)

[2] W. B. Henley, L. Jabstrzebski, and N. F. Haddad, *Electrochem. Soc. Proc. Vol. 93*, Pennington, N.J. (1993), p. 476

[3] A. M. Goodman, J. Appl. Phys. **32** (1961) 2550

[4] L. Jastrzebski, Mat. Sci. and Engineering, **B4** 113 (1989)

[5] F. Shimura, T. Okui, T. Kusama, J. Appl. Phys. **67**, 7168 (1990)

[6] T. S. Horányi, T. Pavelka and P. Tüttö, Appl. Surf. Science **63**, 306 (1993)

[7] V. Lehmann and H. Föll, J. Electrochem. Soc. **135**, 2831 (1988)

[8] M. L. Polignano, F. Cazzaniga, A. Sabbadini, F. Zanderigo, F. Priolo, Mat. Sci. in Semiconductor Processing **1**, 119 (1998)

[9] K. Nagasawa, Y. Matsushita and S. Kishino, Appl. Phys. Lett. **37**, 622 (1980)

[10] M. L. Polignano, G. F. Cerofolini, H. Bender and C. Claeys, J. Appl. Phys. **64**, 869 (1988)

[11] S. Nadahara, M. Watanabe and K. Yamabe, *Extended Abstracts of the 22nd Conference on Solid State Devices and Materials*, Sendai, Japan, 1990, p. 409

[12] J. L. Benton, P. A. Stolk, D. J. Eaglesham, D. C. Jacobson, J. Y. Cheng, N. T. Ha, T. E. Haynes, S. M. Myers, J. Appl. Phys. **80**, 3275 (1996)

[13] M. L. Polignano, D. Caputo, G. Pavia, F. Zanderigo, Solid-State Phenomena **64**, 413-420 (1998)

[14] M. L. Polignano, M. Brambilla, F. Cazzaniga, G. Pavia, and F. Zanderigo, J. of the Electrochem. Soc., **145**, 1632 (1998)

[15] J. Lagowski, P. Edelman, A. M. Kontkiewicz, O. Milic, W. Henley, M. Dexter, L. Jastrzebski and A. M. Hoff, Appl. Phys. Lett. **63** (1993) 3043

[16] J. Lagowski, P. Edelman, A. M. Kontkiewicz, O. Milic, W. Henley, M. Dexter, L. Jabstrzebski and A. M. Hoff, Appl. Phys. Lett. **63**, 3043 (1993)

[17] S. D. Brotherton, P. Bradley and A. Gill, J. Appl. Phys. **57**, 1941 (1985).

[18] G. Zoth and W. Bergholz, J. Appl. Phys. **67**, 6764 (1990)

[19] G. Ferenczi, J. Boda and T. Pavelka, Phys. Stat. Sol. (a) **94**, K119 (1986)

[20] G. Ferenczi, P. Krispin and M. Somogyi, J. Appl. Phys. **54**, 3902 (1983)

[21] G. Ferenczi, C. A. Londos, T. Pavelka, and M. Somogyi, A. Mertens, J. Appl. Phys. **63**, 183 (1988)

[22] D. K. Schroder and H. C. Nathason, Solid-State Electronics **13**, 577 (1970)

[23] M. Zerbst, Z. Angew. Phys. **22**, 30 (1966)

[24] J. S. Kang and D. K. Schroder, Phys. Stat. Sol. (a), **89**, 13 (1985)

[25] P. Edelman, A. M. Hoff, L. Jastrzebski and J. Lagowski, *Proc. of the SPIE Microelectronic Manufacturing 94 Conference*, Vol. 2337, The International Society for Optical Engineering, Bellingham, Washington, (1994) p. 154

[26] E. Yablonovitch, D. L. Allara, C. C. Chang, T. Gmitter and T. B. Bright, Phys. Rev. Letters **57**, 249 (1986)

[27] H. Föll, Symp. on Advanced Science and Technol. of Silicon Mat., Kona, Hawaii, 1991

[28] M. L. Polignano, G. Ferroni, A. Sabbadini, G. Valentini and G. Queirolo, J. of non-Crystalline Solids, 216, 88 (1997)

[29] M. L. Polignano, F. Cazzaniga, A. Sabbadini, G. Queirolo, A. Cacciato and A. Di Bartolo, Mat. Sci. Engineering **42**, 157, (1996)

ANALYTICAL APPROXIMATIONS FOR THE DISTRIBUTIONS OF SUBSTITUTIONAL TRANSITION METAL DEFECTS IN SILICON FLOAT ZONE CRYSTALS

H. Lemke
Institut für Mikroelektronik und Festkörperelektronik Technische Universität Berlin
Jebensstr. 1, D-10623 Berlin, Germany

It is reported about the formation rate of substitutional transition metal (TM) defects (M_S) in dislocated (D) and dislocation-free (DF) 0.8 inch FZ crystals. Theoretical and experimental results are presented for the metals Cu, Ni, Co, Ag, Pd (group I) and Rh, Au, Pt, Ir (group II). It is shown, that the frozen-in concentration $N_S(0)$ can be determined by solving the Frank-Turnbull rate equation. For a given vacancy distribution $C_V(T)$, characterized by its dependence on temperature, all dopants can be arranged according to their binding energies E_B, which are connected with characteristic temperatures T_c. Calculations of $N_S(0)$ are performed by integration with the steepest descent method for D-crystals considering different cooling rates and additionally for vacancy dominated (V) and interstitial dominated (I) regions of DF-crystals. It is shown, that the void and swirl formation in V- and I-regions have a strong influence on $N_S(0)$ in most cases. From a comparison between theoretical and experimental results the binding energies 2.0 to 2.5 eV (group I) and 2.6 to 3.0 eV (group II) are derived. The reduced concentrations $N_S(0)/N_D$ as observed in V-regions, for example, are in the range between $3 \cdot 10^{-4}$ and 1 for $E_B = 2$ up to 3 eV, where N_D is the total TM doping concentration.

1. INTRODUCTION

In previous work /1-3/ it was reported about the distributions of substitutional transition metal (TM) defects in silicon. These investigations were performed on 0.8 inch FZ crystals grown with the velocity V = 3.4 mm/min, where the metal doping was incorporated during crystal growth. Respecting several features of the dopants, it was suitable to distinguish between the groups Cu, Ni, Co, Ag, Pd (I) and Rh, Au, Pt, Ir (II).
It could be shown that the formation rate of substitutional TM defects (M_S) is often markedly different in dislocation-free (DF) crystals compared to that in dislocated (D) crystals. In the case of Rh for example, nearly uniform M_S distributions were observed in D-crystals, whereas for DF-crystals a strongly decreasing M_S concentration with increasing distance from the seed cone was found. A first interpretation of the M_S distributions for dopants of group II was presented in /4/. Using an analytical approximation for the temperature field it could be shown that the growth parameter P = V/G is a strongly decreasing function with increasing distance from the seed cone, where G denotes the temperature gradient at the melt/solid interface. With analytical approximations for the distributions of intrinsic point defects /5/ it was demonstrated that the self-interstitials are the dominating defects in the main parts of the DF-crystals (I-region). As a consequence the concentration of both vacancies and substitutional TM defects is low. In contrast to that, the same parts of D-crystals are vacancy dominated (V-region), because the influence of self-interstitials is markedly diminished by recombination on dislocations. It was pointed out in /4/ that for each dopant a

characteristic temperature range $T < T_c$ exists, which is important for the adjustment of the frozen-in concentration $N_S(0)$.

For the calculation of this concentration it is necessary to solve the Frank-Turnbull (FT) rate equation /6/, which describes the interaction between interstitial metal atoms M_i and vacancies. This equation is part of a coupled system of differential equations considering the recombination of intrinsic point defects and their interaction with sinks and sources in the bulk. In order to simplify the calculations it was assumed in /4/ that the emission rate in the FT-equation can be neglected in the range $T < T_c$. But in contrast to this assumption it will be shown in the present work that for nearly all crystal growth conditions the emission rate is most important for the adjustment of the frozen-in concentration $N_S(0)$.

In this new theoretical analysis the temperature dependence of the vacancy concentration $C_V(T)$, capture coefficient $r_c(T)$ and mass action constant $K(T)$ is approximated by an Arrhenius function characterized by the energy H_V, W and E_B, respectively. The general solution of the FT-reaction will be represented by an integral, which is used as starting point for several analytical approximations. It will be shown that for $H_V > E_B$ and $H_V < E_B$ the emission and the capture rate, respectively, is important for the adjustment of the concentration $N_S(0)$.

The occupation state $y_o = N_S(0) / N_D$, with N_D being the total TM doping concentration, can be generally calculated by integration with the steepest descent method (SDM). But for $y_o << 1$ the result of integration can be also expressed by the Gaussian Gamma function. The theory enables a reliable definition of the characteristic temperature T_c, which depends mainly on the binding energy E_B but weakly also on the quotient $r_c(\infty)$ / $[\eta_m (H_V+W)]$, where η_m denotes the cooling rate at the melting point. It is shown that the formation of the frozen-in concentration takes place in a temperature interval of about 100 K below this temperature, which is labelled in the following by T_c^+. By asymptotic expansion it can be shown that the solution y(T) for $T > T_c^+$ is approximately equal to that at thermal equilibrium. For the calculation of the concentration y_o it is necessary to know the vacancy distribution $C_V(T)$ in the temperature range $T < T_c^+$. For D-crystals the vacancy concentration can be approximated by the thermal equilibrium concentration C_V^{equi}. In the case of DF-crystals, on the other hand, different vacancy distributions have to be considered, as will be shown in section 3.1. For the selection of the appropriate distribution it is necessary to consider besides intrinsic recombination also the defect agglomeration in the I-and V-regions.

Several comparisons between theoretical and experimental results as presented in section 3.4 enable a first determination of the capture coefficients and binding energies of the different dopants. It will be shown that the binding energies for the metals of group I are in the range from 2.0 to 2.5 eV, whereas those of the group II are between 2.6 and 3.0 eV.

2. CRYSTAL GROWTH CONDITIONS

For the experimental investigations 0.8 inch crystals with a length L = 10 - 15 cm were prepared with the FZ technique using the growth velocity V = 3.4 mm/min. The experimental details of this work has been published already in /1-3/. Since the solid/melt interface during crystal growth was nearly flat, the temperature field can be approximately calculated by solving the one-dimensional heat conduction equation as demonstrated in /4/. Behind the seed cone the temperature gradient G_m for $T = T_m$ increases quickly in the first crystal part with a length of about 2 cm. The resulting growth parameter $P = V / G_m$ is shown in Fig.1. Using the relation (6) (s. sect.3.1) the critical growth parameter P_{crit} can be calculated according to $P_{crit} = 2.26 \cdot 10^{-5}$ cm^2 / (Ks). The diagram in Fig.1 shows that each

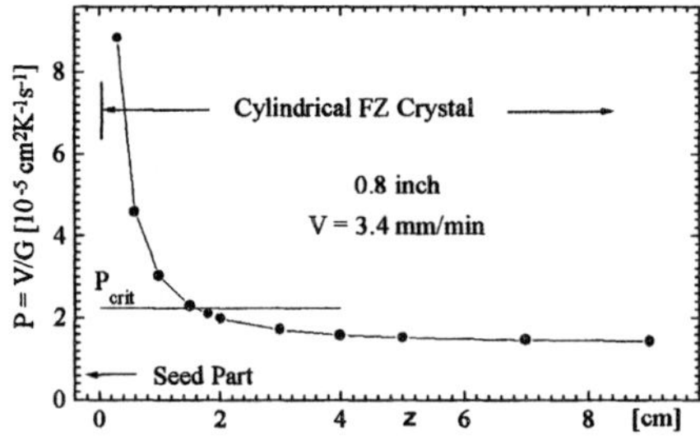

Fig.1. Dependence of the growth parameter P on the axial position for a 0.8-inch crystal.

crystal contains a short vacancy (V) dominated region characterized by $P > P_{crit}$ and a long interstitial (I) dominated region with $P < P_{crit}$. The range of stationary crystal growth starts evidently near the position z = 5 cm. The calculation of the temperature gradient for this range results in $G_m(z > 5cm)$ = 390 K/cm /4/. Near the starting point of the TM doping at about $z = z_A = 0.5$ cm, on the other hand, the temperature gradient is markedly lower with $G_m(z_A) = G_m(z_{stat})/3$.

For the calculation of defect reactions it is further necessary to know the cooling rate $-dT/dt = \eta(T) = V \cdot G(T)$ in a broad temperature range. Neglecting the temperature dependence of the heat conductivity and the influence of growth rate, the cooling rate is given by $\eta(T) = V \cdot G_m(T/T_m)^2$ /4/.

Hence, the main parts of the crystals grown under stationary conditions are characterized by $\eta_m \approx 2$ K/s, whereas near the position z = 0.5 cm the cooling rate is about $\eta_m \approx 0.7$ K/s. Since the growth process was finished by quickly separating the crystal from the feed rod, the cooling rate is markedly increased at the crystal end. In this crystal part the highest cooling rate is about $\eta_m = 20$ K/s at z = L. For the interpretation of the experimental results observed in these regions it is necessary to know the temperature field T(z), with z being the distance to the melt/solid interface.

For a suffiently long crystal the heat conduction equation can be exactly solved by /4/

$$-z(T) = \frac{1}{b}\left(\frac{1}{T} - \frac{1}{T_m}\right) + \frac{a}{b^2}\ln\frac{a - b/T}{a - b/T_m} \tag{1}$$

$$a = \sqrt{\frac{\sigma\sigma_r}{Rk_0}} \quad , \quad b = \frac{c_p \rho V}{k_o} \tag{1a}$$

where z = distance to the melt/solid interface, $k = k_0/T$ = approximate heat conductivity, $\sigma = 5.7 \cdot 10^{-12}$ W/(cm^{2}K^4) = Stefan-Boltzmann constant, $\sigma_r = 0.57$ = emissivity, $c_p = 0.922$ J/(gK) = specific heat, $\rho = 2.33$ g/cm^3 = density, R = crystal radius, V = growth velocity, $k_0 = 360$ W/cm^2.

3. THEORETICAL–EXPERIMENTAL RESULTS

3.1 Distributions of Intrinsic Point Defects in Dislocation-Free Crystals

The formation rate of substitutional TM defects via the Frank-Turnbull reaction is determined by the vacancy concentration in dependence on the growth conditions and temperature. Recently it was reported that the distributions of intrinsic point defects can be described by simple analytical approximations of the transport equations /5/. In the case of high recombination between intrinsic point defects the product of the concentrations of vacancies C_V and self-interstitials C_I is equal to that at thermal equilibrium according to

$$Q(T) = C_V C_I = C_V^{equi} C_I^{equi} = C_{Vm} C_{Im} \exp\left[- \frac{H}{kT_m}\left(\frac{T_m}{T} - 1\right)\right] , \qquad (2)$$

where the equilibrium concentrations C_V^{equi} and C_I^{equi} in the temperature range near T_m are determined by: $H_V = 3.2$ eV = formation energy of vacancies, $H_I = 4.2$ eV = formation energy of self-interstitials, $H = H_V + H_I$, k = Boltzmann constant, $C_{Vm} = 11.7 \cdot 10^{14}$ cm^{-3} = concentration of vacancies at $T=T_m$, $C_{Im} = 9.6 \cdot 10^{14}$ cm^{-3} = concentration of self-interstitials at $T=T_m$.
The point defect data used in this section were taken from /7/. Based on these data main features of the microdefect formation in CZ crystals could be successfully explained /7/.
Using relation (2) the distribution of vacancies and self-interstitials can be written in the form

$$C_V(T) = C + \sqrt{C^2 + q\,Q(T)} \qquad (3)$$

$$qC_I(T) = -C + \sqrt{C^2 + q\,Q(T)} \qquad (4)$$

where

$$C = \left(C_{Vm} - C_{Im} q\right)/2 \qquad (5a)$$

$$q = \frac{P + P_I}{P + P_V} \qquad (5b)$$

$$P_V = \frac{HD_{Vm}}{2kT_m^2} , \qquad P_I = \frac{H\,D_{Im}}{2\,kT_m^2} \qquad (5c)$$

with $D_{Vm} = 4.3 \cdot 10^{-5}$ cm^2/s = diffusivity of vacancies at $T=T_m$, $D_{Im} = 3.8 \cdot 10^{-4}$ cm^2/s = diffusivity of self-interstitials at $T=T_m$, $P = V / G_m$ = growth parameter, V = growth rate, G_m = temperature gradient at $T=T_m$.
The different growth regions are characterized by the sign of the parameter C

$C > 0$	V - Region	$P > P_{crit}$
$C < 0$	I - Region	$P < P_{crit}$

The boundary condition C = 0 is fulfilled for $P = P_{crit}$. Using this relation and considering (5a) up to (5c), Voronkov's formula /8/ for the critical growth parameter can be derived

$$P_{crit} = P_I \frac{C_{Im} - \left(D_{Vm} / D_{Im}\right) C_{Vm}}{C_{Vm} - C_{Im}} \qquad (6)$$

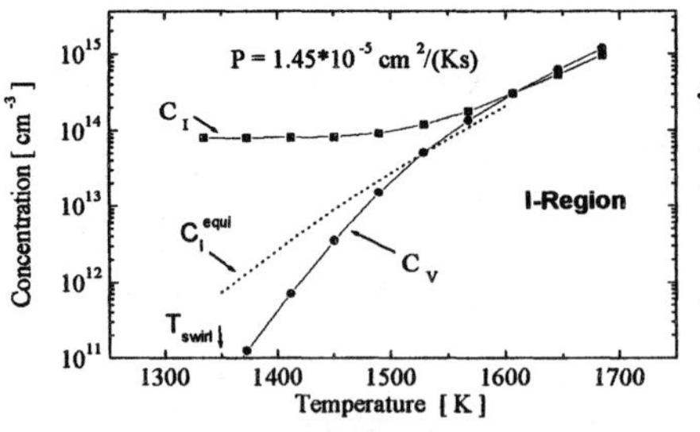

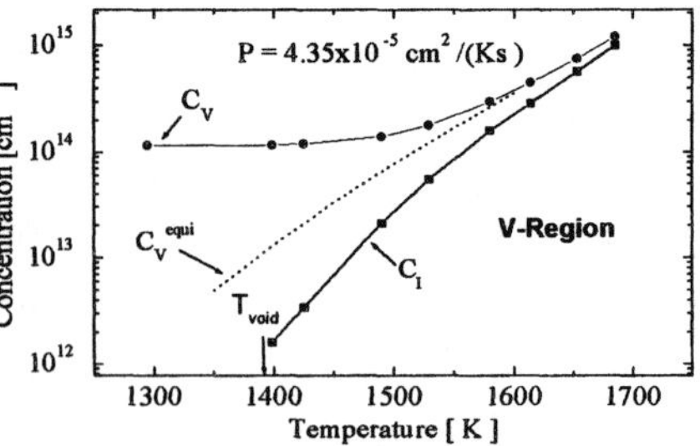

Fig.2. Dependence of the vacancy and self-interstitial concentration on temperature in a 0.8-inch crystal for stationary growth conditions.

Fig.3. Dependence of the vacancy and self-interstitial concentration on temperature for the 0.8-inch crystal as in Fig.2 in a V-region near to the seed cone.

The distributions of intrinsic point defects according to relations (3) and (4) are shown in Fig.2 for a crystal region, which was grown under stationary conditions. The dominance of self-interstitials causes evidently a strongly decreasing vacancy concentration below T = 1550 K. As will be shown in section 3.4, this temperature range is important for the formation of substitutional defects in crystals doped with metals of the group II. In the case of the other metals , a changed vacancy concentration caused by the swirl formation has to be considered. A comparison between the concentrations C_I and C_I^{equi} shows that supersaturation with $C_I > 100 \times C_I^{equi}$ exists in the range $T < T_{swirl} = 1350$ K, where swirl formation can be expected. This formation temperature is consistent with experimental results for 1 inch FZ crystals as published by Roksnoer et al. /9/.

The distributions of intrinsic point defects in the V-region, near the starting point of the TM-doping, is shown in Fig.3. The dominating vacancy concentration enables here a high formation rate of substitutional TM-defects. But as will be demonstrated in section 3.4, for the interpretation of the experimental results it is necessary to know the vacancy distribution in the range of microdefect formation. The diagram shows that supersaturation with $C_V > 10 \times C_V^{equi}$ exists in the temperature range $T < T_{void} = 1390$ K. Detailed theoretical and experimental investigations on the void formation process were recently performed by Sinno et al. for CZ crystals in /10/. According to these investigations the concentration of isolated vacancies during void formation can be approximated by the relation $C_V = 10 \times C_V^{equi}$, which will be used below.

In the case of swirl formation, on the other hand, the approximation $C_V = C_V^{equi}$ has been tried in the temperature range $T < T_{swirl}$. The same vacancy distribution is applied for the interpretation of the experimental results observed in dislocated crystals. But in contrast to DF-crystals, this relation can be used in the whole temperature range $T < T_m$.

3.2 Frank-Turnbull Reaction

The formation of substitutional defects M_S during crystal growth is governed mainly by the Frank-Turnbull reaction, which describes the interaction between interstitial metal atoms M_i and vacancies. Introducing the occupation state $y(T) = N_S(T) / N_D$ and the cooling rate $\eta(T) = -dT/dt$, the rate equation is given by

$$\eta(T)\frac{dy}{dT} = r_c(T)K(T)y - r_c(T)C_V(T)(1-y) \quad (7)$$

with

$$K(T) = N_G \exp[-E_B/(kT)] \quad , \quad C_V(T) = C_{V0} \exp[-H_V/(kT)] \quad (7a)$$

$$r_c = r_{c0} \exp[-W/(kT)] = [D_i + D_V] \times \lambda \quad , \quad (7b)$$

where N_D = total metal concentration, K = mass action constant, E_B = binding energy of the metal atom on the lattice site, T = temperature, k = Boltzmann constant, N_G = concentration of interstitial lattice sites, C_V = concentration of vacancies, H_V = energy characterizing the temperature dependence of C_V (T), r_c = capture coefficient, W = energy characterizing the transport and reaction of the interstitial metal defects, D_i = diffusivity of the interstitial defects, D_V = diffusivity of the vacancies, λ = reaction length.

The solution of the linear differential equation (7) considering the initial condition $y(T_i) = y_i$ can be written in the form

$$y = e^{F(T)}y_i + e^{F(T)}\int_{T}^{T_i} e^{-F(t)} \frac{r_c(t)C_V(t)dt}{\eta(t)} \quad (8)$$

with

$$F(T) = \int_{T_i}^{T} \frac{[K(t)+C_V(t)]r_c(t)dt}{\eta(t)} \quad . \quad (8a)$$

Using the approximation $\eta(T) = \eta_m (T/T_m)^2$ (s. sect. 2) the function F(T) can be exactly calculated. A general calculation of the solution in (8) for all parameter combinations is not possible. But special solutions exist, which will be presented in the Appendix.
A detailed inspection of the integral in (8) shows, that the solution is mainly determined by the temperature range around the maximum of the integrand at $T = T_{max}$. This important parameter can be approximately determined by the equation

$$G' = \frac{d}{dT}\left[F(T) + \frac{H_V + W}{kT}\right] = 0 \quad . \quad (9)$$

If the integrand is approximated by a bell shaped function, the width of the integrand is determined by

$$\Delta T = \sqrt{2/G''} \quad , \quad (9a)$$

which enables the introduction of two characteristic temperatures according to

$$T_c^+ = T_{max} + \Delta T \qquad \text{and} \qquad T_c^- = T_{max} - \Delta T.$$

For the case $y \ll 1$, equations (9) and (9a) can be exactly solved by

$$T_c^{+(-)} = T_{max}\left[1+(-)\frac{\sqrt{2}\,kT_{max}}{\sqrt{(E_B+W)(H_V+W)}}\right] \quad (10)$$

$$kT_{max} = \frac{E_B+W}{\ln\left[\frac{r_{c0}N_G kT_m^2}{\eta_m(H_V+W)}\right]} \quad . \quad (10a)$$

The parameter T_{max} is therefore mainly determined by the binding energy, but depends also logarithmicly on the capture coefficient, cooling rate and vacancy distribution. Generally, the whole temperature range can be divided into three parts:
Range I : $T > T_c^{+}$. Besides a short range connected with the initial condition, the solution is asymptotically equal to that at thermal equilibrium

$$y(T) \approx y^{equi} = \frac{C_V(T)}{K(T)+C_V(T)} \quad . \quad (11)$$

Range II : $T_c^{-} < T < T_c^{+}$. This range with a typical width of about 100 K is most important for the formation of the frozen-in concentration y_0. For the calculation of y_0 it is necessary to know the vacancy distribution in this temperature range.
Range III : $T < T_c^{-}$. In this range the concentration y(T) is nearly constant and equal to that at T = 0.
For further estimations it is useful to transform the variable T and the basic parameters according to

$$x = \exp\left[-\frac{W}{kT_m}\left(\frac{T_m}{T}-1\right)\right] \quad (12)$$

$$r_c = r_{cm}\cdot x \quad , \quad K = K_m\cdot x^{\alpha} \quad , \quad C_V = C_{Vm}\,x^{\beta} \quad (12a)$$

$$\alpha = E_B/W \quad , \quad \beta = H_V/W \quad .$$

Using the initial condition y(T_m)= 0 the solution (8) is then transformed into

$$\frac{y(x)e^{-F(x)}}{(1+\beta)B} = \int_x^1 e^{-F(t)}t^{\beta}dt = \int_0^1[.....]-\int_0^x[.....] \quad (13)$$

with

$$F(x) = Ax^{1+\alpha} + Bx^{1+\beta} \quad (13a)$$

$$B = \frac{r_{cm}kT_m^2C_{Vm}}{\eta_m W(1+\beta)} = a\frac{C_{Vm}}{1+\beta} \quad , \quad A = \frac{r_{cm}kT_m^2K_m}{\eta_m W(1+\alpha)} = a\frac{K_m}{1+\alpha} \quad . \quad (13b)$$

For all parameter specifications needed in this work the limit x = 1 in (13) can be replaced by x = ∞ . In the temperature range I with x > x (T_c^{+}) the integral can be asymptotically expanded. The solution in first approximation is equal to y^{equi} already cited in (11). In the temperature ranges II and III, on the other hand, the solution can be easily determined by series expansion of the second integral in (13) according to

$$y(x) = e^{F(x)}\left[y(0) - Bx^{\beta+1} + AB\frac{\beta+1}{\alpha+\beta+2}x^{\alpha+\beta+2} + \frac{B^2}{2}x^{2(\beta+1)} + \ldots \right] \quad . \qquad (14)$$

For the interpretation of the experimental results only the knowledge of the frozen-in concentration y(0) is important. In the case y << 1 this parameter can be directly calculated with the relation (13).
Considering equation (7) the second term in (13a) can be evidently neglected . In this case the frozen-in concentration can be easily calculated with the Gaussian Γ-Function according to

$$y(0) = \frac{1+\beta}{1+\alpha} B \int_0^\infty e^{-At} t^{\frac{\beta-\alpha}{1+\alpha}} dt \qquad (15)$$

$$y(0) = \Gamma\left(1 + \frac{1+\beta}{1+\alpha}\right) \frac{B}{A^{\frac{1+\beta}{1+\alpha}}} \quad . \qquad (15a)$$

Generally, the frozen-in concentration can be determined with the steepest descent method (SDM) where the integrand is approximated by a bell shaped function. Using the integral in (8) and the relations (9) and (9a) leads to the result

$$y(0) = \sqrt{2\pi}\frac{kT_m R_{max}}{\sqrt{(H+W)(E+W)+(H-E)kT_m R_{max}}} \exp\left[-\frac{H+W}{E+W} + \frac{(H-E)kT_m}{(H+W)(E+W)} R_{max} \right] \qquad (16)$$

with

$$E = E_B, \quad H = H_V, \qquad R_{max} = \frac{T_m}{\eta_m} r_c(T_{max})\, C_V(T_{max}) \quad , \qquad (16a)$$

where the characteristic temperature T_{max} has to be determined with the transcendental equation (9). The formula (16) includes also the special case $H_V = 0$, whereas a similar relation, which can be derived from (13), only for $H_V > 0$ holds.
Figure 4 shows three exact solutions for the Frank-Turnbull equation as presented in the Appendix. The functions were calculated for the binding energy $E_B = 2.5$ eV (W = 0.5 eV) and activation energies $H_V = 1$, 2.5 and 5.5 eV characterizing the temperature dependence of the vacancy concentration. The curve 1 demonstrates the temperature dependence of the occupation state y for a slowly decreasing vacancy concentration as can be observed in a V-region (s. Fig.3). The curve 3, on the other hand, shows the typical behavior for a strongly decreasing vacancy concentration which can be recognized in an I-region (s. Fig.2). The marked characteristic temperatures were calculated with the relations (10) and (10a). All characteristic temperature regions as discussed above can be clearly recognized. The indicated adjustment region caused by the initial condition is with $\Delta T \approx$ 0.01 K obviously narrow. In a broad temperature region below the melting point the

occupation state is equal to that at thermal equilibrium. Below the characteristic temperature region the frozen-in state can be finally recognized. The distribution 1 and 3 are mainly determined by the capture and emission rate, respectively. As a consequence the influence of the reaction parameter a on both distributions is different. If the parameter is lowered, for example, the frozen-in concentration y(0) decreases and increases for the distributions 1 and 3, respectively. According to relation (13b) such a decrease can be initiated, for instance, by a higher cooling rate. Only in the special case $\alpha = \beta$ the occupation state is independent on the reaction parameter.

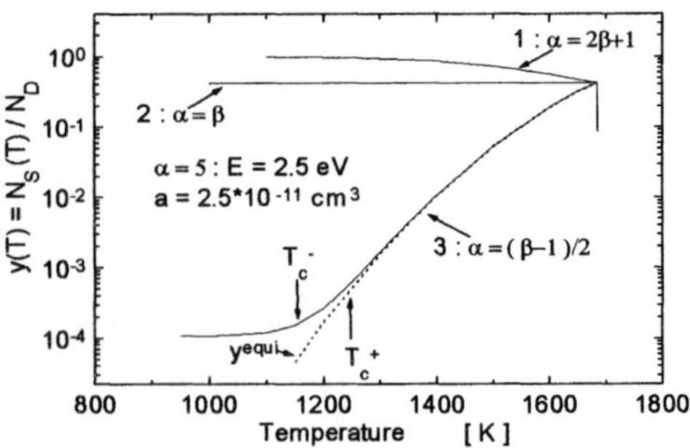

Fig.4. Three exact solutions of the Frank-Turnbull equation (7) as presented in the Appendix. The marked characteristic temperatures on the curve 3 were calculated with the relation (10).

3.3 Axial Distributions of Substitutional Transion Metal Defects

In this section a short review about the distribution of substitutional TM-defects as observed in dislocated and dislocation-free FZ crystals will be given. Detailed experimental results have already been published in /1-3/. The measurements of the concentrations N_S were performed with Deep Level Transient Spectroscopy taking into account known complexes with hydrogen and interstitial TM defects/1-3/. The transition metals with the valence $N > 8$ and their main data as used in this work are listed in Table 1. The segregation coefficients k, except those of Ag and Rh, are known from published data about solid solubilities /11/ and liquidus curves /12/. The segregation coefficients for Ag and Rh were temporarily proposed taking into account the systematic trend in the 3d and 5d group /13/.

M	k	G[mg]	N_D[cm^{-3}]	T_s[K]	y_{stat}, DF	y_{stat}, D	D_m [cm^{2}s^{-1}]	W_M [eV]
Au	$5.5\cdot10^{-5}$	1.75	$7.3\cdot10^{13}$	980	0.75	0.97	$\approx2\cdot10^{-5}$	0.29
Ag	$1.5\cdot10^{-4}$	15	$3.1\cdot10^{15}$	1300	$7.0\cdot10^{-3}$	$3.5\cdot10^{-3}$	?	?
Cu	$4.0\cdot10^{-4}$	20	$2.5\cdot10^{16}$	1030	$1.2\cdot10^{-3}$	$6.0\cdot10^{-4}$	$7.5\cdot10^{-5}$	0.20
Pt	$3.5\cdot10^{-5}$	3	$8.1\cdot10^{13}$	1020	0.58	0.92	$\approx2\cdot10^{-5}$	?
Pd	$9.0\cdot10^{-5}$	6	$7.7\cdot10^{14}$	1000	$3.0\cdot10^{-2}$	$2.5\cdot10^{-2}$	$6.5\cdot10^{-5}$	0.22
Ni	$1.5\cdot10^{-4}$	20	$7.7\cdot10^{15}$	930	?	$5.2\cdot10^{-4}$	$8.0\cdot10^{-5}$	0.47
Ir	$2.5\cdot10^{-6}$	10	$2.0\cdot10^{13}$	1050	0.4	1.2	$\approx2\cdot10^{-6}$	?
Rh	$5.0\cdot10^{-6}$	8	$5.9\cdot10^{13}$	1000	$7.0\cdot10^{-3}$	0.45	?	?
Co	$1.5\cdot10^{-5}$	25	$9.6\cdot10^{14}$	1300	?	$5.5\cdot10^{-3}$	$8.0\cdot10^{-5}$	0.37

Table 1. Typical doping concentrations N_D and reduced concentrations y_{stat} of substitutional TM defects for stationary growth conditions. Solubility and diffusivity data as reviewed in /11/, but for Ir in /14/.

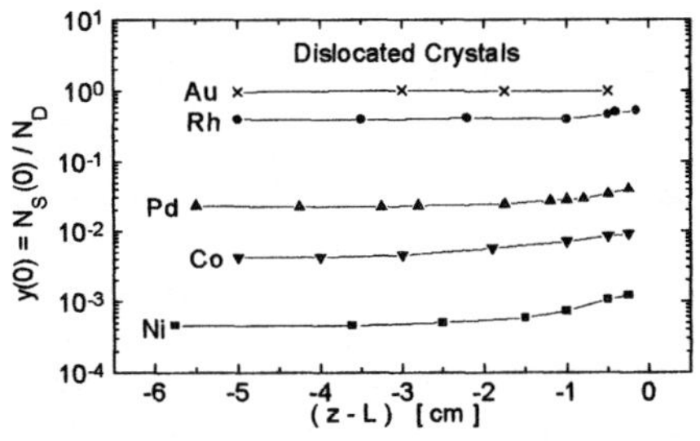

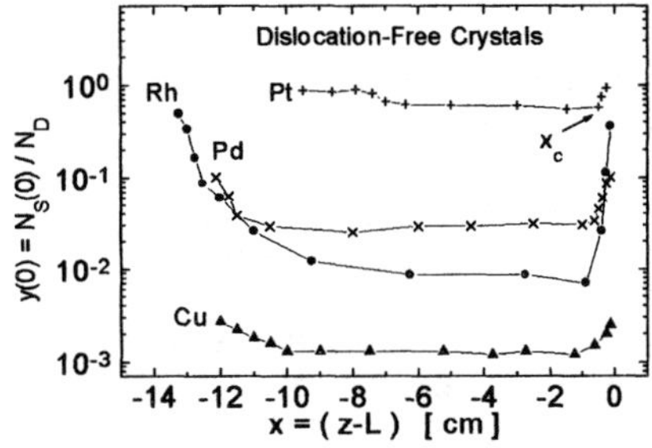

Fig.5. Some axial distributions of substitutional TM defects as observed in dislocated crystals. L = crystal length.

Fig.6. Some axial distributions of substitutional TM defects as observed in dislocation-free crystals. L = crystal length.

Based on these data the total doping concentration N_D has been calculated using the listed weigths G of the doping pellets and the melt volume of 4 cm^{-3}. Using these concentrations and published data for the solid solubilities /11,14/ the temperature range for supersaturation $T < T_S$ has been determined for the dopants. The reduced concentrations of substitutional TM defects observed for stationary growth conditions are listed in the next rows for DF- and D-crystals. In the case of Ni and Co, these data are not known at present for DF-crystals. The diffusivities $D_m = D(T_m)$ and migration energies W_M have been taken from /11/, but for Ir from /14/. The cited diffusivities for Au, Pt, Ir are only rough estimates derived from indirect measurements.

Some characteristic axial distributions of substitutional TM defects in D-crystals are shown in Fig.5. For the crystals with a length L = 10 – 15 cm only the last parts of the distributions including the quickly cooled crystal ends are drawn. The dislocation density in the shown crystal regions is about (1 to 2) × $10^4 cm^{-2}$. Near to the seed cone the dislocation density is about one order of magnitude lower. As a consequence, for the dopants of group I moderately increased concentrations N_S can be observed in these crystal parts. For Au, and similar also for Pt and Ir, the distributions are flat indicating with $y_{stat} \approx 1$ a nearly complete formation of substitutional defects. An increase of the concentration due to the cooling rate can be observed for the metals Rh up to Cu in the range 20 up to 200 %. Generally, the doping behavior between Cu and Ni and also that between Co and Ag is not markedly different.

Some axial distributions of substitutional TM-defects for DF-crystals are shown in Fig.6. In the range of stationary crystal growth a marked decrease of the concentration can be recognized for Pt and also for Au and Ir (s. Table1). This decrease is especially strong in Rh-doped crystals. For the metals Cu up to Pd (group I), on the other hand, the concentrations in these crystal parts are comparable to those in D-crystals. But in the V-regions near z = L and near $z = z_A$ a clearly different behavior can be recognized for DF-crystals. The steep increase of the concentrations in both regions cannot be observed in dislocated crystals.

3.4 Theoretical Interpretation of the Experimental Results

For the interpretation of the experimental results it is necessary to specify the reaction parameter a in the relations (13b). In the following diagrams the values $a = a_o$ and $a = 0.1 \times a_o$ with $a_o = 2.5 \cdot 10^{-11}$ cm^3 are used. For the cooling rate $\eta_m = 2$ K/s and reaction energy W = 0.5 eV the reaction strength at the melting point is therefore given by $r_{cm} \approx 10^{-13}$ cm^3 s^{-1} ($a = a_o$). Since the metals of group I have diffusivities in the range $D_m \approx 10^{-4}$ cm^2 s^{-1} (s. Table1) the reaction length in (7b) has to be chosen according to $\lambda = 10^{-9}$ cm. An important information about the binding energies of several dopants can be derived from the experimental results for D-crystals as shown in Fig.5. Using the steepest descent method according to the relations (16) and (16a) the concentrations y_0 were calculated for the cooling rate $\eta_1 = 2$ K/s ($z = z_{stat}$) and for $\eta_2 = 20$ K/s ($z = L$) considering the vacancy distribution C_V^{equi}. The dependence of the quotient $q = y_o(\eta_1) / y_o(\eta_2)$ on the binding energy is demonstrated in Fig.7 (solid line). For low binding energies this function can be also calculated with the Gamma function in (15a), which is shown in Fig.7 as dotted line. This relation is independent on the reaction parameter a, which can be concluded from the analytic relation according to

$$E_B = \frac{H_V - \rho W}{1+\rho} \quad , \quad \rho = -\frac{\ln q}{\ln(\eta_2 / \eta_1)} \quad . \tag{17}$$

The experimental data shown in Fig.7 enable therefore a direct determination of the binding energies. A comparison with the theoretical relation leads to the energies 2.04 (Ni), 2.08 (Cu), 2.28 (Co), 2.33 (Ag), 2.45 (Pd) and 2.65 eV (Rh), which will be used in the following diagrams as temporary data. Figure 8 shows a comparison between experimental and theoretical results for the concentrations $y(\eta_1)$ of substitutional TM defects for the range of stationary crystal growth. The calculations were performed with the steepest descent method (solid line) and with the Gamma function (dotted line) using the relation $a = a_0$ mentioned above. The experimental results for D-and DF-crystals were taken from the distributions in the Figs.5 and 6 or from Table 1. The results for the 5d metals Au, Pt, Ir are shown only for the D-crystals using the binding energies as determined below. In the case of D-crystals a fair agreement between theory and experiment can be stated for all metals. For the metals of group I and for Rh a strong increase of the occupation state with increasing binding energy can be recognized. For the 5d-metals an occupation state near 1 is observed, which is also in accordance with the theory. In the case of DF-crystals the

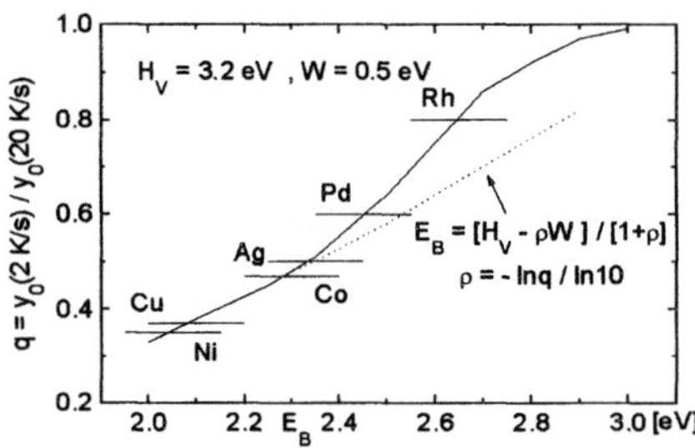

Fig.7. Theoretical and experimental results for the quotient $q = y_o(\eta_1) / y_o(\eta_2)$ characterizing the influence of different cooling rates in dislocated crystals. $\eta_1 = 2$ K/s, $\eta_2 = 20$ K/s ($a = a_0$).

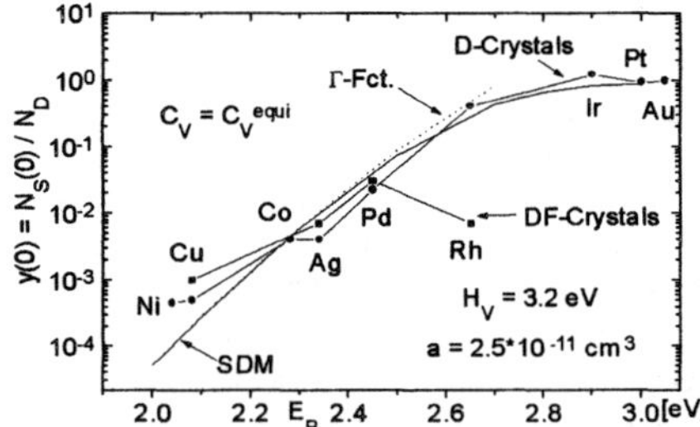

Fig.8. Comparison between theoretical and experimental concentrations $N_S(0)$ for D-crystals and for the I-regions of DF-crystals.

observed M_S concentrations for the metals of group I are similar to those in D-crystals. But the experimental result for Rh deviates strongly from the theoretical relation. This behavior can be easily understood considering the change of point defect distributions caused by the swirl formation in the I-regions of DF-crystals. Due to the low binding energies the characteristic temperatures for the metals of group I are located in the range $T_c^+ < T_{swirl}$. As a result the vacancy concentration is equal to that in thermal equilibrium. For the metals of group II, on the other hand, $T_c^+ > T_{swirl}$ is fulfilled, where the vacancy distribution is determined by intrinsic recombination according to relation (3).

The interpretation of the experimental results for these metals is shown in Fig.9. The calculations were performed with the parameter $a = 0.1\times a_o$ regarding the markedly lower diffusivities of these metals (s. Table1). The vacancy concentration was derived from the relation (3) by expansion of the root, which results in $C_V = qQ(T)/(2|C|)$. Considering relations (5a)-(5c) and the growth parameter cited in Fig.2 this vacancy distribution is characterized by $H_V = 7.4$ eV and $C_{Vm} = 1.4\times10^{16}$ cm^{-3}. The comparison with the experimental results shows that Au and Pt have binding energies near 3 eV and Ir near 2.9 eV. The value 2.65 eV for Rh is evidently in agreement with the results found for dislocated crystals (s. Fig.7). Figure 9 shows additionally the results for dislocated crystals using the same reaction parameter. The concentration of substitutional TM defects is obviously higher in D-crystals, which is in agreement with the theoretical interpretation. This deviating behavior of the dopants in D- and DF- crystals has already been described in /2/.

Figure 10 shows a comparison between experimental and theoretical results for the V-regions of DF-crystals. Due to the lower cooling rate at the position $z = z_A$ the calculation was performed with the reaction parameter $a = 3a_o$. Using thc vacancy concentration $C_V = 10\times C_V^{equi}$ (s. sect.3.1) the estimated temperatures T_c^+ for all dopants are located in the range $T_c^+ < T_{void}$, which is evidently self-consistent with the selected distribution. Hence, the adjustment of the frozen-in concentrations $N_S(0)$ near the seed cone takes place in the temperature range of void formation. The theoretical relations were calculated with the steepest descent method (solid line) and with the Gamma function (dotted line). The experimental concentrations marked in Fig.10 correspond for each dopant to the first measuring point on the left in Fig.6. The diagram shows obviously a fair

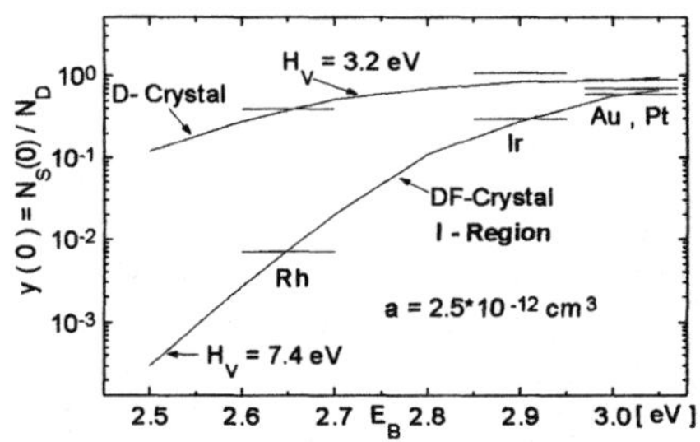

Fig.9. Comparison between theoretical and experimental concentrations $N_S(0)$ for the dopants of group II. The diagram demonstrates the influence of strongly different vacancy distributions as exist in D- and DF-crystals.

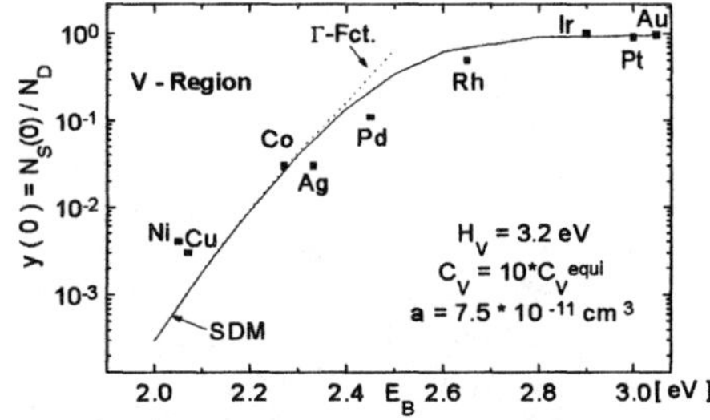

Fig.10. Comparison between theoretical and experimental concentrations $N_S(0)$ for all dopants in the slowly cooled V-regions of DF-crystals.

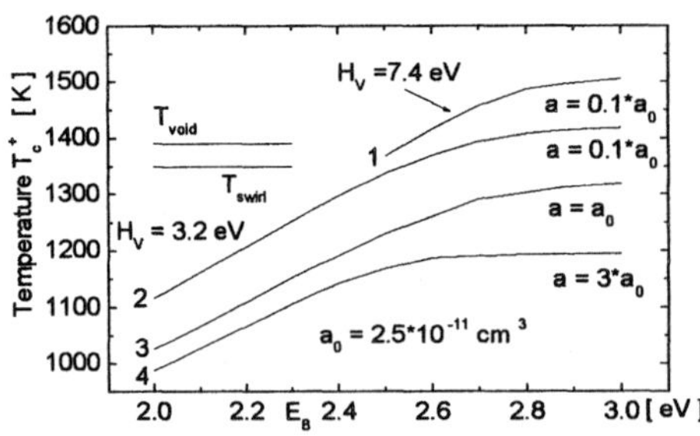

Fig.11. Characteristic temperatures for the frozen-in processes of the TM dopants as shown in Figs. 8, 9, 10.

agreement between the experimental and theoretical results also for the V-regions.

Figure 11 shows the characteristic temperatures for the frozen-in processes of substitutional TM defects as demonstrated in the Figs. 8, 9 10. Using the relation (1) these temperatures can be transformed into characteristic distances to the melt/solid interface. The lowest temperatures (curve 4), located well below T_{void}, can be recognized for the frozen-in processes in the V-regions (s.Fig.10). The curve 3 shows the temperatures, which are important for the results as demonstrated in Fig.8. In the case of Co (E_B=2.28 eV), for instance, the temperature T_c^+ is about 1140 K, which corresponds to a distance of about x_c = 3 cm. The distribution for Co in Fig.5 shows, that the influence of the increased cooling rate at the crystal end can be indeed observed in this distance range. Taking into account a lower reaction rate for Rh (curve 2), a similar evaluation leads to T_c^+ = 1380 K or x_c = 1.2 cm, which is also consistent with the results in Fig.5. The highest temperatures exist for the frozen-in processes in the I-regions of DF-crystals (s.Fig.9). For the 5d metals Au, Pt the characteristic temperature is about 1510 K or x_c = 0.6 cm. The distribution for Pt in Fig.6 confirms this theoretical prediction. A comparable agreement exists obviously also for Rh with T_c^+ = 1440 K and x_c = 0.9 cm. A comparison of the data in Fig.11 with the temperatures T_S in Table 1 shows further that the relation $T_S < T_c^+$ is fulfilled for nearly all experimental situations. Hence, as assumed in relation (7), the influence of metal precipitation on the frozen-in process can be indeed neglected.

CONCLUSION

It was reported about the formation rate of substitutional transition metal defects M_S , which are formed during the growth of silicon FZ crystals. Theoretical and experimental results were presented for the metals Cu, Ni, Co, Ag, Pd (group I) and Rh, Au, Pt, Ir (group II). It was shown that the grown-in concentrations $N_S(0)$, which are observed in dislocated crystals and in the V- and I-regions of dislocation-free crystals, can be interpreted with approximate solutions of the Frank-Turnbull rate equation. For the theoretical analysis the temperature dependence of the vacancy concentration $C_V(T)$, capture coefficient $r_c(T)$ and mass action constant K(T) was approximated by an Arrhenius function characterized by the energy H_V, W and E_B, respectively. Based on an integral representation of the solution the reduced concentrations $y(0) = N_S(0)/N_D$ were calculated with the steepest descent method and for $y \ll 1$ also with the Gamma function. For each dopant the frozen-in process takes place in a characteristic temperature range $T < T_c^+$, which is mainly determined by the binding energy E_B but weakly also by the quotient $r_c(\infty)$ / $[\eta_m (H_V+W)]$, where η_m denotes the cooling rate at the melting point. In the range $T > T_c^+$, on the other hand, the concentration y(T) is equal to that at thermal equilibrium. It was shown that the experimental results for all dopants observed on dislocated crystals can be consistently interpreted with the assumption $C_V(T) = C_V^{equi}$, where C_V^{equi} is the vacancy concentration at thermal equilibrium. The same is true for the I-regions of dislocation-free crystals, whenever $T_c^+ < T_{swirl}$ holds, which is fulfilled for all dopants of group I. In the range of swirl formation $T < T_{swirl}$ the vacancy concentration is markedly different compared to that in the range $T > T_{swirl}$. From a comparison between theoretical and experimental results it could be concluded that the dopants of group I have binding energies between 2.0 and 2.5 eV. In the range $T > T_{swirl}$ the vacancy concentration decreases strongly with decreasing temperature, which is connected with a lowered formation rate of substitutional defects for dopants of group II. It was shown that this behavior can be interpreted with binding energies between 2.6 and 3.0 eV. Based on these energies the frozen-in concentrations can be tentatively interpreted with a capture coefficient of about $r_c(\infty) = 3 \cdot 10^{-12}$ cm^3s^{-1}. The formation rate of substitutional defects in the V-regions of DF-crystals is also influenced by the defect agglomeration. It was shown that the frozen-in concentrations can be consistently explained with a vacancy distributions according to $C_V(T) = 10 \times C_V^{equi}$, which holds in the range of void formation.

ACKNOWLEDGEMENTS

The experimental investigations were performed at the -Institut für Mikroelektronik und Festkörperelektronik der Technischen Universität (TU-Berlin)- and at the -Institut für Kristallzüchtung (IKZ-Berlin)- in cooperation with Wacker Siltronic AG (Burghausen). The author would like to thank B. Hallmann (IKZ) for preparing the doped FZ crystals during the past decade. The extensive technical assistance in sample preparation by G. Harms, J. Colmsee, A. Eckert, B. Tierock (TU-Berlin) and by H. Baumüller, B. Lux and Th. Wurche (IKZ-Berlin) is gratefully acknowledged.

REFERENCES

[1] H. Lemke , in *Materials Science Forum* Vol. 196-201, 683 (1995)

[2] H. Lemke , in *High Purity Silicon IV* , p. 272 (1996), ed. by C. L. Claeys , P. Rai - Choudhury , P. Stallhofer , J. E. Maurits , The Electrochem. Soc., Pennington , N. J.

[3] H. Lemke , W. Zulehner and B. Hallmann , in *Semiconductor Silicon* , p. 572 (1998), ed.by H. R. Huff , U. Gösele , H. Tsuya , The Electrochem. Soc. , Pennington , N. J.

[4] H. Lemke and W. Zulehner, Physica B 273-274, 398 (1999)

[5] H Lemke and W. Südkamp, phys. stat. sol. (a) 176, 843 (1999)

[6] F. C. Frank, D. Turnbull, Phys. Rev. 104, 617 (1956)

[7] T. Sinno , R. A. Brown , W. von Ammon and E. Dornberger, J. Electrochem. Soc. 145, 302 (1998)

[8] V. V. Voronkov , J. Crystal Growth 59 , 625 (1982)

[9] A. J. R. de Kock, P. J. Roksnoer and P. G. T. Boonen, J. Crystal Growth 30, 279 (1975)

[10] T. Sinno, E. Dornberger, R. A. Brown, W. von Ammon, F. Dupret, Mater. Sci. Eng. R. 28, 149 (2000)

[11] W. Schröter and M. Seibt, in *Properties of Crystalline Silicon* p.543 (1999), ed. by R. Hull, Short Run Press Ltd., Exeter

[12] R. P. Elliot, Constitution of Binary Alloys, First Suppliment, Mc. Graw-Hill, New York (1965)

[13] H. Lemke, in *Semiconductor Silicon*, p. 695 (1994) ed. by H. R. Huff, W. Bergholz, K.Sumino, The Electrochemical Society, Pennington, NJ

[14] S. Obeidi and N. A. Stolwijk, Phys. Rev. B 64, 113201 (2001)

APPENDIX

Exact Solutions of the Frank -Turnbull Rate Equation

For special relations between the energy parameters α and β the integration in (13) can be exactly performed. Introducing for

$$\alpha = 2\beta+1$$

the transformation u(x) with $u_m = u(1)$

$$u(x) = \sqrt{A}\left[x^{1+\beta} + \frac{B}{2A}\right] \quad , \qquad \text{(A1)}$$

the solution can be expressed with the error function ϕ(u) according to

$$y(x) = \frac{\sqrt{\pi}B}{2\sqrt{A}} e^{u^2}\left[\Phi(u_m) - \Phi(u)\right] \ . \qquad \text{(A2)}$$

For

$$\beta = 2\alpha + 1$$

the transformation u(x)

$$u(x) = \sqrt{B}\left[x^{1+\alpha} + \frac{A}{2B}\right] \qquad \text{(A3)}$$

leads otherwise to the solution

$$y(x) = 1 - e^{-\left(u_m^2 - u^2\right)} - \frac{\sqrt{\pi}A}{2\sqrt{B}} e^{u^2}\left[\Phi(u_m) - \Phi(u)\right] \ . \qquad \text{(A4)}$$

A simple solution follows for

$$\alpha = \beta$$

$$y(x) = \frac{B}{A+B}\left[1 - e^{-(A+B)(1-v)}\right] \ , \quad v = x^{1+\alpha} \ . \qquad \text{(A5)}$$

GETTERING AND LIFETIME ENGINEERING IN SILICON WAFERS

S. Martinuzzi and O. Palais
University of Marseille 13397 Marseille Cedex 20 - France

Gettering techniques must be applied to silicon wafers and epitaxial layers to meet the main requirement of the SIA Roadmap: metallic impurity concentrations below 10^{10} cm^{-3}. In this paper several gettering techniques are described, like polysilicon backside layer, enhanced precipitation of oxygen, porous silicon layer, highly doped substrate for epitaxial layers and proximity or lateral gettering. To evaluate the effectiveness of these techniques, only minority carrier lifetime measurements can be used as they can detect recombining impurities in the 10^9 to 10^{10} cm^{-3} range. Lifetime measurements by means of contactless techniques, like the microwave photoconductance decay or the microwave phase shift, are well accepted to day and lifetime scan maps are obtained, which can be converted into impurity concentration maps. Combined with minority carrier diffusion length scan maps, the lifetime maps leads to the mapping of minority carrier mobility scan maps.

1. INTRODUCTION

Gettering techniques are used to remove fast diffusers, like iron, copper, nickel, from processed semiconductor materials and to trap them into gettering sites with the aim of improving the lifetime of minority carriers in active regions which contain the devices or which participate in their working.

Different gettering techniques are used depending on the type of devices to be made. The bulk of the silicon wafer may be inactive, like in integrated circuits, or active, like in power devices, solar cells or photodiodes. The case of the epitaxial layer is to be taken into account separately.

Gettering results from two basic mechanisms called segregation and relaxation (eventually enhanced by self-interstitial injection) [1-4], which work together or separately. Relaxation gettering is defined as a non equilibrium mechanism, occuring during the cooling-down of the wafers at the end of a high temperature processing step. It results from heterogeneous precipitation at imperfections or defects (gettering sites) due to a supersaturation of contaminants.

Segregation gettering is rather an equilibrium process based on the enhancement of the solubility of metallic contaminants in the gettering region. Such gettering occurs during the cooling down of the wafers after annealing at high temperature ($T \geq 900°C$) (so that the impurities are dissolved and diffuse through the wafer).

The gettering techniques differ depending on the localisation of gettering sites. Internal gettering occurs when these sites are located in the bulk, while when these sites are located at external surfaces external gettering occurs. Another possibility is the proximity gettering occuring when the sites are created in the vicinity of devices in integrated circuits.

Phosphorus diffusion near the surface of the wafer, from a $POCl_3$ source, is the best example of external gettering technique, well known and used since 1950. It uses the above mentioned basic mechanisms to shrink impurity microprecipitates, to push substitutional impurities in interstitial ones and to trap these impurities in the highly doped N^+ region. Such trapping is due to the increase of solubility, to the decrease of diffusivity of metallic atoms and to the formation of dislocation arrays around SiP precipitates [2] which appear at the surface. Moreover the formation of the SiP precipitates injects in the silicon bulk a high density of self-interstitials which participate in the kick-out and precipitate shrinking phenomena.

Another well known example of external gettering is the Al-Si alloying, when a thick Al layer (> 1 μm) deposited on one side of the silicon wafer is annealed at temperatures T higher than 570°C. At such temperatures the alloy is liquid and in the liquid phase the solubility of metallic impurities is 100 to 10^3 times higher than in solid silicon, and reaches few percents in concentration [5 ; 6 ; 7]. A strong segregation of fast diffusers can occur, with the proviso that these impurities are interstitially dissolved.

Both the preceding techniques works very well, especially for bulk devices like solar cells, but they are invasive and are not easy to use in the case of integrated circuit devices, because phosphorus diffusion creates N^+ layers on the front and back side of the wafer and because deep Al-Si spikes are formed when the Al-Si structure is maintained , after alloy formation, at temperatures higher than 900°C for a few hours (such annealing is needed in order to dissolve microprecipitates as there is no injection of self-interstitials).

Both the preceding external gettering techniques have demonstrated their ability to improve solar cells made from imperfect silicon but have to be replaced by non invasive techniques, i.e. techniques which do not introduce impurity atoms, and which do not cover the entire surfaces of the wafer.

Internal gettering techniques have been well developed in the case of oxygen rich Czochralski grown (Cz) wafers, based on the formation in the bulk of oxygen related microprecipitates and denuded zones close to the surfaces by the so-called "high-low-high" processes. Metallic impurities are trapped at the interface between the precipitates and the silicon matrix, especially at associated crystallographic defects (dislocation loops, stacking faults). However, it has been shown that precipitation of contaminants at oxygen related precipitates, at a given temperature, only occurs when the metal concentration exceeds the solubility limit. This requirement strongly limits the efficiency of internal gettering at low contamination levels. Notice that in FZ wafers like in epitaxial layers, the oxygen concentration is below the solubility limit and gettering sites cannot be created in the bulk.

Other techniques must be employed, like the formation of buried nanovoid bands or of patterned nanovoids bands, the deposition of a polysilicon layer on the back side, the formation of a porous silicon layer on the back surface, the use of highly doped substrates for epitaxial layers, and the use of nitrogen to enhance internal gettering.

The use of gettering techniques to clean silicon wafers so that impurity concentrations decrease below 10^{11} cm^{-3} calls for new analytical techniques which are sufficiently sensitive, which do not destroy the wafers and which allow concentration mapping along the surface of the wafer. The measurement of the lifetime of minority carrier (τ) applied to processed wafers appears to be the most likely method to achieve this concentration, irrespective of the doping level. The lifetime is obtained by contactless techniques using either microwave [8] or optical [9] probes when the wafers are excited by near infrared radiations. These techniques are based on the photoconductance decay or on the phase shift [10 ; 11], provided the surfaces are passivated by means of passivation techniques which do not modify the properties of the investigated material. Obviously the best solution could consist in combining gettering, surface passivation and bulk lifetime measurement.

In contrast to analytical techniques whose the signal amplitude is proportional to the impurity concentration, the lifetime value is inversely proportional to this concentration. Consequently the evaluation of τ is an efficient material and process cleanliness monitor.Lifetime engineering in turn provides the evaluation of the surface recombination velocity. From the values of minority carrier diffusion length (L) and lifetime, the mobility of minority carriers can be obtained.

The present paper gives recent results obtained by means of the promising above-mentioned gettering techniques and describes the potentialities of lifetime engineering in the detection and identification of metallic impurities at very low concentrations.

2. EXPERIMENTAL TECHNIQUES

2.1. Gettering techniques

Polysilicon backside external gettering. This technique is used successfully in device fabrication to minimize the concentration of metallic impurities, Cu and Ni and also Fe atoms are removed although the gettering kinetics of Fe is controlled by the diffusion of interstitial Fe atoms to the gettering sites. These gettering sites are located in a polycrystalline silicon deposited on the back side of a wafer whose front side is devoted to the preparation of IC devices. Shabani et al. [12] have voluntarily contaminated such structures by iron in-diffusion from the surfaces at 900, 1000 and 1050°C and then quenched them to room temperature. The contamination level is in the range 10^{12} to 10^{13} cm^{-3}. Gettering occurs during annealing at 800°C for 1 to 6 hours, and proceeds by relaxation during the cooling-down, provided the temperature is sufficiently high in order that Fe atoms reaches the polysilicon layer.

Internal gettering by enhanced oxygen precipitation. When the oxygen concentration is higher than 10^{18} cm^{-3} in Cz wafers, oxygen related precipitates may be formed which can induce internal gettering. For oxygen concentrations of up to $6x10^{17}$ cm^{-3}, precipitation is negligible and bulk microdefect densities (BMD) sharply decrease in polished as well as in epitaxial wafers.

For 300 mm wafers, used as epitaxial substrates, all kinds of back side layer must be omitted, as these types of treatments have to be applied after a double side polishing process. Consequently internal gettering is required for 300 mm wafers, provided no crystal originated particle (COP) and surface localized light scatterers (LLS) are formed in the near surface region.

Such internal gettering results from an enhanced precipitation of oxygen which can be obtained by two methods :

(i): the use of a thermal pretreatment

(ii): the co-doping with nitrogen in concentrations around 10^{14} cm^{-3}.

Method (i) was applied [13] to wafers containing $5\text{-}6x10^{17}$ cm^{-3} of oxygen atoms. It added two thermal steps prior to the epitaxial deposition process : rapid thermal annealing at 1200°C followed by additional annealing at 750°C for 5h. Bulk micro-defect densities in the range $3x10^4/cm^2$ to $10^5/cm^2$ were observed. These defects survived to the subsequent epitaxial deposition step and could be used for internal gettering.

Method (ii) less expensive than (i), results from the co-doping with small size atoms during the crystal growth. Co-doping with 10^{14} cm^{-3} nitrogen atoms was found efficient for oxygen concentrations in the range $5\text{-}6x10^{17}$ cm^{-3}, with the further advantage of the reduction of the sizes of COPs as well as of LLS on the wafer surface. After nitrogen doping the size of COPs was found around 50 nm instead of 150 nm. [13 ; 14].

Porous silicon layer. This technique is an extension of the polysilicon back side gettering as it uses the segregation of impurities at the interface of a porous silicon layer. The effective surface of the voids in the porous silicon is 100-1000 times larger than that of the wafer surface and gettering effects are enhanced dramatically [15].Porous layers, 1 to 40 μm thick, can be formed by anodisation in HF based electrolytes, and the effective surface of the voids in them is in the range 100 to 400 m^2/cm^3 depending on the porosity.

Like in Fig.1a, the porous layer may be formed on the back side of wafers used for power bipolar devices and solar cells. For integrated circuits created in an epitaxial films the porous layer can be buried beneath the epitaxial film, and very close to the active region, as shown in Fig.1b. Moreover the porous layer may also be discontinuous, as in Fig.1c, in the form of a pattern of isolating regions for VLSI devices. (The pattern is realized in the N^+ wafer before epitaxy).
Such buried porous silicon layers can getter impurities during the epitaxy and also during all the subsequent processing steps.

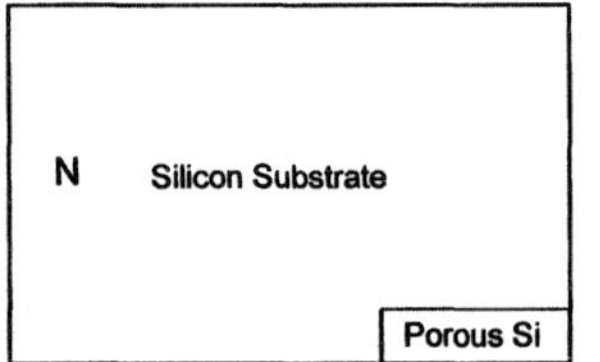

Fig.1a. Backside porous silicon layer used for the gettering of impurity from the silicon substrate in ref.15.

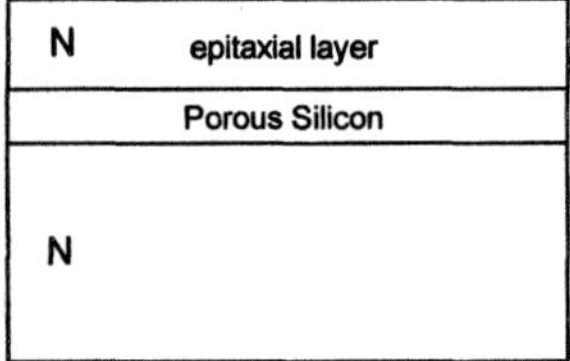

Fig.1b. Continuous frontside porous silicon layer beneath an epitaxial layer (ref.15)

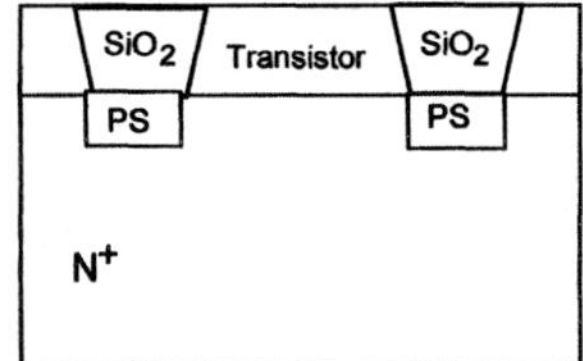

Fig.1c. Discontinuous porous silicon (PS) layer beneath the isoplanar dioxde isolation of bipolar transistors (from ref. 15).

Gettering by a nanocavity band. Proximity gettering could result from the creation of gettering sites close to the working regions of wafers or epitaxial layers by implantation of helium or hydrogen ions. Such implantations at high doses ($\geq 3.10^{16}$ cm^{-2}) form gas bubbles as the solubility of these gases is low in silicon ($< 10^{16}$ cm^{-3}).

After subsequent annealing at temperatures higher than 400°C for hydrogen and 700°C for helium nanocavities are formed, and the gases out-diffuse.
As the ions are concentrated around a depth called the projected range Rp, which depends on the ions species and on the ion energy, a band of cavities is formed beneath the implanted surface around Rp.
The internal surfaces of the cavities, probably covered with unsaturated bonds, act as efficient trapping sites for metallic atoms [16]. Moreover, gettering also occurs when the wafers are annealed during processing steps like oxidation and nitridation. Contamination by fast diffusers could be markedly reduced as these atoms may be trapped in the nanocavity band. The cavity region can be accurately located in wafers or in epitaxial layers by controlling the implanted area and the ion energy. Effects of back side external gettering, internal gettering and also proximity gettering can thereby be obtained.

However, in an epitaxial layer, the cavity band must be located at such a depth that the space charge region developed by the potential barrier which surrounds the cavity layer [17] does not interact with the device region. As the doping level of the epitaxial layers is generally up to around 10^{15} cm^{-3} , high ion energy is required, in the MeV range.

Gettering by highly boron doped substrate. Such a technique is particularly suited to remove iron atoms from an epitaxial layer deposited on a highly boron doped silicon substrate. Iron is known to be a major impurity in the semiconductor device industry, because it can be introduced in the bulk of silicon wafers during high temperature processing steps. In some type of devices, epitaxial layers moderately doped by boron are deposited on highly boron doped substrates. It has been observed that after annealing Fe gettering occurs and was evaluated by measuring the decrease of Fe concentration in the epitaxial layer [18]. This is essentially due to the increase of Fe solubility in silicon when the boron concentration is higher than 10^{18} cm^{-3}.

One extension of this technique is the implantation of boron ions in an epitaxial layer or in a wafer in order to create a region of accumulation of Fe-B complexes around Rp. Benton et al. [19] have shown that a buried region of high boron concentration provides enhanced gettering of Fe in oxygen rich wafers, which can be explained by electronic interactions between interstitial Fe and the dopant boron ions.

Other transition metals can be gettered by this technique and Table III gives some literature data [19 ; 20] concerning such fast diffusers which can form metal-boron pairs.

Lateral gettering. The standard gettering techniques are not effective for thin film-silicon-on-insulator (TFSOI) substrates, due to the presence of a buried oxide, which tends to prevent the migration of fast diffusers from the top TFSOI layer into the bulk substrate. This implies that impurities remain in the active area of the device structure and the gate oxide integrity is degraded. Grown-in dislocations and surface micro-roughness were not found to be the main causes of oxide failure while metallic impurities appeared to be certainly involved in it. Gettering is useful provided the gettering sites are be placed in the TFSOI layer rather than in the silicon substrate. Lateral gettering has been proposed [21] for metal-oxide-semiconductor devices (MOS) and in this technique gettering sites are formed close to the MOS devices. Hong et al. [22] have proposed a more effective lateral gettering which improves significantly the gate oxide integrity. Fig. 2 describes the proposed structures.
Crystallographic defects are introduced in TFSOI by Si ion implantation (175 keV – 10^{15} cm^{-2}). The 100 nm TFSOI layer is completely amorphized and although lateral

recrystallisation occurs at high temperatures, defects remain in the center of the implanted region and act as gettering sites.

Table I. Experimentally determined properties of fast diffusers ($D > 10^{-6}\ cm^2s^{-1}$ at 1100°C) in silicon (from Ref. 19 and 20).

Metal	Charge state	Activation Energy (eV)	Pair	Charge state	Solubility 1100°C (cm^{-3})	Activation Energy (eV)
Co_s	 A	E_c - 0.41 E_c - 0.21	Co-B ?	0/+	$4x10^{15}$	E_v + 0.10
Cr_i	0/+D	E_c - 0.22	Cr_i-B	0/+	$3x10^{14}$	E_v + 0.28
Cu_s	A AA	E_c - 0.16 E_v + 0.41	Cu_i-B	0/+	$8x10^{17}$	E_v + 0.22
Mn_i	+/++DD 0/+D -/0A	E_v + 0.25 E_v + 0.42 E_v + 0.11	Mn_i-B	0/+	$3x10^{15}$	E_v + 0.55
Fe_i	0/+D	E_v + 0.43	Fe_i-B	0/+	$3x10^{15}$	E_v + 0.10
N_i	A A	E_c - 0.43 E_v + 0.22	Ni_i-B	0/+	$5x10^{17}$	E_v + 0.14
with D : donor level ; A : acceptor level ; AA : double acceptor ; i : interstitial ; s : substitutionnal ; E_c and E_v : energy levels of the conduction and valence bands.						

The defects could be located in the future source and drain regions, provided they do not contact the gate edge. Gettering occurs during temperature ramp-up and stabilization, from 750 to 900°C.

The quality of gate oxides can be evaluated by ramp-voltage breakdown and Fowler-Nordheim current voltage.

Notice that a similar structure could be obtained by creating nanocavities in the lateral gettering regions, using He ions in place of silicon ones.

2.2. Minority carrier lifetime measurement

These measurements are generally done today by contactless techniques by means of microwaves or optical probes and the results are usually displayed as color maps for any size wafers. The main advantage of contactless techniques is that they do not need any kind of structure, like p-n junctions or metallic ohmic contacts. They can be used in line to control the cleanliness of the processing steps used in the device industry. Details of the different contactless techniques can be found in ref.8 and 23.

Such techniques are well suited to control the improvement of the electrical properties of the materials by gettering techniques because the lifetime is strongly dependent on impurity concentrations. In addition, the discovery of the ion-boron pair formation and dissociation and its exploitation by Zoth and Bergholz [24], have made it possible to identify and determine the iron density in p type wafers. Other metals like chromium, or even those included in table III can also be considered.

The bulk recombination lifetime τ_b may be given by :

$$\tau_b = (\sigma\, v_{th} N t)^{-1} \qquad (1)$$

with σ the minority carrier capture cross section, v_{th} the thermal velocity and N_t the defect density.

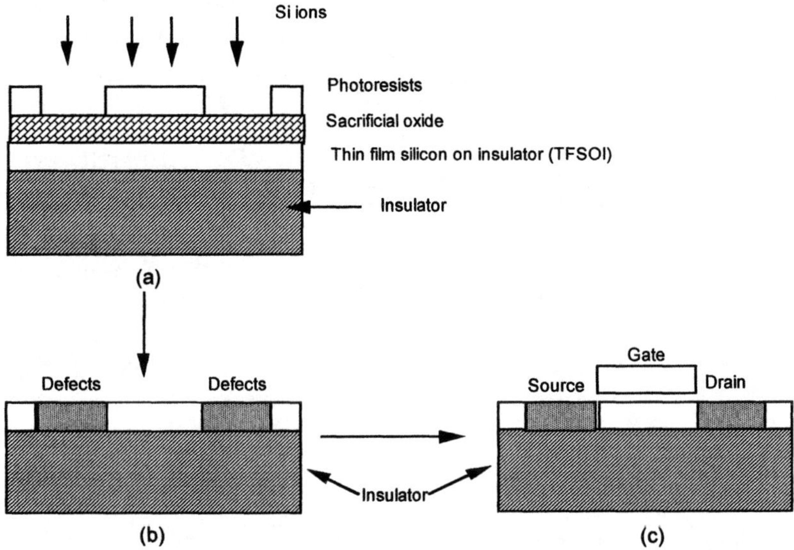

Figure 2. Fabrication of MOS devices with lateral gettering : (a) localized amorphisation of TFSOI by Si ion implant ; (b) formation of crystallographic defects ( ) gate oxide growth and MOS structure with impurities far from channel region (from ref. 22).

The surfaces play an even greater role when the bulk lifetime increases through the surface recombination velocity S and in epitaxial layers their influence is critical because the layers are typically much thinner than the minority carrier diffusion length.

The measured lifetime is an effective lifetime τ_{eff}, given by :

$$\tau_{eff}^{-1} = \tau_b^{-1} + \tau_s^{-1} \quad \text{and} \quad \tau_s = \frac{t}{2S} \qquad (2)$$

with τ_s the surface lifetime and t the sample thickness of the sample.

To obtain τ_b it is necessary to control S by means of surface passivation techniques in which the material is not modified by high temperature annealing, although thermal oxidation is one of the conventional processing steps of the device fabrication. Fig.3 shows the computed variations of τ_{eff} with S for various τ_b.
However the injection level must also be taken in account. It is frequently reported that τ_b increases with the injection level, this is true for deep impurity energy levels, but the opposite can also be true for those close to the band edges, as shown in Fig.4.

Referring to table I, one can assume that τ_b decreases with the injection level in Fe-B pair-containing samples, while τ_b increases in Fe_i containing ones. In Cr contaminated samples τ_b increases slightly, while it increases strongly in oxygen precipitate-containing samples [25].

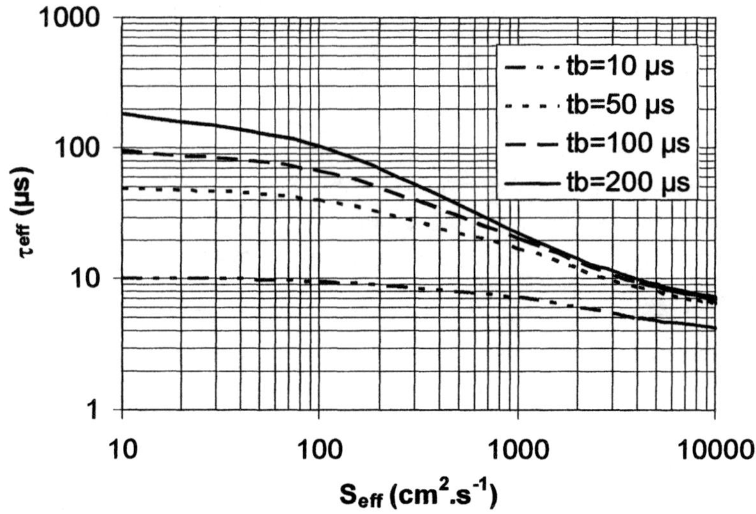

Fig.3. Computed variations of effective lifetime τ_{eff} with surface recombination velocity when τ_b varies in the range 10 to 200 μs as indicated on the curves.

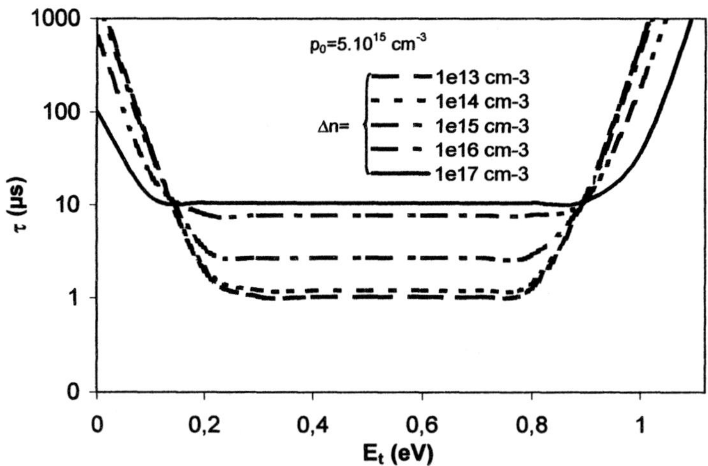

Fig.4 Variation of the lifetime versus deep level energy for various injection levels referred by Δn.

As the photoconductance decay technique is frequently used, we have chosen to focus the present paper on few other techniques.

The phase shift (PS) technique This technique uses a microwave probe and laser light excitation [11 ; 26 ; 27].

A microwave beam from a Gunn diode at 9.4 GHz frequency is directed onto the wafer through a circulator and a coaxial cable while excess carriers are created by a laser diode (λ =940 nm). The laser diode beam intensity is modulated by a sine wave form function generator.

The phase shift between the optical excitation and the microwave reflected power is measured. As the excitation is modulated by small amplitude sine waves, the technique is a quasi steady state one, and its main advantage is that it allows work at practically constant injection level. Moreover, trapping effects which can distort the measured lifetime values are reduced.

The relationship between phase shift ϕ and lifetime τ_b as a function of S (of both surfaces) and of the excitation frequency f is given by equation (15) of reference [10] and expression (3) of reference [11]. The phase shift ϕ is a function of several parameters :

$$\phi = F(S_f ; S_b ; D ; \tau_b ; \alpha ; W ; f) \qquad (3)$$

with S_f and S_b the recombination velocity of the front (illuminated) and back surface, respectively, W the thickness of the wafer , α the exciting light absorption coefficient.

If three parameters are well-known (w, α and f), others are not, but varying the modulation frequency f can help to determine the unknown values of S_f, S_b and D. This is another advantage of the PS technique.

However for passivated surfaces and low excitation frequencies [8], the relationship (3) is simplified and becomes :

$$-tg\phi \cong \omega\tau_{eff} \qquad (4)$$

A lifetime scan map can be obtained if the laser beam is focused (spot diameter ≅ 50 µm) and microwaves are directed through a thin coaxial cable, while the samples are moved by computer controlled step motors, sliding above the end of the coaxial cable. A complete description of such a technique has been given in ref.26.

Optical detected photoconductance decay or phase shift Warta [8] has proposed a technique called MFCA for modulated free carrier absorption which uses an infrared light probe (in place of the microwave one) whose absorption is modulated by the excitation beam. This technique allows measurement up to high doping levels, in the 10^{17} cm^{-3} range. It is based on the variation of the optical absorption coefficient in near infrared light of free carriers, whose concentration is modified by the excitation beam. Lifetime scan maps can also be obtained. The excitation beam is produced by a AsGa laser diode and the wavelength of the optical probe is greater than 1.1 µm.

Resonance coupled photoconductive decay This advanced technique recently proposed by Ahrenkiel and Johnston [28], is also a contactless one and uses an optical excitation and a microwave probe. However it works linearly at frequencies below 1 MHz and over three orders of magnitude of carrier concentrations. Measurements can be done over many decades of injection level allowing the use of injection-level spectroscopy. The technique has been successful in measuring different materials from small area thin films to 350 µm thick samples.

2.3. Minority carrier diffusion length (L) measurement and mapping techniques

Such techniques are not contactless and are based on steady state photocurrent. However they can bring confirmations of the lifetime investigations and the combination of L and τ_b mappings leads to the mapping of the mobility of minority carriers (μ).

The diffusion length L of minority carriers is determined by correlating the spectral variation of the steady state short-circuit photocurrent of (MIS) diodes with the spectral variation of α. The diodes are prepared by deposition of a semitransparent 100 Å thick aluminium layer (electron gun evaporation) on the illuminated surface.

Scan maps of local values of L are obtained using the previously described light beam induced current technique (LBIC) [29] from which lifetime scan maps can be deduced if the diffusion coefficient D is known.

L can also be determined by contacless techniques like the surface photovoltage as reported by Schroder [30].

2.4. Surface passivation

It is necessary to passivate the surfaces in order to measure a lifetime as close to τ_b as possible. High temperature techniques (thermal oxidation ; silicon nitride deposition) are well known, however the material properties may be modified during the annealing.

Another efficient technique consists in immersing the investigated wafer in a hydrofluoric acid solution [31], which drastically reduces S whatever the doping level of the wafer and the injection level (S decreases to 40 $cm.s^{-1}$). Immersion in an ethanol-iodine solution also gives good results [32]. Unfortunately S varies with time, which impedes the use of scanning techniques and the ethanol-iodine solution must be replaced by an aqueous solution of iodine polyvidone which works well for 10 to 12 hours without noticeable variations of S [27].

Notice that it was observed that phosphorus diffusion close to the surface from a $POCl_3$ source reduces the influence of S and may be combined either with plasma enhanced chemical vapor deposition (PECVD) deposition of a hydrogen rich silicon nitride layer on the N^+ layer, or with the immersion in the iodine solution.

3. RESULTS

3.1. Gettering engineering

External polyback side gettering [12] under slow ramp down cooling conditions is able to decrease the iron concentration below 8 x 10^{12} cm^{-3} in substrates as well as in epitaxial layers.

The increase of the internal gettering in oxygen poor Cz wafers by co-doping with nitrogen appears to be very efficient as the density of bulk microdefects (BMD) is in the range 10^4 to 10^6 cm^{-2} when the dissolved oxygen concentration varies between 6.5 to 7.5 x 10^{17} cm^{-3}.

The potential getter efficiency is very high since more than 90 % of Ni atoms and more than 95 % of copper atoms are removed from a contaminated crystal.

Similar iron atom gettering has also been verified. The results obtained by Gräf et al. [13] on 300 mm epi pp$^-$ wafers confirm that sufficient gettering capability is achievable. The co-doping of the epi substrates by nitrogen appears the most cost-effective approach to enhance gettering capabilities.

Porous silicon is also highly efficient and the gettering of Au and Cu atoms works through the wafer back side or by the front side, underneath an epitaxial layer. Au and Cu atoms are accumulated in the porous silicon which acts as a sink for such metallic impurities [15]. Generation lifetime is increased by 6 times in the epitaxial layer deposited on a N$^+$ substrate with a porous silicon back side layer, after annealing at 1100°C for 30 min.

Nanocavities produced by He or H ion implantation are also of great interest. It was demonstrated that the cavity band traps metallic atoms like Cu, Ni, Co, Fe and Au [33 ; 34]. Monolayer coverage of the cavity walls was observed, consistent with a cavity saturation near 5.10^{15} cm^{-2} for Cu, Ni, Co and Fe [35]. A band of cavities created beneath a surface protects the wafer bulk from a metallic contamination [36].

A highly boron doped region can also concentrate metallic impurities thanks to the enhancement of the solubility which depends on the boron concentration when this concentration is higher than 10^{18} cm^{-3}. Such segregation gettering works well during the cooling down of the samples [18] and Benton et al. [19] have reported a partition coefficient as high as 10^6 at a P$^+$P interface for iron. Boron implantation gettering is more effective than gettering produced by silicon implantation damages. This technique makes a large contribution to the effective gettering of Fe by PP$^+$ epitaxial silicon wafers and is expected to operate for Cr an Mn atoms. When the B ion energy is in the MeV range and the ion dose is higher than 10^{14} cm^{-2}, the total concentration of iron included in Fe-B pairs and in interstitial iron is found to be below 10^{10} cm^{-3}.

The results of Benton et al. [19] showed that a separate contribution of B implantation doping and of oxygen precipitates occur in the P$^+$ implanted wafers. This is due to the positive charge of iron interstitial atom when the Fermi level is below the trap level of Fe_i in silicon, i.e. $E_v + E_t = 0.39$ eV.

Lateral gettering is also a technique of choice for the improvement of epitaxial layers and of gate oxide integrity formed on TFSOI substrates [22]. As a result of such gettering the average breakdown voltage of MOS capacitors is improved, with much thighter breakdown distribution, and becomes very close to the values obtained when bulk substrates are used to make these devices. Lateral gettering also leads to the elimination of I-V humps. Such improvements are certainly related to the gettering of metallic impurities and Fe lateral gettering was reported in ref.21.

3.2. Lifetime engineering

Provided the surfaces are sufficiently well passivated, the effective lifetime is close to the bulk lifetime, which is strongly related to the impurity concentration N_t.

The contactless techniques, like the microwave phase shift [37], allows the detection of impurity concentration in the range 10^{10} to 10^9 cm^{-3}. Fig.5 shows the variation of the phase shift ϕ versus N_t, for a metallic impurity with an effective capture cross section $\sigma = 10^{-14}$ cm^2 for 3 values of the surface recombination velocities 10, 500 and 5000 cm.s^{-1}. The computed curves saturate for $N_t \cong 10^9$ cm^{-3}.

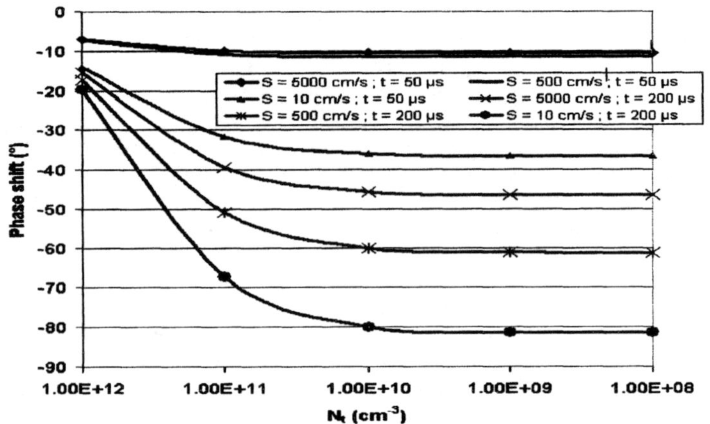

Fig.5 Phase shift induced by a sample containing a metallic impurity, versus impurity density N_t, for an effective capture cross section $\sigma = 10^{-14}$ cm^2, two τ values and for three surface recombination velocities: 10 cms^{-1}, 500 cms^{-1} and 5000 cms^{-1}. ($D = 30$ cm^2s^{-1}, $W = 300$ µm, $f= 6000$ Hz).

Such a sensitivity is irrespective of the doping level and can be achieved in samples doped to 10^{16} cm^{-3} range, which is not possible by means of deep level transient spectroscopy.

Measuring the lifetime can also allow the evaluation of N_t and the identification of metallic species. The case of Fe is well known in p-type silicon. In boron doped wafers, Fe-B pairs are formed and practically all the iron atoms which have not precipitated are included in such pairs. These pairs can be dissociated by short annealing at 210°C [24] or by increasing the injection level [39], and interstitial Fe_i atoms are released. The difference in the capture cross sections of Fe_i and Fe-B induces a decrease of the bulk lifetime, after the dissociation. The interstitial concentration of iron (which has formed pairs) can be evaluated from equation (5):

$$[Fe] = K \left[\frac{1}{\tau_{Fe_i}} - \frac{1}{\tau_{FeB}}\right] \text{ cm}^{-3} \qquad (5)$$

with K a coefficient depending on the doping level and on the rate of FeB pair dissociation due to the annealing.

The determination of the coefficient K is of paramount importance, because even at low injection level, τ_{FeB} depends strongly on the dopant concentration, as shown by Fig.6. This figure gives the computed variations of the lifetime associated with FeB and Fe_i, where p_o and Δp are, respectively, the doping level and the excess carrier concentration for various p_o

From the Fig.6 results and using experimental data obtained by DLTS, the variation of K versus the doping level can be computed. It is given in Fig.7, for a given injection level. It appears that we must distinguish between two various

doping ranges: K is negative for low doping level ($\tau_{Fei} < \tau_{FeB}$) and K is positive ($\tau_{Fei} < \tau_{FeB}$) for high doping level [37].
Between these two ranges there is an indetermination, due to the recombination activities of Fe_i and FeB, which are approximately the same. Experimental data obtained from lifetime measurements before and after FeB pair dissociation lead to the experimental values represented by black dots in Fig.7.

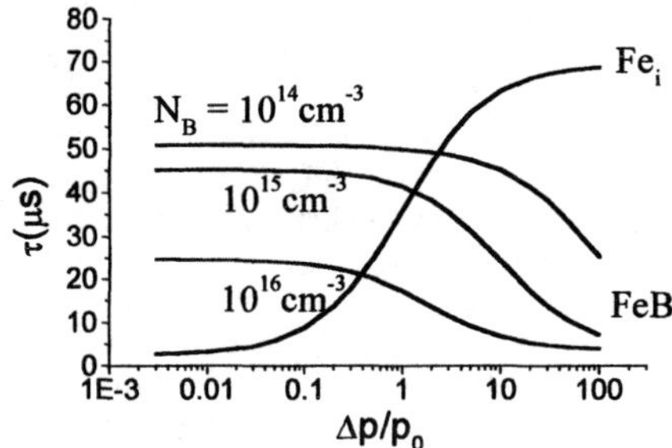

Fig. 6 : Calculated lifetime dependence versus excitation level for FeB pairs and Fe_i, in samples with various dopant concentrations and $[Fe_i] = 1x10^{12}cm^{-3}$.

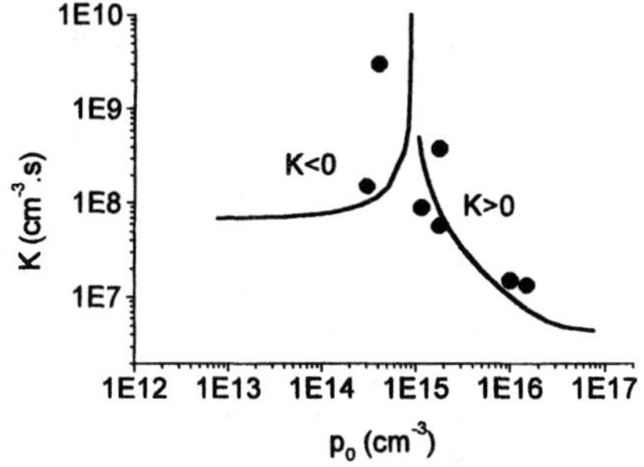

Fig.7 : variation of K vs doping level for a given injection level of $4x10^{14}cm^{-3}$. Dots correspond to experimental measurements and straight lines to computed values

The iron concentration map of a multicrystalline silicon (mc-Si) sample doped to 3×10^{16} cm^{-3} ($K = 52 \times 10^6$ cm^{-3} s), previously gettered by phosphorus diffusion was obtained by this method [37]. In the initial wafer, iron concentration was in the 10^{14} cm^{-3} range, and was not easily evaluated because several other impurities like Cu and Cr contaminated the material. After the gettering step, metallic impurity concentrations decrease, the lifetime is neatly increased making easy the determination of iron concentration (in fact, the interstitial iron Fe_i associate to boron atoms).

Scan maps of τ_{Fei} and τ_{FeB} allow to map concentration of Fe, with a sensitivity limit in the 10^{10} cm^{-3} range and, thanks to the microwave phase shift technique, with a spatial resolution of 50 μm [37].
Obviously the sensitivity limit depends on the presence of other recombination centres, like oxygen precipitates in Cz silicon [38].

Rein et al. [40] have proposed a lifetime spectroscopy technique based on temperature dependent lifetime measurements which allows a direct determination of the energy level of the recombination centres which controled τ_b. By means of this technique it is possible to detect and identify impurities like titanium and molybdenum, and also metastable defects in boron doped Cz silicon.

The combination of L and τ_b scan maps allows the mapping of minority carrier diffusion coefficient D (or mobility μ). For a Cz monocrystal an homogeneous scan is obtained, as expected, with D_n close to 30 $cm^2.s^{-1}$, which is a corrected value in a crystal doped to 10^{15} cm^{-3} by boron.

For multicrystalline silicon wafers boron doped to 2 x 10^{16} cm^{-3} and previously gettered by phosphorus diffusion (and whose the surfaces are passivated by this diffusion), the scan map of D exhibits the same features as those displayed by the lifetime and diffusion length scan map. In the homogeneous grains, D reaches 29 $cm^2.s^{-1}$ like in a monocrystal, while it decreases strongly at grain boundaries and dislocation clusters [41 ; 42].

Such a variation of D (or μ) for minority carriers validates the technique of resistivity mapping used by Barrance Diaz et al. [43] to describe the features of defects in multicrystalline wafers.

CONCLUSION

As the new generation of devices requires that the impurity level in active regions be below 10^{10} cm^{-3}, gettering techniques must be used, although their cost represents 10 % of the total integrated circuit manufacturing costs.

In this paper we have described some gettering techniques whose the engineering can satisfy the above mentioned requirement, like polysilicon back side external gettering, enhanced oxygen precipitation, internal gettering, porous silicon layer external and internal gettering, nanocavity band gettering, gettering by highly boron doped substrates and lateral gettering.
Each technique possesses its own engineering and its own field of application and every silicon integrated circuit processing line favors the choice of a particular gettering technique.

However, to evaluate the gettering efficiency at the required level of impurity only the lifetime engineering has the capability to give a clear answer up to 10^9 cm^{-3}. Moreover, in some cases, it is also possible to identify impurities which remain in the active regions. This is particularly true for Fe atoms, but it is not precluded that other metallic contaminants may also be identified.
This is why we have also described some contactless lifetime measurement techniques, and especially the phase shift one, which are particularly suited for impurity mapping.

Although there are still many unknowns about the recombination phenomena and the gettering effects, the field of the present paper is likely to be an important subject of investigation for the years to come.

The complementarity of gettering and lifetime engineerings makes them essential to the production of semiconductor devices.

REFERENCES

[1] W. Schröter, M. Seibt, D. Gilles, Mat. Sciences and Technology, vol.4, 539 (1991).

[2] A. Ourmazd, W. Schröter, Appl. Phys. Lett. 45, 781 (1984).

[3] S.A. Mc.Hugo, H. Hieslmair in Wiley Encyclopedia of Electrical Electronics Engineering, J.G. Webster Ed., p. 388, John Wiley Inc. N.Y. (1999).

[4] J.S. Kang, D.K. Schroder, J. Appl. Phys. 65, 2974 (1989).

[5] M. Appel, J. Hanke, R. Schindler, W. Schröter, J. Appl. Phys. 76, 4432 (1994).

[6] S. Martinuzzi, I. Périchaud, J.J. Simon, Appl. Phys. Lett. 70, 2744 (1997)

[7] R. Grafiteanu, U. Gösele, T.Y. Tan, Material Research Society Symposia Proceedings (MRS ed., Pittsburgh, PA, USA, p.297(1995).

[8] J.W. Orton, P. Blood in The Electrical Characterisation of Semiconductor : Measurements of Minority Carrier Properties, pp. 71-122, ed. by Academic Press, London, pp.71-122(1990).

[9] W. Warta, Solar Energy Mat. and Solar Cells 72, 389 (2002).

[10] M. Orgeret, J. Boucher, Rev. de Phys. Appliquée, 13, 29 (1978).

[11] A. Schönecker, J.A. Eikelboom, A.R. Burgers, P. Lölgen, C. Leguijt, W.C. Sinke, J. of Appl. Phys. 79, 1497 (1996).

[12] M.B. Shabani, Y. Shiina, Y. Shimanuki, Proceedings of te 6th International Symposium "High Purity Silicon VI" ed. by C.L. Claeys, P. Rai-Choudhury, M. Watanabe, P. Stallhoer and H.J. Dawson, p. 305, The Electrochemical Soc. Proc. Series, Pennington N.J. (2000).

[13] D. Gräf, U. Lambert, R. Schmolke, R. Wahlich, W. Siebert, E. Daub, W.V. Ammon, in ref.12, p.219.

[14] D. Yang, R. Fan, Y. Shen, D. Tian, L. Li, D. Que, in ref.12 p.357.

[15] G. Lamedica, M. Balucani, A. Ferrari, V. Bondarenko, V. Yakovtseva, L. Dolgyi, Solid State Phenomena, 82-84, p.405 (2002).

[16] S.M. Myers, D.M. Follstaedt, D.M. Bishop and J.W. Medernach in Semiconductor Silicon1994, H.R. Huff, W. Bergholz and K. Sumino Eds, p. 808, The Electrochemical Soc. Proceedings Series, Pennington, N.J.-USA, (1994).

[17] C.H. Seager, S.M; Myers, R.A. Anderson, W.L. Warren and D.M. Follstaedt, Phys. Rev. B, 50, 2458 (1994)

[18] M.B. Shabani, Y. Shiina, S. Shimanuki, F.G. Kirscht, Solid State Phenomena 82-84, 331(2002).

[19] J.L. Benton, P.A. Stolk, D.J. Eaglisham, D.C. Jacobson, J.Y. Cheng, J.M. Poate, N.T. Ha, T.E. Haynes, S.M. Myers, J. Appl. Phys. 80, 3275 (1996).

[20] K. Graff, Metal Impurities in Silicon Device Fabrication, Springer Ed., NY (1995).

[21] S. Koveshnikov, A. Agarwal, K. Beaman, G.A. Rozgonyi, Solid State Phenomena 47-48, 183 (1995).

[22] S.Q. Hong, T. Wetteroth, H. Shin, S.R. Wilson, D. Werho, T.C. Lee, D.K. Schroder, Appl. Phys. Lett. 71, 3397 (1997).

[23] Recombination lifetime measurements in silicon, D.C. Gupta, F.R. Bacher, W.M. Hughes, eds, ASTM-STP 1340, West Conshohocken, PA-USA, (1998).

[24] G. Zoth and W; Bergholz, J. Appl. Phys. 67, 6764 (1990).

[25] D.K. Schröder in ref.12, p.365.

[26] O. Palais, J. Gervais, L. Clerc, S. Martinuzzi, Rev. of Scientific Instruments, 70, 4044 (1999).

[27] O. Palais, J. Gervais, E. Yakimov, S. Martinuzzi, Eur. Phys. J. AP 10, 157(2000)

[28] R.K. Ahrenkiel, S.W. Johnston, Solar Cells and Solar Energy Materials 55,59 (1998).
[29] M. Stemmer, S. Martinuzzi, Inst. Phys. Conf. Serv. 135, 239 (1993).
[30] D.K. Schroder, Mat. Science and Engineering B91-92, 196 (2002).
[31] E. Yablonowitch, T. Gmitter, Appl. Phys. Lett. 49, 587 (1986).
[32] T.S. Horanyi, T. Pavelka, P. Tüttö, Appl. Surf. Science 63, 306 (1993).
[33] S.M. Myers, D.M. Bishop, D.M. Follstaedt, Mater. Res. Soc. Symp. Proc. 316, 33 (1994).
[34] V. Raineri, Solid State Phenomena, 57-58, 43 (1997).
[35] S.M. Myers, Nucl. Instr. and Meth. in Phys. Res. B106, 379 (1995).
[36] I. Périchaud, E. Yakimov, S. Martinuzzi in ref.12, p.341.
[37] O. Palais, E. Yakimov, S. Martinuzzi, Mat. Science and Engineering B91-92, 216 (2002).
[38] O. Porrini, P. Tessariol in ref.12, p.236.
[39] D. Mac Donald, A. Cuevas, J. Wong-Leung, J. Appl. Phys. 89, 7932 (2001).
[40] S. Rein, T. Rehrl, J. Isenberg, W. Warta, S.W. Glunz, Proc. of 16th European Photovoltaic Solar Energy Conf. Glasgow, 1476, (2000)
[41] D. Sontag, G. Hahn, P. Geiger, P. Fath, E. Bucher, in ref. 12, p.533.
[42] O. Palais, L. Clerc, A. Arcari, S. Martinuzzi, Proc. of E-MRS Conf. 2002 Symp. E, to be published.
[43] M. Barrance Diaz, W. Koch, C. Hässler, H.C. Brantigam, in ref. 9, p.473.

SENSITIVE COPPER DETECTION IN P-TYPE CZ SILICON USING μ-PCD

H.Väinölä, M. Yli-Koski, A. Haarahiltunen, J. Sinkkonen
Helsinki University of Technology, Electron Physics Laboratory, P.O. BOX 3000, 02015-HUT, Finland

Conventional microwave photoconductive decay method in conjunction with high-intensity bias-light has been used to study out-diffusion and precipitation of copper in *p*-type silicon. The high-intensity light is used to activate the precipitation of interstitial copper, which follows Ham's kinetics and results in a distinctive decrease in the excess carrier lifetime. Results indicate that Cu concentration well below 10^{12} cm^{-3} can be detected by this method. It is demonstrated that positive corona charge can be used to prevent out-diffusion of interstitial copper, while negative charge enables copper to freely diffuse to the wafer surfaces. It was observed that precipitation rate of copper increases significantly when the bias-light intensity is raised above certain critical level. In addition, the copper precipitation rate was discovered to be much higher in samples, which have internal gettering sites. These findings suggest that *i*) high-intensity light reduces the electrostatic repulsion between positively charged interstitial copper ions and copper precipitates enabling copper to precipitate in wafer bulk even at low concentration level *ii*) during high-intensity illumination oxygen precipitates provide effective heterogeneous nucleation sites for copper.

INTRODUCTION

Copper is a harmful impurity in modern IC technology as it degrades gate oxide integrity and causes increased junction leakage. To be able to detect copper contamination by contactless minority-carrier lifetime methods is an attractive idea, as measurements are fast, non-destructive, and sensitive to measured defect concentrations, making them a suitable characterization tool for the industrial environment. However, conventional recombination studies show weak dependence of low copper concentration on minority carrier lifetime in *p*-type silicon [1].

Recently, it was observed by surface photovoltage method (SPV) that optical activation drastically enhances recombination activity in *p*-type silicon, which contains copper [2,3]. Bazzali *et al.* [4] have reported similar behavior of enhanced recombination activity of copper in silicon in presence of small oxygen precipitates. By combining these observations, the authors have discovered that microwave photoconductive decay (μPCD) method in conjunction with high-intensity bias-light can be used to detect copper contamination in *p*-type silicon, which contains small oxygen precipitates [5].

One of the objectives of this paper is to gain understanding about the origin of the above-mentioned increase in recombination activity of copper. Previously, it has been observed that in *p*-type wafers, precipitation of copper takes place in wafer bulk only when copper concentration exceeds approximately 10^{16} cm^{-3} [6]. This has been explained by the following electrostatic model. As Cu_i^+ ions act as shallow donors in *p*-type silicon, the Fermi level approaches the conduction band edge with increasing copper concentration. When the Fermi level is raised above $E_C - 0.2eV$, which is known to be the electroneutrality level of copper precipitates, the charge state of copper precipitates changes from positive to neutral or negative diminishing the electrostatic repulsion between positively charged interstitial copper ions and copper precipitates. Instead, in *n*-type silicon copper can precipitate easily as Fermi level is close to the electroneutrality level of copper precipitates even at low copper concentration level [6]. Similarly, we suggest that electron Fermi level can be raised in *p*-type silicon by optical activation. This affects the electrostatic repulsion according to the above electrostatic model and results in an enhancement in copper precipitation rate.

In order to confirm the above argument, we begin this paper by studying how the generated excess-carrier density affects the precipitation rate of copper. Then we study the effect of external corona charge on the out- and in-diffusion of interstitial copper ions. Finally, we estimate the sensitivity of copper detection of the method used.

EXPERIMENTAL

Sample Preparation

Samples used in this study are single side polished 100 mm Czochralski grown silicon wafers with 525 μm thickness. The resistivity in boron-doped wafers is 10 Ωcm and in-doped wafers 3 Ωcm. The initial interstitial oxygen concentration is about 13 ppma, measured by Fourier transform infrared spectroscopy (FTIR) according to the ASTM F 121-83 standard method. All wafers were pre-oxidized at 1050°C for 15 minutes to dissolve the grown-in precipitate nuclei, *i.e.,* to diminish the effect of thermal history of the wafer.

Small oxygen precipitates were grown to some of the wafers by two-step thermal anneal. The nucleation of oxygen precipitates was made in nitrogen ambient at 625°C for 4 hours. This was followed by the growth of precipitates in oxygen ambient at 800°C for 4 hours. This procedure should result in oxygen precipitate density about 10^9 cm^{-3} [7], which is small enough that precipitates should not have an excessive effect on minority carrier recombination by themselves.

Copper was chemically deposited on the oxide-covered wafer surfaces by immersing them in diluted copper-sulfate solution. Copper in-diffusion was performed in nitrogen ambient at 800°C for 20 minutes through the silicon dioxide layer, which was used to diminish unintentional iron contamination.

Description of Measurement Method

In the experiments we use a method, recently developed by the authors, in which the microwave photoconductive decay (μPCD) method is used in conjunction with high-intensity bias light [5]. In *p*-type silicon the high-intensity light activates the precipitation of interstitial copper in the presence of small oxygen precipitates. The precipitation follows Ham's kinetics and results in a distinctive increase in the recombination rate, as copper precipitates are shown to be very effective recombination centers [8]. The change in recombination lifetime is clearly detectable even with low copper concentration.

We use 1 $\mu C/cm^2$ positive corona charge to prevent out-diffusion of copper in *p*-type silicon. If positive corona-charge is omitted, impact of high-intensity light may be very different. The method is applicable also in case when copper has completely out-diffused to the wafer surfaces, as we have demonstrated [5] that it is possible to in-diffuse copper back into wafer bulk from the oxide interface by changing the polarity of surface corona charge from negative to positive.

The minority-carrier recombination lifetime was measured at high injection level using Semilab's WT-85 XL measurement equipment. Optical activation was performed by external 973.5 nm bias-light, whose intensity was manually adjusted. Recombination lifetime measurements were carried out at constant excess-carrier concentration $\Delta n = 10^{16}$ cm^{-3}. Figure 1 shows an example of the effect of optical activation, with 29 W/cm^2 intensity, on the excess-carrier recombination lifetime in *p*-type silicon, which had copper concentration about $2 \cdot 10^{12}$ cm^{-3}. It can be seen that measured high-injection level lifetime $\tau(t)$ decreases from 5 ms to 150 μs during high-intensity illumination in a wafer, which contains small oxygen precipitates. The decrease in lifetime is much weaker in a homogenized wafer. The reason for this difference will be discussed later in this paper.

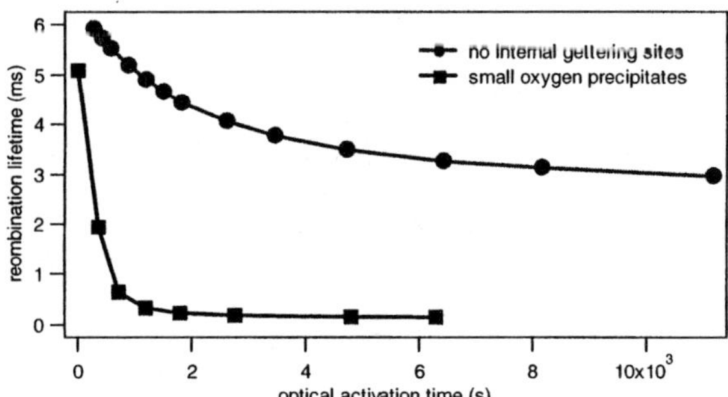

Figure 1. Effect of optical activation on the excess-carrier recombination lifetime in p-type silicon with and without oxygen precipitates.

Similar measurements were made on reference wafers, with and without oxygen precipitates, which were not intentionally copper contaminated. No degradation in lifetime was observed in these wafers, which supports the idea that the observed change in optically activated recombination lifetime is related to precipitation of copper.

The measured excess-carrier lifetime $\tau(t)$ is converted to optically activated excess-carrier recombination rate $R(t)$ as follows

$$R(t) = \Delta n[\tau^{-1}(t) - \tau^{-1}(0)], \tag{1}$$

where $\tau(0)$ is the measured lifetime without optical activation, *i.e.*, at $t = 0$. Therefore, by applying Ham's law [9] and assuming that optically activated recombination rate R is proportional to the total amount of precipitated copper, the following equation can be used to fit the experimental data

$$R(t) = R_{final}(1 - \exp(-t/\tau_{Cu})), \tag{2}$$

where τ_{Cu} is the time constant of precipitation. It is determined as follows

$$\tau_{Cu} = (4\pi n r_0 D)^{-1}, \tag{3}$$

where n is the effective precipitation density, r_0 is the effective radius of the precipitates and D is the diffusion constant of the interstitial copper. The measured data can be well explained by Ham's law, which gives the precipitation rate of copper as a function of optical activation time.

It is worth mentioning that the effect of optical activation on the recombination lifetime is permanent. Therefore, it was not possible to measure the same point more than once. For example, in copper out-diffusion study, the optical activation had to be done at several separate points on a wafer.

RESULTS AND DISCUSSION

Effect of optical intensity on recombination rate

Optically activated recombination rate was measured as a function of optical activation time with different bias-light intensities. The incident optical power was varied between 4.5 W/cm^2 and 46 W/cm^2. We determined the time constant τ_0 for each intensity by fitting Equation (2) to the measurement points. Figure 2 shows both the experimental and fitted data at two different light intensities in a wafer, which contains small oxygen precipitates.

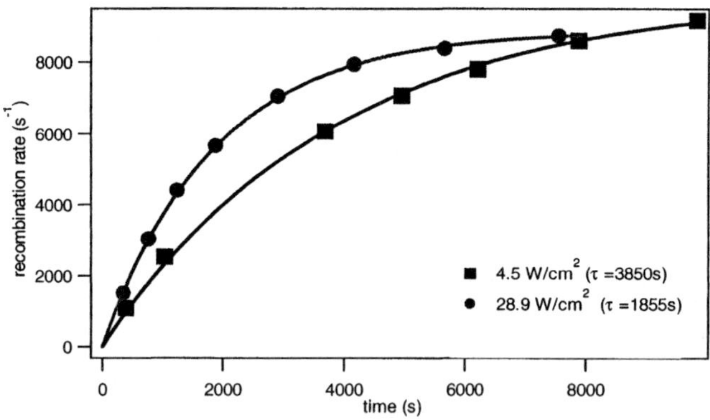

Figure 2. Optically activated recombination rate as a function of activation time at two different incident optical intensities.

It can be seen from Figure 2 that there is a clear difference in time constants between these two intensities. This result can be explained by different precipitation rates of copper, *i.e.*, in case of higher optical intensity precipitation of copper is faster and the corresponding time constant is smaller. Figure 2 shows also that the recombination rate saturates close to the same level in the end of the optical activation regardless of the optical intensity. This could indicate that the saturation level is proportional to the copper concentration as the measurements were made on the same uniformly copper-contaminated wafer.

To further study the influence of the bias-light intensity on the time constant of the optically activated recombination, we found it instructive to calculate the quasi-Fermi levels for electrons with different optical intensities. Calculations were made with commercial PC1D simulation software [10]. To obtain the time constant at different Fermi level positions, we measured the optically activated recombination rate with different optical intensities. Figure 3 shows the results of the obtained time constant of the recombination rate as a function of quasi-Fermi level for electrons E_{fn}.

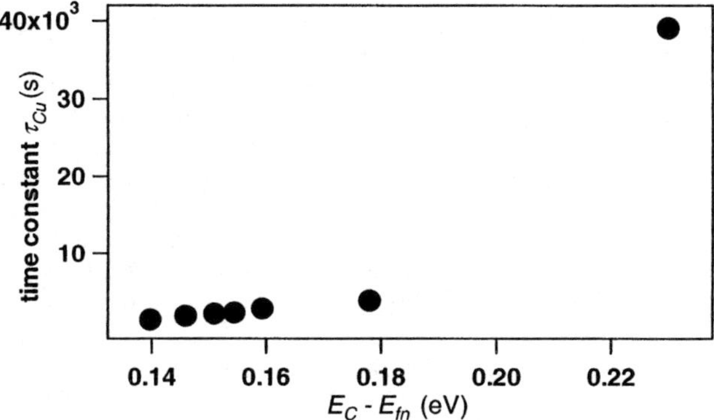

Figure 3. Time constant of the recombination rate as a function of electron quasi-Fermi level in wafers with oxygen precipitates.

We can see from the above figure that precipitation rate of copper depends on the electron Fermi level. Therefore, it is reasonable to expect that the change in recombination activity is due to the electrostatic effects, which are affected by the illumination. C. Flink *et. al.* [11] have reported that without optical activation precipitation of copper is very slow in the bulk of the wafer at low copper concentration, mainly due to the electrostatic repulsion between the positively charged copper precipitates and the interstitial Cu_i^+ ions. However, during illumination the quasi-Fermi levels for electrons raises towards the conduction band edge, which may change the charge state of some defect level(s) thus increasing the recombination activity.

First, we studied the energy range (0.14 – 0.18) eV to examine if the defect level E_c – (0.15 to 0.16) eV, related to the interstitial copper [12], affects the precipitation rate of copper. Figure 3 shows no significant change in recombination activity around this energy level. In addition, if the charge state of interstitial copper ions had an impact on the increased precipitation rate of copper, the precipitation should not follow the Ham's law as the precipitation is limited by the reaction - not by the diffusion. The observed change in recombination lifetime is exponential (Fig. 1), suggesting that the precipitation is rather diffusion limited.

Next, copper precipitates and their possible change in charge state during illumination was studied. It is reported that copper precipitates homogeneously forming copper silicides Cu_3Si, which are positively charged with electroneutrality level at approximately $E_c - 0.2$ eV [13]. Therefore, we measured the time constant with such a low intensity that the electron Fermi level is well below the electroneutrality level of copper silicides (E_c-E_{fn} = 0.23 eV). We observed that the time constant was considerably higher, around 39000 s, which indicates very low copper precipitation rate. Figure 3 shows that the increase in precipitation rate is very strong between energies (0.18-0.23) eV. This result indicates that there is a significant, probably step-like, increase in the copper precipitation rate between energies 0.18-0.23 eV, which supports the idea of the change in the charge state of copper precipitates from negative to positive or neutral.

The same measurements were made on the reference wafers, which went through only the first homogenization anneal, *i.e.*, in those wafers oxygen precipitation density was extremely low. It is seen from Figure 1 that oxygen precipitates have a significant effect on the precipitation rate of copper. The decrease in the time constant, i.e. the enhancement in the precipitation rate, is explained by the increase in the effective density of precipitation sites (Eq. (3)). In samples without intentionally formed internal gettering (IG) sites copper precipitates slowly, as the effective density of precipitation sites is small. In samples, which contain internal gettering sites, the copper precipitation rate is much higher as there are heterogeneous nucleation sites present. This suggests that oxygen precipitates, indeed, can act as effective sinks for copper.

Although the decrease in lifetime due to optical activation was observed in wafers, which did not have intentionally grown IG sites, the samples with IG sites are far more interesting from practical point of view. If the above method will be used to detect copper contamination, the required measurement time is considerably shorter in wafers, which have IG sites. Figure 1 shows that the required measurement time to obtain the saturation level is about 30 minutes in samples with IG sites. The corresponding time in homogenized samples is over 2 hours.

Out-diffusion of copper

When positive corona charge was deposited on the wafer surface the decrease in lifetime after optical activation was the same regardless of the elapsed time after deposition of corona charge. This indicates that the out-diffusion of copper is successfully eliminated by positive corona charge. Instead, by depositing negative corona charge the out-diffusion of copper to the wafer surfaces is enabled. By assuming that negative corona charge allows copper to freely out-diffuse to wafer surfaces and using out-diffusion kinetics based on Fick's law, we can estimate the concentration of copper [Cu] in wafer bulk as a function of time

$$[\mathrm{Cu}] = [\mathrm{Cu}]_0 \exp(-t/\tau_{diff}), \tag{4}$$

where $[Cu]_0$ is the initial copper concentration. Time constant of the diffusion, τ_{diff}, is

$$\tau_{diff} = d^2 / (\pi^2 D_{Cu_i}), \tag{5}$$

where d is the thickness of the wafer and D_{Cu_i} the diffusion constant of the interstitial copper ions.

We studied the out-diffusion of copper by depositing positive corona charge on both sides of the p-type silicon wafer. Then we measured the saturation value of the optically activated recombination rate at constant bias-light intensity (29 W/cm^2) as a function of elapsed time after deposition of corona charge. As optically activated recombination rate, measured at the same bias-light intensity, is proportional to the amount of precipitated copper in wafer bulk, we get as a result copper concentration (in arbitrary units) as a function of diffusion time. Equation (4) can then be used to fit the experimental data to find the time constant τ_{diff}. Figure 4 presents the experimental results as well as the fitted curve.

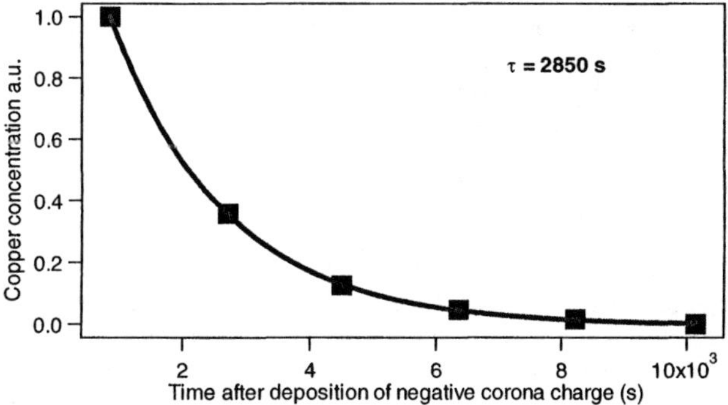

Figure 4. Copper concentration in wafer bulk as a function of time after deposition of negative corona charge.

The fitted curve was weighted by the three last measurement points as it can be assumed that the copper distribution profile is not necessarily constant in the beginning. The obtained time constant of the curve is 2850 s, which corresponds the diffusivity $9.8 \cdot 10^{-8}$ cm^2s^{-1}. This is in excellent agreement with room temperature diffusivity value of interstitial copper in this material ($9.7 \cdot 10^{-8}$ cm^2s^{-1}) [14]. We have thus shown that the decrease in optically activated recombination rate after deposition of negative corona charge is due to the out-diffusion of interstitial copper atoms.

Sensitivity of the method

We used n-type wafers as a reference to get a rough estimate for the amount of copper in wafer bulk. It is known that copper reduces the minority carrier lifetime in n-type wafers even in case of quite low copper contamination level [2,15]. However, we did not detect any decrease in lifetime in n-type wafers after intentional copper contamination. The charge carrier lifetime values were in the order of 4 to 5 ms both before and after contamination. This suggest that the copper surface contamination value is less than 10^{12} cm^{-2} [15]. The low copper content was supported by the observation of small increase in the lifetime of p-type wafers after contamination. This passivation effect of copper has been observed earlier with low copper concentration [1].

In order to get a better estimate for the amount of copper in the samples, adsorbed copper concentration at the oxide-covered wafer surface was quantified by vapor phase decomposition total reflection x-ray fluorescence analysis (VPD/TXRF), using $HF\text{-}H_2O_2$ containing scanning liquid. The copper concentration was measured to be $1 \cdot 10^{11}$ Cu-atoms/cm^2 before in-diffusion and $4 \cdot 10^{10}$ Cu-atoms/cm^2 after in-diffusion anneal. Assuming that no out-diffusion takes place, we can calculate that during in-diffusion only 60% of the surface copper diffuses into wafer bulk. This results in bulk copper concentration of $2.3 \cdot 10^{12}$ cm^{-3} in 525 μm thick wafer. Figure 1 shows that the lifetime decreases from 5 ms to 150 μs with this $2.3 \cdot 10^{12}$ cm^{-3} copper concentration. However, the detection limit is far below this concentration. For example, based on the out-diffusion studies, when less than 10% of the initial copper concentration is left in wafer bulk due to out-diffusion, the decrease in lifetime is still clearly detectable. Thus, we suggest that copper concentration well below 10^{12} cm^{-3} can be detected by this method. To obtain the exact sensitivity of this method, other independent method should be used to determine the quantitative amount of copper present in samples.

CONCLUSIONS

Recent studies of copper precipitation and out-diffusion in p-type silicon show that below a critical contamination the copper predominantly diffuses out to the wafer surface, while for higher copper concentrations it mainly precipitates in the bulk [6,11]. The behavior is explained by the electrostatic interaction between the positively charged interstitial copper and the copper precipitates. We have shown that copper can precipitate in wafer bulk far below this critical concentration under high-intensity bias light and that positive corona charge can be used to prevent its out-diffusion.

In addition, studies of internal gettering of copper have indicated that oxygen precipitates are relatively inefficient as sinks for copper [16]. However, our experiments show that under high-intensity illumination oxygen precipitates provide effective heterogeneous nucleation sites for copper. This could find application in developing gettering for copper.

It has been reported that a surface band bending repulses positively charged copper ions, which retards the out-diffusion of copper [11]. We have shown that by depositing negative corona charge on p-type silicon wafer the copper ions can freely diffuse to the wafer surfaces.

Based on the studies of out-diffusion of copper we estimate that copper concentration of about 10^{11} cm^{-3} can be detected by this method. To be able to use this method as a quantitative method for determining the copper concentration in p-type silicon wafers, the measurement results should be calibrated with some other quantitative method, for example transient ion drift (TID).

As copper precipitation follows Ham's kinetics, we expect that the precipitation rate is dependent on oxygen precipitate size and density. We will be applying this technique in the future for more detailed studies of the role of oxygen precipitates to get some data related to intrinsic gettering efficiency. In addition, more work concerning the copper concentration sensitivity will be performed.

ACKNOWLEDGEMENTS

Comments and suggestions given by Olli Anttila are gratefully acknowledged. The authors acknowledge also the financial support from the Finnish National Technology Agency, Academy of Finland, Okmetic Oyj, Micro Analog Systems Oy and VTI Technologies Oy.

REFERENCES

[1] A.A. Istratov and E.R. Weber, *J. Electrochm. Soc.*, **149**, G21 (2002).
[2] W.B. Henley, D.A. Ramappa, and L. Jastrezbski, *Appl. Phys. Lett.*, **74**, 278 (1999).
[3] D.A. Ramappa, *Appl. Phys. Lett.*, **76**, 3756 (2000).
[4] A. Bazzali, G. Borionetti, R. Orizio, D. Gambaro, and R. Falster, *Mater. Sci. Eng.* **B36**, 85 (1996).
[5] M. Yli-Koski, M. Palokangas, A. Haarahiltunen, H. Väinölä, J. Storgårds, H. Holmberg, and J. Sinkkonen, *J. Physics C.*, in press.
[6] R. Schadeva, A.A. Istratov, and E.R. Weber, *Appl. Phys. Lett.*, **79**, 2937 (2001).
[7] P. F. Wei, K.F. Kelton, and R. Falster, *J. Appl. Phys.* **88**, 5062 (2000).
[8] A.A. Istratov, H. Hielsmair, T. Heiser, C. Flink, E.R. Weber, W. Seifert, and M. Kittler, in *Proceedings of the 7th Workshop on the Role of Impurities and Defects in Silicon Device Processing*, edited by B.L. Sopori, (NREL, Copper Mountain, CO, 1997), p. 158.
[9] F.S.Ham, J. Phys. Chem. Solids, **6**, 335 (1958).
[10] D.A. Clugston and P.A. Basore, *Proceedings of 26th IEEE Photovoltaic Specialists Conference*, (Anaheim, USA, 1997), p. 207.
[11] C. Flink, H. Feick, S.A. McHugo, W. Seifert, H. Hielsmair, T. Heiser, A.A. Istratov, and E.R.Weber, *Phys. Rev. Lett.* **85**, 4900 (2000).
[12] A.A. Istratov and E.R. Weber, *Appl. Phys. A*, **66**, 123 (1998).
[13] A.A. Istratov, H. Hedemann, M. Seibt, O.F. Vyenko, W. Schroeter, T. Heiser, C. Flink, H. Hielsmair, and E.R. Weber, *J. Electr. Chem. Soc.* **145**, 3889, (1998).
[14] A.A. Istratov, C. Flink, H. Hielsmair, E.R. Weber, and T. Heiser, *Phys. Rev. Lett.* **81**, 1243 (1998).
[15] M. Miyazaki, *Recombination lifetime measurements in silicon* (ASM STP 1340), 294 (1998).
[16] A.A. Istratov, R. Sachdeva, C. Flink, S. Balasubramanian, and E.R. Weber, *Solid State Phen.* **82-84**, 323 (2002).

Topographic change at the SiO2/Si interface with multilayer oxidation for various temperatures

Daisuke Hojo, Norio Tokuda, and Kikuo Yamabe
Institute of Applied Physics
University of Tsukuba
Tsukuba, Ibaraki, 305-8573, Japan
Phone: +81-298-53-6937 Email: daisuke@esys.tsukuba.ac.jp

Oxidation temperature dependence of the topographic change at SiO_2/Si interfaces was investigated using an atomic force microscope (AFM). The vestiges of two-dimensional-oxide-island growth were observed at the interfaces. It was observed at the interfaces that the vestiges such as dark holes or islands observed at a lower oxidation temperature were significantly smaller than that at a higher oxidation temperature. The number of the vestiges was higher at a lower oxidation temperature. These results may indicate that oxidation stress affect oxidation.

KEYWORDS: silicon, silicon dioxide, interface, oxidation, step, terrace, layer-by-layer, oxidation stress, roughness, etching

Introduction

To realize high reliable thin oxide with low leakage current, Si surface topography should be atomically controlled.[1)] The authors also reported that the SiO_2/Si interfaces with wide step/terrace structures was roughened during thermal oxidation.[2)] It was found that the SiO_2/Si interfaces consisted of up to four atomic layers separated one another by 0.3 nm at an oxidation temperature of 1050°C in order to exclude oxidation stress as much as possible.[3)] These results provided evidence of breaking the layer-by-layer fashion at the SiO_2/Si interfaces. In the layer-by-layer model, there are only two layers to be observed at the interfaces. On the other hand, it is well known that a trench corner is rounded or sharpened by the thermal oxidation depending on the oxidation conditions.[4)] The results of the trenched surfaces indicated that the Si convex corner had become sharper during low temperature oxidation because of oxidation stress.

In this study, temperature dependence of the topographic change at the SiO_2/Si interfaces was investigated using an atomic force microscope (AFM) after removing the SiO_2 films.

Experimental

P-type, (111) oriented, Czochralski grown Si wafers in the $(11\bar{2})$ direction at an off-angle of 2'±1' with a resistivity of 10-20Ωcm were used. These are then etched, using ultra low dissolved oxygen pure water (LDW) to form the wide terrace/step structures on the Si wafer surface. The LDW was obtained by adding ammonium sulfite monohydrate as reported by Matsumura et al.[5] for 1 hour. Subsequently, these wafers were treated by a modified RCA cleaning treatment. These wafers were terminated by hydrogen using diluted hydrofluoric acid (DHF) to prevent the growth of chemical oxide.[6,7] The Si surfaces were thermally oxidized both at 1050°C in 2% dry O_2 diluted with argon gas and at 800°C and 600°C in 100% dry O_2. The SiO_2/Si interface topographies were observed by tapping mode AFM after removing SiO_2 film by DHF. It was confirmed that Si surfaces were not roughened through the DHF etching within our experimental error.

Results and Discussion

A typical initial topographic AFM image of the Si surface after forming the wide terrace/terrace structures is shown in Fig. 1(a). The Si surface before oxidation had clear wide terrace/step structures. Each step height was about 0.3 nm corresponding to the

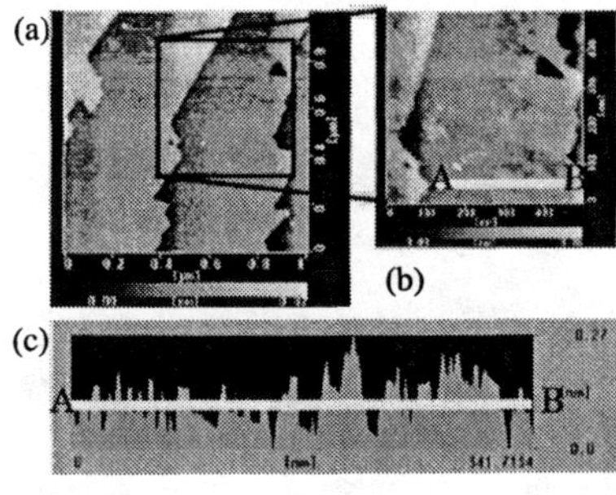

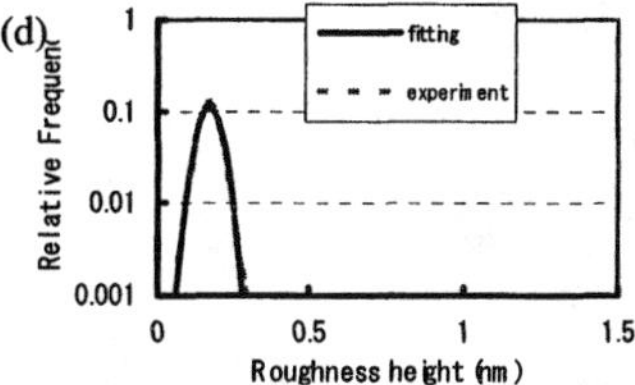

Fig. 1(a) An AFM image of initial Si(111) surface with the area of 1μm×1μm. (b) Magnified image magnified the terrace. (c) The cross-sectional image shows roughness was within 1 atomic layer. (d) The histogram shows the roughness distribution fitted in single Gaussian distribution.

atomic step height in (111) orientation. The magnified image is shown in Fig. 1(b). Fig. 1(c) is a cross-sectional view. As shown in Fig. 1 (d), the roughness height distribution on this terrace fitted to single Gaussian distribution with full width at half maximum (FWHM) of about 0.08 nm. This result indicates that the terrace surface is atomically flat. The FWHM value is the experimental error in our AFM measurement. This statistical method quantitatively shows that roughness is within an atomic layer.

A topographic image of an SiO_2/Si interface after oxidation at 1050°C for 5 min in the diluted O_2 ambient is shown in Fig. 2(a). The thickness of the SiO_2 was 4.2 nm by ellipsometry. The configuration of the step edges seemed to be preserved at the interfaces.[8, 9] The magnified image in Fig. 2(b) shows that dark holes are distributed over the whole range of the interface. From the cross-sectional view in Fig. 2(c), the depth of the dark holes was about 0.3 nm[10]. The oxidation gets ahead of a main interface, most highly occupied area, by one atomic layer at the dark holes.

Strictly speaking, in the layer-by-layer fashion, only two atomic layers should be occupied. In this figure, however, not only dark holes but also islands were observed. The height of these islands was 0.3 nm as shown in Fig. 2(c). A histogram was plotted to evaluate roughness more quantitatively as shown in Fig. 2(d). From this histogram, the roughness height distribution on this terrace was decomposed into three Gaussian peaks. Those peaks were separated one another by 0.3 nm. Therefore, the Gauss1 peak in Fig. 2(d) is due to the dark holes and the Gauss3 is due to the islands. The peak intensity of each Gaussian distribution, which is determined by the product of its peak height and its FWHM, can be attributed to an occupied area ratio of each Si atomic layer.

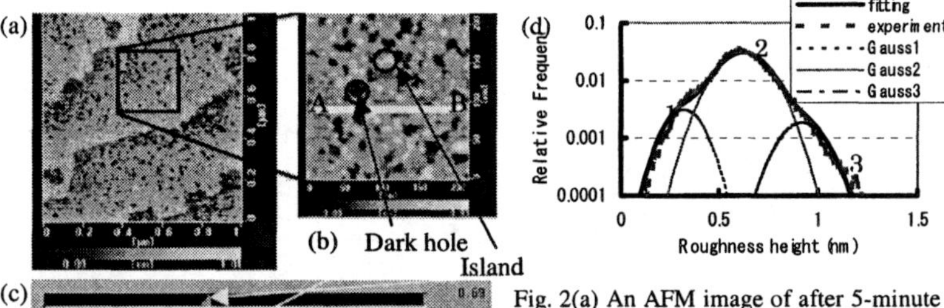

Fig. 2(a) An AFM image of after 5-minute oxidation of the SiO_2/Si interface with the area of 1μm×1μm. (b) There are dark holes and white dots around in the magnified image (c) The cross-sectional image shows the depth of dark holes were 0.3nm and the height of white dots were 0.3nm. (d) The histogram shows the oxidation front consisted of three atomic

The FWHM of each Gaussian distribution was about 0.2 nm, which is larger than that of pre-oxidation surface shown in Fig. 1(d). The reason for this is that an AFM cantilever had a finite radius. Its influence strongly appeared around the dark holes and the islands. The occupied area ratio for the dark holes was 7 %, 87 % for the main interface and 6 % for islands.

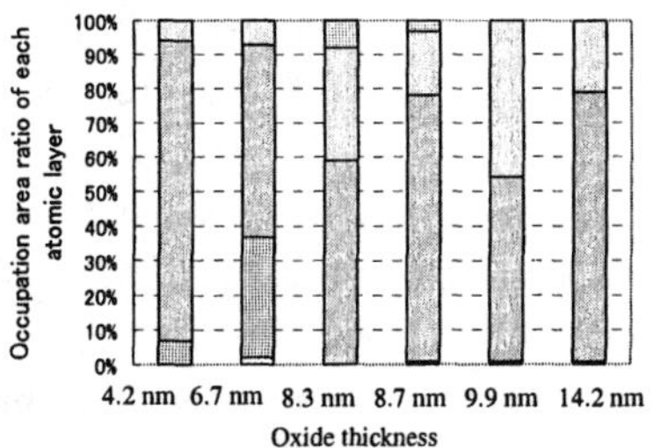

Fig. 3 Occupation area ratio of each atomic layer, such as dark holes or islands, was shown for each oxide thickness. The SiO_2/Si interfaces consist of at least three atomic layers for every oxide thickness up to oxide thicknesses of 14.2

Figure 3 shows the occupation area ratios of each atomic layer, such as dark holes or islands for various oxide thicknesses at 1050°C. The SiO_2/Si interfaces consist of at least three atomic layers for every oxide thickness up to oxide thickness of 14.2 nm. These occupation area ratios of each atomic layer show snapshots of the oxidation progress at the SiO_2/Si interfaces. It means that the thermal oxidation shrinks the islands on the main interface, while it creates and grows the dark holes there. Therefore, whether the interfaces consist of three or four atomic layers seems to be the difference of oxidation phase. However, it was interesting that as the oxide thickness increases, the total ratios of atomic layers other than highly occupied two atomic layers decreased gradually. The oxidation seems to become layer-by-layer-like[11-13] as the oxide thickness increases.

Next, the interface at a lower oxidation temperature was investigated. Fig. 4(a) is a

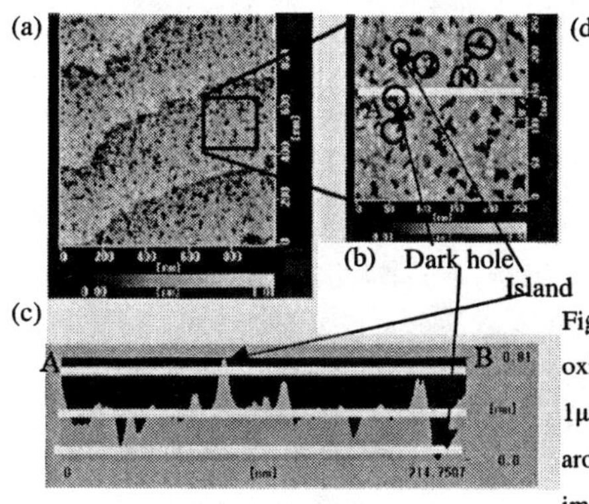

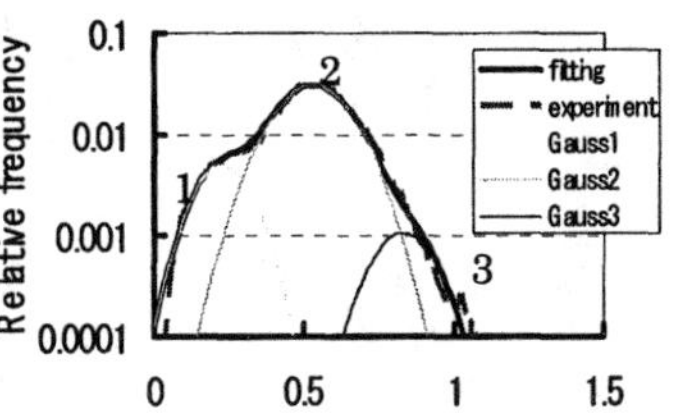

Fig. 4(a) An AFM image of after 800℃5-minute oxidation of the SiO_2/Si interface with the area of 1μm×1μm. (b) There are dark holes and white dots around in the magnified image (c) The cross-sectional image shows the depth of dark holes were 0.3nm and the height of white dots were 0.3nm. (d) The histogram shows the oxidation front consisted of three

topographic image of an SiO_2/Si interface after oxidation at 800°C. The thickness of the SiO_2 was 5.7 nm. The configuration of the step edges seemed to be preserved no matter what an oxidation temperature was. A magnified image in Fig. 4(b) shows islands are distributed over the whole range of the interface as observed at the interface of 4.2 nm-thick-oxide at an oxidation temperature of 1050°C. The number of the islands is higher and the size of those is smaller. The height of these islands, as shown in Fig. 4(c), was also 0.3 nm. As shown in Fig. 4(d), roughness distribution is decomposed into three peaks. The occupied area ratio of the dark holes was about 10 % (Gauss1), 87 % for the main interface (Gauss2) and 3 % for the islands (Gauss3). The FWHM of each Gaussian distribution became larger than that of Gaussian distributions at a higher oxidation temperature. This reason is that there are many small islands distributed over the whole range of the interface, then, influence of the radius of AFM cantilever more strongly appeared.

Figure 5(a) shows an AFM image of the interface after 600°C oxidation. A thickness of the film was 2.0 nm. A magnified image in Fig. 5(b) shows that both holes and islands are also distributed over the whole range of the interface. The height of these islands, as shown in Fig. 4(c), was also 0.3 nm. As shown in Fig. 5(d), a height histogram is decomposed into five peaks. The occupied area ratio of dark holes was about 7% (Gauss1), 74% for the main interface (Gauss2), 16% for islands (Gauss3) and 3% for towers (Gauss4) on the islands from the intensities. The occupied ratio of the peak higher than tower layer was less than 1% (Gauss5). Although the oxide thickness

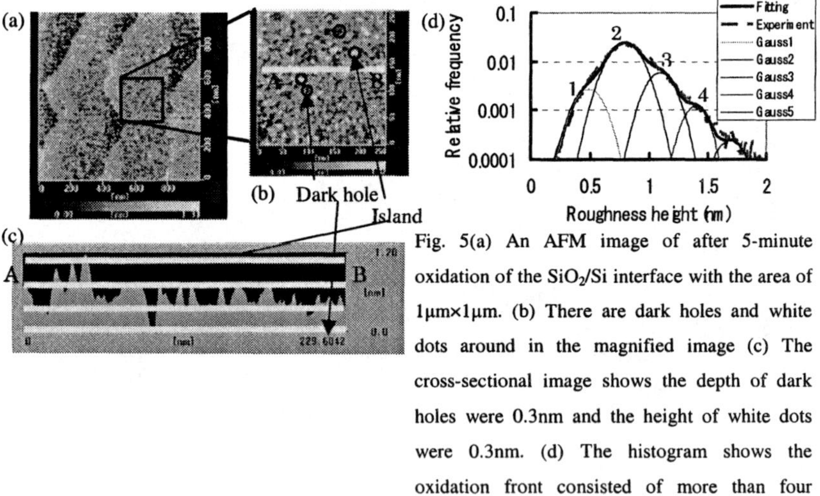

Fig. 5(a) An AFM image of after 5-minute oxidation of the SiO_2/Si interface with the area of 1μm×1μm. (b) There are dark holes and white dots around in the magnified image (c) The cross-sectional image shows the depth of dark holes were 0.3nm and the height of white dots were 0.3nm. (d) The histogram shows the oxidation front consisted of more than four

is thin, the interface was considerably rough. The numbers of dark holes, islands and towers at 600°C are more than that at 800°C. As far as we investigated, the mean separation distance among vestiges such as dark holes or islands increases with oxidation temperature.

In the oxidation, the dark holes are created on the atomistically flat Si surface and they are laterally stretched. The grown holes start to connect with the neighborhood holes. Furthermore, as the oxidation progresses, the islands are formed. This process is repeated one after another at the new interfaces. Figure 6 schematically shows a microscopic aspect of the generation of the holes and the lateral growth of them at an SiO_2/Si interface. That is, the hole growth depends on the oxidation rate in parallel to the interface, oxidation rate A. On the other hand, the hole generation rate is attributed to the oxidation rate in perpendicular to the interface, oxidation rate B. A ratio of these two oxidation rates determines the interface topography. Precisely, the oxidation rate B is divided into two parts in (111) orientation, which are from on-top layer to back-bond layer and from back-bond to on-top. These two differences are beyond the resolution in our AFM measurements. The difference between these two oxidation rates seems to depend on the oxidation conditions, such as oxidation temperature. The number of dark holes is closely related to this two oxidation rates because it decides the mean separation distance between dark holes. The islands and the towers are the same as the holes. The temperature dependence of the interface topography described above indicates that the oxidation rate A is more rapidly decreased with increasing oxidation temperature than the oxidation rate B. This temperature dependence suggests that the oxidation stress is one of the key factors to describe this phenomenon. We can consider that the oxidation stress suppresses the oxidation rate A, while it promotes the oxidation rate B, because the generation of the dark holes on the next atomic layer is easier than the growth of them on same atomic layer under the high oxidation stress at the interface. Based on this presumption, a reason why oxidation at 1050°C approached to the layer-by-layer with increasing oxide thickness is described as follows. As the oxidation progresses, a net oxidation rate reduces. Slow oxidation rate relaxes the oxidation stress at the interface at high oxidation temperature. So, the oxidation rate A is restored and the oxidation rate B gets relatively smaller. This condition is met with layer-by-layer oxidation. Although the oxidation rate is smaller at 800ºC than at 1050ºC despite diluted O_2 gas, the number of the vestiges was significantly large. This result reveals that the probability of attacking the flat interfaces with oxidation at a higher oxidation temperature is lower than at a lower oxidation temperature. The oxidation stress is more relaxed at 1050˚C than at 800˚C and at 600˚C. This result provides evidence that the oxidation stress affects the probability of attacking the flat interfaces.

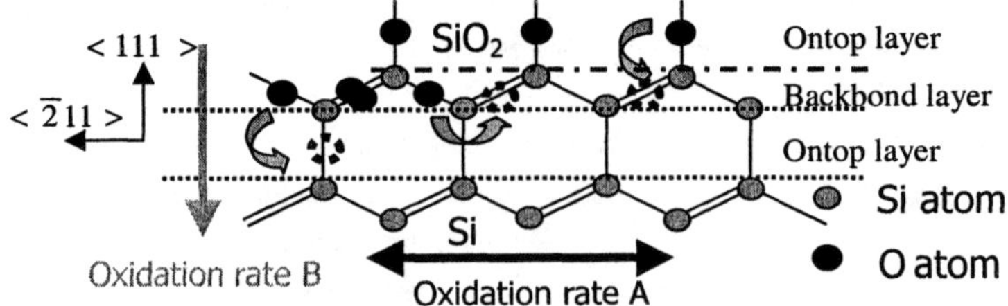

Fig. 6 Schematically microscopic aspect of the generation of the holes and the lateral growth of them at an SiO_2/Si interface. The growth and generation rates of dark holes depends on the oxidation rate A in parallel to the interface and the oxidation rate B in perpendicular to it, respectively. Precisely, the "oxidation rate B" is divided into two parts; from on-top layer to backbond layer and from back bond to on-top.

Conclusion

The interface topographic change was investigated using AFM. It was found that topographic change at the interface depended on an oxidation temperature. This difference may derive from whether oxidation stress is relaxed or not at the interfaces.

References

1) M. Murata, N. Tokuda, D. Hojo and K. Yamabe: Thin Solid Film **414** (2002) 56
2) N. Tokuda, M. Murata, D. Hojo and K. Yamabe: Jpn. J. Appl. Phys. **40** (2001) 4763
3) D. Hojo, N. Tokuda and K. Yamabe: Jpn. J. Appl. Phys. **41** (2002) L505
4) K. Yamabe and K. Imai: IEEE Trans. Electron Devices **ED-34** (1987) 1681.
5) H. Fukidome and M. Matsumura: Jpn. J. Appl. Phys. **38** (1999) L1085.
6) G. S. Higashi, Y. J. Chabal, G. W. Trucks and K. Raghavachari: Appl. Phys. Lett. **56** (1990) 656.
7) M. Sakuraba, J. Murota and S. Ono: J. Appl. Phys. **75**(1994) 3701.
8) V. Tsai, X. S. Wang, E. D. Williams, J. Schneir and R. Dixon: Appl. Phys. Lett. **71**(1993) 1495.
9) M. Suzuki, Y. Homma, Y. Kudoh and N. Yabumoto: Jpn. J. Appl. Phys. **32** (1993) 1419.
10) H. Watanabe, T. Baba and M. Ichikawa: Jpn. J. Appl. Phys. **39** (2000) 2015.
11) H. Watanabe, K. Kato, T. Uda, K. Fujita and M. Ichikawa: Phys. Rev. Lett. **80** (1998) 345.

12) H. Watanabe, T Baba and M. Ichikawa: Appl. Phys. Lett. **74** (1999) 3284
13) N. Miyata, H. Watanabe, and M. Ichikawa: Phys. Rev. B **58** (1998) 13 670.

DIODE ANALYSIS OF DEEP SUBMICRON CMOS P-WELL IMPLANTATION DAMAGE

A. Poyai[1,2], E. Simoen[1], C. Claeys[1,2] and R. Rooyackers[1]
[1]IMEC, Kapeldreef 75, B-3001 Leuven, Belgium
[2]also at E.E. Dept., KU Leuven, Kasteelpark Arenberg 10, B-3001 Leuven, Belgium

ABSTRACT

This paper derives the carrier generation and recombination lifetime, and the surface generation velocity from a dedicated analysis of the current- and capacitance-voltage characteristics of n^+-p-well shallow junctions. The selected deep boron implantation energy and dose were compatible with deep submicron CMOS technology. It is demonstrated that a higher leakage current and yield loss are obtained for a lower implantation energy and higher dose. Studying diodes with different geometries assesses the impact of the implantation energy and dose on the bulk, the perimeter and the corner leakage current components. Anomalous effects related to defect-assisted hole trapping in the p-well and to shallow-trench isolation (STI) related corner leakage current are illustrated.

INTRODUCTION

The scaling of the feature size in ultra large scale integrated (ULSI) technologies includes amongst others a tailoring of the substrate carrier concentration, which peak value increases for every next CMOS technology node. A high substrate doping concentration is required in order to control the short-channel effect and is also useful in latch-up suppression [1]. At the same time, a lower ion density near the channel is necessary in order to lower the Coulombic scattering and to enhance the inversion layer mobility. These requirements can be met by using a so-called retrograde well doping profile, which is usually fabricated by combining a low-energy with a high-energy ion implantation. This is followed by a dopant activation anneal, which should also reduce the implantation damage. However, it has been observed frequently that in the case of a high-energy boron implantation, the residual damage is difficult to remove by annealing [1-5]. Generally, silicon-interstitial related extended defects, ranging from small clusters to dislocation loops have been found after processing, which may affect the leakage current of junctions made in the well [6-9]. In addition, these extended defects may act as a source of interstitials during subsequent processing, resulting in so-called transient-enhanced diffusion (TED) of dopants [10].

The trend to lower the thermal budget in deep submicron CMOS and to reduce the deep well implantation energy raises considerable concern. This may not only result in insufficient damage removal but also brings the defective zone, which shifts with the projected range (R_p) of the implantation, closer to the junction. If these defects are located too close to the junction they can impact the diode current-voltage (*I-V*) characteristics. Therefore, the aim of the paper is to use p-n junction diodes as an assessment tool for the study of the impact of the deep boron implantation energy and dose on the electrical properties of a retrograde p-well, compatible with deep submicron CMOS technology. This is achieved by combining *I-V* and capacitance-voltage (*C-V*) characteristics.

EXPERIMENTAL DETAILS

Different geometry shallow n^+-p diodes (as indicated in Table I) compatible with submicron CMOS technology have been processed on 200 mm p-type substrates. A shallow trench isolation (STI) depth of 400 nm was used. A retrograde p-well was obtained by combining a different deep (as described in Table II) and a fixed shallow (35 keV, 1.3×10^{13} cm^{-2}) boron ion implantation, followed by a dopant activation anneal (10 min, 850°C). The n^+ region was created by an arsenic ion implantation (50 keV, 4×10^{15} cm^{-2}) and anneal (10 s, 1100°C). This was followed by Co-silicidation with a titanium capping layer (12/8 nm) with a maximum anneal temperature of 850°C. The automatic current measurement was performed at –2V, which is 10 % above the nominal operation bias of a 0.18 μm technology (1.8V). Per wafer 38 dies have been stepped automatically, allowing a statistical analysis of the leakage current (yield, wafer maps). Additional manual measurements provide more detail and should in principle enable to extract lifetime values, possible defect parameters, bulk versus perimeter versus corner components, etc. The static *I-V* characteristics of the diodes were measured on wafer level, in the dark, and in the range of –2V to +1V. The bias was applied to the back p-type substrate and the current was measured at the grounded top n^+ and bottom contact. Further, *C-V* characteristics have been measured on the same diodes at a frequency of 100 kHz, in order to assess the doping density and the depletion width. The latter parameter can be used to transform the reverse bias axis into a depth axis, so that one can easily obtain profiles of the electrical parameters.

Table I. Description of the different diode geometries used in the analysis. Indicated are also the width and length of one rectangle in the meander (ND) structure, and of one square in the square (SQ) structure.

Diode	Area (*A*) (μm²)	Perimeter (*P*) (μm)	Number of corners (N_C)	Width (μm)	Length (μm)	P/A (μm^{-1})
SQ	4999696	8944	4	2236	2236	0.001789
ND1	200000	400160	320	1.0	2500	2.0008
ND3	199680	200928	1248	2.0	320	1.00625
ND4	187500	45000	2400	12.5	25	0.24
ND5	2496000	407160	1248	12.5	640	0.163125
ND6	2500000	204000	320	25	1250	0.0816
ND7	2496000	215280	1248	25	320	0.08625
ND8	2480000	229400	2480	25	160	0.0925
ND9	5000000	404000	320	25	2500	0.0808
ND10	5000000	431250	2500	25	320	0.08625
ND11	4960000	229400	1240	50	320	0.04625

Table II. Description of the split batch silicon diodes for investigating the residual p-well implant damage induced leakage current. All wafers received the same implantation anneal (10 min at 850°C and 10 s at 1100°C).

Wafers	Energy (keV)	Dose (cm^{-2})
D3-D4	180	$1.2x10^{13}$
D5-D6	180	$8x10^{12}$
D9-D10	120	$1.2x10^{13}$

The same analysis approach has previous been used to study the impact of the thermal budget on the junction leakage current in a 0.18 and 0.13 μm CMOS technology [11]. It was observed that a lower thermal budget results in a higher doping density and a higher leakage current. Both the diffusion and the generation components of the area current increased due to the presence of residual implantation damage.

RESULTS AND DISCUSSION

Figure 1 gives the *I-V* characteristics of different geometry n^+-p-well junction diodes for the 180 keV, $1.2x10^{13}$ cm^{-2} p-well split. The *I-V* characteristics of large-area (SQ) and large-perimeter (ND1) diodes exhibit a normal behaviour (no current at zero bias). The *I-V* characteristics of the corner-rich diodes (ND4) show a clearly abnormal behaviour (current flow at zero bias). The origin of this phenomenon is related to the high number of corners (Table I). This is inferred from the fact that the ND4 diode has the highest number of corners per area and perimeter ratio (Fig. 2). This abnormal behaviour is also found in other corner-rich diodes (such as ND3) but the effect is smaller than for ND4 due to a lower corner current contribution to the total leakage current.

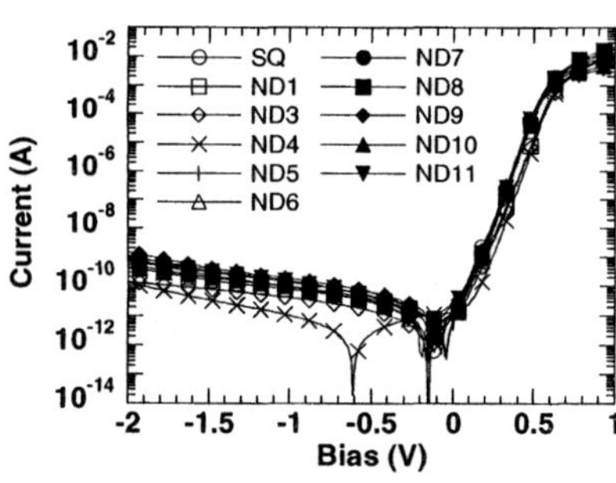

Fig. 1 Current-voltage (I-V) characteristics of different geometry n^+-p-well junction diodes for the 180 keV, $1.2x10^{13}$ cm^{-2} p-well split.

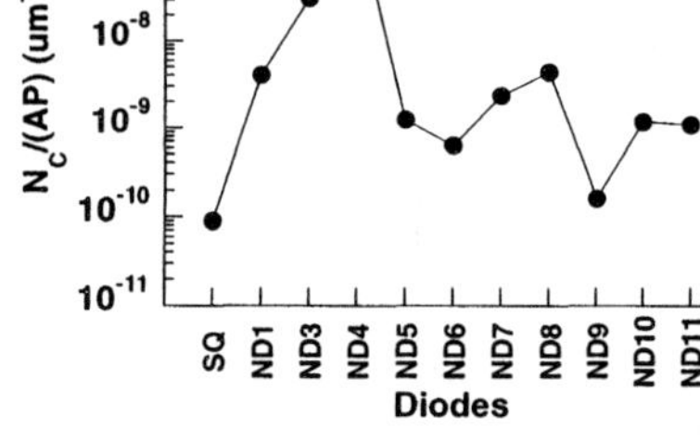

Fig. 2 $N_C/(AP)$ for different geometry n^+-p-well junction diodes for the 180 keV, $1.2x10^{13}$ cm^{-2} p-well split.

Combining a number of SQ and ND diodes allows to separate the volume or bulk current density [J_A (A/cm^2)] which scales with the diode area (*A*), from the perimeter density [J_P (A/cm)], assumed proportional with the perimeter (*P*). In principle, also a corner component [J_C (A/corner)] can be separated. More details on the analysis procedure have been published before [12]. The separation of the geometrical current components is based on the assumption that the total reverse current (I_R) is a linear combination of the different components, given by:

$$I_R = AJ_A + PJ_P + N_C J_C + I_{par} \quad (1)$$

where I_{par} represents a parasitic sample-independent system leakage.

By considering I_R/A versus P/A, as shown in Fig. 3, one can check whether I_R in Fig. 1 is a linear combination of the different geometrical current components as given by Eq. (1). A constant value J_A should be obtained in case that I_R is dominated by the area component, while a linear relationship with slope J_P is found when the peripheral component dominates I_R. Figure 3 indicates that I_R of SQ is governed by the area component. I_R of ND1 is dominated by the peripheral component. I_R/A of ND4 and ND10 is lower and of ND8 is higher than the expected line. ND10 has the same area, about 3 % higher perimeter and 77 % larger number of corners than ND9 (Table I). A lower I_R/A for ND10 than for ND9 is thought to be related to the corner component J_C, which has opposite sign compared with the area and peripheral current, according to the separation based on Eq. (1). The different current densities are shown in Fig. 4. A lower I_R/A for ND4 than the expected line is related to the corner component. This also confirms that the abnormal behaviour of ND4 in Fig. 1 is due to the corner-rich structure. A more in-depth discussion of Fig. 3 has recently been published by the authors [11]. Based on that analysis it was decided to use SQ, ND1 and ND4 as representative structures for the study of the effect of the thermal budget on the area, the peripheral and the corner component, respectively.

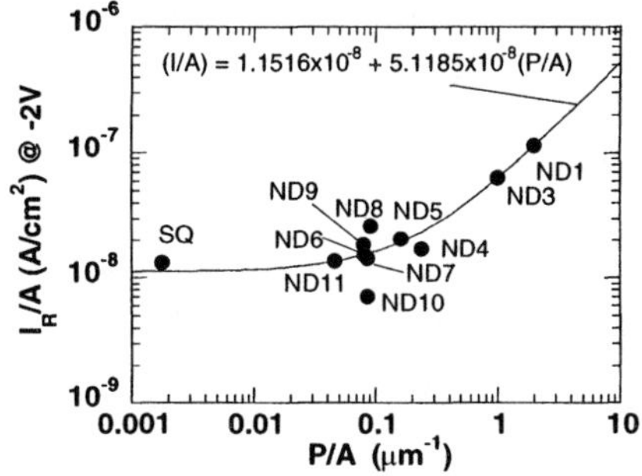

Fig. 3 The I_R/A versus P/A of different geometry n^+-p-well junction diodes for the 180 keV, $1.2x10^{13}$ cm^{-2} p-well split.

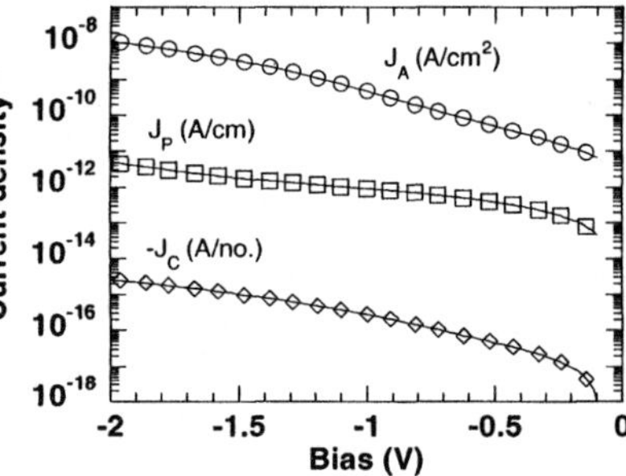

Fig. 4 Different current components of different geometry n^+-p-well junction diodes for the 180 keV, $1.2x10^{13}$ cm^{-2} p-well split.

Figure 5 shows the cumulative probability of SQ, ND1 and ND4 for different p-well conditions (Table II). In all components, the highest leakage current is observed for the lowest implantation energy, while the dose has a lesser impact. The same goes for the yield, where the best values are observed for $8x10^{12}$ cm^{-2} @ 180 keV. Analyzing Fig. 5, one can conclude that the area leakage current density has the most pronounced impact on the overall leakage current. Therefore, the area (J_A) and peripheral (J_P) current density have been calculated from Eq. (1) by combining at least 4 different diodes. The result is shown in Fig. 6. The highest J_A and J_P are found for the 1.2x10^{13} cm^{-2} @ 120 keV condition, while the lowest J_A and J_P are for the 8x10^{12} cm^{-2} @ 180 keV diodes.

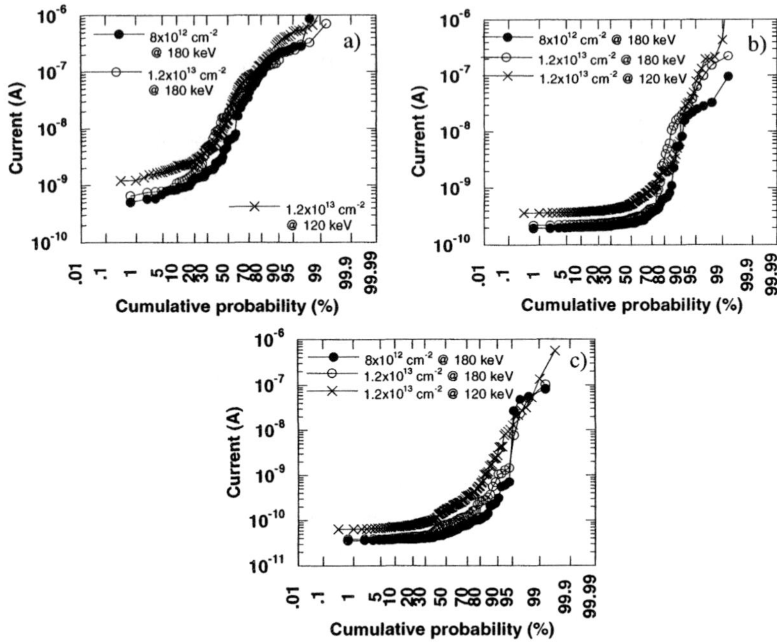

Fig. 5 The cumulative probability of a) SQ, b) ND1 and c) ND4 for different p-well implantations.

It should be noted that J_A consists of the area diffusion (J_{dA}) and generation (J_{gA}) current density, while the surface generation current normally dominates J_P [12,13]. J_{dA} depends on the doping density and the recombination lifetime, while J_{gA} relates to the depletion width (W_A) and generation lifetime (τ_g). W_A is also a function of the doping density. Since the devices are fabricated in a highly doped p-well, the electric field can also enhance the generation current [14,15]. The electric field is inversely proportional with the depletion width [16]. Both recombination and generation lifetime are strongly defect dependent [17]. Therefore, the higher J_A (Fig. 6a) for different implantation conditions may be due to a lower doping density and/or lower generation and recombination lifetimes, while J_P (Fig. 6b) may relate to the surface generation velocity. The origin of the higher leakage current is further investigated.

The carrier concentration derived from the *C-V* characteristics, by taking the series resistance into account [18], is shown in Fig. 7a. The highest carrier concentration is obtained for the 1.2x10^{13} cm^{-2} @ 120 keV condition, while the lowest one is found for the 8x10^{12} cm^{-2} @ 180 keV diodes. The higher carrier concentration of diodes fabricated by $1.2x10^{13}$ cm^{-2} @ 120 keV compared with the $1.2x10^{13}$ cm^{-2} @ 180 keV condition is due to the lower implantation energy. The higher carrier concentration of diodes fabricated by $1.2x10^{13}$ cm^{-2} @ 180 keV compared with $8x10^{12}$ cm^{-2} @ 180 keV is due to the higher implantation dose. The experimental profiles are in qualitative agreement with TSUPREM4 simulations, as shown in Fig. 7b. A higher doping concentration from TSUPREM4 than the one derived from *C-V* characteristics is found. Moreover, a more pronounced difference in doping concentration for the different cases is found in the measurements, compared with the simulation. This discrepancy is due to the fact that the simulation does not take into account the full thermal budget similar to the wafers. Anyway, they show the same trend.

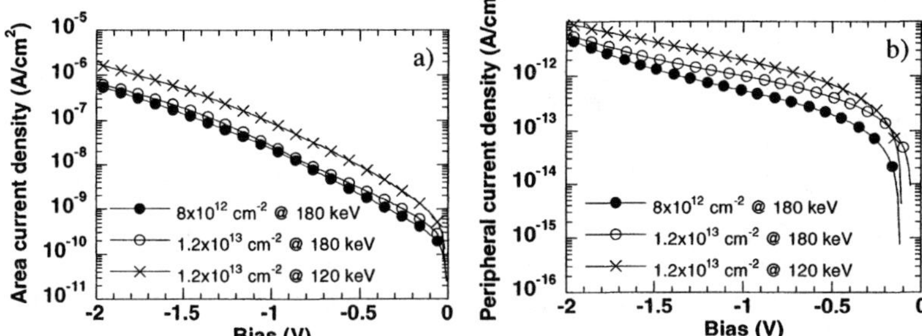

Fig. 6 Area current density (a) and peripheral current density (b) versus reverse bias for different p-well implantations.

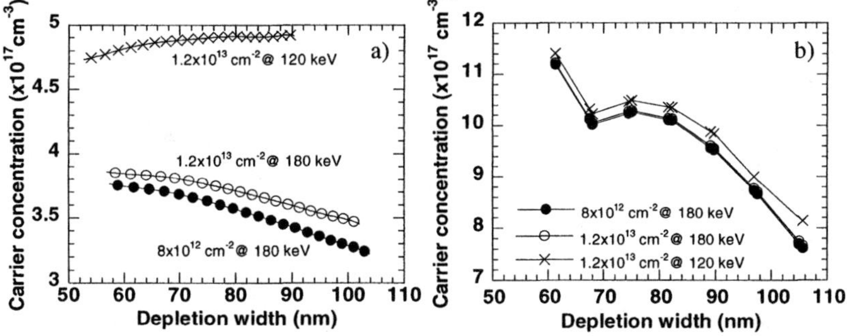

Fig. 7 a) Experimental and b) simulated p-well profile for different p-well implantations.

The area depletion width (W_A) (Fig. 8) is calculated from the area capacitance (C_A), using $W_A = \varepsilon_{Si}/C_A$, where ε_{Si} is the silicon permittivity. C_A is obtained from the capacitance by combining at least 4 different diodes [18]. The smallest W_A is found for the $1.2x10^{13}$ cm^{-2} @ 120 keV case due to the highest doping density (Fig. 7a). From this it is concluded that the higher doping density and smaller W_A can not explain the high J_A observed in Fig. 6a. Therefore, only alternative explanation relies on a reduction of the generation and/or recombination lifetime.

For junctions fabricated in a highly doped p-well, one has to account for the high electric field (F_m). This field effect can be derived for example from Fig. 6a, which shows a nearly exponential increase of the leakage current density with higher reverse bias. In addition, the carrier generation from deep levels takes place only in the generation width (W_{gA}) [19]. W_{gA} in Fig. 8 is calculated from W_A by using

$$W_{gA} = \sqrt{W_A^2 - W_{A0}^2} \tag{2}$$

where W_{A0} is the depletion width at zero bias.

The effect of the electric field can be studied through the generation current enhancement factor (Γ) and is given by [20]

$$\Gamma = \frac{d\left(J_{gbA}/W_{gA}\right)}{dF_m} \tag{3}$$

The enhancement factor, corresponding with the Poole-Frenkel (PF) effect (Γ_{PF}) [14], a combination of the PF effect and phonon-assisted tunneling (PAT) in a Coulombic potential well ($\Gamma_{PF}+\Gamma_{Coul}$), PAT in a Dirac potential well or trap-assisted tunneling (TAT) ($1+\Gamma_{Dirac}$) [15] is compared with the empirical factor calculated from Eq. 3, in Fig. 9. Also shown in that figure is F_m versus depletion width. A higher Γ and F_m are found for a p-well implantation with 120 keV than with 180 keV. The observed Γ values do not agree with the theoretical ones expected for PF, TAT or PF combined with phonon-assisted tunneling in a Coulombic potential well for a single defect energy level. The different Γ for different p-well implantation energies also points towards the presence of different defect behaviour (concentration and/or energy level).

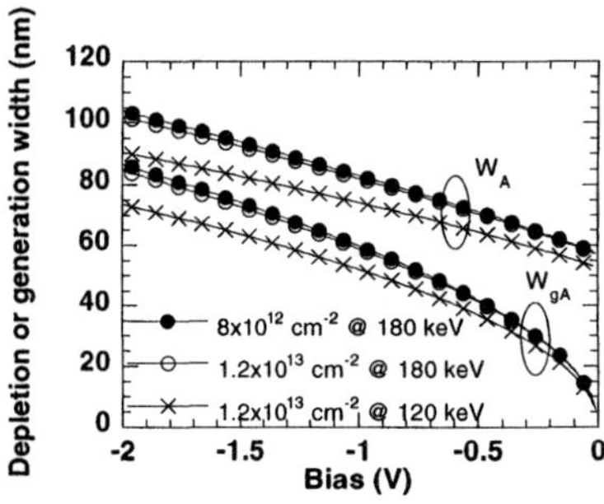

Fig. 8 Depletion (W_A) and generation (W_{gA}) width versus reverse bias for different p-well implantations.

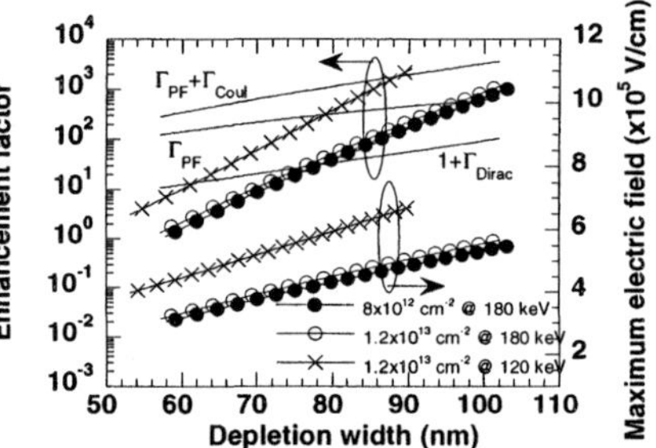

Fig. 9 Enhancement factor corresponding with the Poole-Frenkel (PF) effect (Γ_{PF}), a combination of the PF effect and phonon-assisted tunneling (PAT) in a Coulombic potential well ($\Gamma_{PF}+\Gamma_{Coul}$), PAT in a Dirac potential well or trap-assisted tunneling (TAT) ($1+\Gamma_{Dirac}$) and calculated using Eq. 3. The maximum electric field (F_m) versus depletion width for different p-well implantations is also shown.

Taking both Γ and W_{gA} into account, the generation lifetime, as shown in Fig. 10, can be calculated from

$$\tau_g = \frac{qn_i W_{gA} \Gamma}{J_A - J_{dA}} \tag{4}$$

where q is the electronic charge and n_i the intrinsic carrier concentration

A lower τ_g is observed for a lower p-well implantation energy (120 keV), while it is only slightly lower for a higher p-well implantation dose. It can be derived from Fig. 10 that the p-well implantation with the 1.2x10^{13} cm^{-2} @ 180 keV condition produces more electrically active defects than for the 8x10^{12} cm^{-2} @ 180 keV case due to a higher implantation dose. The 1.2x10^{13} cm^{-2} @ 120 keV p-well implantation causes more defects than the 1.2x10^{13} cm^{-2} @ 180 keV condition owing to the lower implantation energy. It is also shown in Fig. 10 that two minima in τ_g occur. For example, they occur around 65 and 95 nm below the junction for the p-well implantation with the 1.2x10^{13} cm^{-2} @ 180 keV condition. It is tempting to associate these minima with the damage produced by the shallow and the deep boron implantation, respectively. A projected range (R_P) of 100 and 480 nm for the shallow (35 keV) and deep (180 keV) boron implantation is expected. Adding an estimated junction depth of 100 nm yields a depth of around 165 and 195 nm for the first and second τ_g minimum. The first τ_g minimum is close to the expected position for extended defects induced by the shallow boron implantation. The second minimum τ_g is maybe due to the deep boron implantation, as reported by Hsu et al. [21] who observed by Scanning Electron Microscopy dislocation loops starting around 60 % of R_P. It is also found that the second τ_g minimum moves closer to the junction by reducing the p-well implantation energy.

The generation lifetime depends on the trapping level(s) in the bandgap and the corresponding density. This indicates that a different τ_g may relate to a different energy level and/or density. The effective activation energy (E_T) can be calculated from [17]

$$\tau_g = 2\tau_r \cosh\left(\frac{E_T - E_i}{kT}\right) \tag{5}$$

where τ_r is the recombination lifetime, E_i is the intrinsic concentration, k is the Boltzmann's constant and T is the absolute temperature.

Normally, τ_r can be calculated from J_{dA}, if the total diffusion current can be approximated by the one from the lowly doped region [22]. However the diffusion current from the lowly and highly doped region are comparable in shallow junctions. As shown elsewhere, there exists, however, a fairly simple straightforward way to extract both lifetimes simultaneously from a single *I-V* curve [23], as shown in Fig. 11. The method relies on the forward recombination current density, which can be represented by:

$$\frac{qn_i W_A}{J_{rbA0}} = 2\tau_r \exp\left(\frac{qV}{2kT}\right) + \tau_g \tag{6}$$

where V is the forward bias.

The saturation area bulk recombination current density (J_{rbA0}) can be obtained from the area bulk recombination current density (J_{rbA}) by dividing with [exp(qV/kT)-1]. The average τ_r, τ_g and $|E_T-E_i|$ are summarized in Table III. The average τ_g is in good agreement with the one obtained from the generation current (Fig. 10). The average τ_r is very similar in all cases, which can be interpreted as that σN_T is not different. A lower $|E_T-E_i|$ is found for the lower p-well implantation energy (120 keV), while it is not significantly influenced by the p-well implantation dose. This indicates that a higher J_A (Fig. 6a) for the lower p-well implantation energy (120 keV) is due to a higher Γ and a lower $|E_T-E_i|$ making the created defects more efficient for leakage current generation. It should be remarked that a single slope of the plot qn_iW_A/J_{rbA0} versus exp$(qV/2kT)$, as shown in Fig. 11, and in agreement with τ_g derived from forward and reverse I-V characteristics indicates an absence of the Schottky effect due to silicidation [24].

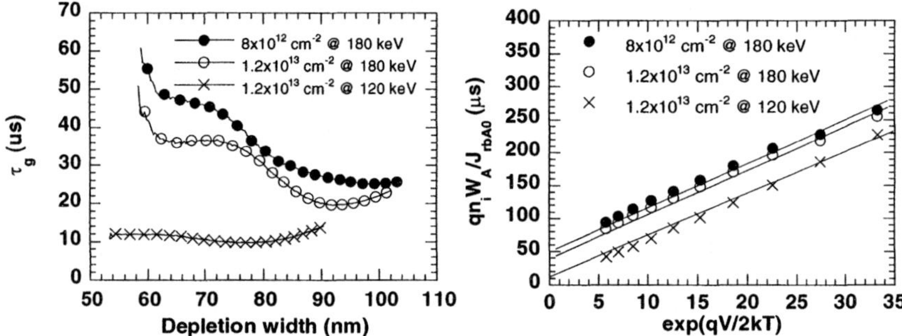

Fig. 10 Generation lifetime (τ_g) versus depletion width (W_A) for different p-well implantations.

Fig. 11 qn_iW_A/J_{rbA0} versus exp$(qV/2kT)$ for different p-well implantations.

Table III The recombination (τ_r) and generation (τ_g) lifetimes obtained from the recombination current and $|E_T-E_i|$ derived from the lifetime ratio for different p-well implantations.

p-well	τ_g (μs)	τ_r (μs)	$\|E_T-E_i\|$ (meV)
8x10^{12}, 180 keV	46	2.2	78.0
1.2x10^{13}, 180 keV	44	2.2	77.3
1.2x10^{13}, 120 keV	12	2.1	44.4

It has been mentioned that the residual defects present after a deep boron ion implantation show a significant electrical activity, which causes a hump in the hole current (bottom current) [25]. By reducing the p-well implantation energy the hump moves closer to the junction as shown in Fig. 12. This demonstrates that the defects induced during the p-well implantation move closer to the junction, in agreement with what can be expected from the lowering of R_P, the shift of the second τ_g minimum (Fig. 10) and the literature [21]. As there exists a tendency to lower the deep boron ion-implantation energy for future CMOS technologies, this could be a potential source of device leakage.

As shown in Fig. 6, both J_A and J_P increase with a reduction of the energy and an increase dose of the p-well implantation. A higher J_A is mainly due to a reduction of τ_g (Fig. 10), while a higher J_P is caused by a higher surface generation velocity (Fig. 13). The surface generation velocity (S_{geff}) is calculated by taking the peripheral generation width (W_{gP}) into account and using

$$S_{geff} = \frac{(J_P - J_{dP})}{qn_i W_{gP}} \tag{7}$$

A higher S_{geff} is found for the lower p-well implantation energy (120 keV) and the higher p-well implantation dose. This could be due to a higher lateral doping density and/or a higher interface trap density [26]. A higher S_{geff} with increasing depletion width relates to a higher stress along the sidewall of the STI [27].

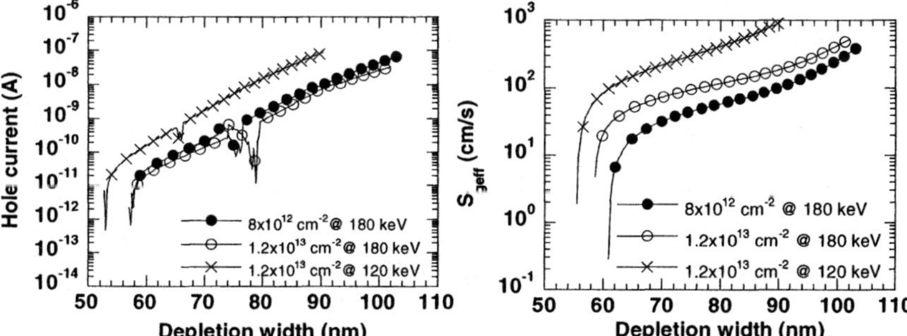

Fig. 12 Hole current versus depletion width for SQ diodes with different p-well conditions.

Fig. 13 Effective surface generation velocity (S_{geff}) versus depletion width for different p-well implantations.

CONCLUSION

The residual defects present after a deep boron ion implantation due to the peak displacement damage near the projected range show significant electrical activity, which affects the *I-V* characteristics of p-well diodes. These defects can cause an increase of the leakage current of the area, the peripheral and the corner components. A higher leakage current occurs when the defects move closer to the junction by reducing the implantation energy. This is the case for state-of-the-art CMOS technology. These defects cause an undesirable hump in the hole current due to hole trapping.

ACKNOWLEDGEMENTS

The Authors are grateful to the members of the CMOS team for helpful discussions on the technology. A. Poyai is indebted to the Thai government for his scholarship supported through the National Science and Technology Development Agency (NSTDA) of Thailand.

REFERENCES

[1] K.K. Bourdelle, Y. Chen, R.A. Ashton, L.M. Rubin, A. Agarwal, and W.A. Morris, *IEEE Trans. Electron Devices*, **48**, 2043 (2001).
[2] H. Sayama, M. Takai, Y. Yuba, S. Namba, K. Tsukamoto, and Y. Akasaka, *Appl. Phys. Lett.*, **61**, 1682 (1992).
[3] J.Y. Cheng, D.J. Eaglesham, D.C. Jacobson, P.A. Stolk, J.L. Benton, and J.M. Poate, *J. Appl. Phys.*, **80**, 2105 (1996).
[4] W.-C. Hsu, M.-C. Chen, and M.-S. Liang, *J. Electrochem. Soc.*, **147**, 3111 (2000).
[5] H. Cerva, E. Hammerl, R. Lemme, U. Schwalke, K. Wangemann, and G. Zoth, in *Third International Symposium on Defects in Silicon*/1999, T. Abe, W.M. Bullis, S. Kobayashi, W. Lin, and P. Wagner, Editors, PV 99-1, p. 55, The Electrochemical Society Proceeding Series, Pennington, NJ (1999).
[6] H. Sayama, M. Takai, Y. Akasaka, K. Tsukamoto, and S. Namba, *Jpn. J. Appl. Phys.*, **28**, L1673 (1989).
[7] D.C. Jacobson, A. Kamgar, D.J. Eaglesham, E.J. Lloyd, S.J. Hillenius, and J.M. Poate, *Nucl. Instrum. Meth. Phys. Res.*, **B 96**, 416 (1995).
[8] K.K. Bourdelle, S. Chaudhry, and J. Chu, *IEEE Trans. Electron Devices*, **49**, 521 (2002).
[9] A. Poyai, E. Simoen, C. Claeys, R. Rooyackers, and G. Badenes, *J. Electrochem. Soc.*, **48**, G507 (2001).
[10] C.S. Rafferty, H.-J. Gossmann, A. Kamgar, D.C. Jacobson, E.J. Lloyd, S.J. Hillenius, H.-H. Vuong, J. Becerro, H.M. Vaidya, S.A. Lytle, M.J. Thoma, and H.S. Luftman, in Techn IEDM 1996, p. 791(1996).
[11] A. Poyai, E. Simoen, C. Claeys, R. Rooyackers and A. Redolfi, in *Microelectronics Technology and Devices – SBMicro 2002*, N.I. Morimoto, R.P. Ribas, and P. Verdonck, Editors, PV 2002-8, p. 213, The Electrochemical Society Proceeding Series, Pennington, NJ (2002).
[12] A. Poyai, E. Simoen, C. Claeys, R. Rooyackers, G. Badenes, and E. Gaubas, in *High Purity Silicon VI*, C.L. Claeys, P. Rai-Choudhury, M. Watanabe, P. Stallhofer and H.J. Dawson, Editors, PV 2000-17, p. 403, The ElectrochemSoc Proc Ser, Pennington, NJ (2000).
[13] S.M. Sze, John Wiley & Sons, New York, 1981.
[14] G. Vincent, A. Chantre, and D. Bois, *J. Appl. Phys.*, **50**, 5484 (1979).
[15] G.A.M. Hurkx, H.C. de Graaff, W.J. Kloosterman, and M.P.G. Knuvers, *IEEE Trans. Electron Devices*, **39**, 2090 (1992).
[16] M.J.J. Theunissen and F.J. List, *Solid-St. Electron.*, **28**, 417 (1985).
[17] D.K. Schroder, *IEEE Trans. Electron Devices*, **44**, 160 (1997).
[18] A. Poyai, E. Simoen, and C. Claeys, *Appl. Phys. Lett.*, **80**, 1192 (2002).

[19] A. Czerwinski, E. Simoen, A. Poyai, and C. Claeys, in *Third International Symposium on Defects in Silicon*, T. Abe, W.M. Bullis, S. Kobayashi, W. Lin, and P. Wagner, Editors, PV 99-1, p. 88, The Electrochem Soc Proc Ser, Pennington, NJ (1999).
[20] A. Poyai, E. Simoen, and C. Claeys, *IEEE Trans. Electron Devices*, **48**, 2445 (2001).
[21] W.-C. Hsu, M.-S. Liang, and M.-C. Chen, *J. Electrochem. Soc.*, **149**, G184 (2002).
[22] J. Vanhellemont, E. Simoen, C. Claeys, *Appl. Phys. Lett.* **66**, 2894 (1995).
[23] A. Poyai, E. Simoen, C. Claeys, E. Gaubas, A. Huber, and D. Gräf, Paper presented at the *E-MRS Spring Symp.* on *Advanced Characterisation of Semiconductor Materials and Devices*, Strasbourg (Fr.), 18-21 June, 2002.
[24] J.-S. Park, D.-K. Sohn, J.-U. Bae, C.-H. Han, and J.-W. Park, *IEEE Trans. Electron Devices*, **47**, 994 (2000).
[25] A. Poyai, E. Simoen, and C. Claeys, *Appl. Phys. Lett.*, **78**, 949 (2001).
[26] W.D. Eades and R.M. Swanson, *J. Appl. Phys.*, **58**, 4267 (1985).
[27] S.M. Hu, *J. Appl. Phys.*, **67**, 1092 (1990).

NEW METHOD FOR ACCURATE DETERMINATION OF THE ELECTRIC-FIELD ENHANCEMENT IN JUNCTIONS – THEORETICAL MODEL AND APPLICATION TO STI DIODES WITH HIGH FIELDS

A. Czerwinski
Institute of Electron Technology, Al. Lotnikow 32/46, 02-668 Warszawa, Poland

E. Simoen,[1] A. Poyai,[1,2] and C. Claeys[1,2]
[1]IMEC, Kapeldreef 75, B-3001 Leuven, Belgium
[2] E.E. Dept., KU Leuven, Kasteelpark Arenberg 10, B-3001 Leuven, Belgium

The origin of p-n junction reverse current is investigated by two new methods based on the analysis of the leakage activation energy. These studies reveal significant leakage components that strongly increase with the junction voltage, which were indistinguishable before. They are the local Schottky-current in state-of-the-art silicided Shallow Trench Isolation (STI) or Local Oxidation of Silicon (LOCOS) junctions and the peripheral diffusion current in small-size LOCOS junctions. The overall area of the silicide-related Schottky (or Shannon) contacts is determined. The above currents must be accounted for, when using the second method, which deals with the tunneling current, the main reason of the remaining leakage. It reveals a large enhancement of the local electric field inside STI p-n junctions, yielding fields greatly overcoming the one due to ionized impurities only.

INTRODUCTION

A huge number of p-n junctions is present in state-of-the-art CMOS ULSI integrated circuits (ICs). Consequently, the main IC parameters, like the leakage and standby power limits and DRAM retention time, are determined by the p-n junction reverse (leakage) current (I_R). As the IC integration density continues to increase, the problem of leakage becomes more serious [1, 2]. Many techniques can be used to analyze the leakage and the related lifetime [3], e.g. noncontact and nondestructive techniques, like microwave photoconductance decay (μPCD) and surface photovoltage (SPV). However, these methods investigate the sample volume over a distance of typically tens to hundreds of microns, inappropriate in the case of small-size junctions in modern ICs, where the leakage originates mostly from junction peripheries i.e. edges and corners [4, 5]. The peripheral leakage is measured on devices after all processing steps, when the methods based on current-voltage (I-V) measurements are listed among the best [2, 3]. This leakage is used for its own diagnostics. The article presents new methods based on I-V measurements, namely on analyses of the leakage activation energy.

EXPERIMENTAL

Shallow n^+-p diodes compatible with submicron (0.18 μm) CMOS technology have been processed on Czochralski (Cz) 150 mm p-type substrates. Shallow trench isolation (STI) was applied. A retrograde p-well was obtained by a deep (200 keV, 1.2×10^{13} cm^{-2}) and a shallow (55 keV, 1.5×10^{13} cm^{-2}) boron ion implantation, followed by a dopant activation anneal (10 min, 850°C). The estimated projected range (Rp) was

~ 0.5 μm and ~ 0.2 μm below the silicon surface, for the deep and shallow boron implantation, respectively. The n^+-region was obtained by an arsenic ion implantation (70 keV, $4x10^{15}$ cm^{-2}) and subsequent anneal (10 s, 1100°C). A junction depth around 0.1 μm is expected. The back-end of the process consists of a cobalt silicide with titanium capping layer, a TEOS Inter-Metal Dielectric layer and an Al-Si-Cu metallization. The maximum silicidation temperature was 700°C.

Reference diodes were used, which were manufactured on Czochralski (Cz) grown wafers with high (HO) initial oxygen concentration, internally gettered (IG) using a high-low-high temperature cycle and surrounded by a local oxidation of silicon (LOCOS) isolation. No p-well was fabricated in this case. The n^+-region was obtained by an arsenic ion implantation (70 keV, $3x10^{15}$ cm^{-2}) and subsequent anneal (10 s, 1100°C + 30 min, 800°C). A standard Al metallization without a silicidation was applied [6].

Junctions with different geometry were studied. The area (A), the perimeter (P) and the number of corners (N_C) of the measured test structures: large area (SQ1) and medium area (SQ2) rectangles and meander (ME1) junctions were: A = 0.1, 0.001, 0.001 cm^2, P = 1.3, 0.13, 8.04 cm, and N_C = 4, 4, 320, respectively. With the total reverse current I_R being a linear combination of the different components, given by [4]:

$$I_R = AJ_A + PJ_P + N_C J_C + I_{par} \qquad (1)$$

where J_A (A/cm^2), J_P (A/cm), and J_C (A/corner) are the planar, perimeter and corner current densities and I_{par} is a parasitic component, a combination of the currents measured for a pair of structures allows to leave components of one type and remove these of another type. The subtraction of SQ2-structure current from SQ1-structure current leaves mainly the planar leakage, with P/A ratio = 11.8, while the subtraction of SQ2-structure current from ME1-structure current leaves the peripheral leakage only, related to junction edges and corners, with theoretically infinite value of P/A.

The static current-voltage (I-V) characteristics of the diodes were measured in the dark for the junction-voltage (V_R) range from 0.1 to 6 V for STI and to 10 V for LOCOS diodes. Temperature-dependent measurements were done in the range from 15 to 120°C, allowing extraction of the reverse-current activation energy. High-frequency (100 kHz and 1 MHz) capacitance-voltage (C-V) measurements yielded in the low-doped part of junctions an impurity concentration N_A of about 6×10^{14} cm^{-3} for LOCOS and 2.5×10^{17} cm^{-3} for STI diodes.

ACTIVATION ENERGY

The activation energy (E_a) is a useful parameter that can be derived from I-V measurements. It is found from temperature-driven changes of the reverse current I_R, assuming a definition of E_a as follows:

$$I_R(T) \propto \exp(-E_a/kT)] \qquad (2)$$

with k the Boltzmann constant, and T the absolute temperature. The semi-logarithmic Arrhenius plot is used to find a linear dependence between the logarithm of reverse current measured at one reverse bias for different temperatures, and (1/kT) values on the horizontal axis [7].

For the pure thermal-generation processes, described by the Shockley-Read-Hall (SRH) model, or in the case of the diffusion current of an ideal p-n junction, no direct impact of the electric field (F) on the activation energy exists. Even then, the leakage-current activation energy may vary with the junction voltage, for example due to an increase of the junction depletion-region width, so that the region and the traps responsible for the carrier generation are changing. For other leakage mechanisms, e.g. the band-to-band or trap-assisted tunneling, or when the generation is enhanced by the Poole-Frenkel effect, there is a strong dependence of the current activation energy on the electric field. In some other cases, as for the reverse current in a Schottky junction, the activation-energy dependence on the electric field and junction voltage is small [8].

It is often observed that $E_{aDIF} \approx E_g$ and $E_{aGEN} \approx E_g/2$, where E_{aDIF} and E_{aGEN} are the activation energies for $I_{R,DIF}$ and $I_{R,GEN}$, while E_g is the silicon bandgap (1.12 eV at 300 K). In fact, when considering the measured leakage currents, their dependence on temperature is not purely exponential but, e.g., a T^3 prefactor exists for the diffusion current [9]. It shifts the "as measured" (effective) E'_a value (further labeled with the apostrophe) in comparison with the "pure" E_a of the exponential part. For example E_{aDIF} is increased by 0.08-0.09 eV in the temperature range 273-373 K, to E'_{aDIF} = 1.21-1.22 eV, a value which is often reported in the literature [10, 11]. When the origin of the measured current is known, such a shift of the "as measured" E'_a value can be accounted for, and E_a may be obtained. However it may lead to wrong E_a values, when this origin is wrongly attributed, so it may be better to consider E'_a values.

In the literature, the simple Arrhenius plot representation at a fixed reverse bias is generally followed to extract the activation energy, even when the tunneling origin of the measured leakage current is fully realized [12, 13]. Only in a few cases the impact of the electric field in the junction on the activation energy of the leakage current has been studied to some detail, for standard p-n diodes [14, 15], or as a function of the gate voltage for gated diodes [16]. Fitting procedures have been applied to the leakage-current curves to determine the energy level of the dominant generation-recombination (G-R) centers [14, 15], to derive the donor- or acceptor-like nature of these centers [15, 16] or to detect the conditions in the junction, like the local electric-field enhancement due to defects and the value of mechanical-stress [15]. E_{aGEN} values, often considered equal to $E_g/2$, in fact are given by $E_{aGEN} = E_g/2 + (E_T - E_i)$ [17], when the impact of electric field can be neglected, where E_T and E_i are the trap and the intrinsic energy-levels, respectively. Under the influence of the electric field in the depletion region, E_{aGEN} decreases, first towards $E_g/2$, while for very high field conditions, the values as low as 0.15-0.2 eV have been reported. Such low activation energies are seen also for peripheral and planar leakages (with the constant $I_{R,DIF0}$ component already removed) in Fig. 1.

Instead of using the traditional Arrhenius representation, combined with curve fitting, one can rely on a recently established analytical approach for the derivation of the activation energy characteristics $E_a(V_R)$, dependent on junction voltage. The method requires the combination of two I-V curves at closely spaced temperatures T and T+ΔT. $E'_a(V_R,T)$ characteristics are then determined for a reverse current I_R by:

$$E'_a \equiv [-1/I_R(T)]\times[dI_R/d(1/kT)] \qquad (3)$$

with $\Delta T = 10°C$, and the temperature stability ± 0.1°C. The activation energy E'_{aQG} is determined from the remaining part ($I_{R,QG}$) of the overall (peripheral or planar) leakage after removing the saturation current $I_{R,DIF0}$. Sometimes, this $I_{R,QG}$ current has been determined and its activation energy was calculated, but only the method, which will be further extended here, allows to determine its real physical components and corresponding activation energies. The label "QG" for this remaining leakage $I_{R,QG}$ means here that it is a "quasi-generation" current, i.e. it is typically (and often wrongly, as will be shown) attributed to the generation current. But the true generation current $I_{R,GEN}$, which corresponds to E'_{aGEN}, is in fact only a part of this $I_{R,QG}$ leakage.

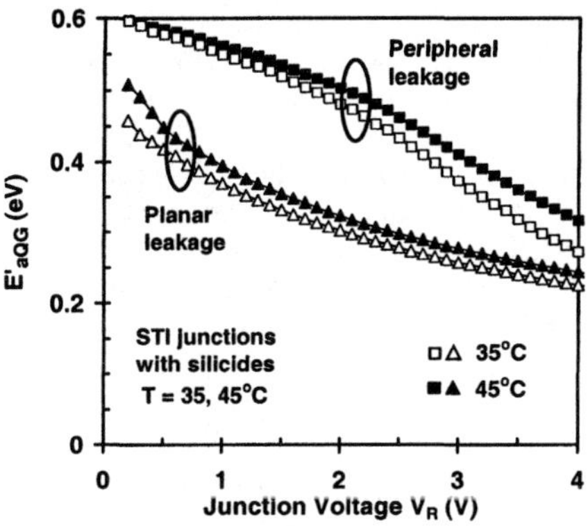

Figure 1. Activation energies E'_{aQG} of a STI diode, which were determined for the peripheral and the planar leakage from their "quasi-generation" part $I_{R,QG}$, remaining after the subtraction of the saturation current $I_{R,DIF0}$.

This so-called dual-temperature method has been successfully applied to demonstrate the increase of the junction diffusion current ($I_{R,DIF}$) with increasing reverse bias, i.e., the function $I_{R,DIF}(V_R)$ [9]. Usually $I_{R,DIF}$ is assumed to be a constant so-called saturation-current ($I_{R,DIF0}$), independent of V_R, which value is determined by the linear extrapolation of the total current at low bias to zero value of the generation current ($I_{R,GEN}$) [18]. However for the peripheral part of the leakage an increase of $J_{R,DIF}$ with V_R due to the lateral extension of the depletion region has been found [9]. Also for the planar leakage the $J_{R,DIF}$ component may vary with V_R, e.g., for IG or epitaxial wafers [18], which however is not the case for the STI junctions studied here.

In the following, if not mentioned, peripheral and planar leakages will be considered separately, i.e. "the total current" will mean either total peripheral or planar leakage, etc, and for simplicity this peripheral- or planar-type of the leakage will be mentioned only within the text and not in the current-labels. For the peripheral leakage the mechanical stress is stronger and also the surface generation at the silicon-oxide interface occurs [9, 19]. This leakage dominates in state-of-the-art small-size junctions with high P/A ratio [4, 5], and is mostly responsible for the summed (peripheral and planar together) leakage of junctions being components of VLSI and ULSI ICs.

A new method is given, to consider two physical leakage-current components, characterized by different activation energies, which may depend on the reverse bias. More specifically, in addition to the generation current component $I_{R,GEN}$ a second voltage-dependent leakage-current ($I_{R,ADD}$) is considered which origin is a priori not known and characterized by unknown activation energy (E'_{aADD}). The knowledge of the activation energy $E'_{aADD}(V_R)$ may allow to specify the origin and source of this excessive leakage, while its value $I_{R,ADD}(V_R)$ enables to assess the relative importance of this component (and the necessity for its suppression), when compared with the total leakage. Also the real values of $I_{R,GEN}$ and E'_{aGEN} can be obtained that way. It should be remarked that the derived analysis is performed after subtraction of the diffusion current contribution, assumed constant here. It is found, that even at high temperatures the contribution of the saturation $I_{R,DIF0}$ current is negligible for the peripheral leakage (below 4% at V_R = 2 V even for 70°C), while for the planar leakage it is also insignificant for temperatures much higher than room temperatures (below 10% at V_R = 2 V even for 50°C), but as this contribution increases with the temperature, it becomes already significant at high temperatures (above 30% at V_R = 2 V for 70°C).

The previously proposed method for solving this problem, which is also based on the activation-energy analysis, assumed [9]: (1) the same origin of $I_{R,DIF0}$ and $I_{R,ADD}(V_R)$ currents, (2) a constant and known $E'_{aADD}(V_R)$ energy, and (3) a high $I_{R,DIF0}/I_R$ ratio. These assumptions were reasonable in the case of diffusion current increasing with V_R, analyzed in [9], however they are not generally applicable, e.g. for the considered STI diodes. The new method allows to derive the above variables also in the case of a priori unknown values. For the general case of $I_{R,QG} = I_{R,GEN} + I_{R,ADD}$ leakage, Eq. (3) gives:

$$E'_{a,QG} = (I_{R,GEN}\, E'_{aGEN} + I_{R,ADD}\, E'_{aADD}) / (I_{R,GEN} + I_{R,ADD}) \quad (4)$$

The derivation of $E'_{a,QG}$ leads after transformation to:

$$dE'_{aQG}/d(1/kT) = I_{R,ADD}\times(E'_{aADD} - E'_{aQG})^2 / (I_{R,QG} - I_{R,ADD}) \quad (5)$$

$$I_{R,ADD} = I_{R,QG}\times[dE'_{aQG}/d(1/kT)] / \{(E'_{aADD} - E'_{aQG})^2 + [dE'_{aQG}/d(1/kT)]\} \quad (6)$$

When the E'_{aADD} value is known (as in [9] for the diffusion current), Eq. (6) already yields $I_{R,ADD}(V_R)$ for each voltage V_R. But in the general case of an unknown $E'_{aADD}(V_R)$, a set of equations (5) and (6) with two dependent unknowns $I_{R,ADD}$ and E'_{aADD} results. They can be found when Eq. (6) is applied also for a second temperature T_2. With the use of Eq. (6) two values $I_{R,ADD}(T_1)$ and $I_{R,ADD}(T_2)$ are obtained from the experimental results measured at two temperatures T_1 and T_2, and from an initial (input) guess for E'_{aADD} assumed for each V_R. From these $I_{R,ADD}(T_1)$ and $I_{R,ADD}(T_2)$ values, an "experimental" (output) $E'_{aADD}(V_R)$ is determined, using simply Eq. (3). The calculation procedure stops when both input and output $E'_{aADD}(V_R)$ values coincide.

This method applied for non-silicided LOCOS Cz-Si HO-IG junction gives almost uniform E'_{aADD} values close to 1.2 eV, as shown in Fig. 2. It confirms the attribution of $I_{R,ADD}$ to the diffusion current, as assured also by results of the high-temperature method

[9]. An E'_{aADD} for each V_R has been fitted independently, so if an almost constant value results for all V_R, then it is not an assumption.

For STI silicided junctions also almost constant E'_{aADD} values in the range 0.55-0.6 eV are revealed for both, peripheral and planar leakages (Figs. 3, 4). Such E'_{aADD} values and their weak dependence on V_R, while $I_{R,ADD}$ significantly increases with V_R, suggest the relation of the excess $I_{R,ADD}$ leakage to the silicidation process. For silicided p-n junctions small regions of silicide penetration are often reported, which results in an excessive leakage [20-24].

The leakage due to the silicidation process behaves as a Schottky or quasi-Schottky (Shannon) junction [20], depending on the length of the silicide penetration and the junction depth. The slope of the extracted $I_{R,ADD}$ leakages versus 1/kT, as shown in Fig. 5 from $I_{R,ADD}$ transformation to the form of the Schottky-junction reverse-current [8], is equal to ϕ_{B0} = 0.55-0.6 eV, while the overall area of the local Schottky contact A_S, found as the extrapolation of the plot to 1/kT = 0, is in the range 0.1-0.2 μm^2. A similar approach was used for the forward current in [21], which however is inapplicable here. From Fig. 5 follows that A_s is larger for the peripheral region than for the planar one.

The resulting effective activation energy is increased in comparison with the barrier height by the presence of T^2 prefactor in the Schottky leakage-current. On the other hand, the dependence of the current on electric field lowers this effective value. For the local field in the range of several tenth of MV/cm, this difference between the effective activation energy and the barrier height may be reduced almost to zero because of the opposite effects of T^2 prefactor and field.

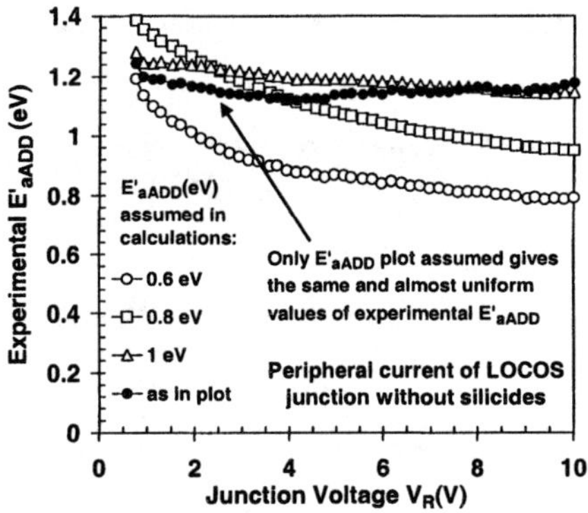

Figure 2. Activation energies $E'_{aADD}(V_R)$ of the voltage-dependent leakage component $I_{R,ADD}(V_R)$ of Cz-Si HO IG LOCOS diode, determined for the peripheral leakage.

When silicide penetrations are in direct contact with the low-doped p-type substrate or well-region, a localized Schottky junction occurs. The barrier heights for contacts to low-doped p-Si are reported to be from 0.42 or 0.43 eV [21] to 0.48 eV [22] for $CoSi_2$ contacts. It points rather to a $CoSi_x$ composition of the penetrations [23]. They cause very high leakage density, e.g. 23 000 times higher locally than for the standard p-n junction, as found in [20]. When the formed silicide penetrations are not in direct contact with the low-doped part of junction, but instead with the depletion region of junction, then a higher barrier-height is reported, e.g. 0.86 eV for $CoSi_2$/p-Si (low doped) [24]. Leakage densities for such a quasi-Schottky (Shannon) junction are lower then, but have been still found in the range e.g. 220-3500 times higher than for the standard p-n junction [20]. The magnitude of this ratio depends on the quality of the "ideal" p-n junctions (apart from the localized Schottky contacts), and in the case of very good quality p-n junctions it may be even higher. The ratio depends also on the junction voltage and temperature, as seen in Fig. 6.

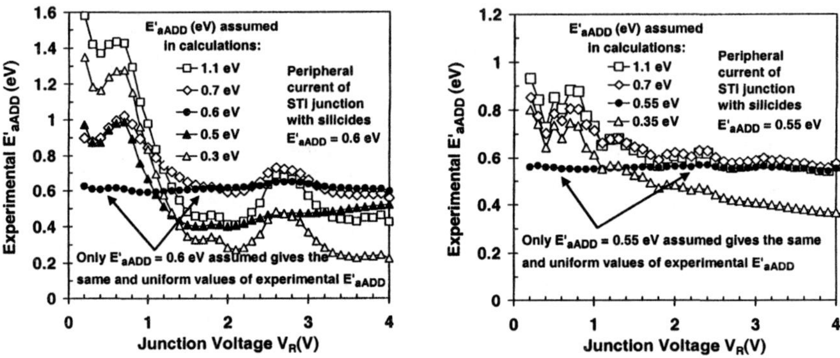

Figure 3. Activation energies $E'_{aADD}(V_R)$ of the voltage-dependent leakage component $J_{R,ADD}(V_R)$ for two different STI diodes, determined for the peripheral leakage.

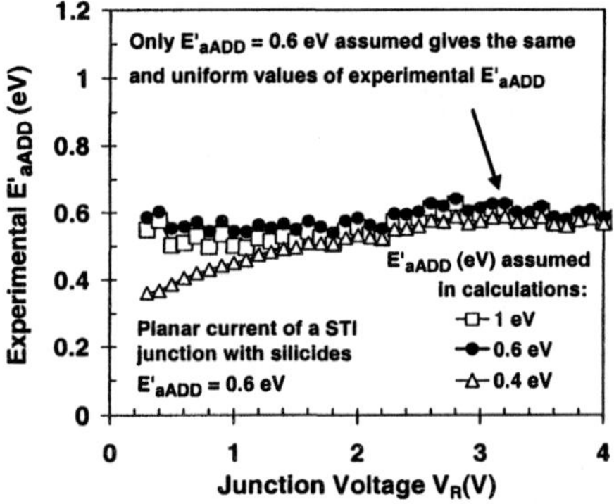

Figure 4. Activation energies $E'_{aADD}(V_R)$ of the voltage-dependent leakage component $J_{R,ADD}(V_R)$ for STI diode, determined for the planar leakage.

The obtained results indicate that silicide penetrations form local Schottky or Shannon contacts in the measured silicided junctions.

Although the presence of various leakage mechanisms is often suggested, previous quantitative characterizations of such excessive leakage [13, 20, 24] concerned the case of its domination in the overall leakage, so that the whole leakage was related to silicide penetration. The present method deals with the case, when the quality of silicidation process is higher, and the leakage related to silicide and the overall area of localized Schottky junctions may be smaller. The density of silicide penetration (Fig. 5) - expressed as the overall area of Schottky contacts divided by the area of p-n junction - is several hundred times lower than that depicted in [13]. Although the area of Schottky contacts was in [13] in range of 1.25 μm^2, the sizes of junctions were much smaller there.

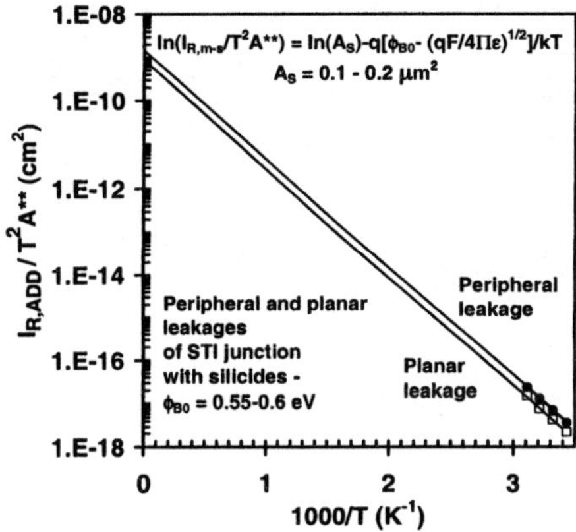

Figure 5. Plot of $I_{R,ADD}/T^2A^{**}$ versus 1000/T for peripheral and planar leakages of a silicided STI junction, considered to be the current of the Schottky junction ($I_{R,m-s}$) (in form presented in [8]), for the extraction of Schottky contact area A_S and of the slope of curves, which gives the Schottky barrier height ϕ_{B0}. This plot corresponds to one value of the electric field F at $V_R = 2$ V. A** is the effective Richardson constant.

As visible in Fig. 6, the peripheral leakage is constituted mainly of the $I_{R,ADD}$ leakage for the low-voltage range up to about 2 V, which is more and more important with the decrease of supply voltage for scaled ICs. For higher V_R and stronger electric field the $I_{R,GEN}$ contribution is increasing faster. For the planar leakage, the $I_{R,ADD}$ component is also strong, although not dominant. The contribution of $I_{R,GEN}$ is generally higher for lower temperatures. The STI silicided junctions show also a predominant impact of $I_{R,ADD}$ in the whole voltage range studied for the peripheral current in the temperature range of 70-80°C, typical for DRAM operation. For planar junctions in this temperature range the saturation diffusion current $I_{R,DIF0}$ dominates.

The vulnerability of junction peripheral regions to the creation of silicide penetration, supposedly corresponds to critical conditions (also the mechanical stress) at junction edges and corners close to the isolation, where the silicide layer is also present, besides the planar region.

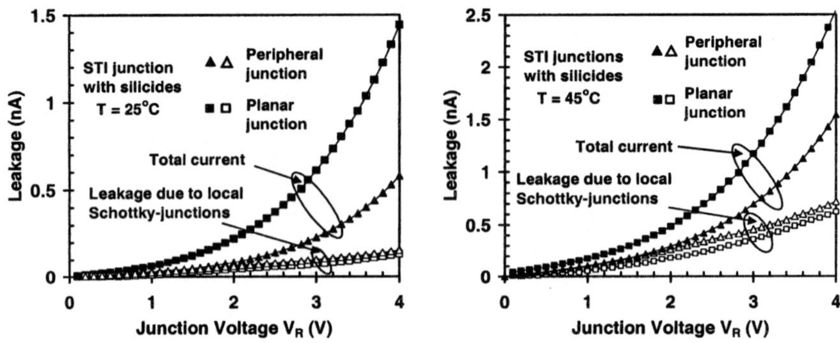

Figure 6. The contribution of additional leakage $I_{R,ADD}$ due to Schottky junctions to the total peripheral or planar leakage in STI diode (with $I_{R,DIF0}$ current also accounted for) at 25 and 45 °C.

METHOD FOR ELECTRIC-FIELD ENHANCEMENT EXTRACTION

Many attempts have been made to reveal the quantitative dependence of the junction leakage on the electric field [15, 25-27]. An obstacle is the poor knowledge of the real (local) field F_{LOC}, enhanced in comparison with the field F_d existing only due to the ionized impurities in the junction depletion region. The enhancement occurs e.g. in the vicinity of defects [15, 25, 26]. Here, a way is pointed out allowing a better estimation of this local field.

The crucial fact for the proposed method, is that when considering the dependence of the leakage-current activation-energy on the electric field F_{LOC}, the minimum of $|dE_a/dF_{LOC}|$ occurs for a value $F_{LOC} = F_{LOC,CRIT}$, which is close to 0.8-1 MV/cm (Fig. 7). The plot is based on the theory, which describes the carrier emission-coefficient enhancement due to electric field for either Coulombic or Dirac wells, i.e., for trap-assisted-tunneling (TAT) generated by defects where the dominant emission process occurs either with or without Poole-Frenkel effect, respectively [14-16, 25]. The plot in Fig. 7 is derived from the temperature-driven theoretical changes of the emission coefficient for various values of the electric field. Because F_{LOC} is proportional to the voltage V_R, this $|dE_a/dF_{LOC}|$ minimum is visible as a minimum in $|dE_a/dV_R|$ plot for a specific $V_R=V_{R,CRIT}$, and constitutes a distinctive mark that the specific local field $F_{LOC,CRIT}$ occurs there. This specific $F_{LOC,CRIT}$ (and also $V_{R,CRIT}$) is only weakly dependent on the values of the energy barrier E_{IB} for carrier emission and of the carrier effective-mass m*.

The generally applied procedure involves the numerical derivative of the experimental E'_{aQG} versus V_R followed by a determination of $V_{R,CRIT}$ corresponding with the minimum. For example such shape was found for the measured peripheral or planar leakage-currents (with $I_{R,DIF0}$ saturation current subtracted) of STI diodes, with $V_{R,CRIT}$ in the range from 2.2 to 2.6 V for various samples (about 2.6 V in Fig. 8). Because this minimum corresponds to a field $F_{LOC,CRIT} \approx$ 0.8-1 MV/cm, while the maximum field F_d (due to ionized impurities only) at $V_{R,CRIT}$ can be calculated, the local-field enhancement $\eta = F_{LOC,CRIT}/F_d$ can be estimated [15].

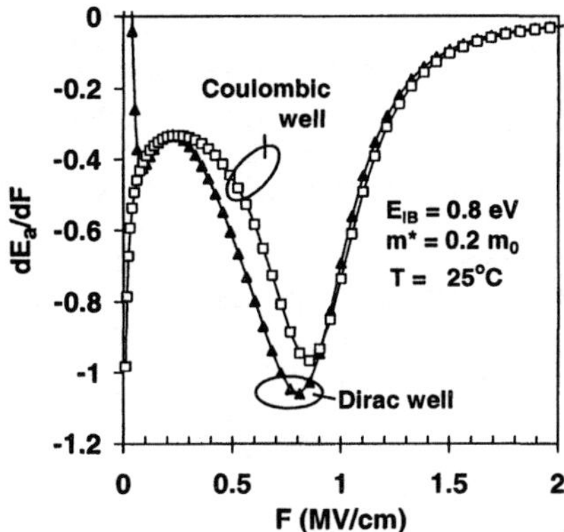

Figure 7. Theoretical values of dE_a/dF for the field-enhanced carrier-emission, calculated for a barrier height E_{IB} = 0.8 eV. The minimum corresponds generally to a specific local field $F_{LOC,CRIT}$ in the range 0.8-1 MV/cm for the trap-assisted tunneling either with or without Poole-Frenkel effect (i.e with a Coulombic or Dirac well, respectively).

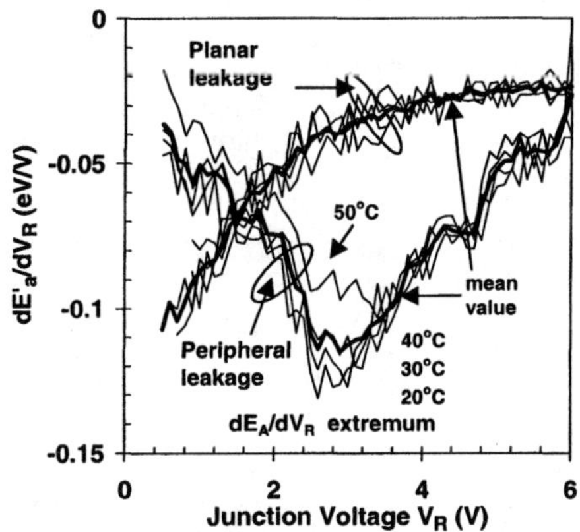

Figure 8. Experimental values of dE'_a/dV_R, which illustrate the idea of method. The same $V_{R,CRIT}$ was revealed for peripheral leakage at different temperatures. Although the presence of a significant $I_{R,ADD}$ in the considered leakage could shift $V_{R,CRIT}$ value, the real local electric-field enhancement is estimated to be not less than 2.

However, the results of method for voltage-dependent leakage-components extraction, which was presented before, show, that in the case of significant $I_{R,ADD}$ leakage, dE'_{aQG}/dV_R is not equal to dE'_{aGEN}/dV_R, while only the latter corresponds to the theoretical dependence and to a $F_{LOC,CRIT}$ value.

In the low-voltage range where $I_{R,ADD}$ constitutes a large part of $I_{R,QG}$ (Fig. 6), the accurate derivation of dE'_{aGEN}/dV_R is impossible, because $I_{R,GEN}$ is determined with low relative accuracy, as a difference of two closeby values. For higher V_R the relative $I_{R,GEN}$ accuracy is better, however because the leakage current $I_{R,ADD}$ and its activation energy E'_{aADD} are determined from derivatives, their characteristics are not smooth, oppositely to the results obtained directly from the measurements. The derivative dE'_a/dV_R requires the use of smooth $I_R(V_R)$ characteristics. Unfortunately, with the measurement setup used, the results are too scattered to give an accurate $V_{R,CRIT}$. This $V_{R,CRIT}$ value is however found not bigger than 2 V. This is concluded from the fact, that the shape of E'_{aGEN} versus V_R plot in the whole $V_R > 2$ V range is concave, i.e. typical for the $V_R > V_{R,CRIT}$ behavior, in contrast to the expected convex shape for $V_R < V_{R,CRIT}$. Also the values of E'_{aGEN} are significantly below the range where $F_{LOC,CRIT}$ is present for TAT (which is about 0.3-0.4 eV). For a doping $N_A = 2.5\times10^{17}$ cm^{-3} in STI diodes, the field F_d of ionized impurities at $V_R = 2$ V is equal to about 0.45 MV/cm. When comparing it with $F_{LOC,CRIT} = 0.8$-1 MV/cm, the electric-field enhancement coefficient can be estimated $\eta \geq 2$. With this electric-field enhancement, for higher voltages not only TAT but also band-to-band tunneling (BBT) may occur.

Strong enhancements of the local electric field in p-n junctions have been reported for the vicinity of various types of defects [15, 26, 27]. An enhancement coefficient close to 2 may be related, e.g., to oxide precipitates. However the highest enhancements are related to the vicinity of sharp edges of metallic (or silicide) precipitates [26]. Therefore, a possible explanation for the electric-field enhancements may also be related to the presence of silicide penetrations, which sharp edges may cause a large field enhancement in their vicinity. In conclusion, these silicide penetrations may be the origin of both: a relatively important Schottky current and the largely enhanced tunneling current.

CONCLUSION

Dedicated methods have been proposed, which are based on the analysis of the leakage-current activation energy. They allow to extract voltage-dependent leakage components as well as the enhancement of the local electric-field in p-n junctions. In small-size p-n junctions of ULSI ICs manufactured with the use of silicidation and STI processes, a marked presence of Schottky leakage currents and of tunneling currents due to an enhanced local electric field was revealed. The Schottky (or Shannon) current is attributed to the presence of silicide penetrations, as suggested by the activation energy of the leakage in the range 0.55-0.6 eV. Its relative contribution to the overall leakage is high or significant even in the case of a high-quality silicidation process, i.e., even for a low density of the silicide penetrations. The presence of silicide penetrations may be also a possible reason of the tunneling-current increase due to the observed electric-field enhancement.

ACKNOWLEDGMENT

A. Czerwinski acknowledges partial financial support by the State Committee for Scientific Research, Poland, under Grant 7 T11B 084 20.

REFERENCES

[1] K. Kim, C.-G. Hwang, and J. G. Lee, IEEE Trans. Electron Devices, **45**, 598 (1998).

[2] H. Uchiyama, K. Matsumoto, T. Mchedlidze, N. Nisimura, and K. Yamabe, J. Electrochem. Soc., **146**, 2322 (1999).

[3] J. E. Park, D. K. Schroder, S. E. Tan, B. D. Choi, M. Fletcher, A. Buczkowski, and F. Kirscht, J. Electrochem. Soc. **148**, G411 (2001).

[4] A. Czerwinski, E. Simoen, C. Claeys, K. Klima, D. Tomaszewski, J. Gibki, and J. Katcki, J. Electrochem. Soc., **145**, 2107 (1998).

[5] H.-D. Lee, S.-G. Lee, S.-H Lee, Y.-L. Lee, and J.-M. Hwang, Jpn. J. Appl. Phys., **37**, 1179 (1998).

[6] J. Vanhellemont, E. Simoen, A. Kaniava, M. Libezny, and C. Claeys, J. Appl. Phys., **77**, 5669 (1995).

[7] A. Poyai, E. Simoen E, C. Claeys, A. Czerwinski, and E. Gaubas, Appl. Phys. Lett. **78**, 1997 (2001).

[8] S. M. Sze, *Physics of Semiconductor Devices,* p. 281, Wiley Interscience, New York (1981).

[9] A. Czerwinski, E. Simoen, A. Poyai, and C. Claeys, J. Appl. Phys., **88**, 6506 (2000).

[10] C. T. Wang, Solid-State Electron., **20**, 967 (1977).

[11] H. Aharoni, T. Ohmi, M.M. Oka, A. Nakada, and Y. Tamai, J. Appl. Phys., **81**, 1270 (1997).

[12] T. Hamamoto, S. Sugiura, and S. Sawada, IEEE Trans. Electron Devices, **45**, 1300 (1998).

[13] H.-D. Lee, IEEE Trans. Electron Devices, **47**, 762 (2000).

[14] M. J. J. Theunissen and F. J. List, Solid-State Electron., **28**, 417 (1985).

[15] A. Czerwinski, Appl. Phys. Lett., **75**, 3971 (1999).

[16] M. Rosar, B. Leroy, and G. Schweeger, IEEE Trans. Electron Devices **47**, 154 (2000).

[17] D.K. Schroder, IEEE Trans. Electron Devices, **29**, 1336 (1982).

[18] Y. Murakami and T. Shingyouji, J. Appl. Phys., **75**, 3548 (1994).

[19] A. Czerwinski, E. Simoen, and C. Claeys, Appl. Phys. Lett., **72**, 3503 (1998).

[20] E. C. Jones and N. W. Cheung, J. Vac. Sci. Technol., B **14**, 236 (1996).

[21] Q. Wang, C. M. Osburn, and C. A. Canovai, IEEE Trans. Electron Devices, **39**, 2486 (1992).

[22] B. S. Chen and M. C. Chen, IEEE Trans. Electron Devices, **43**, 258 (1996).

[23] K.-I. Goto, A. Fushida, J. Watanabe, T. Sukegawa, Y. Tada, T. Nakamura, T. Yamazaki, and T. Sugii, IEEE Trans. Electron Devices, **46**, 117 (1999).

[24] G. P. Schwartz and G. J. Gualtieri, J. Electrochem. Soc., **133**, 1266 (1986).

[25] G. Vincent, A. Chantre, and D. Bois, J. Appl. Phys., **50**, 5484 (1979).

[26] D. A. Kirkpatrick, A. Mankofsky, and K. T. Tsang, Appl. Phys. Lett., **60**, 2065 (1992).

[27] K. Ohyu, M. Ohkura, A. Haraiwa, and K. Watanabe, IEEE Trans. Electron Devices, **42**, 1404 (1995).

DEFECT REACTIONS OF COPPER IN SILICON

S. Knack[1], J. Weber[1], S.K. Estreicher[2]
[1] TU Dresden, 01069 Dresden, Germany
[2] Texas Tech University, Lubbock, TX , USA

Copper doped samples were studied by capacitance spectroscopy and photoluminescence. Experiments on the formation of defects are presented which allow an identification of the number of Cu atoms involved in the defect. The microscopic identity of the 1.014-eV-photoluminescence line in Cu doped Si was reconfirmed and the Cu_ICu_S pair model revisited. The formation of the Cu^* defect is favored by low copper concentrations or the presence of dislocations in the material.

INTRODUCTION

As a fast diffuser and an element that can occupy both interstitial and substitutional lattice sites, copper can form various defects and defect complexes in silicon (1,2,3). It has the ability to passivate shallow acceptors, such as boron, gallium, aluminum or indium, by forming neutral pairs. At room temperature, these are not stable and the copper will either diffuse out to the surface or form precipitates.

When copper has been incorporated on the substitutional lattice site, additional interstitial copper atoms preferably bind to it, forming Cu_ICu_S pairs (4-6). This defect gives rise to strong luminescence at liquid-helium temperature with a zero-phonon line (ZPL) at 1.014 eV and has a deep level at 0.1 eV above the valence band. A very similar defect has been named Cu^*, with luminescence band at 1.3 μm (ZPL at 0.943 eV) and a corresponding deep level at E_V + 0.17 eV. The Cu_ICu_S pair defect has been re-interpreted in the light of a recent PL-study as a defect containing a single Cu atom (7,8).

In this paper, we describe experiments on the formation of two copper defects, which give information on their microscopic identity. Using capacitance spectroscopy the temperature and concentration dependence of the defects were studied. We find that the formation of the Cu_ICu_S defect is possible even at room temperature. The formation of the Cu^* defect is favored by low copper concentrations or the presence of dislocations in the material.

EXPERIMENTAL

The defect reactions of copper and hydrogen were studied in two different sets of samples. (A) Copper was dissolved in the melt of silicon and incorporated into the crystal during the floating zone growth process. The boron concentration of the p-type crystals was about 2 x 10^{15} cm^{-3}. A second p-type crystal was Cu-doped with the same technique but contained grown-in dislocations. The boron concentration for the dislocated crystal was about 1 x 10^{14} cm^{-3}. (B) A set of p-type floating zone silicon with a boron concentration of 1 x 10^{14} cm^{-3} was copper implanted with various doses at room temperature and afterwards annealed at 700 °C for 30 min.

For copper indiffusion at room temperature a piece of pure copper wire (99,999%) was dissolved in a mixture of $HF:HNO_3$ (1:6). Before PL or DLTS measurements samples were either etched in this copper saturated solution or in the same mix of acids without any copper.

Photoluminescence (PL) was excited with the 514.5 nm line of an argon-ion laser and measured at 4.2K with a liquid nitrogen cooled germanium detector with standard lock-in technique. The setup used for the Deep-Level Transient Spectroscopy (DLTS) measurements was a home-built system using analogue filtering of the transients and a liquid helium cryostat for measurements down to 40K.

RESULTS

Figure 1 shows DLTS spectra of the dislocation-free silicon crystal and the PL-spectra from this sample is depicted in figure 2a. The zero phonon luminescence line at 1.014 eV is well known and has been extensively studied before (4-8). In the following we will call this luminescence center Cu_{PL} following the nomenclature of (7) and likewise denote the DLTS-center at E_V +0.1 eV Cu_{DLTS}..

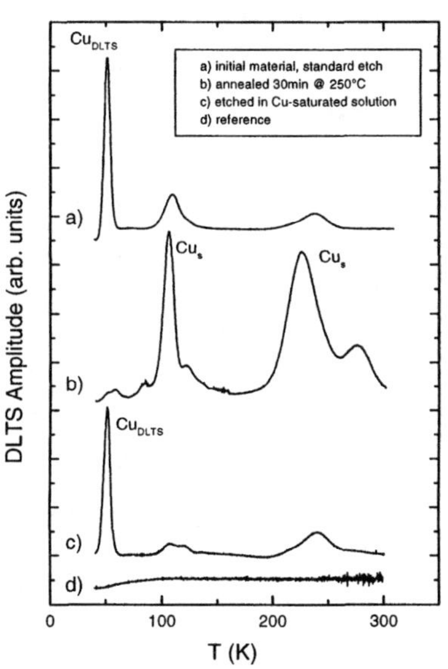

Figure 1: DLTS-spectra of the melt-doped dislocation-free Si sample, a) spectra measured after standard etch, b) spectra taken after an additional annealing step at 250 °C for 30 min, c) spectra taken after etching of the annealed sample in a copper solution, d) reference sample containing no substitutional copper after etching in a copper solution.

In the as-grown dislocation-free samples these are the dominant centers in PL as well as in DLTS. In the dislocated material neither the PL- (fig. 2b) nor the DLTS signal (fig. 3a) can be found. Instead the levels for substitutional copper (9,10) and copper-hydrogen complexes (11) were measured with DLTS, while the photoluminescence showed a broad line at 0.807 eV which is related to dislocations (12).

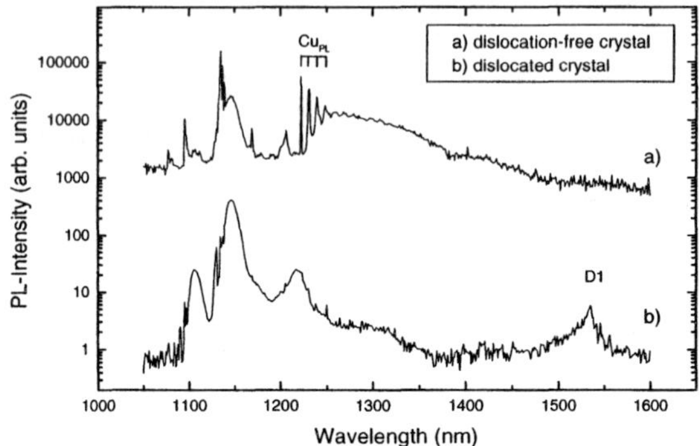

Figure 2: Photoluminescence spectra of the dislocation-free crystal showing the Cu_{PL} signal (a) and the dislocated crystal with the D1 line attributed to dislocations (b).

In the following we describe some simple annealing and diffusion experiments performed with the dislocation-free samples : Annealing at 250°C for 30min almost completely removes Cu_{DLTS} , instead the levels of substitutional copper and of the copper-hydrogen complexes are found (fig. 1b). In a next step the annealed sample is etched in a copper-containing solution, which again restores the signal Cu_{DLTS} (fig. 1c). In a reference sample of FZ silicon with a similar boron-concentration but no Cu-doping no defects could be detected after etching in the same Cu-containing solution, (fig. 1d).

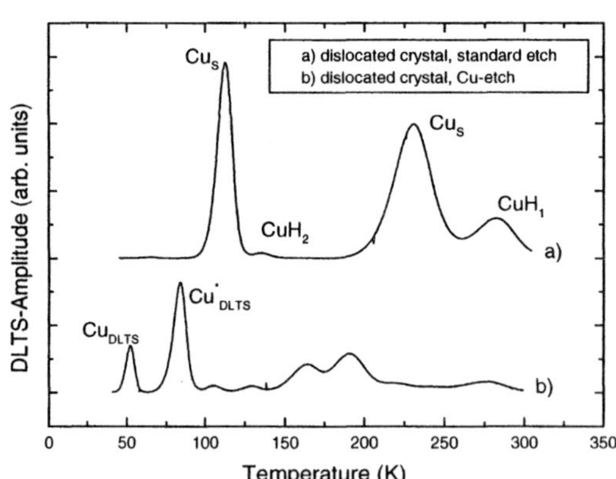

Figure 3: DLTS-measurements of the melt-doped, dislocated crystal, a) spectra taken after standard etch, b) spectra after etching in copper solution.

In the dislocated sample the transformation from the substitutional Cu-defect to Cu_{DLTS} can also be seen after etching with a Cu-containing solution (spectra 3b), but is not as efficient as in the dislocation-free material. The dominant DLTS peak comes from a level at E_V + 0.185 eV and is labeled Cu^*_{DLTS}. The depth distribution of the defects in this sample was measured by DLTS and is shown in figure 4. The two levels Cu_{DLTS} and Cu^*_{DLTS} have a constant concentration near the surface and fall off towards the bulk in a parallel behavior. The substitutional copper defect shows a contrary behavior with very low concentrations near the surface and starting to rise at the depth where the two other defects are falling off in concentration.

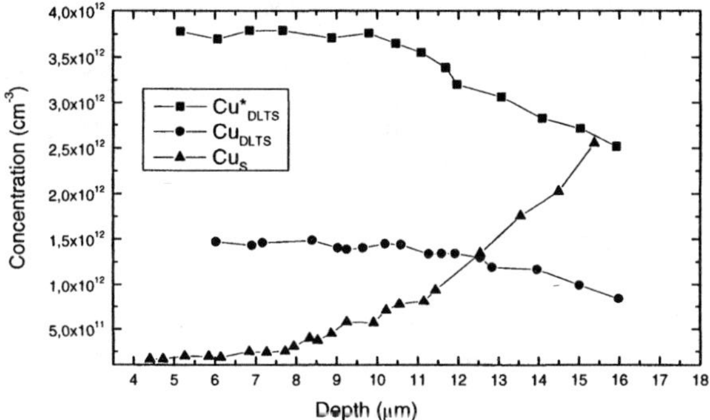

Figure 4: Depth-profiles of the major levels in spectrum Fig. 3b). The distribution of the Cu_{DLTS} and Cu^*_{DLTS} levels are shown as well as the level of sub-stitutional copper.

The Cu^*_{DLTS} level together with Cu_{DLTS} is also found in the Cu-implanted samples that have been annealed at 700 °C for 30 min (fig. 5). Photoluminescence measurements on the implanted samples detected the features of both Cu_{PL} and a system of lines reported by Henry and co-workers in (13,14) and labeled Cu^*.

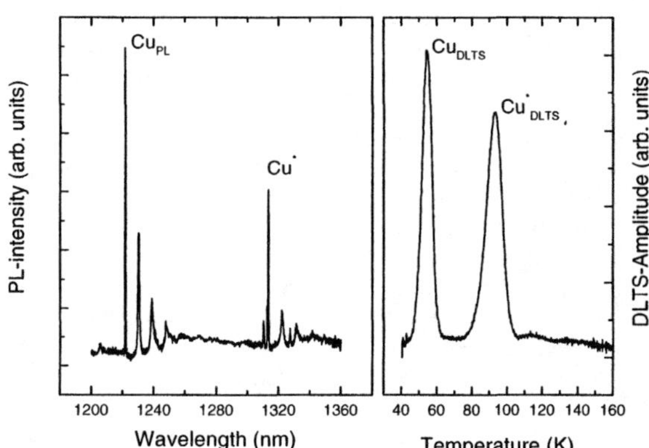

Figure 5: PL and DLTS data of a Cu-implanted sample after annealing at 700 °C for 30 min and rapid quenching to room temperature.

The material of figure 1 was subjected to a series of anneals at different temperatures: first the samples were annealed at 400°C for half an hour after which no Cu-related luminescence could be detected. Afterwards the samples were annealed for 5 minutes at 800, 900 and 1000 °C, respectively, and rapidly quenched to room temperature. The resulting luminescences is shown in figure 6. As expected, the Cu_{PL} signal is increasing with the temperature of the anneals. The Cu* luminescence does not increase in intensity and becomes hidden by a strong background luminescence.

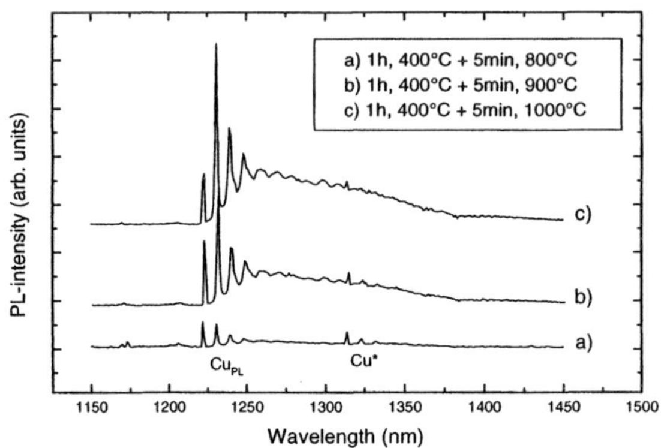

Figure 6: Formation of the Cu* center in melt-doped samples for different annealing conditions. The spectra were measured at 4.2 K.

DISCUSSION

The annealing and diffusion experiments show that the Cu_{DLTS} can be transformed into the defect attributed to substitutional Cu_S (9,10). The PL measurements show an analogue behavior of the Cu_{PL} defect. Cu_S cannot be detected in luminescence, but the experiments are consistent with the former assignment of Cu_{DLTS} and Cu_{PL} to the same copper center (6). The annealing experiments show that $Cu_{DLTS,PL}$ is thermally less stable than the substitutional copper. At room temperature, in-diffusing interstitial copper is very efficiently bound to Cu_S and restores $Cu_{DLTS,PL}$. On the other hand, the defect can not be formed at room temperature in samples where no Cu_S is present before the copper in-diffusion. These results clearly suggest $Cu_{DLTS,PL}$ to be a copper-pair defect consisting of one substitutional and one interstitial copper atom.

In reference (7) Nakamura *et al.* observed a linear dependence of the Cu_{PL} signal on the total copper concentration. Using calibrated copper solutions, they achieved controlled surface contamination of silicon wafers with different, but very low Cu-concentrations. They annealed the silicon samples for 30 min at 700 °C, thus achieving different concentrations of copper well below the solubility limit at the temperature of annealing. The intensity of Cu_{PL} then showed a linear dependence on the incorporated copper concentration, which lead them to suggest the defect to consist of only one Cu atom. Nevertheless, also in terms of the pair model can the linear concentration dependence of the $Cu_{DLTS,PL}$ defect be understood. We will discuss this in the following.

The law of mass action gives a quadratic dependence on the total copper concentration for the reaction:

$$Cu + Cu \rightarrow Cu_{pair} \qquad [1]$$

But this chemical equation does not describe the formation of the defect correctly. The defect formation during growth or annealing, as evidenced by the above described experiments, is a two step process which can be described as

$$Cu_I + V \rightarrow Cu_S \quad \text{at temperature } T_1 , \qquad [2]$$

$$Cu_I + Cu_S \rightarrow Cu_ICu_S \quad \text{at temperature} T_2 < T_1 , \qquad [3]$$

The high stability of Cu_S is predicted by calculated binding energies:

$Cu_I + V \rightarrow Cu_S + 2.71$ eV (ab initio Hartree-Fock (3))

$Cu_I + V \rightarrow Cu_S + 2.78$ eV (density-functional theory (21))

For fast quenching after the annealing, one can set T_1 as the temperature of the anneals and T_2 as room temperature. At the elevated temperature, only substitutional copper is formed as the copper-pair is not stable. The concentration of Cu_S is thus determined by equation (2). In the limit of low copper concentration one would therefore expect

$$[Cu_S] = k\, [Cu_{tot}] , \qquad [4]$$

with $k < 1$. When quenched to room temperature, the remaining interstitial copper is still mobile and moves freely until trapped by some defect. In dislocation-free material, the main sink for Cu_I is substitutional copper as copper-boron pairs are not stable at room temperature and the small fraction of Cu_I lost to the surface can be neglected. Thus, if $k > 0.5$, i.e., if the concentration of substitutional copper is higher than the concentration of the remaining interstitial copper, then

$$[Cu_ICu_S] = [Cu_I] = (1\text{-}k)\, [Cu_{tot}] , \qquad [5]$$

otherwise

$$[Cu_ICu_S] = [Cu_S] = k\, [Cu_{tot}] . \qquad [6]$$

Depending on whether more or less than half of the copper has been quenched into the substitutional lattice site, the formation of the pair-defect will be limited either by the concentration of the interstitial or of the substitutional species. As in the dislocation-free crystals no substitutional copper could be detected in the as-grown state, it is to be assumed that $k < 0.5$. In both cases, the concentration dependence would be linear as observed in the experiment of Nakamura et al. (7,8).

For higher copper concentration, the solubility of substitutional copper at 700 °C would be reached and therefore

$$[Cu_S] = \text{const.} = [Cu_ICu_S] \qquad [7]$$

as no additional Cu_S can be formed at room temperature. Thus, for higher concentrations, a saturation of the copper-pair concentration would be expected as is experimentally observed. It is important to note that the solubility of *substitutional* copper determines the saturation of the pair-formation not the much higher total solubility which is dominated by the interstitial species.

The results from the melt-doped silicon samples support this analysis. In dislocation-free silicon, the copper-pair defect is found after crystal growth. This situation changes in dislocated material, where the dislocations are an efficient trap for the interstitial copper, leaving no free interstitial copper at lower temperatures for the formation of the copper pair. Thus, only the isolated substitutional copper is found in the dislocated material after crystal growth. From this discussion follows that the observed concentration dependence is consistent with the model of a Cu_ICu_S defect.

	1000 K	1500 K	references
$[Cu_I]$	1×10^{16} cm^{-3}	1×10^{18} cm^{-3}	APL A30, 1 (1983)
[V]	7×10^{7} cm^{-3}	8×10^{13} cm^{-3}	PRL 56, 2195 (1986)
[V]	4×10^{6} cm^{-3}	1.4×10^{13} cm^{-3}	PSS b222, 219 (2000)
H_{Cu}	2.7 eV		PRB 60, 5375 (1999)
H_{Cu}	2.2 eV		APL 77, 2142 (2000)

Table 1: Parameters for the temperature-dependence of the copper-pair formation: Concentrations of interstitial copper and vacancy concentrations at 1000K and 1500K and the formation enthalpy H_{Cu} of substitutional copper.

In an earlier study a quadratic dependence of the PL-intensity on the Cu-concentration was found (5). Here, Cu was evaporated onto the sample surface. During annealing at different temperatures, ranging from 700 °C to 1200 °C, the solubility limits were reached. This is in contrast to the other study with fixed-temperature anneals and low copper concentrations. This experiment is more difficult to analyze within my model. The equilibrium concentrations of interstitial copper (15,16) and the silicon vacancy (17,18), as well as the formation enthalpy for substitutional copper (9,19) have to be known, as the law of mass action has to be applied to the reaction [2] at different temperatures:

$$[Cu_S]_T \propto [Cu_{tot}]_T \, [V]_T \, e^{-(H/kT)} . \qquad [8]$$

From this, the concentration of the copper-pair can then be calculated according to the reaction [3].

The values found in literature show a large scatter. They are summarized in table 1. The absolute concentration calculated with different sets of values varies by about an order of magnitude. Yet, for the analysis of the concentration dependence only the exponent of concentration dependence is used:

$$[Cu_ICu_S]_{T=700°C} / [Cu_ICu_S]_{T=1200°C} = ([Cu_{tot}]_{T=700°C} / [Cu_{tot}]_{T=1200°C})^X . \qquad [9]$$

From the different sets of values that can be found in the literature one obtains

$$[Cu_ICu_S] \propto [Cu_{tot}]^{2.2\pm0.2} , \qquad [10]$$

which is very close to the experimental finding of

$$[Cu_{PL}] \propto [Cu_{tot}]^{2.0\pm0.2} \quad . \qquad [11]$$

Thus, together with previous experimental results, the original assignment of the $Cu_{DLTS,PL}$ signal to a Cu_iCu_S pair can be confirmed. The existence of such a defect is supported by molecular dynamic calculations (21). The main properties of the Cu_iCu_S pair have been calculated by theory and matched to the experimental results from DLTS and PL measurements. The calculated binding energy of 1.16 eV (21) is close to the measured value of 1.02 eV (20). The vibrational properties of the defect have been calculated using a new theoretical approach and pseudo-local modes have been found (21) which are close in energy to the phonon replicas seen in the luminescence spectra.

A single bond-centered copper atom proposed by Nakamura as the microscopic structure (7) is found to be highly unstable in these calculations. While the Si-H bonding length (~0.15 nm) which served as analogy for proposing this model is small enough to fit within two silicon nearest neighbors (Si-Si bond length ~0.235 nm), the Cu-H bond length (~0.24nm) so big that large lattice relaxations would be necessary to accommodate it on the bond-centered site. These outward relaxations require too much energy for this structure to be stable. The proposal of bond-centered copper can be ruled out on other grounds as well: if it were an energetically stable site, interstitial copper should be captured to these sites which contradicts the fact that the mobile interstitial copper has been observed at room temperature in TID experiments (22).

There is less experimental data on Cu* than there is for the Cu_iCu_S defect. The microscopic identity of this center cannot yet be established. Interstitial copper itself was first proposed as a possible candidate. This would account for the observed tetrahedral symmetry (14). Yet, there seem to be a number of problems with this assignment: interstitial copper should out-diffuse or precipitate at room temperature (23) which contradicts the long-term stability of Cu*. There have been reports that interstitial copper can be stabilized in silicon at room temperature in low concentrations when there is no copper-silicide defining the boundary conditions on the surface (24). This might be a possible explanation for the low intensities of Cu* observed in experiments.

From experimental observations it cannot be ruled out that Cu* is due to a complex of Cu with another impurity. Cu* has been observed in a number of different materials and circumstances. Also, experiments that we performed with the co-diffusion of Cu with iron - another common impurity - failed to increase the intensity of the Cu* luminescence. Further work is needed to clarify the nature of the defect.

SUMMARY

In the light of the annealing and diffusion experiments on copper doped silicon samples, the concentration dependence of the 1.014 eV luminescence line and the E_V + 0.1 eV DLTS center is carefully analyzed and the originally proposed model of an Cu-pair defect revisited.

The microscopic origin of Cu^* and Cu^*_{DLTS} is still not resolved. The luminescence feature with a zero-phonon line at 943.7 meV is found in samples with low Cu contamination and has tetrahedral symmetry (13,14). It is shifted in energy by -70 meV from the line at 1.014 eV. The fact that it shows up together with the DLTS-level at E_V + 0.185 eV, which is likewise shifted by approximately +80 meV from the Cu-pair at E_V + 0.100eV suggests that the DLTS peak and the photoluminescence line are both due to the same defect which shows remarkable similarity to the Cu-pair defect.

ACKNOWLEDGMENTS

The authors want to thank T. Heiser for helpful discussions. The technical assistance of B. Koehler and S. Behrendt is gratefully acknowledged. R. Botha is thanked for his help with photoluminescence measurements. The growth of the FZ samples was perfectly performed by H. Riemann from the Institute for Crystal Growth, Berlin. The work of SK was supported by the Deutsche Forschungsgemeinschaft under project WE 1319/9, JW was supported by the BMBF (05 KK1ODA/2) and SKE was supported by a grant from the R.A. Welch foundation and a contract from the National Renewable Energy Laboratory.

REFERENCES

1. A. Mesli, T. Heiser, Phys. Rev. B. **45,** 11632 (1992).
2. A. A. Istratov and E.R. Weber, Appl. Phys. A: Mater.Sci. Process. **66**, 1 (1998).
3. S.K. Estreicher, Phys. Rev. B **60**, 5375 (1999).
4. N.S. Minaev, A.V. Mudryi, V.D. Tkachev, Sov. Phys. Semicond. **13**, 233 (1979).
5 J. Weber, H. Bauch, R. Sauer, Phys. Rev. B **25**, 7688 (1982).
6. H.B. Erzgraeber, K. Schmalz, J. Appl. Phys. **78**, 4066 (1995).
7. M. Nakamura, S. Ishiwari, A. Tanaka, Appl. Phys. Lett. **73**, 2325 (1998).
8. M. Nakamura, Appl. Phys. Lett. **73**, 3896 (1998).
9. H. Lemke, phys. stat. sol. (a) **95**, 665 (1986).
10. S.D. Brotherton et al., J. Appl. Phys. **62**, 1826 (1987).
11. S. Knack, J. Weber, H. Lemke, Physica B**273-274**, 387 (1999).
12. R. Sauer, J. Weber, J. Stolz, E.R. Weber, K.-H. Kuesters, H. Alexander, Appl. Phys. A **36**, 1 (1985).
13. K.G. McGuigan, M.O. Henry, E.C. Lightowlers, A.G. Steele, M.L.W. Thewalt, Solid State Comm. **68**, 7 (1988).
14. K.G. McGuigan, M.O. Henry, M.C. Carmo, G. Davies, E.C. Lightowlers, Mat. Sci. Engr. B**4**, 269 (1989).
15. J.D. Struthers J. Appl. Phys. **27**, 1560 (1956).
16. R.N. Hall, J.H. Racette, J. Appl. Phys. **35**, 379 (1964).
17. P.B. Rasband, P. Clancy, M.O. Thompson, J. Appl. Phys. **79**, 8998 (1996).
18. R. Falster, V. V. Voronkov, F. Quast, Phys. Stat. Sol. (b) **222**, 219 (2000).
19. U. Wahl, A. Vantomme, G. Langouche, J.P. Araujo, L. Peralta, J.G. Correia, Appl. Phys. Lett. **77**, 2142 (2000).
20. A.A. Istratov, H. Hieslmair, C. Flink, T. Heiser, E.R. Weber, Appl. Phys. Lett. **72**, 474 (1998).
21. S.K. Estreicher, D. West, J. Goss, S. Knack, J. Weber, submitted to Phys. Rev. Lett.
22. T. Heiser, A. Mesli, Appl. Phys. A **57**, 325 (1993).
23. C. Flink, H. Feick, S.A. McHugo, W. Seifert, H. Hieslmair, T. Heiser, A.A. Istratov, E.R. Weber, Phys. Rev. Lett. **85**, 4900 (2000).
24. A.A. Istratov, H. Hieslmair, C. Flink, T. Heiser, E.R. Weber, Appl. Phys. Lett. **71**, 2349 (1997).

Photoluminescence Intensity Analysis In Application To Contactless Characterization Of Silicon Wafers

A. Buczkowski, B. Orschel, S. Kim, S. Rouvimov, B. Snegirev[(a)], M. Fletcher, F. Kirscht[(a)]

SUMCO USA
1351 Tandem Avenue, N.E.
Salem, OR 97303, USA

[(a)] formerly with SUMCO USA

Photoluminescence (PL) signal has been modeled under conditions of steady-state, depth-dependent excitation, ignored surface recombination and neglected carrier diffusion. In spite of serious model limitations, the experimental data of PL versus resistivity collected in the 0.0009 to 20 Ωcm range showed a reasonably good agreement with the simulation output. PL signal reaches maximum within 0.01 to 0.1 Ωcm range. At lower resistivity, the signal strongly depends on doping level and it can be used for doping striation monitoring. At higher resistivity, PL depends on applied excitation levels and concentration of deep recombination centers (material contamination). Within entire range, the PL can be used for imaging of extended defect, such as dislocations and precipitations.

INTRODUCTION

Photoluminescence (PL) technique in application to non-contact characterization of silicon materials has been employed for a long time[1]. In particular, analysis of PL spectra for deep-level assisted photoluminescence at low, liquid helium temperature was found to be very effective for defect identification[2, 3]. Both, time-resolved[4, 5] as well as steady state[6] measurement conditions have been applied for PL signal recording. In spite of PL advantages, PL utilization in industrial practice remains rather limited, most likely due to inconveniences related to liquid He handling and unavailability of commercially built high throughput equipment. Recently, a PL-based tool operating at room temperature, known under the brand name SIPHER[7] has been introduced. Full wafer, band-edge PL intensity maps as well as micro-zoom PL images down to one-micron resolution are available as the tool output. With the SIPHER, contactless characterization of structural defects[8] and mapping of contamination[9] became functional within a wide range of material resistivity, not assessable before by common tools like Photoconductance Decay (PCD) or Surface Photovoltage (SPV). However, the tool sensitivity for defect or contamination detection is not yet fully recognized. In this work, theoretical simulation of PL intensity is performed. The simulation takes into account the tool specific design and aims at gaining better insights into the tool output. Three basic recombination mechanisms, radiative recombination competing with multi phonon Shockley-Read-Hall (SRH) and Auger recombination, were assumed to control the signal level.

In order to model the signal intensity, lifetime dependence on injection level was imperative due to extremely high excitation levels exercised in the tool. In calculations, a steady-state case, with generation rate balanced by recombination process without diffusion have been postulated. Carrier diffusion was neglected due to very low lifetime, observed despite of low defect density and low doping. This low lifetime is a result of contribution of Auger recombination to the lifetime at high injection level.

For illustration, typical examples of tool application are provided, including a high-resolution mapping of doping, analysis of precipitation density and contrast, in particular after decoration with metals. Examples of extended defect in active regions of fully processed device wafers are also shown.

THEORETICAL ANALYSIS

Theoretical analysis of photoluminescence intensity presented in this paper aims at gaining better insights into the specific output of the SIPHER instrument. Therefore, when the physics model of the tool operation is constructed, the distinctive tool design is taken into consideration. The SIPHER uses laser light for semiconductor excitation. The light is focused to a micrometer spot on the semiconductor surface by a system of lenses. Due to high intensity of light applied, and because of light focusing, it is expected that very high concentration of free carriers is generated at the semiconductor surface. The carriers, in the band-to-band recombination process, produce photoluminescence radiation, which is collected by the same optical system as used for the excitation. During PL measurements, samples (silicon wafers) are placed on a motorized chuck and are moved in x-y directions to enable wafer mapping. Either full wafer mapping or high-resolution PL imaging can be performed, depending on settings of the chuck axis movement. Due to the specific design of the sample excitation path, modeling of the PL signal intensity requires high excitation levels to be taken into account. At such conditions, no analytical solution can be derived from the continuity equation. Although in general, a numerical analysis can be performed, accepting some simplifications enables easy, yet powerful solution of the continuity equation. This solution can be used for semi-quantitative interpretation of the experimental data. Three major model assumptions are made (i) "steady state" condition is applied, and (ii) spatial carrier flow (diffusion) and (iii) surface effects are ignored. With respect to (i), the condition seems to be reasonably well fulfilled, as the PL signal acquisition is done within milliseconds, while carrier recombination lifetime is in the nanosecond range. The condition (ii) is more controversial, because at recombination lifetimes on the order of nanoseconds, diffusion lengths are still in the micrometer range and, therefore, are comparable with or larger than the generation spot (about 1-2 μm). It is estimated that by neglecting the carrier diffusion, the injection levels may be overestimated by two or more orders of magnitude (as diffusion "dilutes" the actual carrier density). The condition (iii) seems to be reasonably well satisfied if the samples are passivated, for example by oxidation.

In order to model PL intensity, one can begin with the common continuity equation.[10] For the case of holes, it can be written:

$$q^{-1}\nabla I_p + g_E - \frac{dp}{dt} = (r - g) \tag{1}$$

where:
q is the electronic charge, I_p is the hole current density (due to carrier diffusion), g_E is the carrier external (optical) generation rate, p is the free carrier concentration, t is time, and (r- g) is the internal net recombination rate.
If "steady state" condition (i) is assumed, the dp/dt term can be neglected, i.e. dp/dt = 0. The condition (ii) allows a further simplification, $q^{-1}\nabla I_p = 0$ leading to:

$$G_E = R_E \equiv \frac{\Delta n_e}{\tau(\Delta n_e)}, \qquad at \quad \Delta n_e = \Delta p_e, \tag{2}$$

where Δn_e is the excess carrier concentration at equilibrium, τ is the recombination lifetime, G_E is the generation rate at steady state conditions equal to the net recombination rate, R_E.

The generation rate is assumed to be depth-dependent:

$$G_E(z) = -\alpha_L \, NF \exp(-\alpha_L \, z), \tag{3}$$

$$\alpha_L \, [cm^{-1}] = \left(\frac{84.732}{\lambda \, [\mu m]} - 76.417\right)^2 \tag{4}$$

where α_L is the absorption coefficient, λ the excitation laser wavelength, and NF the photon flux. The introduction of depth-dependent generation rate enables simulation of PL intensity dependence on laser wavelength. The SIPHER instrument uses two different laser wavelengths for excitation, 0.532 µm and 0.827 µm.

The presence of three basic recombination lifetime components is assumed for lifetime calculations: radiative recombination (τ_{rad}) lifetime, competing with multi phonon Shockley-Read-Hall (τ_{SRH}), and Auger (τ_A) recombination lifetime:

$$\frac{1}{\tau} = \frac{1}{\tau_{rad}} + \frac{1}{\tau_A} + \frac{1}{\tau_{SRH}} \tag{5}$$

where

$$\tau_{SRH} = \frac{\tau_p \left(n_0 + n_1 + \Delta n_e\right) + \tau_n \left(p_0 + p_1 + \Delta n_e\right)}{p_0 + n_0 + \Delta n_e} \tag{6}$$

$$\frac{1}{\tau_p} = \frac{1}{\sigma_p \, v_{th} \, N_T}, \qquad \frac{1}{\tau_n} = \frac{1}{\sigma_n \, v_{th} \, N_T}, \tag{7}$$

$$n_1 = N_c \exp\left(\frac{E_T - E_c}{kT}\right), \qquad p_1 = N_v \exp\left(\frac{E_T - E_c}{kT}\right) \tag{8}$$

$$N_c = 6.2 \times 10^{15} \, T^{1.5}, \qquad N_v = 3.52 \times 10^{15} \, T^{1.5}, \qquad v_{th} = 1.93 \times 10^7 \sqrt{T/300} \tag{9}$$

$$\tau_A = \frac{1}{C_n\left(n_0^2 + 2\, n_0\, \Delta n_e + \Delta n_e^2\right) + C_p\left(p_0^2 + 2\, p_0\, \Delta n_e + \Delta n_e^2\right)} \tag{10}$$

$$\tau_{rad} = \frac{1}{B\left(p_0 + n_0 + \Delta n_e\right)} \tag{11}$$

Based on Eqs. 2-11, at a given generation rate G_E (related to power of laser diode used in the PL instrument), the corresponding equilibrium Δn_e, called also the injection level, can be found as the root of the Eq. 12:

$$F(\Delta\, n_e) \equiv G_E - \frac{\Delta\, n_e}{\tau(\Delta n_e)} = 0 \tag{12}$$

The injection level depends on material parameters such as doping (resistivity), defect and contamination properties, and generation rate. In light of G_E dependence on depth, it is imperative that the carrier concentration Δn_e and all lifetime components are dependent on depth, too. Therefore, when referring to those parameters, a distinction between the local and average (depth-integrated) parameter values should be made. In the following part, if not explicitly stated, the average values are considered.

When the injection level is determined from Eq. 12, the local PL intensity *PLD(z)*, total internal efficiency η_{eff} and external *PL* intensity can be calculated as follows:

$$PLD\,(z) = \frac{1}{NF} R_{rad} = \frac{1}{NF} \frac{\Delta\, n_e}{\tau_{rad}(\Delta n_e)}, \qquad \eta_{eff} = \sum_z PLD(z)\,\Delta\, z, \quad PL = C \times NF \times \eta_{eff} \tag{13}$$

where C is a constant dependent on the instrument detector/ amplifier sensitivity and optics system design.

SIMULATION RESULTS

In order to simulate the PL signal intensity a number of parameters related to material, defects, doping and instrument have to be assumed as the input. Quantitative levels of these parameters used for calculations have been summarized in Table 1. For conversion of doping density (primarily used in modeling) into resistivity, the doping versus resistivity conversion formula following ASTM standards[11] is used.

Table 1. Quantitative levels of parameters used for PL signal modeling.

Material[1, 12]	Defects	Doping	Instrument
$T = 300K$	N_T - variable	N_D - variable	$\lambda = 0.532$ µm, 0.827 µm
$C_n = 2.8\times10^{-31}$ cm^6/s	$\sigma_n = 1\times10^{-14}$ cm^3/s	ρ (N_D) - variable	NF - variable
$C_p = 1.0\times10^{-31}$ cm^6/s	$\sigma_p = 1\times10^{-14}$ cm^3/s	n-type	C = 1
$B = 1.0\times10^{-14}$ cm^3/s	$E_T = 0$ eV		
$E_c = 0.56$ eV			
$Ev = -0.56$ eV			
$kT = 0.026$ V			

The results of PL signal simulations are illustrated in the next several figures. Figure 1 illustrates the depth-dependence of injection level, $\Delta n_e(z)$, and related local photoluminescence

PLD(z) for two different lasers (penetration depths) used in SIPHER. This data shows that over 90% of the total PL signal ("depth of information") is generated within about 3 μm and 25 μm for short (0.532 μm) and long (0.827 μm) wavelength laser, respectively.

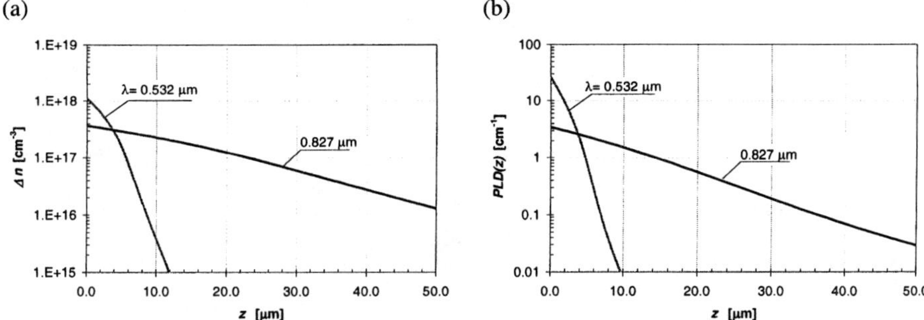

Figure 1. Depth-dependence of (a) injection level, $\Delta n_e(z)$, and (b) local photoluminescence *PLD(z)* under $NF=10^{20}$ cm^{-2}s^{-1} excitation. $N_D=10^{17}$ cm^{-3}, and $N_T=10^{12}$ cm^{-3}.

Obviously, higher excitation levels *NF* lead to higher injection levels $\Delta n_e(z)$, see Fig. 2. The $\Delta n_e(z)$ = f (*NF*) dependence becomes nonlinear, especially at low doping. At typical excitation levels, i.e. laser power density, the "high injection" state is achieved for lightly doped samples, while for heavily doped material, the "low" injection state is still prevalent.

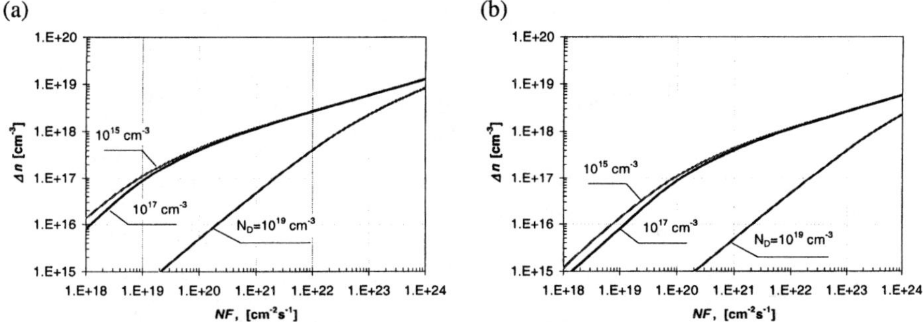

Figure 2. Impact of the *NF* excitation on the average injection level, Δn_e, with doping concentration N_D as a parameter. $N_T=10^{12}$ cm^{-3}; (a) 0.532 μm and (b) 0.827 μm excitation laser.

Furthermore, the excitation level affects the recombination lifetime observed in the material and contribution of each lifetime component to the lifetime total value, see Fig. 3. At medium doping levels, as the excitation level increases, a switch from predominant SRH to Auger recombination mechanism is observed. The radiative recombination component is also dependent on excitation level but overall the radiative lifetime is very long (as one would expect for an indirect bandgap semiconductor) and therefore it has no influence on the total lifetime.

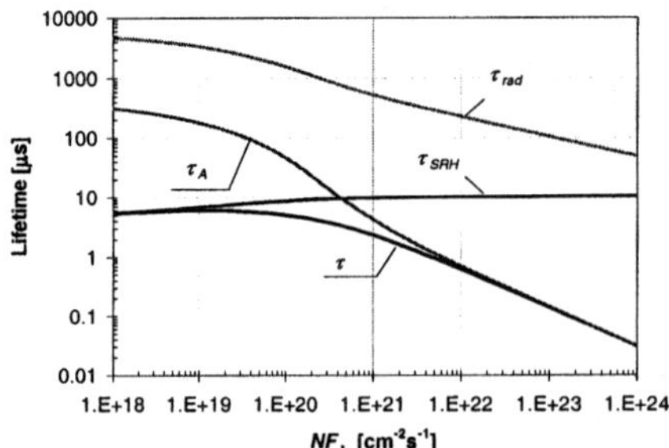

Figure 3. Impact of the *NF* excitation on the average lifetime components. $N_D = 10^{17}$ cm^{-3}, $N_T = 10^{12}$ cm^{-3}, $\lambda = 0.532$ µm.

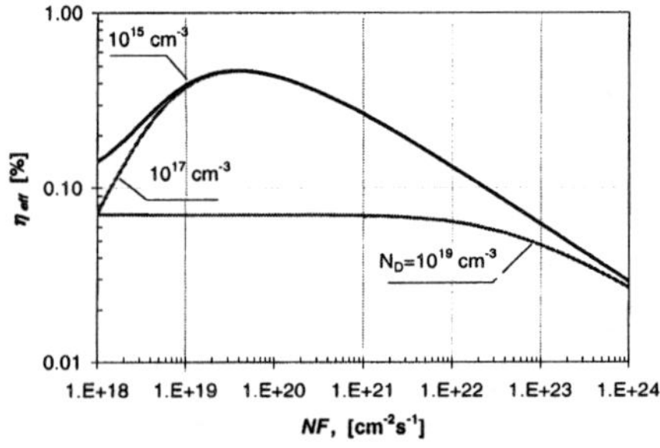

Figure 4. Impact of the *NF* excitation on PL efficiency with doping concentration N_D as a parameter. $N_T = 10^{12}$ cm^{-3}, $\lambda = 0.532$ µm.

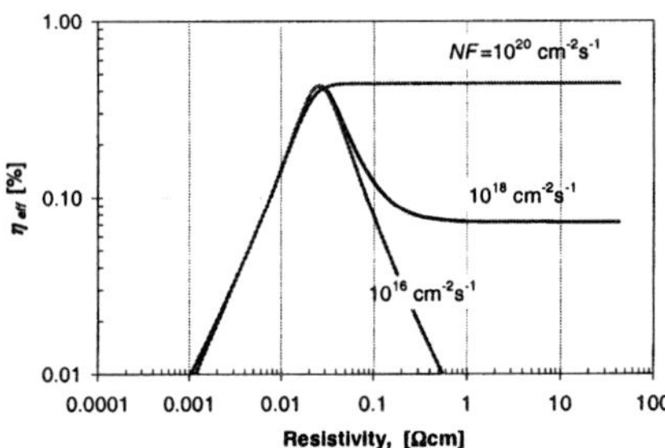

Figure 5. Dependence of PL efficiency on resistivity with excitation level *NF* as a parameter. $N_T = 10^{12}$ cm^{-3}, $\lambda = 0.532$ µm.

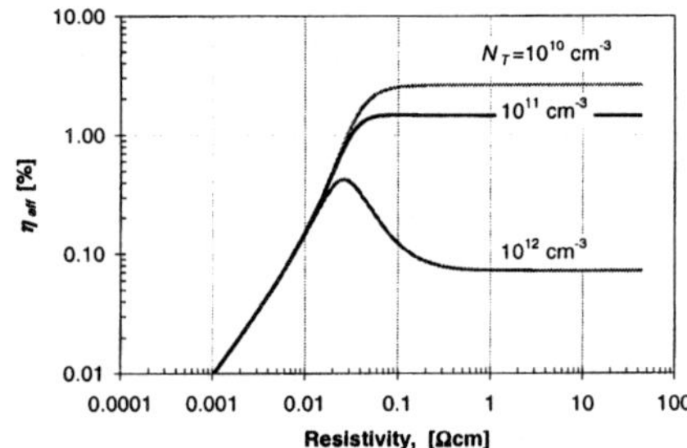

Figure 6. Dependence of PL efficiency on resistivity with deep center concentration N_T as a parameter. $NF = 10^{18}$ cm^{-2}s^{-1}, $\lambda = 0.532$ µm.

It is worth noticing that the internal efficiency of photoluminescence η_{eff}, defined as the ratio of generated to absorbed photons, may have a maximum as a function of excitation level, compare Fig. 4. This maximum occurs when the ratio of radiative recombination rate to SRH and Auger rates reaches a maximum. The η_{eff} efficiency can be relatively high (exceeding 1%) under certain conditions of doping level and excitation, see simulations illustrated in Figs. 5 and 6. The modeling shows also that at given excitation and material purity (N_T), the PL intensity reaches a maximum for material of certain resistivity (doping level), usually within 0.01 to 0.1 Ωcm range.

The analysis of simulation outcome leads to the following observations:

(1) At high excitation levels, the recombination lifetime in the material is ruled by Auger process, even at low doping levels. The radiative recombination lifetime is very long and its impact on the total lifetime is negligible at any excitation/doping/contamination conditions.

(2) Due to low lifetime, the effects of carrier diffusion are limited. As the result, by changing wavelength of the excitation laser, the depth of information can be modified.

(3) In heavily doped materials, PL is primarily a function of doping. The PL signal intensity can be used for analysis of doping striations. Major "lifetime killers" such as precipitations, if able to compete with the Auger component, may give rise to PL contrast.

(4) In lightly doped materials, the PL intensity is not sensitive to doping, but it is controlled by contamination and excitation levels. The sensitivity for contamination ($\Delta PL/\Delta N_T$) is reduced at high excitation and low N_T. Therefore, the detection of bulk contamination at levels below 10^{10} cm^{-3} may be questionable.

EXPERIMANTAL RESULTS

Model Verification

In order to verify the basic model concept, a variety of n-type samples doped with Arsenic, Antimony and Phosphorus with resistivity ranging over four orders of magnitude, from about 1 mΩcm to over 10 Ωcm, has been evaluated. To minimize the effects of major carrier loss due to diffusion towards the surface, prior to the evaluation, the samples were thermally oxidized in a diffusion furnace for surface passivation. For PL measurements, the 0.827 μm fixed power laser was used. The experimental and model-based fitted data is illustrated in Fig. 7. In spite of the "crude" approach used for modeling, a reasonably good agreement between experimental and modeled data was obtained. One major initial discrepancy was noticeable: the experimental PL peak position was shifted towards lower resistivity, as compared to simulation data given in Fig. 6. To correct this discrepancy, the Auger recombination parameters used for modeling had to be lowered by over an order of magnitude, as compared to the published values[1, 12]. Furthermore, the excitation level ($NF = 3\times10^{19}$ $cm^{-2}s^{-1}$) obtained by fitting was lower than expected from the laser power by about three orders of magnitude. These discrepancies are believed to result from disregarding the carrier diffusion process in the model. Therefore, the fitted NF, C_n and C_p parameters should be considered "effective", i.e. incorporating the effects of neglected carrier diffusion.

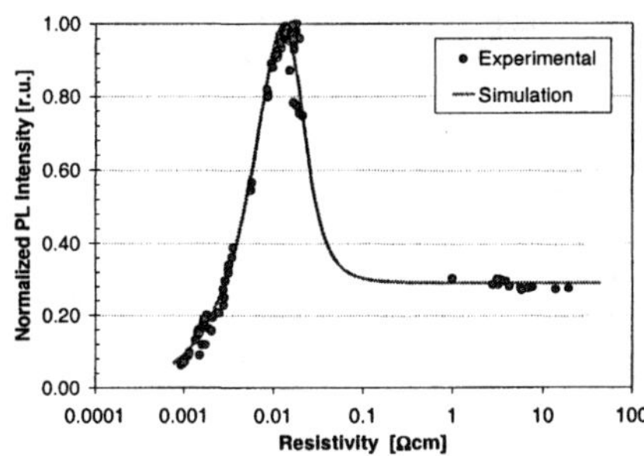

Figure 7. Plot of experimental (dots) and simulated (solid lines) photoluminescence versus resistivity for n-type silicon material (oxide passivated samples). Fitting parameters: $N_T = 4\times10^{11}$ cm^{-3}, $NF = 3\times10^{19}$ $cm^{-2}s^{-1}$, $C_n = 0.07 \times 10^{-31}$ cm^6/s, $C_p = 0.1 \times 10^{-31}$ cm^6/s.

Applications

One of major advantages of the PL tool, as compared to more classical lifetime-based instruments such as PCD or SPV, is PL "zooming" capability, i.e. acquisition of images within a wide range of resolution. The "zooming" capability is illustrated in Fig. 8, where PL images of solar grade, large grain polycrystalline EFG material are shown.

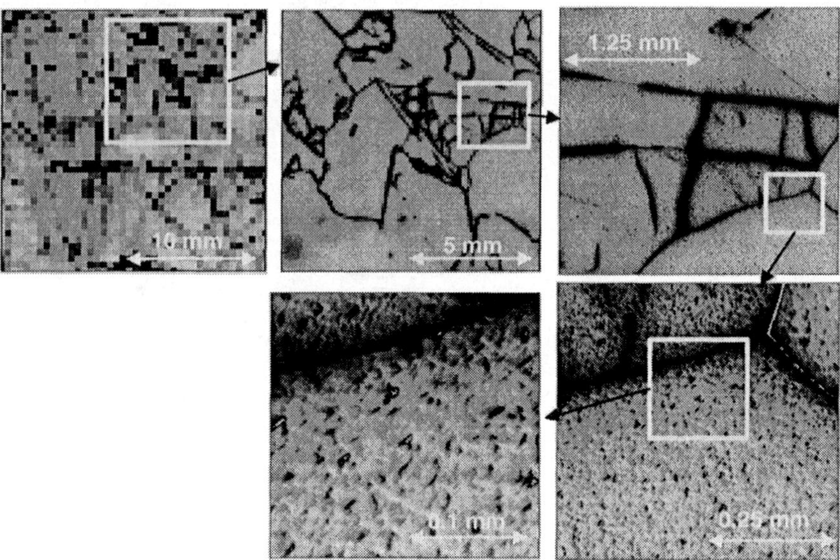

Figure 8. PL image "zooming" capability for large grain polycrystalline EFG material.

Heavily doped substrates, epitaxial and lightly doped wafers can be evaluated at full wafer size. Fig. 9 shows an example of a 200 mm epitaxial wafer measured twice with two different lasers. Complementary half-wafer images, each taken at different wavelength are aligned

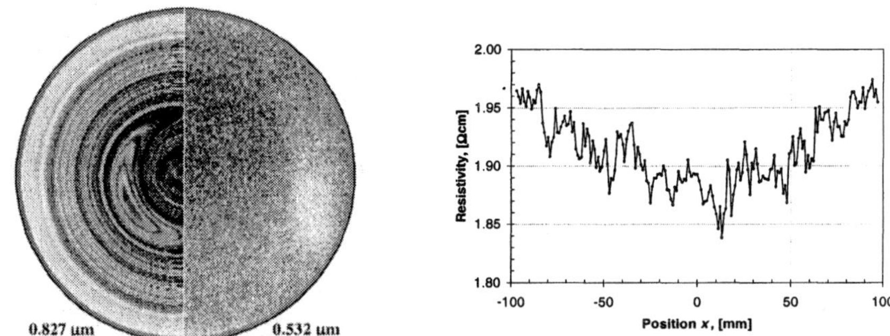

Figure 9. Detection of doping striations in 200 mm, <100>, n/n$^+$ epitaxial wafer. (a) full wafer map and (b) resistivity profile based on the resistivity versus PL transformation formula.

side-by-side to illustrate the impact of laser penetration on the information depth. At short wavelength, the signal is primarily dependent on lightly doped epitaxial layer quality, while at long wavelength the information is related to heavily doped substrate (Fig. 9(a)). For the heavily doped substrate, the modeling data indicates that the observed contrast variation represents doping striations. Thus, based on the relationship of PL on resistivity, the PL map can be transformed into a high-resolution resistivity image enabling detailed analysis of resistivity profiles (Fig. 9(b)).

For lightly doped materials, no impact of doping is observed; instead, the material quality/ contamination and injection level come into effect. Figure 10 shows two superimposed images of a lightly doped wafer with a fragment of OISF ring; lifetime map taken with PCD and PL intensity image obtained with SIPHER. The same magnification (despite of different resolution) was used for both images. It is evident that the PL contrast is much smaller than the PCD contrast, compare 2% and 30%, respectively. The difference in contrast is due to impact of injection level: the injection level in PL is a few orders of magnitude higher than in PCD.

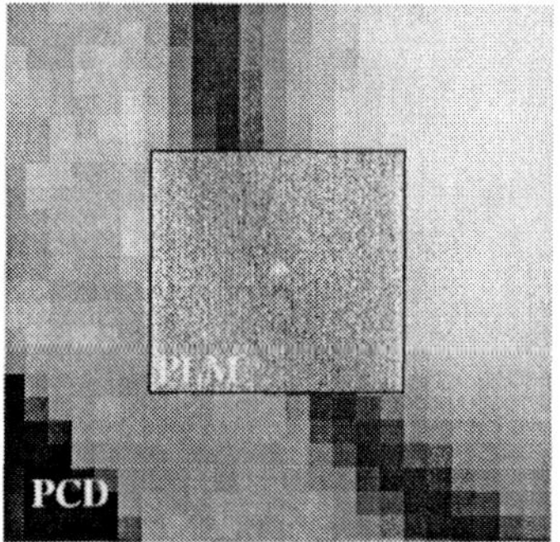

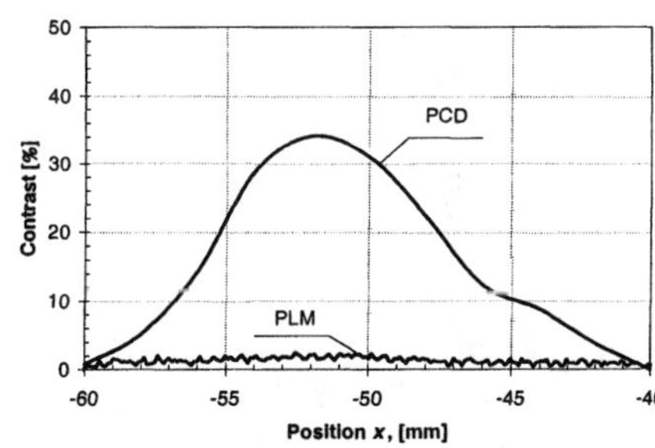

Figure 10. OISF ring detection in lightly doped polished wafer; (a) superimposed PL and PCD images and (b) PL and PCD contrast associated with the OISF ring.

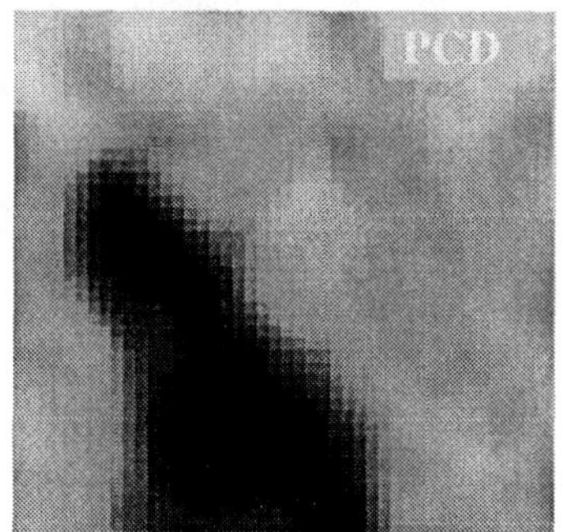

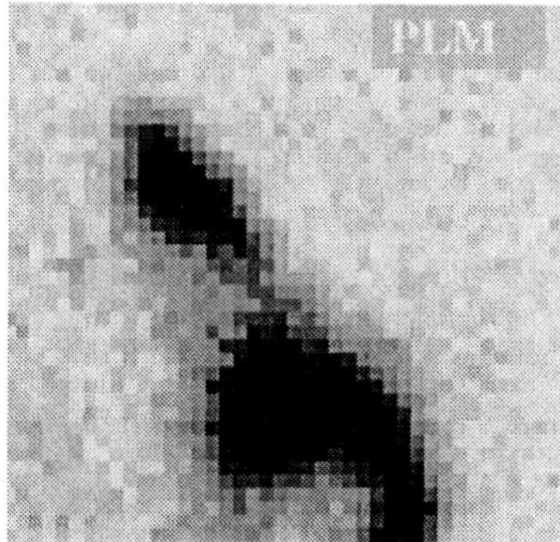

Figure 11. Detection of metal contamination in-diffused into the wafer bulk: (a) PCD lifetime and (b) PL intensity maps.

As the result of high injection, the weak defects, which are relatively "clean", do not show much higher electrical activity than the background. For strong recombination centers such as related to contamination by heavy metals (like Fe), both techniques PL as well as PCD show similar strong contrast, compare Fig. 11. The advantage of high PL image resolution is quite noticeable despite of the same 500 μm step used for PCD and PL images. The higher resolution of PL is not a matter of difference in typical image raster, rather because of limited carrier diffusion and better focusing of excitation beam.

The application of PL imaging at enhanced resolution is illustrated in Figs. 12 and 13. Figure 12 shows misfit dislocations (MD) in Si/Si(Ge)/Si heterostructure. Both, spatial density of MD (400 cm^{-1}) and MD electrical recombination "strength" (contrast=30% for 0.532 μm laser) can be quantified. In addition, the shape of the MD peaks gives a hint on the material diffusion length (DL), about 20 μm in this case. Roughly, such DL corresponds to 200 ns lifetime. In the image, a superposition of contrast due to MD and doping striation is also noticeable.

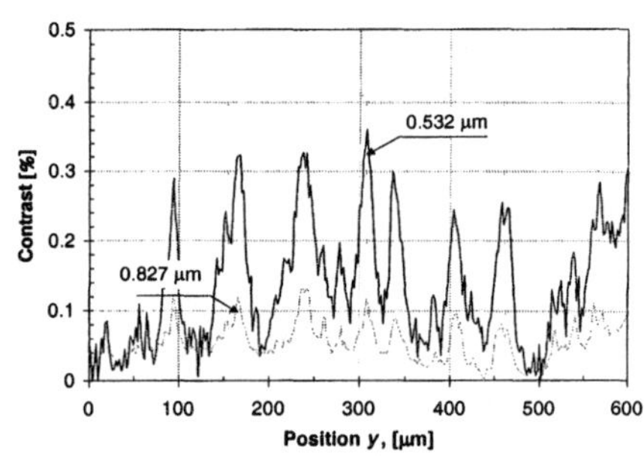

Figure 12. Detection of misfit dislocations (MD) in Si/Si(Ge)/Si heterostructures; (a) 1×1 mm^2 PL image taken with 0.827 μm laser and (b) *y*-scan contrast.

A typical image of oxygen precipitations is illustrated in Fig. 13. The aerial density of bulk micro defects in this sample can be quantified as about 2×10^5 cm^{-2}. In this particular sample the precipitates were decorated with Nickel. The metal decorated precipitates are "electrically" very active, here with contrast exceeding 60%. The diffusion length and lifetime can be estimated to be 4 μm and 8 ns, respectively.

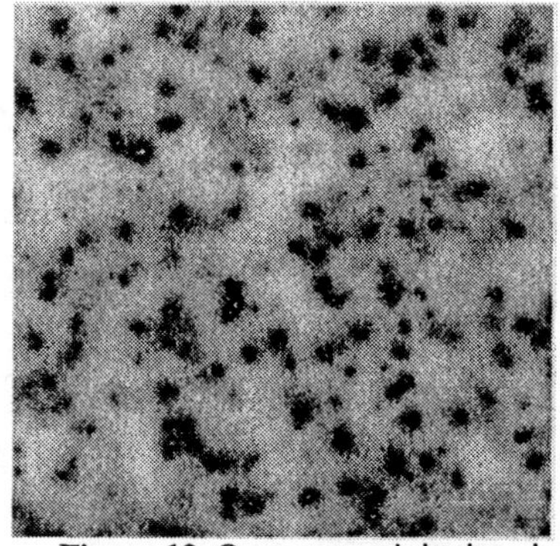

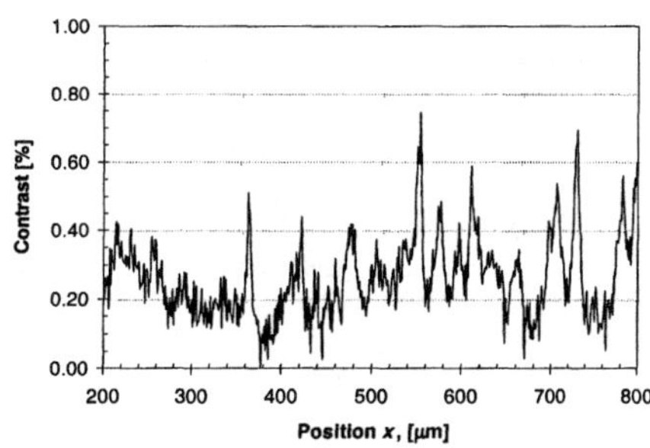

Figure 13. Oxygen precipitations in lightly doped, p- material, decorated with Ni: (a) 250×250 μm^2 PL image taken with 0.532 μm laser, and (b) *y*-scan contrast.

CONCLUSIONS

PL signal simulation based on assumptions of steady state conditions and at neglected carrier diffusion seems to be a reasonable approach for PL modeling, especially, if samples with a passivated surface are considered. The possibility of neglecting diffusion comes from the fact that very high excitation levels are applied in the tool used. At such excitation levels, Auger process dominates recombination leading to recombination lifetime in the nanosecond range.

Relatively good agreement between simulated and experimental data was achieved. The data shows that the PL signal reaches a maximum within 0.01 to 0.1 Ωcm resistivity range. At the lower resistivity end, the PL signal depends primarily on resistivity and the tool becomes very useful for analyzing doping uniformity and striations. At the high resistivity end, the PL signal is primarily dependent on the concentration of deep levels and the tool can be used for defect and contamination analysis.

For lightly doped materials, a strong impact of excitation level on the PL output is also anticipated. Within the entire resistivity range, at very high concentration of defects such as dislocations and oxygen or metal precipitates, SRH recombination may effectively compete with Auger recombination. Besides defect concentration, the recombination activity of specific defects will impact this competition.

REFERENCES

1. D. Schroder, *Semiconductor Material And Device Characterization*, John Wiley&Sons, Inc., New York (1990)
2. S. Ibuka, M. Tajima, H. Takeno, M. Warashina, T. Abe, K. Nagasaka, Jpn. J. Appl. Phys., **36**, L494-L497, (1997)
3. K. H. Cho, J. Y. Lee, E. K. Suh, K. W. Kim, Jpn. J. Appl. Phys., **36**, 1-5 (1997)
4. S. Ibuka, M. Tajima, Jpn. J. Appl. Phys., **39**, L1124-L1126 (2000)
5. V. Y. Timoshenko, J. Rappich, T. Dittrich, Jpn. J. Appl. Phys., **36**, L58-L60, (1997)

6. M. Warashina, M. Tajima, Jpn. J. Appl. Phys., **35**, 120-123 (1996)
7. SIPHER, Bio-Rad Semiconductor Systems Division (recently Accent Optical Technologies)
8. F. Kirscht, B. Orschel, S. Kim, S. Rouvimov, B. Snegirev, M. Fletcher, M. Shabani and A. Buczkowski, *Application of Room-Temperature Photoluminescence For Characterizing Thermally Processed CZ Silicon Wafers* in Defect and Impurity Engineered Semiconductors and Devices III, Ed. S. Ashok, J. Chevallier, N.M. Johnsonn, B.L. Sopori, H. Okushi, MRS, **719**, (2002)
9. S. Westrate, V. Higgs, Solid State Technology, **45**, 57-58 (2002)
10. J. Blakemore, *Semiconductor statistics*, Dover Publications, Inc. New York 1987
11. Annual Book of ASTM Standards, F 723-99, *Conversion Between Resistivity And Dopant Density for Boron-Doped, Phosphorus-Doped, and Arsenic-Doped Silicon*, **10.05**, 275-291 (2000)
12. H. K. Jung, K. Taniguchi, C. Hamaguchi, Jpn. J. Appl. Phys., **34**, 3054-3058, (1995)

Local Electrical Characteristics of Cu-Contaminated SiO_2 Thin Films

Norio Tokuda[1], Takahiro Kanda[1,*], Satoshi Yamasaki[2], Kazushi Miki[3], and Kikuo Yamabe[1]
1 Institute of Applied Physics, University of Tsukuba, Tsukuba, Ibaraki 305-8573, Japan
2 Advanced Semiconductor Research Center-AIST, Tsukuba, Ibaraki 305-8562, Japan
3 Nanotechnology Research Institute-AIST, Tsukuba, Ibaraki 305-8562, Japan
*Present address: Komatsu Electronic Metal Co., Ltd., Miyazaki, Miyazaki 889-1693, Japan

e-mail : tokuda@esys.tsukuba.ac.jp

Dielectric degradation of intentional Cu contaminated SiO_2 film on Si substrate using conducting atomic force microscopy and metal-oxide-semiconductor capacitors was investigated. Comparison of both measurements clarified that local oxide leakage currents increased not at Cu particle positions but at the random points except for Cu particles. Under Cu particles, the thick SiO_2 film was grown. Combination of the electric characteristics and total reflection x-ray fluorescence analyses indicated that high density Cu atoms on SiO_2 surface induced dielectric breakdown and low density Cu inside SiO_2 film induced high leakage current in low electric field range.

1. Introduction

With continuous decreasing feature size of Metal-Oxide- Semiconductors (MOS) devices, decreasing in oxide thickness of MOS systems is inevitable. At the same time, the scaling of devices to submicron dimension has resulted in the performance constrained by interconnect RC delay. The interconnect resistance can be lowered by choosing alternate conductor materials such as Cu, Ag or other new materials. Such materials have still low purities such as 99.9% or 99.99%. Not only the impurities but also the main elements lead to degradation of the device performance. Especially, heavy metal contaminants are one of the most serious causes of degraded device performance[1)]. Degradation mechanism by the impurity inside the SiO_2 films has been evaluated by the dielectric characteristics of MOS capacitors[2-8)]. Microscopic or physical mechanism, however, is not obvious. Further to introduce the new materials, it is essential to understand the degradation mechanism by the metal elements inside the SiO_2 films.

Cu and Au have larger electronegativity than Si. In a diluted HF (DHF) immersion before a gate oxidation, such a metal can be adsorbed on the bare Si surface[9-11)]. It is well known that even a trace amount of metallic impurities on a Si surface can drastically degrade the electronic performance of semiconductor devices such as increase in inverse leakage current at the p-n junction, and acceleration of the carrier lifetime deterioration. Especial increase in the gate oxide leakage current and decrease in the gate oxide breakdown voltage are serious. In conventional evaluation of electrical characteristics of an MOS capacitor, it is impossible to evaluate the local dielectric degradation. Using a conducting atomic force microscopy (CAFM), it is possible to simultaneously observe both topographic image of ultrathin SiO_2 surface and the corresponding local oxide leakage current distribution[12-16]. That is, we directly evaluated a correlation between the metallic particles and the two-dimensional leakage current distribution using the CAFM.

2. Experimental

The Si substrates used in this study were p-type (111) oriented Czochralski grown Si wafers with a resistivity of 10-20 Ω·cm. The Si wafers were annealed at 1200 °C in hydrogen ambient for 1 hour to form step/terrace structure on the surfaces. The Si wafers were cleaned by a modified RCA method[17], followed by hydrogen termination using diluted HF (DHF) solution to prevent the chemical oxide growth during ultra pure water (UPW) rinse and air exposure[18,19]. After the cleaning procedure, some of the wafers were intentionally contaminated by DHF solution (HF=0.5%) immersion for 10 min. At that time, DHF solution was contaminated with $CuCl_2$ concentration of 1 ppm in volume. The Cu contaminated wafers were rinsed with the UPW for 10 min. The quantitative chemical analyses of the metallic contamination on the wafer surfaces were performed by means of total reflection x-ray fluorescence (TXRF). Angle, excitation voltage and current of an incident x-ray were 0.05 degree, 40 kV and 40mA, respectively. A detection limit of Cu on the Si surface by this method was about 3×10^9 atoms/cm^2. The surface Cu concentrations of reference non-contaminated wafers and the contaminated wafers were less than $1\text{x}10^{10}$ atoms/cm^2 and about $3\text{x}10^{14}$ atoms/cm^2, respectively. The contaminated wafers were thermally oxidized in dry O_2 ambient at 600°C for 20 min and at 850°C for 30 min. Their oxide thicknesses were 1.3 nm and 8.2 nm by ellipsometry. Some of the Cu contaminated oxide was immersed in sulfuric acid/hydrogen-peroxide mixture (SPM) solution to remove the adsorbed Cu atoms. The Al film was evaporated on the 8.2-nm-thick SiO_2 film. The gate electrodes of MOS capacitors were delineated by a photolithography technique and a wet etching. The oxides grown on a backside of the wafers were removed completely by clean DHF solution. These MOS capacitors were used to measure oxide leakage current vs. gate voltage (J-V) characteristics of the 8.2-nm-thick SiO_2 film. On the other hand, both the topographic and the leakage current images of the 1.3-nm-thick silicon oxide were obtained simultaneously by the CAFM with an Au-coated conductive tip. The force constant and the resonant frequency of the cantilever were 0.14 N/m and about 12 kHz, respectively. The sample bias voltage during the CAFM scanning was kept constant at -3.0 V.

3. Results and Discussion

Figure 1 shows J-V characteristics of two kinds of the 8.2-nm-thick gate oxide measured using the MOS capacitor structures. One of them was thermally grown on the Cu contaminated Si surface. The other was immersed in the SPM solution to remove the residual Cu atoms on the gate oxide. A broken line in Fig.1 represents a J-V curve of the reference oxide without the intentional Cu contamination. The intentional Cu contamination clearly lowered the breakdown voltage of the gate oxide. This is consistent with experimental results described in other manuscripts[5]. In addition, increase in the leakage current can be seen in the voltage range of -3 to -6 V, comparing with the leakage current of the reference oxide. The characteristics of the oxide immersed in the SPM solution were also shown in Fig.1. It was found that the SPM immersion avoided the low voltage dielectric breakdown events of the Cu contaminated gate oxide. After the SPM immersion, however, increase in the leakage current was left in the voltage range of -3 to -6V, in comparison with that of the referenced oxide. This will be discussed below.

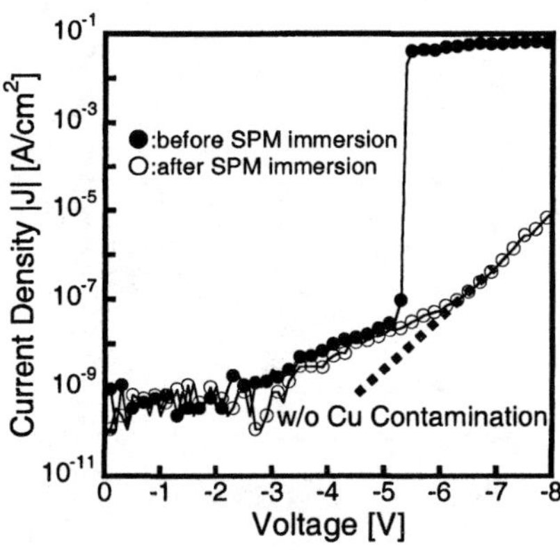

Fig. 1. J-V characteristics of MOS capacitors with 8.2-nm-thick gate oxides. The gate oxides were thermally grown on Si wafers contaminated by immersing in DHF solution contaminated with Cu 1 ppm for 10 min. Some of the contaminated oxides were immersed in SPM solution to dissolve the Cu atoms on the oxide. A dashed line indicated the leakage current of the reference oxide without the intentional Cu contamination.

These results indicate that the Cu contamination degraded the oxide dielectric characteristics. It was speculated that the Cu contamination degraded the dielectric characteristics of the SiO_2 at the adsorbed particles. That is, the oxide might be thinned at the particle positions[3,20] or the dielectric resistivity might be degraded. It is, however, not clear yet where the dielectric characteristics of the gate oxide were locally degraded. To clarify this, we observed a relationship between the oxide leakage current distribution and the oxide surface topography using the CAFM.

Figures 2(a) and (b) show the surface topography image and the leakage current distribution of the Cu contaminated SiO_2 by means of the CAFM, respectively. The scanning area of the CAFM is 500 nm × 500 nm. In Fig. 2(a), five lines of atomic step edge in the surface topography are indicated by arrows. Each height of the atomic step was 0.3 nm. It is known that Cu particles are formed on Si surface after immersion in DHF solution with Cu ions[9-11]. We observed the particles both on the Cu contaminated Si surface before thermal oxidation and on the Cu contaminated SiO_2 surface after thermal oxidation. Contrast in the current image of Fig. 2(b) indicates the leakage current range of 0 to -10nA. Dark points indicated positions with the high leakage current. Although not shown in a figure, the dark points were hardly observed in the CAFM image of the reference SiO_2 surface. In Fig. 2(b), however, a large number of the dark points were randomly observed in the whole region of the Cu contaminated SiO_2 surface except for the Cu particles. It has been

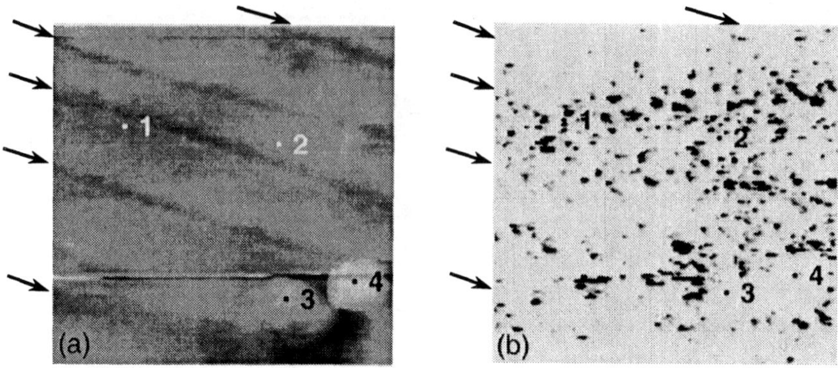

Fig. 2. (a) A topographic AFM surface image of an oxide thermally grown on contaminated Si(111). A scanning area is 500 × 500 nm^2. (b) A current image obtained simultaneously with (a) using the CAFM. The sample bias was kept at – 3 V for (a) and (b). Color contrasts in Figs. 2(a) and (b) indicate the height range of 0 to 2.0 nm and the leakage current range of 0 to 10nA, respectively. In the current image (b), the high current density region is represented by dark dots.

reported that the leakage currents selectively flow at the step edge, when the SiO_2 film thickness is about 1 nm[14-16]. In Fig. 2(b), however, the oxide leakage current at the step edge was not higher than that within the terraces. The high leakage currents randomly flowed irrespectively of the step edge. This result can be explained as follows. In ref.15, the increase in the oxide leakage current was observed for 1.0nm-thick oxide, while it was not observed for 1.4nm-thick oxide. The oxide thickness in this experiment is 1.3 nm and too thick. In addition, the Cu may selectively adsorb the atomic step edge only under a restricted condition beyond our understanding.

Two Cu particles were observed at the points 3 and 4 on thermal oxide surface after the Cu contamination as shown in Fig. 2(a). The leakage current at the points 3 and 4 in Fig. 2(b) hardly flowed. We confirmed that before the thermal oxidation, the high current flowed over the whole region of the Cu contaminated Si surface including the Cu particles. These results indicate that the thermal oxidation suppressed the oxide leakage current at the Cu particles.

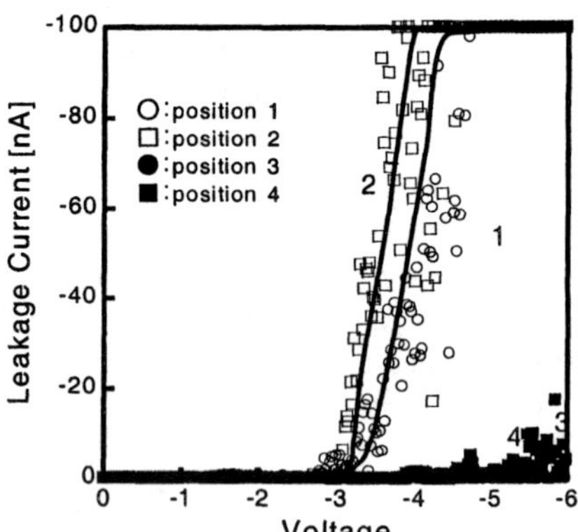

Fig. 3. Local I-V curves measured by using cantilever of CAFM fixed at four points indicated in Fig. 1(a).

Figure 3 shows local oxide leakage current vs. sample bias voltage (I-V) curves measured at the four positions in Fig. 2 by CAFM. At the points 1 and 2, the oxide leakage

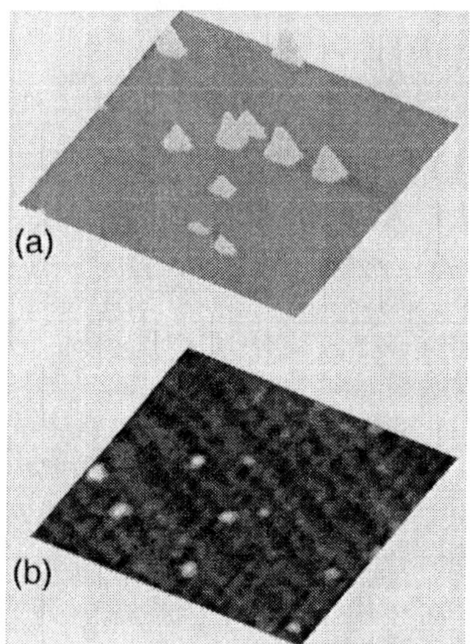

Fig. 4. A three-dimensional AFM surface image of Cu contaminated oxide (a) before an immersion in SPM and (b) after the immersion in SPM. A scanning area is 2 × 2 μm². The maximum of the z-axis is 100 nm.

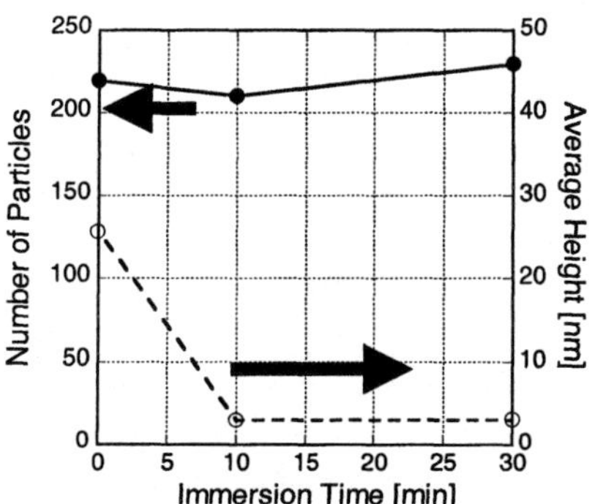

Fig. 5. The number and the average height of the particles as a function of the immersion time in SPM.

Fig. 6. A topographic AFM surface image of Cu contaminated oxide after an immersion in DHF solution. A scanning area is 2 × 2 μm². Dark spots indicate pits.

currents rapidly increased at around -3 V. On the other hand, the leakage current at the points 3 and 4 was lower than 20nA over the whole range of the sample bias voltage. It has been reported that the oxidation of silicon is accelerated by the Cu particles[21] and Cu_3Si[22,23]. The thick SiO_2 film of which the growth rate was accelerated by Cu would suppress the oxide leakage current. We suggest that it becomes three-layer structure in the particles: copper oxide layer, SiO_2 layer that was thickened by Cu, and Si substrate. This detail is discussed below. It is well known that copper metal as well as copper oxide dissolve in acid solution. If the top of the particles consists of copper metal or copper oxide, the particles should become smaller after an immersion in acid solution, such as SPM. Figure 4(a) and (b) show three-dimensional AFM images of the topography of the Cu contaminated SiO_2 surface before and after the immersion in SPM, respectively. The scanning area of the AFM is 2 μm × 2 μm. From the figures the average height of the particles became lower by the immersion in SPM. Figure 5 shows both the number and the average height of the particles in the area of

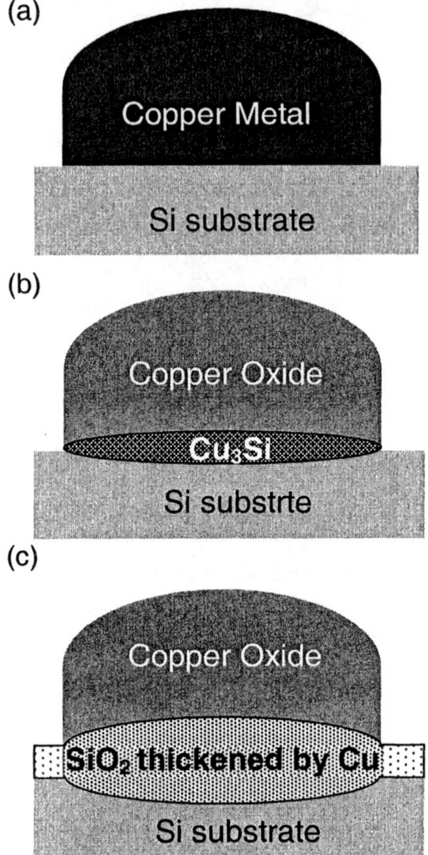

Fig. 7. Mechanism of accelerated silicon oxidation at the Cu particle in thermal oxidation process. (a) as-Cu deposited in DHF solution. (b) Cu_3Si was formed at the interface. (c) The Cu3Si completely reacted the oxidant to form SiO_2 and Cu.

10 μm × 10 μm as a function of the immersion time in SPM. After the immersion in SPM, although the number of the particles was almost constant, the average height of the particles became lower from 28 nm to 3 nm. Figure 6 shows an AFM image of the topography of the Si surface after an immersion in DHF solution. All particles disappeared after etching oxides. Instead some pits were observed after the immersion in DHF solution. It can be considered that a silicon dioxide or a silicate is grown. But it has been reported that Cu do not silicate with sintering at 1150°C[24]. Therefore the particles are considered silicon dioxides that were thickened by Cu. These suggest that the thick SiO_2 film growth under the Cu particles suppressed the oxide leakage current. Figure 7 shows the flow of forming the underlying thick SiO_2. After immersing the Si wafer in DHF solution with Cu 1 ppm, copper deposits on the Si substrate as shown in Fig. 7(a). The sample was conveyed to a thermal furnace in dry O_2 ambient at 850°C. It has been reported that Cu_3Si formed at Cu/Si interface at 200°C[22] and Cu was oxidized to form Cu_2O below 250°C and CuO above 275°C[25], respectively. Immediately the sample was conveyed to the thermal furnace, Cu was oxidized at the surface of the particles and changed to Cu_3Si at the interface of the particles, as shown in Fig. 7(b). And Si also was oxidized at the Si surface except for the particles. The Si oxidation is catalyzed by the Cu_3Si. It was reported that the SiO_2 can be grown to a thickness of 1-2 μm at room temperature over a period of a few days. The increase of thickness of the Cu_3Si film stops because an oxidant reaches to the interface. The oxidant reacts with the Cu_3Si to form SiO_2 and Cu according to: $Cu_3Si + 2O \rightarrow SiO_2 + 3Cu$. The Cu_3Si changes to SiO_2 and Cu, and Cu and Si are oxidized to form copper oxide and silicon oxide, respectively. Finally, as shown in Fig. 7(c), the particles are composed of following three layers: copper oxide layer, SiO_2 layer that was thickened by Cu, and Si substrate.

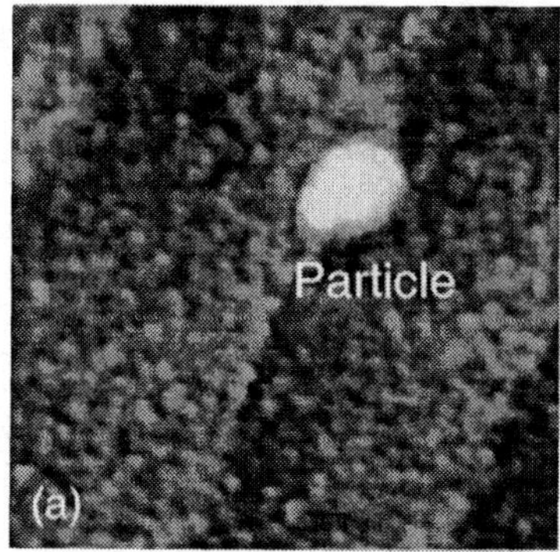

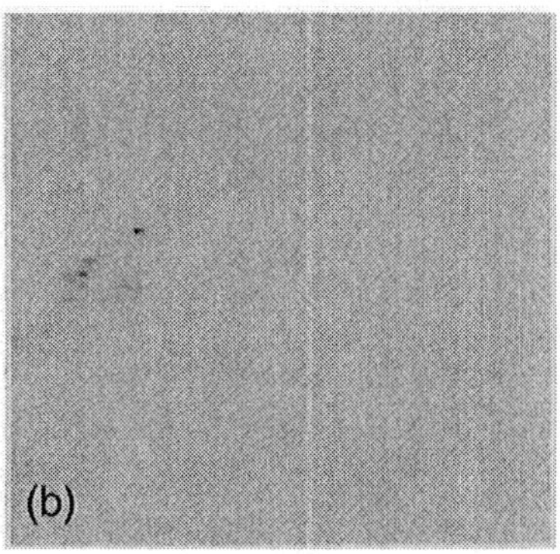

Fig. 8. (a) A topographic AFM surface image of Cu contaminated oxide after SPM immersion for 10 min. A scanning area is 500 × 500 nm^2. (b)A current image obtained simultaneously with (a) using the CAFM. The sample bias was kept at – 3 V for (a) and (b). Color contrasts in Figs. 2(a) and (b) indicate the height range of 0 to 2.0 nm and the leakage current range of 0 to 10nA, respectively. The dark points with high leakage current were hardly observed over the whole region of Cu contaminated SiO_2 surface after SPM immersion.

Figure 8(a) shows an AFM image of the topography of the Cu contaminated SiO_2 surface after the immersion in the SPM. The scanning area is 500 nm × 500 nm. A height range of 0 to 2.0 nm is represented with color contrast. Before the immersion in SPM, a large amount of the dark points with the high oxide leakage current was randomly observed in the region except for the Cu particles in Fig.2 (b). After the immersion in SPM, however, the dark points were hardly observed over the whole oxide surface region in Fig.8 (b). The current image was similar with that of the reference SiO_2 surface. It was reported that Cu positive ions were injected into a SiO_2 film under bias temperature stress of MOS capacitors with Cu electrodes[4]. Therefore, it is considered that the Cu positive ions near the oxide surface were injected into the SiO_2 film and the leakage current flowed through the SiO_2 film under sample voltage application of the CAFM. On the other hand, it is known that Cu on the Si surface is dissolved by immersion in such a strong oxidative solution as SPM[9-11]. It is concluded that dissolution of the Cu on or near the oxide surface by the SPM immersion caused diminishing of the dark points.

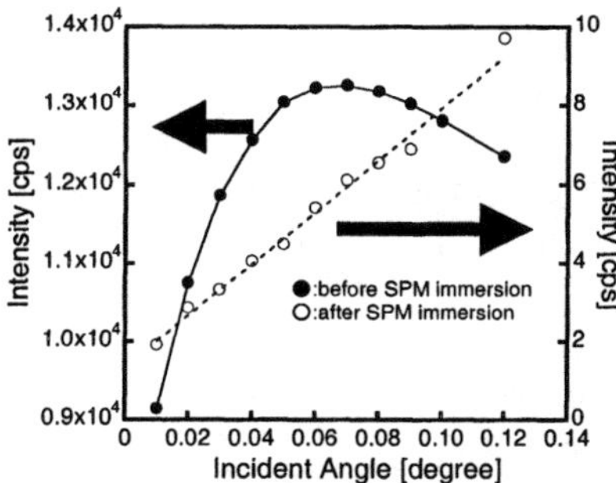

Fig. 9. Cu signal intensity near Cu contaminated oxide surface as a function of x-ray incident angle of TXRF. Intensities before and after SPM immersion are compared.

Figure 9 shows the signal intensity of the Cu atoms of the Cu contaminated SiO_2 films as a function of the x-ray incident angle in the TXRF measurement. In this figure, both the signal intensities of the SiO_2 films before and after the SPM immersion are shown. Before the SPM immersion, the maximum of the Cu intensity near 0 degree means that most of the Cu atoms were on the oxide surface. After the SPM immersion, the maximum of the Cu signal intensity disappeared. Correspondingly, the dark points diminished in Fig. 8(b). The intensity monotonically increased with increasing x-ray incident angle, that is, detection depth. This result indicates that the low-density Cu atoms after the SPM immersion were approximately uniformly distributed in the depth direction of the SiO_2 films except for the surface. It is known that the oxide charges or traps can create a local energy variation inside the oxide[26]. Therefore, it may be presumed that this residual of the Cu atoms inside the SiO_2 films caused the increase in the oxide leakage current in the voltage range of -3 to -6V.

4. Conclusions

We investigated the dielectric degradation of the intentional Cu contaminated SiO_2 film using the CAFM and the MOS capacitors. In conclusion, comparison of both experimental results clarified the following conclusion. That is, the local oxide leakage currents increased not at the Cu particle positions but at the random points except for the Cu particles. From a result in a wet etching, the particles are composed of following three layers: copper oxide layer, SiO_2 layer that was thickened by Cu, and Si substrate. The thick SiO_2 film would suppress the oxide leakage current. It was also found that the low voltage dielectric breakdown of the Cu contaminated SiO_2 films were recovered by the SPM immersion. The SPM immersion and diminished the high oxide-leakage-current positions. The TXRF analyses indicated that the SPM immersion removed the Cu atoms on the SiO_2 surface and left them within the SiO_2 films. These results mean that the high-density Cu atoms on the SiO_2 surface induced the oxide dielectric breakdown and the low density Cu inside the SiO_2 film induced the high leakage current in the low electric field range.

Acknowledgement

The authors would like to thank Dr. Masaharu Watanabe for fruitful discussion and Komatsu Electronic Metal Co., Ltd., Advanced Semiconductor Research Center and Advanced LCD Technologies Development Center Co., Ltd. for sponsoring this work.

References

1) L. Jastrzebski, R. Soydan, B. Goldsmith and J. T. Miginn: J. Electrochem. Soc. 131 (1984) 2944.

2) R. Takizawa, T. Nakanishi, and A. Ohsawa: J. Appl. Phys., 62, (1987) 4933.

3) K. Hiramoto, M. Sano, S. Sadamitsu and N. Fujino: Jpn. J. Appl. Phys., 28 (1989) L2109.

4) G. Raghavan, C. Chiang, P. B. Anders, S.-M. Tzeng, R. Villasol, G. Bai, M. Borh, D. B. Fraser: Thin Solid Films 262 (1995) 168.

5) D.A.Ramappa and W.B.Henley: J. Electrochem. Soc. 146, (1999) 2258.

6) P.-T. Liu, T.-C. Chang, J. C. Hu, Y. L. Yang, and S. M. Sze: J. Electrochem. Soc. 147, (2000)

368.
7) T. Kobayashi, I. Uchiyama, A. Kimura, S. Oka, Y. Kitagawara: J. Crystal Growth 210 (2000) 112.
8) Y. H. Lin, Y. C. Chen, K. T. Chan, F. M. Pan, I. J. Hsieh, and Albert Chin: J. Electrochem. Soc., 148 (2001) F73.
9) T. Ohmi, T. Imaoka, I. Sugiyama, and T. Kezuka: J. Electrochem. Soc., 139, 11, (1992) 3317.
10) T. Ohmi, T. Imaoka, T. Kezuya, K. Takano, and M. Kogure: J. Electrochem. Soc., 140, 3, (1993) 811.
11) H. Morinaga, M. Suyama, M. Nose, S. Verhaverbeke, and T. Ohmi: IEICE Trans. Electron. E79-C, (1996) 343.
12) R. Hasunuma, A. Ando, K. Miki, Y. Nishioka: Appl. Surf. Sci., 159-160 (2000) 83.
13) A. Ando, R. Hasunuma, T. Maeda, K. Sakamoto, K. Miki, Y. Nishioka, T. Sakamoto: Appl. Surf. Sci., 162 -163 (2000) 401.
14) R. Hasunuma, A. Ando, K. Miki, Y. Nishioka: Appl. Surf. Sci., 162-163 (2000) 547.
15) M. Murata, N. Tokuda, D. Hojo and K. Yamabe: Proc. High Purity Silicon, eds. C. L. Claeys et al. (Electrochemical Society, Pennington, 2000) Vol. PV2000-17, p. 425.
16) H. Ikeda, N. Kurumado, K. Ohmori, M. Sakashita, A. Sakai, S. Zaima, Y. Yasuda: Surf. Sci., 493 (2001) 653.
17) W. Kern and D. A. Puotinen: RCA Rev. 31 (1970) 187.
18) G. S. Higashi, Y. J. Chabal, G. W. Trucks and K. Raghavachari: Appl. Phys. Lett. 56 (1990) 656.
19) M. Sakuraba, J. Murota and S. Ono: J. Appl. Phys. 75 (1994) 3701.
20) K. Honda, A. Ohsawa, and N. Toyokura: Appl. Phys. Lett., 45(1984)270.
21) D.Gräf, M.Grundner, L.Muhlhoff and M. Dllith: J. Appl. Phys.,69(1991)7620.
22) J. M. E. Harper, A. Charai, L. Stolt, F. M. d'Heurle, and P. M. Fryer: Appl. Phys. Lett. 56 (1990) 2519.
23) C. S. Liu, L. J. Chen: Thin Solid Films 262 (1995) 187.
24) M. Takiyama, S. Ohtsukba, S. Hayashi and M. Tachimori: Proceedings of 19th workshop on ULSI Ultra Clean Technology, 95(1992).
25) W. Gao, H. Gong, J. He, A. Thomas, L. Chan, S. Li: Mat. Lett. 51 (2001) 78.
26) Y. Taur and T. K. Ning: Fundamentals of Modern VLSI Devices, Cambridge University Press, New York (1998) 85.

INFLUENCE OF IRRADIATION-INDUCED DEFECTS ON THE ELECTRICAL PERFORMANCE OF POWER DEVICES

H.-J. Schulze[1], F.-J. Niedernostheide[1], M. Schmitt[1,2], U. Kellner-Werdehausen[3], G. Wachutka[2]

[1]Infineon Technologies AG, Balanstr. 59, D-81541 Munich, Germany
[2]Physics of Electrotechnology, Munich University of Technology, Arcisstr. 21, D-80290 Munich, Germany
[3]eupec, Max-Planck-Str. 5, D-59581 Warstein, Germany

The electrical properties of high-voltage devices based on silicon can efficiently be tailored by utilizing irradiation-induced defects which modify the charge carrier lifetime and/or the effective doping concentration. Effects of electron and proton irradiation on the charge carrier lifetime and the effective doping concentration are discussed. Special focus is aimed at high proton fluence (10^{12} - 10^{15} cm^{-2}) and its influence on the doping variation of FZ silicon. The effective donor concentration is determined by spreading-resistance measurements and analyzed as a function of the irradiation fluences and the annealing conditions. The lifetime modifications are utilized with a view to optimizing the switching behavior of diodes, thyristors, and CoolMOS™ devices, while doping effects are employed in order to adjust overvoltage protection functions in thyristors.

INTRODUCTION

In this paper, we discuss the application of irradiation techniques for performance tailoring of high-power devices. First, methods for lifetime control are briefly recapitulated and the advantages of irradiation techniques with electrons or light ions over doping techniques with gold or platinum are described. In the main part experimental and simulated results are presented, showing how the specific advantages of electron and proton irradiation can be exploited to optimize the electrical data of power devices such as diodes, thyristors, gate turn-off thyristors and CoolMOS™ devices. To this end, a dedicated choice of irradiation fluences, irradiation energies and annealing temperatures is required.

One of the most important applications of light-ion irradiation in the field of high-power switching devices such as IGBTs (Insulated Gate Bipolar Transistors) or thyristors concerns the adjustment of the charge carrier distribution by the vertically inhomogeneous reduction of the charge carrier lifetime to improve the turn-off behavior while keeping the on-state losses low (1, 2). Often, a combination of electron and light ion irradiation is advantageous (3, 4). In the case of a diode, a lifetime distribution with a double peak is particularly beneficial when a soft reverse recovery behavior and an efficient suppression of the dynamic avalanche is required. Such a distribution can be realized by a two-step ion irradiation with different energies or even by a single-step ion irradiation when a suitable mask is used (5).

Except for the local reduction in the charge carrier lifetime, light-ion irradiation may also be used to change the doping profile in the devices without exposing them to a high-temperature treatment. This is especially important for high-power devices, where low-doped substrates with resistivities of up to 1000 Ωcm are used to achieve high blocking voltages. In proton-irradiated and subsequently annealed silicon, hydrogen-related donor formation, for example, leads to a locally higher doping concentration in an n-type layer and can be used to decrease the blocking voltage of p^+-n junctions. This effect can be utilized for integrating an overvoltage protection function into thyristors (6) or for realizing an n-type buffer layer in high-voltage devices. With a single proton irradiation it is even possible to exploit both effects – local reduction in the carrier lifetime and increase in the doping concentration – simultaneously to realize planar local lifetime-controlled IGBTs with a punch-through structure (7).

METHODS FOR THE REDUCTION OF THE CARRIER LIFETIME IN POWER DEVICES AND THEIR INFLUENCE ON THE DOPING LEVEL

Today, the carrier lifetime τ of power devices is reduced either by incorporation of suitable heavy metals, such as Au or Pt, by diffusion techniques or by irradiation with high-energy particles such as electrons, protons or helium nuclei. All these methods produce defects with deep levels in the energy band gap of Si which then act as recombination centers for free electrons and holes.

For a long time, diffusion of Au has practically been the only method used to reduce the carrier lifetime in high-power devices. Gettering of Au or Pt by defects in the Si crystal or by phosphorous diffusion strongly influences the reduction of the carrier lifetime and often results in laterally inhomogeneous distributions of the carrier lifetime. Furthermore, Au doping produces relatively high leakage currents and, along with it, considerable self-heating effects due to its deep level close to the center of the band gap.

As a result of these disadvantages irradiation methods become more and more important. They have the advantage of not being influenced by gettering effects. Furthermore, they can be applied at the end of the fabrication process. It is thus possible to improve the electrical data after having them measured. Moreover, the resulting leakage currents are relatively low compared with Au-diffused devices.

Vertically inhomogeneous carrier lifetime profiles can be obtained by proton or helium irradiation. When such MeV-particles penetrate into the semiconductor, they lose their kinetic energy by ionization processes and collisions with the target atoms and, thereby, produce primary vacancy-interstitial pairs, which in turn form stable secondary defects like divacancies or complexes with impurities. The energy deposition becomes especially effective when the protons have already lost a great fraction of their energy at the end of their range. The recoiling target atoms also have enough energy to generate further vacancy-interstitial pairs in collision cascades. Therefore, the defect distribution shows a peak near the penetration depth of the protons, whereas between this peak and the surface there is only a small tail of defects. The depth of the defect-rich layer below the Si surface is determined by the energy of the irradiation particles, while the defect concentration and, therefore the recombination rate in this narrow region is defined by the irradiation fluence.

The absorption coefficient of high-energy electrons in Si is much smaller compared to that of ions. Consequently, electron irradiation is suitable for a homogeneous reduction of the carrier lifetime in the vertical direction of the wafer.

While electron irradiation for moderate fluences has no significant effect on the doping level in the power devices, the irradiation with protons can result in the creation of additional donors in the irradiated volume of the wafer [e.g. (8,9)]. Helium irradiation, on the other hand, can effect a partial compensation of the doping level of the n-type starting material.

EXPERIMENTAL DETAILS

Diode and Thyristor Preparation

The starting wafers used for these types of power devices are 4 inch or 5 inch, float zone, n-type Si wafers, which are doped by means of neutron transmutation to obtain a homogeneous resistivity distribution in the n^--base. For blocking voltages between 5 and 8 kV the resistivity ranges from 200 to 500 Ωcm and the wafer thickness from 0.7 to 1.5 mm.

The blocking p-n junctions were realized by an Al vacuum predeposition (10) with a subsequent drive-in step. This reduces process-induced contamination because Al is the p-type dopant with the highest diffusion coefficient and, therefore, the diffusion temperatures can be kept relatively low together with short diffusion times. The n^+-emitter was produced by $POCl_3$-diffusion. The silicon wafers were connected with the molybdenum substrate by the use of a low temperature joining technique (11), which avoids heavy metal contamination induced by alloying techniques.

The lateral distribution of the carrier lifetime in the n^--base was characterized by the Elymat and the microwave photoconductive decay (μ-PCD) method. If the samples were analyzed after doping processes, the resultant doping layers were removed by lapping and etching so that the n^--layer remained. Detailed information about the principles of these techniques can be found in Ref. (12). The vertical doping profiles in the samples were measured by the spreading resistance technique.

The irradiation energies available for electron irradiation were up to 16 MeV. For proton irradiation we used energies of up to 3.5 MeV. To obtain special doping profiles by the application of proton irradiation annealing temperatures of up to 500°C have been applied.

CoolMOS^{TM} Preparation and Operation

CoolMOS^{TM} power MOSFETs with a nominal breakdown voltage of 600 V and a nominal current of 20 A were prepared as described in detail in Ref. (13). Since the CoolMOS^{TM} is a quite recent device concept we briefly describe its operation principle. The structure is very similar to a conventional vertical power MOSFET (Fig. 1).

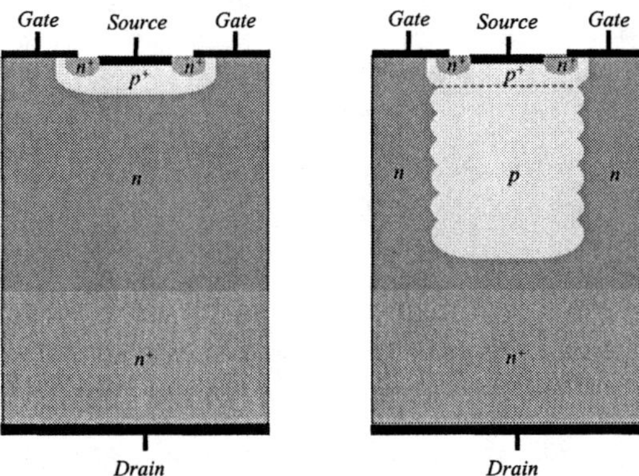

Fig. 1: Structure of a conventional power MOSFET (left) and the CoolMOS^{TM} (right)

In the normal on-state an electron current is sustained between the highly n-doped source region and the drain contact by applying a positive gate bias across a very thin gate oxide. This creates an inversion layer at the semiconductor surface under the gate and forms an conducting channel between the n^+-source and the n-doped epi-layer. In the case of a conventional MOSFET, the on-resistance $R_{(DS)on}$ is dominated by the resistance of the epi-layer, since it is relatively low-doped to accommodate the voltage in the blocking state when the gate is turned off and a high drain-source voltage is applied. In order to increase the blocking capability $V_{BR(DSS)}$, the epi-layer must be made wider and the epi-doping must be reduced. Therefore, the on-resistance is a more than quadratically increasing function of the blocking voltage (14). The CoolMOS^{TM} exhibits two major differences: First, the p-well is extended vertically to the device interior by several 10 µm and the doping concentration of the so formed p-column is controlled in such a way, that it exactly equals the doping concentration of the surrounding n-region. The doping integral along a line perpendicular to the direction of main current flow has to be smaller than the breakthrough charge Q_c (≈ 1 - $2 \times 10^{12} cm^{-2}$ for silicon). Consequently, in the blocking state the whole epi-layer is depleted from mobile carriers at very low bias. We have a superposition of a lateral and a vertical electric field and the resulting field profile is nearly constant along the vertical direction because the opposite charges of the p- and n-regions cancel each other. The second difference is the fact that the doping of the n-epi can now be raised by roughly one order of magnitude without losing the blocking capability. The result is a device whose on-resistance rises approximately linear with the blocking voltage, hence allowing an enormous reduction in the on-resistance with unaltered blocking voltage at the same time. By application of this technique we can combine high doping levels with high blocking voltages, i. e. this principle breaks the $R_{(DS)on}$-$V_{BR(DSS)}$ limit line of conventionally designed silicon MOSFETs.

The described normal operation mode relies on a unipolar effect, i.e. only majority carriers carry the electrical current and the switching speed does not depend on the recombination lifetime of the charge carriers. The situation changes if a negative bias is applied to the drain contact. Then holes are injected from the p-well and electrons from the drain emitter to the region of the lower doped p- and n-regions like in a p^+-p-n-n^+ diode. This internal reverse diode ("body diode") can be utilized in switching topologies that require reverse current flow through the switching transistor to save an additional freewheeling diode. A necessary condition for that is the fast commutation of the diode from the forward state to the blocking state, where all mobile carriers have to be removed from the base region to sustain the applied blocking voltage. In the regular CoolMOS™ this process runs rather slowly because the p-n junction deeply extends into the bulk region and almost the whole electron-hole plasma has to be removed before a substantial depletion region forms that can sustain the blocking voltage. As a consequence the diode current does not drop to zero instantaneously, but passes a pronounced negative maximum and, thus, produces considerable switching losses.

In order to improve the reverse recovery behavior of the body diode, the completely processed wafer were irradiated with electrons or protons. In the case of electron irradiation, the energy E was 4.5 MeV with the fluence ranging from $1.5 \cdot 10^{15}$ cm^{-2} to $4.5 \cdot 10^{15}$ cm^{-2}. After irradiation they were annealed in an N_2/H_2 atmosphere at T = 220 °C for 10 hours or in an N_2 atmosphere at T = 350 °C for 4 hours.

Proton irradiation was performed from the drain side with energies ranging from 3.75 MeV to 4.14 MeV so that the penetration depth was located in the region of the alternating p- and n-regions, i. e. between 17 and 42 μm under the gate oxide. The fluences ranged between $1 \cdot 10^{11}$ cm^{-2} and $2 \cdot 10^{12}$ cm^{-2}. These devices were also annealed in an N_2/H_2 atmosphere at T = 220°C for 4 hours. For electrical measurements all devices were mounted in TO220 packages.

EXPERIMENTAL RESULTS

Electron Irradiation

The absorption coefficient of electrons is quite small in comparison to that of protons or helium ions. Therefore, not only a single wafer but a stack comprising several wafers in series can be irradiated with electrons, reducing the irradiation cost. Experiments have shown that the electrical data of the devices do not depend on their position in the stack, demonstrating the homogeneity of the carrier lifetime reduction along the stack axis. The following two examples reveal how electron irradiation can effectively be used to optimize the switching behavior of a thyristor and a CoolMOS™. In the first example also the influence on the device characteristic arising from the electron energy and annealing temperature is analyzed.

Thyristor: Optimization of the Q_{rr}-V_T Relation. High-power thyristors are characterized by a high blocking voltage and a high current capability of several kilovolts and kiloamperes, respectively. In the on-state, when a high current is flowing, it is therefore desirable that the voltage drop V_T across the thyristor is as low as possible to keep the power losses small. This requires a large high-injection carrier lifetime. On the other hand, an effective removal of the charge carriers is necessary for turning off the device in order to re-establish its voltage blocking capability and to keep the reverse recovery charge Q_{rr} and with it the turn-off losses small, which can be achieved by a small carrier lifetime. Hence, one has to cope with the well-known trade-off between V_T and Q_{rr} and to find an optimum by properly adjusting the carrier lifetime for the different operation states.

Important defect centers arising from electron, proton or helium irradiation and determining the carrier lifetime are divacancies (VV) and vacancy-oxygen (VO) defects. For electron irradiation the concentration ratio [VO]/[VV] of the generated defects decreases with increasing electron energy. Since the VO defects are most important for the high-injection lifetime, which is important in the on-state, and the VV defects govern the low-injection lifetime, which is relevant during turn-off, the Q_{rr}-V_T trade-off of electron irradiated thyristors can be improved by choosing high-energy electrons. This is demonstrated in Fig. 2 showing the simulated dependence of Q_{rr} on V_T for two different concentration ratios [VO]/[VV]= 4 and 16 corresponding to an electron energy of about 10 and 2.5 MeV. The forward current of 1.5 kA was switched off with a dI/dt rate of -2 A/µs and a reverse voltage of -4 kV was applied to the device. As it is evident from the diagram, the high-energy electron irradiation leads to a reduction of the reverse recovery charge by about 5% for voltage drops around 2.8 V, which is a typical V_T value for high-voltage thyristors. In addition to the simulation results, experimental values for an electron energy of 10 MeV are shown for comparison. The small differences between experimental and simulated values probably result from contact resistances of the devices not taken into account in the simulation.

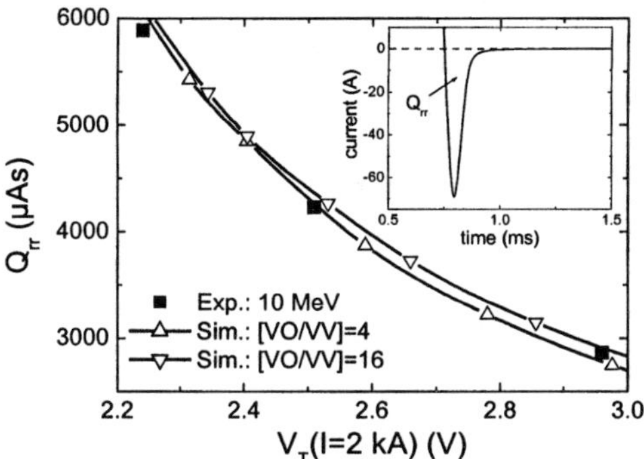

Fig. 2: Simulated Q_{rr}-V_T relation of electron irradiated 8 kV thyristors for two different ratios [VO]/[VV] = 4, 16 and experimental results for an electron energy of 10 MeV. The inset shows how the reverse recovery charge Q_{rr} is determined by integrating the thyristor turn-off current.

Since the VV and VO defects anneal out between 250 and 300 °C and between 300 and 350 °C, respectively (15), all process steps after irradiation should not exceed a temperature of 250 °C to keep the concentration ratio [VO]/[VV] as low as possible.

Electron Irradiation of the Internal CoolMOS™ Body Diode. As a second example for device optimization by electron irradiation we consider CoolMOS™ devices. As already mentioned, the intention is to improve the switching speed of the internal reverse diode.

Experimental results are displayed in Fig. 3, where the turn-off behavior of the internal diode of a CoolMOS™ device irradiated with electrons ($\phi = 4.5 \times 10^{15} cm^{-2}$, E = 4.5 MeV) and annealed after irradiation at T = 350 °C is compared with that of a non-irradiated device. Evidently the reverse recovery time can be drastically reduced, leading to an improved switching speed. The reverse recovery charge Q_{rr} drops by a factor of almost ten from 12.5 μC to 1.3 μC. Moreover, the irradiated device exhibits a lower diode voltage peak as a consequence of the lower reverse current peak.

In Fig. 4 the dependence of Q_{rr} on the electron fluence is shown for two different annealing temperatures. Obviously, Q_{rr} is lower if the devices are annealed at T = 220 °C, because the concentration of defects that can act as recombination centers is still high. But annealing at 350°C can be unavoidable to re-adjust the remaining electrical parameters of the device (16), which still has to work as a regular CoolMOS™ transistor beside the diode action. And here one important parameter is the leakage current in the blocking state, which should be as low as possible. It is found that for the annealing at 350 °C the reverse recovery charge is about three times higher compared to the 220 °C annealing, but the leakage current is about 50 times smaller.

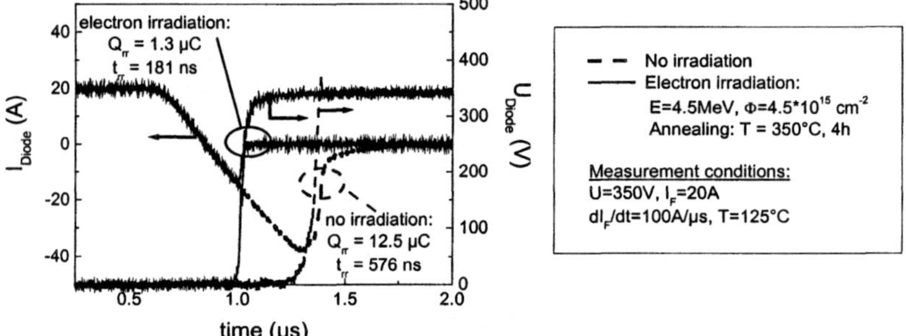

Fig. 3: Turn-off behavior of the 600V-20A-CoolMOS™ body diode for an electron-irradiated and a non-irradiated device

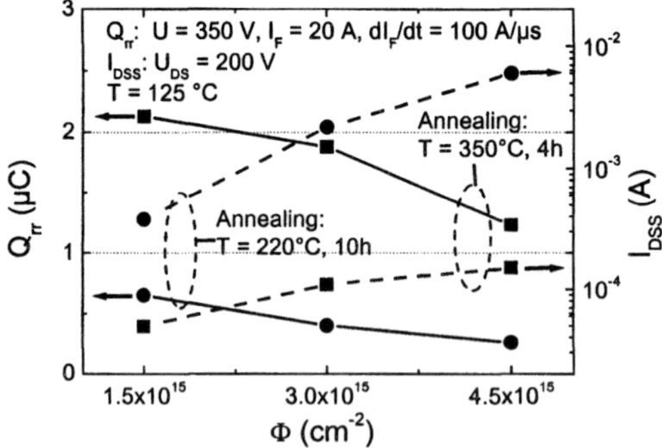

Fig. 4: Q_{rr} and I_{DSS} of the body diode of electron irradiated 600V-20A-CoolMOS™ devices for different fluences and annealing conditions.

The described electrical behavior and the influence of the annealing conditions are based on the microscopic defects that are generated by the high-energetic electrons of the irradiation beam. For the identification of the character of these defects we performed majority carrier DLTS measurements at Schottky diodes, which were processed from the typical CoolMOS™ phosphorus-doped n-epi starting material. To make also the typical p-type material accessible to measurement, wafers were boron-implanted so that the original n-epi was converted to p-type.

Fig. 5 shows DLTS spectra that we obtained from the n-epi sample after electron irradiation and subsequent annealing. For an annealing temperature of 220 °C we identify three well-known peaks from the vacancy-oxygen complex VO at $E_C - E_T = 0.16$ eV and the two levels of the divacancy at $E_C - E_T = 0.22$ eV and $E_C - E_T = 0.45$ eV. The PV-complex that could contribute to the divacancy level at $E_C - E_T = 0.45$ eV, should already have been annealed out at 220 °C. Except for the VO-level another dominant peak appears at $E_C - E_T = 0.32$ eV, which is significantly higher than the VV signal. It may be caused by the V_2O defect or congregations of vacancies of the form V_n, $n \geq 3$ (17-19). In Ref. (17) this defect is detected after electron, proton and helium irradiation, especially in regions of high energy deposition. This corresponds well with the fact that our electron irradiation was performed at high electron fluences, which favors the generation of extended defect complexes due to the high concentration of primary vacancies. Another possible origin of the signal is the VOH-complex which is usually found in proton-irradiated silicon. In our case the hydrogen could be introduced by wet chemical etching in an acid mixture during the preparation of the DLTS samples. This hypothesis is supported by the appearance of a defect level at E_C-0.31eV after electron irradiation and subsequent etching in an acid mixture as reported in Ref. (20).

When measuring the p-type material we had to cope with high leakage currents probably caused by non-optimized electrical contacts. Hence, the DLTS spectrum after electron irradiation and subsequent annealing at 220 °C (Fig. 6) is rather noisy. Despite this fact, there is a clearly visible peak at $E_V + 0.25$ eV accompanied by a small shoulder at $E_V + 0.20$ eV. The latter can be attributed to the donor level of the divacancy (18, 21). The former level can be identified as B_iB_s- or B_iC_S-complexes (18, 21) or a second level of the mentioned VOH complex (22).

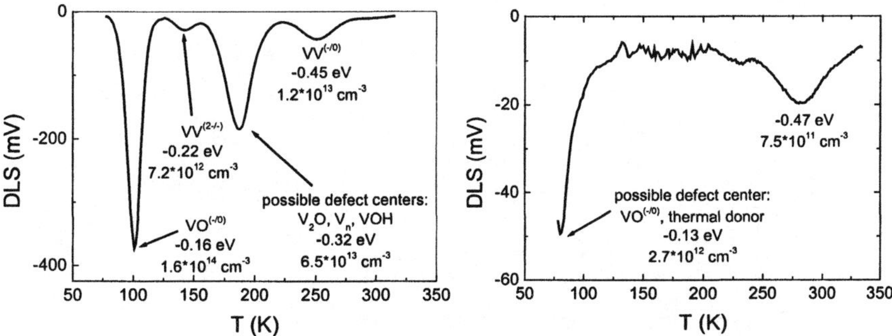

Fig. 5: DLTS spectrum of electron irradiated n-epi Si, E = 4.5 MeV, $\Phi = 1.5 \cdot 10^{15}$ cm^{-2}, annealed at T = 220 °C, 4h (left) and 350 °C, 4h (right)

If the n-type material is annealed at 350°C, the DLTS spectrum changes drastically (Fig. 5). There are only two peaks left at E_C – 0.13 eV and E_C – 0.47 eV, with the first one dominating. The concentration of the defects is approximately two orders of magnitude lower than the concentration after annealing at 220°C. The identification is not completely clear. The peak at E_C – 0.13 eV could be caused by the VO defect or by a thermal donor (17,23). Wondrak (17) detected a similar signal at E_C – 0.14 eV, which grows above 300°C and can be observed also in non-irradiated samples. However, four-point resistance measurements at high-resistivity wafers subjected to the same irradiation and annealing conditions as described above gave no hints for thermal donor formation. In p-type material we could not find any signal after annealing at 350 °C, which could be also an effect of the mentioned high leakage currents.

As described elsewhere (24), we obtained a realistic simulation of the diode turn-off and the leakage current after electron irradiation and annealing at T = 220°C by assuming two dominating defects, the VO-complex $VO^{(-/0)}$ and the single acceptor level of the divacancy $VV^{(-/0)}$ in the simulation model. The divacancy has an energetic level close to the center of the bandgap and, hence, dominates the generation rate in the space charge region. Therefore, it can be extracted from analyzing the measured leakage current when a reverse bias is applied. The VO-concentration is obtained by adjusting the simulation to the measured diode turn-off curves. In this way we achieved a very good agreement with the measured data (24). The extracted concentrations were [VO] = $1.5{\cdot}10^{14}$ cm^{-3} and [VV] = $1.0{\cdot}10^{14}$ cm^{-3}, while the DLTS measurements yielded [VO] = $1.6{\cdot}10^{14}$ cm^{-3} and [VV] = $1.2{\cdot}10^{13}$ cm^{-3}. The large discrepancy between the measured and the simulated VV concentration can have several reasons:

Possibly there are minority carrier traps which could not be observed in our DLTS set-up. A second reason could be that the level E_C - 0.32 eV is the VOH defect. It is reported in Ref. (20) that the concentration of vacancy-related defects like VO or VV decreases in the presence of the VOH complex. Therefore, in the actual device, where no etching is performed after electron irradiation, the VV concentration could be higher than in the DLTS measurement, and this could explain the high measured leakage current.

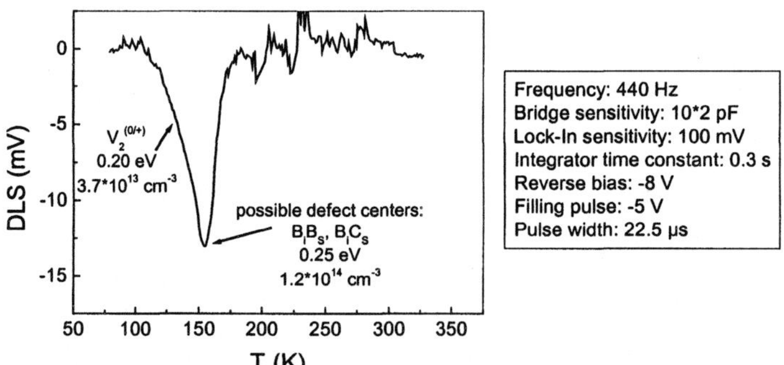

Fig. 6: DLTS spectrum of electron irradiated boron-implanted n-epi Si converted to p-type (E = 4.5 MeV, $\Phi = 1.5{\cdot}10^{15}$ cm^{-2}, annealed at T = 220 °C, 4h)

Apart from these explanations further reasons for the high leakage current are conceivable. Investigations on defects caused by neutron irradiation (25, 26) give evidence for

- the existence of an acceptor level at the center of the bandgap (E_C - E_T $\approx$0.5 eV), which could not be detected unambiguously so far and which is attributed to the V_2O defect.
- an inter-center charge transfer between divacancies in VV-clusters. Especially the reaction VV-VV → VV^+VV^- can result in an enhanced occupation probability of the acceptor state of the divacancy $VV^{(-/0)}$ when the VV density in the VV cluster exceeds 10^{15} cm^{-3}. The clusters could be formed at the end of the trajectories of the recoiling silicon nuclei when an electron has hit them, because there the energy deposition is rather high and concentrated to a small volume.

In our DLTS spectra there is no level near the center of the bandgap, which could explain the missing leakage current, but the level at E_C - 0.32 eV may indicate that there are at least low-density vacancy clusters which could enable the inter-center charge transfer mechanism. For the annealing at T = 350 °C the difference between the measured defect concentration and the electrical data, which reflects a still rather high defect concentration, is even more pronounced. Further experiments are required to clarify this behavior. Nevertheless, electron irradiation seems to be a very suitable choice in order to improve the internal reverse diode of the CoolMOS™.

Proton Irradiation

Non-Uniform Lifetime Profile in the CoolMOS™ Device. The CoolMOS™ was irradiated with protons from the drain side, since this procedure avoids damage effects on the sensitive gate oxide. The penetration depth of the protons under the gate oxide was varied from 17 to 42 µm and the fluence from $1{\cdot}10^{11}$ cm^{-2} to $2{\cdot}10^{12}$ cm^{-2} (Fig. 7).

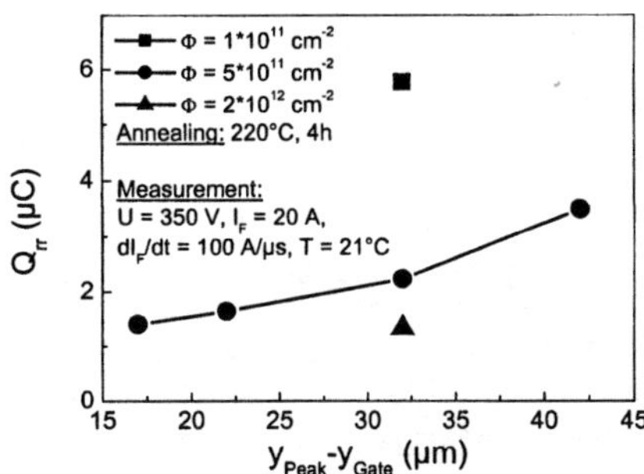

Fig. 7: Q_{rr} of the body diode of 600V-20A-CoolMOS™ devices irradiated by protons from the drain side as a function of the proton penetration depth and fluence

The reverse recovery charge Q_{rr} of the internal reverse diode shows a strong dependence on the proton energy and fluence and reaches values comparable to electron irradiation with subsequent annealing at 350 °C. The reverse recovery charge decreases monotonously when the defect peak approaches the gate oxide. This has two reasons:

- Like in the conventional vertical MOSFET, the electron-hole plasma has first to be removed from the junction under the p^+-well due to the higher electron mobility compared to the hole mobility. This process is accelerated if the defect maximum is located near the p^+-well and, consequently, the reverse recovery current will decrease.
- The fraction of the plasma that is covered by the defect tail becomes larger when the peak approaches the p^+-well. This is especially important in the CoolMOS™ because the whole p-n junction between the compensated regions has to be depleted before at least a small space charge region can build up to sustain the applied bias voltage.

The second statement is supported by device simulations (Fig. 8), which are again based on the two dominating defect levels $VO^{(-/0)}$ and $VV^{(-/0)}$, the concentrations of which were extracted from the leakage currents and diode turn-off curves. The simulation results clearly demonstrate the strong influence of the defect tail. If a tail concentration with a quarter of the peak concentration is assumed, the relation between Q_{rr} and peak depth follows the experimental data. But for a tail concentration with, for example, one tenth of the peak concentration, the Q_{rr} value saturates when the peak reaches the vicinity of the gate oxide.

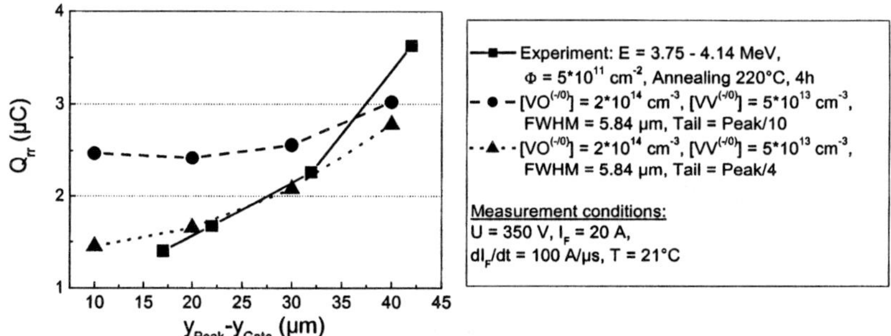

Fig. 8: Simulated reverse recovery charge of the body diode of 600V-20A-CoolMOS™ devices proton-irradiated from the drain side as a function of the proton penetration depth compared to experiments.

Proton-Induced Doping Effects: Analysis of Spreading Resistance Profiles. The second important effect induced by irradiation concerns modifications of the effective doping concentration. Several hydrogen-correlated donor families formed in silicon at annealing temperatures in the range of about 250 - 500 °C were identified in the past. However, the microscopic structure of the corresponding donor centers is not completely clear. In oxygen-rich Cz silicon five shallow thermal donors containing hydrogen labeled STD(H) with binding energies between 35.4 and 37.8 meV are known (27). There is some evidence that the centers C_i-O_i-H and C_i-O_{2i}-H are elements of these family (28, 29). However, a direct experimental proof that the STD(H)s contain carbon is still missing (28). According to recent results three other hydrogen-related donor centers labeled D1-D3 have very similar properties as elements of the STD(H)-family and might be identical with elements of this family (28, 29). In FZ silicon irradiated with neutrons and subsequently subjected to a hydrogen plasma treatment, hydrogen-related donors with ionization energies between 31.8 and 52.5 meV were found after temperature treatments in the range of 250 - 400°C (9, 30). It could be shown that these donors are only formed in the presence of hydrogen and if damage in silicon is sufficiently high (9, 30). Furthermore, the appearance of hydrogen-related double donors (HDDs) has been reported for silicon irradiated with 30 MeV protons (31). This donor family is different from the above mentioned donor types but shows some similarity to the well-known oxygen-related thermal double donors (TDDs), because experimental results suggest that HDDs and TDDs consist of the same core, [110] chains of <001> Si interstitials, surrounded by different shells containing hydrogen and oxygen, respectively (31).

For device applications the effective electrical activity of the donor centers is a very important point, the microscopic structure is of minor importance. Therefore, our investigations focus on spreading resistance (SR) measurements of proton-irradiated FZ silicon and the analysis of the dependence of donor formation on the proton fluence and the annealing conditions.

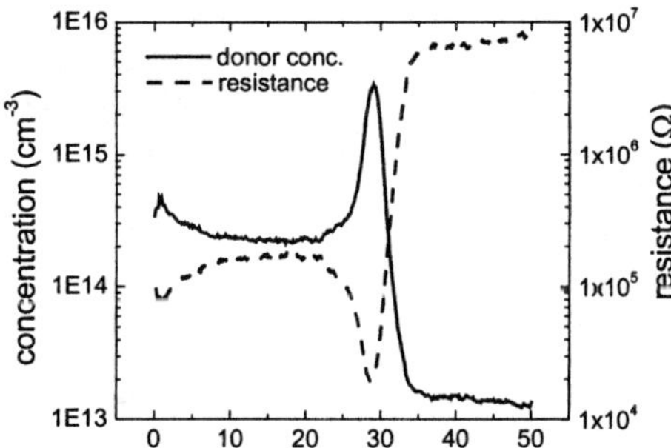

Fig. 9: SR-profile and resulting donor profile of a proton ($E = 1.5$ MeV, $\Phi = 10^{15}$ cm^{-2}) irradiated n-type silicon ($\rho = 210$ Ωcm), annealed at 500 °C for 30 min, the sample was pre-oxidized before irradiation

A typical SR profile of a proton-implanted n-type FZ silicon sample after a 500 °C annealing is shown in Fig. 9. The minimum of the profile coincides with the penetration depth of the implanted 1.51 MeV protons indicating an increased donor concentration in this region. Assuming that the whole measured sample area is of n-type conductivity, the electrically activated donor concentration profile has been calculated from the SR-profile. Furthermore, neither irradiation-induced effects on the charge carrier mobility nor possible charge carrier trapping due to irradiation-induced defects were taken into account. However, at annealing temperatures higher than 400°C, the above-mentioned VO and VV defects are annealed anyway. The peak concentration calculated under these assumptions is about $3.3 \cdot 10^{15}$ cm^{-3} (Fig. 9) and the integrated H-induced donor concentration is about $1.5 \cdot 10^{12}$ cm^{-2}. The latter is close to the breakthrough charge Q_c ($\approx 1.3 \cdot 10^{12}$ - $2 \cdot 10^{12}$ cm^{-2}) which is a measure of the minimum charge necessary to achieve the maximum blocking voltage of a reverse biased p-n junction. The peak has a full width at half maximum (FWHM) of about 2 μm which is comparable to that of the primary defect peak (32). Furthermore, the peak has broadened essentially in the direction towards the sample surface while there is only a small broadening towards the sample interior. This is in agreement with results reported previously (9) and is indicating that donors are created only when hydrogen and irradiation-induced defects are present simultaneously. Because of the high mobility of hydrogen at low temperatures, donor formation is not only effective in the heavily damaged area close to the penetration depth of the implanted ions but - to a certain extent - also in the defect tail resulting in a corresponding donor tail between the peak area and the sample surface.

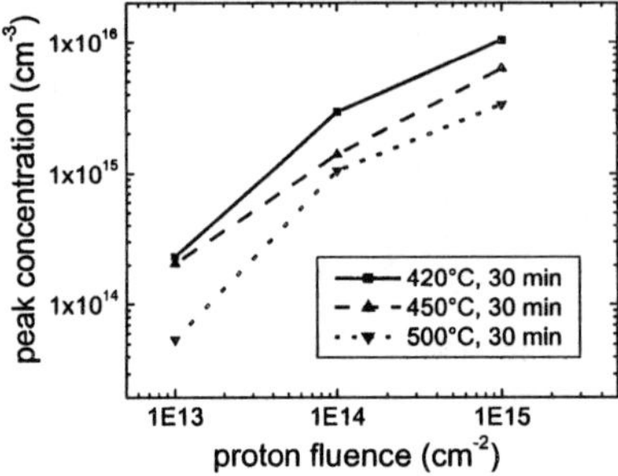

Fig. 10: Peak-donor-concentration vs. proton fluence for different annealing temperatures; $\rho = 210\ \Omega cm$, $E = 1.5$ MeV, samples were pre-oxidized before irradiation

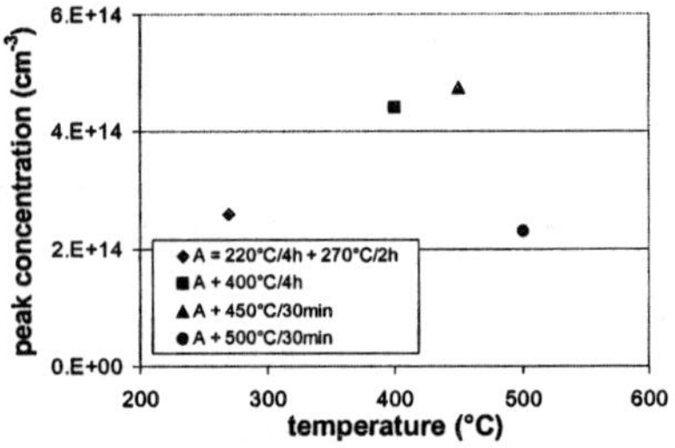

Fig. 11: Peak-donor-concentration vs. annealing temperature; $\rho = 1000\ \Omega cm$, $E \approx 2.1$ MeV, $\Phi = 3 \cdot 10^{13}\ cm^{-2}$, samples were subjected to a complete thyristor fabrication process before irradiation

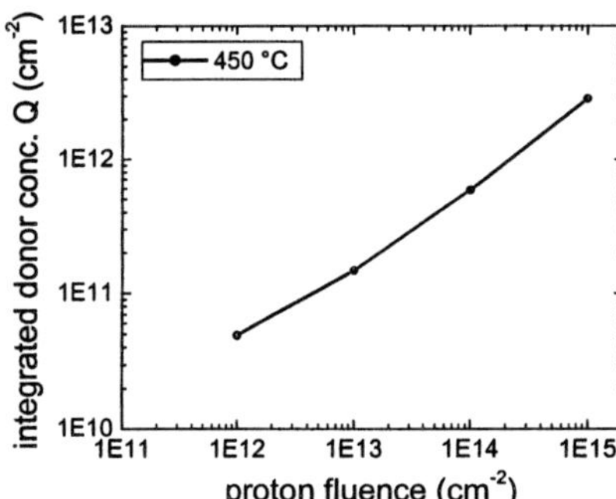

Fig. 12: Integrated donor concentration Q as function of the implanted proton fluence ($E \approx 1.5$ MeV); annealing: 450 °C, 30 min, samples ($\rho = 210\ \Omega cm$) were pre-oxidized before irradiation

Fig. 10 shows that the donor peak concentration increases monotonously with the proton fluence and can be varied in a wide range. Even for moderate fluences (10^{13} - $10^{15}\ cm^{-2}$) relatively high peak donor concentrations (10^{14} - $10^{16}\ cm^{-2}$) can be achieved for an annealing temperature of 420 °C. In Fig. 11 the dependence of the peak concentration on the annealing temperature is displayed. In this case a slightly higher proton energy was used (E = 2.1 MeV). All four samples were annealed first by a two-step annealing (220 °C/4h followed by 270 °C/2h). After that three samples were annealed additionally at higher temperatures. As expected, the peak concentration increases with annealing temperature in the lower temperature range, reaches a maximum between 400 and 450 °C and finally decreases in agreement with the measurements presented in Fig. 10. This reduction of the peak concentration is caused by a broadening of the donor profile (cf. Fig. 9) which extends more and more to the surface with increasing annealing temperature. This broadening can be explained by a stronger diffusion of the implanted hydrogen atoms.

As already mentioned, it is necessary for the realization of field stop layers in power devices that the integrated donor concentration Q of the H-induced donor profile exceeds the breakthrough charge Q_c. In Fig. 12, Q is drawn as a function of the implanted proton fluence (E = 1.51 MeV) for samples annealed at 450 °C for 30 min. The monotonous increase is caused by the increase of the peak concentration as well as by a monotonous broadening of the area around the peak (cf. Fig 9) with increasing proton fluence. For these annealing conditions the proton fluence must exceed a value of roughly $5 \cdot 10^{14}$ cm^{-2} to reach the effective donor charge Q_c. However, for longer annealing times and a lower annealing temperature the minimum proton fluence can be reduced further.

Application of H-Induced Doping Effects: Integrated Overvoltage Protection Function for Thyristors. It is a specific advantage of proton-induced donor formation that only a relatively low thermal budget is necessary for the donor formation, opening the possibility to modify the doping profile when device processing is completed. This is exemplified in the following, where we consider the adjustment of the breakdown voltage of a diode operating as integrated breakover diode (BOD) inside the amplifying gate (AG) structure in a light triggered thyristor. The central part of the investigated structure is schematically drawn in Fig. 13. The BOD is located in the center of the concentric structure and surrounded by a *p*-ring forming the *p*-base of the first amplifying gate of a thyristor structure. The whole thyristor consists of four amplifying gates and a main thyristor which are all arranged concentrically around the central BOD. When a forward blocking voltage is applied to the thyristor structure, the p-n^- junction is biased in reverse direction. Due to the curvature of the p-n^- junction of the BOD its breakdown voltage is reduced compared to the breakdown voltage of the surrounding amplifying gates and the main thyristor. Therefore, avalanche generation will set on first in the BOD area, when the forward voltage exceeds the breakdown voltage of the BOD. The resulting avalanche current flows through the surrounding *p* base towards the main thyristor turning on the first AG, which in turn triggers the second AG and so on until finally the main thyristor turns on safely and non-destructive (33, 34).

By a local proton irradiation in the *n*-region below the central *p*-region the effective donor concentration can be increased locally and, consequently, the breakdown voltage of the integrated BOD can be reduced. The breakdown voltages of the investigated non-irradiated thyristors are around 8.6 kV. A proton penetration depth of about 120 µm corresponding to an acceleration energy of 3.5 MeV was chosen in order to locate the peak area of the proton-induced donor profile in the n^- base but close to the central junction depth. After irradiation with proton fluences ranging between 10^{12} and $5 \cdot 10^{14}$ cm^{-2} the samples were annealed at 470 °C for 1 h. As shown in Fig. 14 the breakdown voltage of the BODs is diminished by about 2.8 kV for the lowest proton fluence and by nearly 7.8 kV for the highest fluence, demonstrating that the breakdown voltage can be efficiently and easily decreased in a wide voltage range after device processing has been completed. This allows a well-defined adjustment of the BOD voltage in the completely processed device.

Finally, it is worth to note that, by using helium ions instead of protons for irradiation, the breakdown voltage can even be increased. Details concerning this subject can be found in Ref. (24).

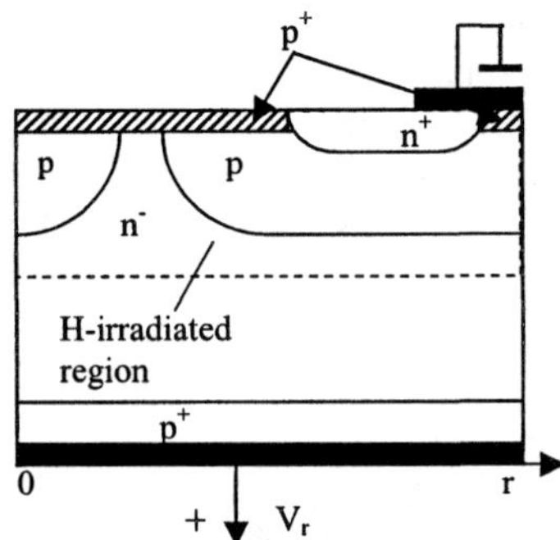

Fig. 13: Schematic drawing of the breakover diode (BOD) and the first amplifying gate of a thyristor

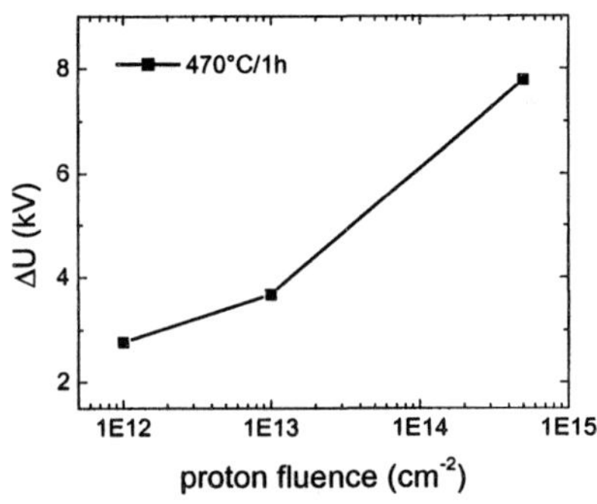

Fig. 14: Reduction of the breakdown voltage V_{br} of the BOD as function of the 3.5 MeV-proton fluence after annealing at 470 °C for 1 h, mean value of four samples

SUMMARY

In order to optimize the electrical performance of high-power devices irradiation-induced effects can be used very effectively by modifying the charge carrier lifetime and the doping profiles. For the fine-tuning of trade-off relations mainly determined by irradiation-induced recombination and generation centers as, e.g., the $Q_{rr}(V_T)$-relation, not only a suitable particle type has to be selected, but also the influence of other parameters determining the ratio of the defect centers generated by the irradiation has to be considered. As illustrative examples we have considered the optimization of the $Q_{rr}(V_T)$-relation of thyristors and the internal CoolMOS^{TM} diode by electron irradiation, where the application of high-energy electrons and high electron fluences, respectively, combined with proper annealing conditions turned out to be favorable. For local donor modifications proton irradiation followed by annealing in the temperature range of 270-500°C was investigated. Spreading resistance and electrical measurements show that, for suitably chosen proton energies, the breakthrough charge of $Q_c \approx 10^{12}$ cm^{-2} can be achieved with proton fluences in the range of 10^{14}-10^{15} cm^{-2} for annealing temperatures varying between 400 and 500 °C. Consequently, field stop layers often used in power devices can be realized with a low temperature budget. Finally, the influence of proton irradiation on the effective doping concentration was used for adjusting the breakdown voltage of breakover diodes integrated in light-triggered high-power thyristors for overvoltage protection.

ACKNOWLEDGMENTS

The authors would like to thank C. Protop (Mettmann) for many helpful discussions, and R. Bommersbach (Infineon Technologies, Munich) for experimental support.

REFERENCES

1. F. Robb, I. Wau, S. Ju, *Proc. ISPSD'1997,* Weimar, p. 251 (1997).
2. T. Nakagawa, K. Satoh, M. Yamamoto, K. Hirasawa, K. Ohta, *Proc. ISPSD'95*, Yokohama, p. 175 (1995).
3. A. Hallen. M. Bakoski, *Solid-State Electronics* **32**, 1033 (1989).
4. J. Vobecký, P. Hazdra, N. Galster, E. Caroll, *Proc. PECM'98*, Prague, p. 1.22 (1998).
5. P. Hazdra, J. Vobecký, N. Galster, O. Humbel, T. Dalibor, *Proc. ISPSD'2000*, Toulouse, p. 123 (2000).
6. P. Voss, German Patent DE3817160 (1989).
7. H. Iwamoto, H. Haruguchi, Y. Tomomatsu, J.F. Delon, E.R. Motto, *Proc. IEEE IAS*, p. 692 (1999).
8. W. Wondrak and D. Silber, *Physica* **129B**, 322 (1985).
9. J. Hartung and J. Weber, *Phys. Rev. B* **48**, 14161 (1993).
10. H.-J. Schulze and Kolbesen, *Solid-State Electronics* **42**, 2187 (1998).
11. H. Schwarzbauer and R. Kuhnert, *IEEE Trans. Industry Applications* **27**, 93 (1991).
12. D.K. Schroder, *Semiconductor Material and Device Characterization*, John Wiley & Sons, New York (1998).
13. G. Deboy, M. Pürschel, M. Schmitt, A. Willmeroth, *Proc. ESSDERC 2001*, Nuremberg, Germany, p. 61 (2001).
14. C. Hu, *IEEE Transaction on Electron Devices* **26**, 243 (1979).
15. G.D. Watkins, *EMIS Datareviews Series*, No. 20, Inspec, London, p. 643 (1991).
16. M. Schmitt, H.-J. Schulze, A. Schlögl, M. Vossebürger, A. Willmeroth, G. Deboy, G. Wachutka, *Proc. ISPSD'02,* Santa Fe, p. 229 (2002).
17. W. Wondrak, *Ph.D. thesis*, J.-W.-Goethe-Universität, Frankfurt a. M., Germany (1985).
18. S. Coffa and F. Priolo, *EMIS Datareviews Series*, No. 20, Inspec, London, p. 746 (1999).
19. L.C. Kimerling, *Inst. Phys. Conf. Ser.* **31**, p. 221 (1977).
20. Y. Tokuda and H. Shimada, *Mat. Res. Soc. Symp. Proc.* **513**, p. 363 (1998).
21. B.G. Svensson, *EMIS Datareviews Series*, No. 20, Inspec, London, p. 763 (1991).
22. O. Feklisova, N.Yarykin, E.B. Yakimov, J. Weber, *Physica,* **B 308-310**, 210 (2001).
23. C.A.J. Ammerlaan, in: Properties of Crystalline Silicon, ed. By R. Hull, *EMIS Datareviews Series*, No. 20, Inspec , London, p. 66 (1991).
24. F.-J. Niedernostheide, M.Schmitt, H.-J. Schulze, U. Kellner-Werdehausen, A. Frohnmeyer, G. Wachutka, *J. Elektrochem. Soc.*, in press (2002).
25. B.C. MacEvoy and S.J. Watts, *Solid State Phenomena* **57-58**, 221 (1997).
26. K. Gill, G. Hall and B. MacEvoy, *J. Appl. Phys.* **82 (6)**, 126 (1997).
27. R.C. Newman, M.J. Ashwin, R.E. Pritchard, J.H. Tucker, *phys. stat. sol. (b)* **210**, 519 (1998).
28. V.P. Markevich, T. Mchedlidze, M.Suezawa, L.I. Murin, *phys. stat. sol. (b)* **210**, 545 (1998).
29. J. Coutinho, R. Jones, P.R. Briddon, S.Öberg, L.I. Murin, V.P. Markevich, J.L. Lindström, *Phys. Rev. B* **65**, 014109 (2001).
30. J. Hartung, *Ph.D. thesis*, University of Stuttgart, Germany (1991).
31. S.Zh. Tokmoldin, B.N. Mukashev, Kh.A. Abdullin, Yu.V. Gorelinskii, B. Pajot; *Materials Science and Engineering* **B71**, 263 (2000).
32. Calculated with SRIM, version 2000.39
33. M. Ruff, H.-J. Schulze, U. Kellner, *IEEE Transactions on Electron Devices* **46**, 1768 (1999).
34. F.-J. Niedernostheide, H.-J. Schulze, U. Kellner-Werdehausen, *Microelectronics Journal* **32**, 457 (2001).

IMPACT OF HIGH-TEMPERATURE ELECTRON IRRADIATION ON THE ELECTRICAL PARAMETERS OF N-TYPE CZ SILICON

V. Neimash, A. Kraitchinskii, M. Kras'ko, V. Tischenko and V. Voitovych
Institute of Physics, National Academy of Science of Ukraine, Kiev, Ukraine

E. Simoen and C. Claeys*
IMEC, Leuven, Belgium
*Also at E.E. Dept, KU Leuven, Kasteelpark Arenberg 10, B-3001 Leuven, Belgium

ABSTRACT

This paper describes the impact of a 1 MeV electron irradiation performed at 450°C on the formation of oxygen thermal donors (OTDs). For that purpose, a dedicated set-up has been constructed whereby the sample heating is performed by the power of the electron beam itself. For a sufficiently high sample thickness, it is possible to derive the effect of both thermal annealing and irradiation at 450°C on the electron concentration by comparing the electrical properties of the exposed front- and the non-irradiated back-side of the n-type Czochralski silicon samples. Besides 4-point resistance probing and Capacitance-Voltage (C-V) measurements, the deep levels associated with the Radiation Defects (RDs) have been assessed by Deep Level Transient Spectroscopy (DLTS). The main conclusions are that electron irradiation enhances considerably the OTD formation rate. This could originate from a radiation-induced vacancy-assisted enhancement of the effective oxygen diffusivity. At the same time, a whole set of RDs have been revealed which can be related to higher order vacancy-oxygen (V_x-O_y) complexes and may also act as additional OTD nulei.

INTRODUCTION

Simultaneous heat treatment and irradiation is known for investigating the interaction between thermal- and radiation defects [1]. Recently, interest for using high temperature particle irradiations as a method of defect and lifetime engineering has emerged [2-4]. The motivation is related to the fact that the stability of the radiation-induced defects at the operation temperature of the devices is limited, if they are created for example by a room temperature exposure. In practice, irradiation temperatures in the range up to 800°C have been studied [2-4]. However, when dealing with Czochralski (CZ) silicon, one has to take into account the agglomeration and precipitation behaviour of interstitial oxygen, leading to the formation of oxygen-related thermal donors (OTDs) when performed in the range 350-500°C [4,5].

One of the issues related to the simultaneous formation of RDs and OTDs is the fact that unlike other shallow donors, OTDs do not interact with radiation defects at room temperature [6]. None of the existing models can explain this feature. Previously, we have attempted to utilize gamma irradiation during a thermal treatment at 450°C to influence the OTD formation [1]. It is a well-known fact that as OTDs grow the corresponding donor activation energy becomes smaller [7]. It turned out that "hot" gamma-irradiation essentially accelerates this process but does not affect the rate of the OTD generation. This fact is at variance with the common conception that OTDs act as an intermediate stage of oxygen precipitation in Si. Behind this work is the authors' strong believe that the comprehension of these "anomalies" in the behavior of OTDs under "hot" irradiation is extremely significant when investigating the practical problems traditionally associated with the radiation hardness and thermal stability of silicon materials. Therefore, the purpose of the work was to obtain information about the nature and rate of radiation-induced defects and OTD generation during a 1 MeV electron irradiation at 450°C of n-type CZ Si.

EXPERIMENTAL

The influence of a 1 MeV electron irradiation at 450°C on the change of the conductivity of n-type Czochralski (CZ) silicon was investigated. The initial parameters of the studied material are indicated in Table I. The carbon content in the samples is lower than 5×10^{16} cm^{-3}. The generation of OTDs is the main reason for the change of the specific electrical resistance during heat treatments at 450°C. Their kinetics strongly depends on the initial interstitial oxygen concentration denoted by $[O_i]$ [5]. The distribution of the oxygen impurities in silicon crystals is usually non-uniform [8]. This is a major obstacle for the study of the kinetics of OTDs, since it is very difficult to select samples with identical oxygen concentration and oxygen distribution on a microscale. The rate of OTD generation also strongly depends on the temperature T of the heat treatment. Furthermore, temporal fluctuations in the electron beam flux can cause considerable temperature variations in an irradiated sample. This means that it is difficult to irradiate a sample under exactly the same conditions (i.e., same $[O_i]$, T,...) as the annealed reference sample. Here we propose an original approach to eliminate these problems, which is schematically represented in Fig. 1.

The n-type Cz silicon samples were diced in a parallelepiped with size 10x10x2 mm^3. The surface of the samples was mechanically and chemically polished. Simultaneous heating and irradiation was achieved by a high-power ($I=8$ $\mu A/cm^2$) beam of high-velocity (1 MeV) electrons. The samples were heated to 450°C by the absorbed radiation energy, in air. The electron beam impinged perpendicularly on one of the main surfaces. In one of the small sides a hole with a diameter of 0.8 and a depth of 5 mm was drilled. A thermocouple was inserted in this hole for measurement of the temperature. It served as a mechanical support too. Total absorption of electrons of such energy in silicon occurs in a depth of approximately 1 mm. However, due to the high thermal conductivity of silicon the temperatures difference between the irradiated front- and non-irradiated back-side of the sample does not exceed 0.3°C. Thus, the OTD generation happened under irradiation in the near surface layer of the exposed edge of the samples (side 2 in Fig. 1) and in a completely similar thermal condition but without irradiation near the surface of the second major edge (1 in Fig. 1). After the hot irradiation, the grown oxide film was removed by HF etching.

The creation of thermal donors (TDs) was monitored via the change of specific electrical resistance by the four point resistance method at room temperature. The resistance of a near surface layer of the sample of width less than 0.1 mm is measured in this way. Additionally, circular Au Schottky barriers have been evaporated permitting high frequency (1 MHz) Capacitance-Voltage (C-V) profiling of the material. The same diodes were used in a Deep Level Transient Spectroscopy (DLTS) study of the created RDs. A Schottky barrier was first applied to the front side (2 in Fig. 1), while an InGa eutectic ohmic contact was made at the back-side. After a full front-side characterization, the contact configuration was reversed to assess the deep levels at the back-side of the samples.

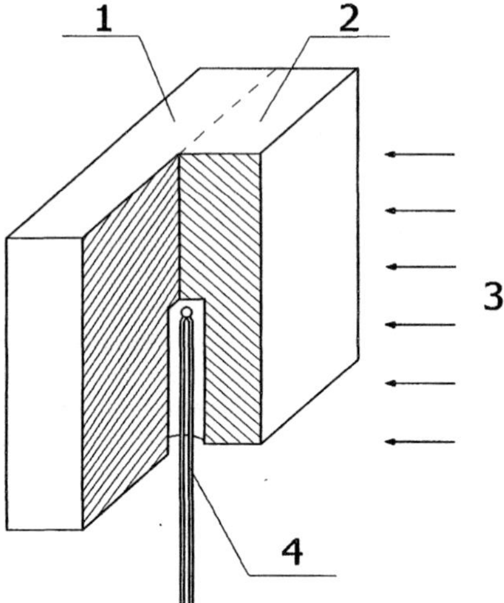

Fig. 1. Schematic sample configuration for the "hot" electron irradiation. An intense 1 MeV electron beam (3) is directed to the front surface (2) of the n-type CZ samples. The back-side (1) only experiences the heating by the electron beam, while a thermocouple (4) monitors its temperature.

RESULTS

The results are summarized in Table 1. The initial concentration of oxygen and the free electron density before and after processing are indicated by N_O, n_0 and n, respectively. t_{HT} is the duration of heat treatment; Φ_e the radiation fluence obtained during the heat treatment; $dn_{TD}/dt = (n - n_0)/ t_{HT}$ is the generation rate of thermal defects of the donor type and assumed to correspond to the oxygen-related thermal donors. Finally, $dn_{RD}/d\Phi$ is the introduction rate of radiation defects of compensating acceptor type. As is clear from the table, in the irradiated part of sample №1 there was a strong reduction of the conductivity. Apparently, it turns out that in this sample the generation of radiation defects that are either of the acceptor type or neutralizing the carrier concentration is happening faster than the generation of OTDs. In samples №2 and №3 the concentration of oxygen is much higher. The electron concentration increases here in comparison with the initial value even at a radiation dose which is 12 times higher than for №1. Hence, in this case the generation of OTDs evolves faster than for the radiation-induced acceptors. The difference between №1, on the one hand, and №2 and №3, on the other can be explained by considering the $\sim[O_i]^{3.5\text{-}4}$ power dependence of the OTD formation [5]. The efficiency of the radiation-induced defect creation is assumed identical in all samples, in a first approximation, because it is primarily determined by the dissociation of Frenkel pairs in independent vacancies and self-interstitial atoms, resulting in the formation of complexes which are stable at the irradiation temperature. The dissociation of Frenkel pairs is determined by the crystal type, the concentration of dopants and by the radiation flux [9], which were identical for all samples.

The carrier concentrations n of Table I were qualitatively confirmed by the C-V measurements. A possible reason for quantitative differences between the two techniques is the depth probed by the measurements.

Table I. Experimental data for the first set of samples studied. The free electron concentration (n) was derived from 4-point resistance measurements.

sample №	N_O $x10^{17}cm^{-3}$	n_O $x10^{13}cm^{-3}$	t_{HT} min	Φ_e $x10^{16}cm^{-2}$	n $x10^{13}cm^{-3}$	dn_{TD}/dt $x10^{10}cm^{-3}s^{-1}$
1	<5	9.1	10	3	0.4	–
1	<5	9.1	10	0	10.8	0.47
2	8	9.4	120	36	9.75	14.5 (17.55)
2	8	9.4	120	0	31.9	1.5
3	9	10.5	120	36	23.1	16.25 (19.25)
3	9	10.5	120	0	46.3	2.5

INTERPRETATION OF THE CARRIER CONCENTRATION DATA

In order to explain the results of Table I, one can consider three variants of the influence of irradiation on the OTD creation:

i) The rate of OTD generation decreases under irradiation;
ii) The rate of OTD generation does not depend on the irradiation;
iii) The OTD generation rate increases under irradiation.

These three different options are further investigated in order see whether or not one of them can explain the experimental observations.

i. Radiation kills the generation of OTDs

Samples such as №1 differ from the other samples by a smaller starting concentration of interstitial oxygen. In Fig. 2 the dependence of the electron concentration change $\Delta n = (n - n_O)$ (due to OTD formation) for samples of this type versus the annealing time at 450°C without irradiation is shown. It is seen that the OTD concentration, accumulated after 10 min is small. This also follows from the change of the electron concentration on the dark side of sample №1 (Table I). Therefore, the decrease of the electron concentration on the irradiated side of sample №1 approximately corresponds to the concentration of radiation-induced acceptors. From these data it is possible to calculate the introduction rate of RDs.

$$dn_{RD}/d\Phi = (n_O - n)/\Phi = 2.9x10^{-3}cm^{-1}. \quad (1)$$

This $dn_{RD}/d\Phi$ is a lower limit, for the case of absence of OTD generation. Actually, OTDs will be created in the irradiated part of the sample too. This can be derived from the data for №2 and №3, where the electron concentration is increased after thermo–radiation treatment. However Eq. (1) can be used for an estimate of the RD concentration in these samples too. Therefore, for the free electron concentration after thermo-radiation one finds:

$$n = n_O + 2n_{TD} - n_{RD} = n_O + 2n_{TD} - \Phi\, dn_{RD}. \quad (2)$$

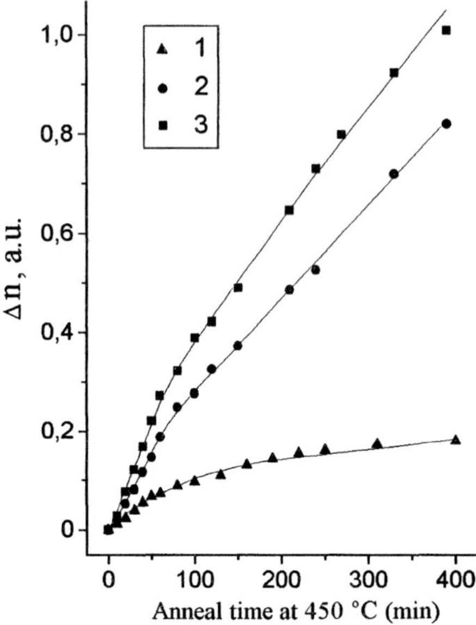

Fig.2 Dependence of the electron concentration change $\Delta n = (n - n_0)$ (due to OTD formation) versus the annealing time at 450°C (without irradiation).

The factor 2 follows from the double donor nature of the OTDs, i.e., at room temperature it is assumed that each OTD center gives up two electrons to the conduction band. The OTD generation rates calculated from Eq. (2) are summarized in Table I. It is clear that the OTD generation in the irradiated part of №2 and №3 is higher, than in the non-irradiated part. It contradicts the assumption about suppression of OTD generation by radiation.

ii. The irradiation does not influence the rate of OTD generation

In this case, using the data for the irradiated part of the sample, we find:

$$dn_{RD} / d\Phi = (n_0 + 2n_{TD} - n)/\Phi. \qquad (3)$$

where n_{TD} can be derived from the data for the not irradiated side of the sample: $2n_{TD} = n - n_0$. Substituting the appropriate values from Table I, one finds for sample №1 $dn_R / d\Phi = 3.5 \times 10^{-3}$ cm^{-1} and for №2 and №3 $dn_{RD}/d\Phi = (6.2 \pm 0.2) \times 10^{-4}$ cm^{-1}. In other words, for №2 and №3 we observe a 6 times lower introduction rate for the RDs than for №1. It contradicts the result of hypothesis i) that $dn_{RD}/d\Phi$ for sample №1 is close to the real value. If $dn_{RD}/d\Phi$ is substituted by 3.5×10^{-3} cm^{-1}, one finds for №2 and №3 the dn_{TD}/dt, values which are indicated in Table I in brackets. It is observed that in this case the TD generation rate in the irradiated part is further increased compared with the non-exposed side. It, therefore, contradicts the supposition that radiation does not influence the rate of the OTD generation.

iii. The rate of OTD generation increases as a result of the irradiation

As the starting value we use the RD introduction rate for the irradiated side of sample №1, given by Eq. (1). In this case dn_{TD}/dt in samples №2 and №3 under irradiation is 6 to 10 times higher than without irradiation. That is in agreement with the above assumption.

The question rises what can cause the acceleration of OTD generation under irradiation? In Fig. 3, the dependence of OTD generation rate on the time of the heat treatment at 450°C is shown for samples such as № 1-3.

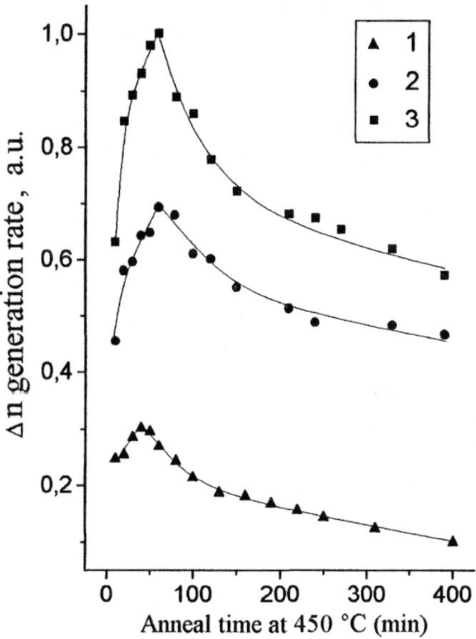

Fig. 3. OTD generation rate versus time of heat treatment at 450 ^{0}C: 1 – sample № 3; 2 – sample № 2; 3 – sample № 1.

It is clear that in the time range used for the thermo-radiation treatment (up to 120 min), dn_{TD}/dt has not reached a steady-state value yet. As was demonstrated in Ref. 10, at this stage the heterogeneous mechanism of OTD generation dominates. Thus the rate of OTD generation can be written as:

$$dn_{TD}/dt = 4\pi D r N_O N_z \, exp(-4\pi D r N_O t) \quad (4)$$

where D is the diffusivity of interstitial oxygen, r the capture radius of oxygen by nuclei, N_z the concentration of nuclei and t the time of heat treatment. It is well-known that under irradiation the diffusivity of oxygen can be modified [11]. The ratio of the OTD generation rates at thermo-radiation $dn_{TD\,(irr)}/dt$ and at pure heat treatment dn_{TD}/dt can be described by:

$$dn_{TD(irr)}/dn_{TD} = (D_{irr}/D) \exp[-(D_{irr} - D)\, 4\pi D r N_O t] \quad (5)$$

At small t the exponential in Eq. (5) is close to 1. Therefore (as a first approximation) the increase of the OTD generation rate under irradiation (as follows from the experimental results) indicates a similar increase of the diffusivity of oxygen.

It is interesting to remark that a pulsed electron irradiation of the samples was made at a duty cycle (ratio of time between pulses to the pulse duration of 2.5 μs) of 1000. It is known that the lifetime of primary vacancies in silicon which are generated by intense electron irradiation hardly exceeds the pulse length [12]. At 450°C the lifetime of vacancies can be only lower than this value. If the free radiation-induced vacancies are the origin of the acceleration of the oxygen diffusion, they influence the movement of oxygen only 1/1000 parts of the time of the thermo-radiation treatment. This suggests that the increase of the diffusivity of oxygen is not one, but in reality 3 orders of magnitude.

RADIATION-INDUCED DEEP LEVELS

In order to support the above interpretation, the properties of the secondary radiation-induced defects created under "hot irradiation" have been investigated by the DLTS technique. Besides the above samples the extra ones listed in Table II with various concentrations of oxygen and phosphor impurities have been investigated. The reason is that the starting resistivity of the material of Table I was too low to assess the OTD donor levels after irradiation, for both sides. As will be shown, compensation by a deep acceptor causes carrier freeze-out in the 77-100 K range. The carbon content in the second set of samples does not exceed $5x10^{16}$ cm^{-3}. The table indicates also both the time of the thermo-radiation treatment (at 450°C for samples №4-6 and at 535°C for sample №7) and the 1 MeV electron fluence.

Table II. Experimental conditions of the second set of n-type CZ material studied.

Sample	N_O $x10^{17}cm^{-3}$	n_0 $x10^{14}cm^{-3}$	t_{HT} min	Φ_e $x10^{16}cm^{-2}$
4	2.0	7.0	10	3
5	9.1	9.1	20	6
6	7.5	219	60	18.0
7	7.0	181	36	18.2

4-6: anneal at 450°C
7: anneal at 535°C

As can be seen from Fig. 4, corresponding with a №3 sample irradiated for 120 min, a whole set of electron traps is revealed by DLTS. They are labeled here E1 to E5, as one can expect that the radiation-induced defects stable at 450°C are different from the ones introduced during a room temperature irradiation [13-14]. Surprisingly, some of the peaks are also observed in the back-side spectrum in Fig. 4, although the concentration is typically one decade smaller. A final puzzling fact is that there is no trace of the deep levels associated with the OTDs, in either side of the samples. This is particularly strange for the non-exposed side. The explanation for this is shown by the Capacitance-Temperature (C-T) graphs of Fig. 5, demonstrating the freeze-out of the material at a deep acceptor level in the range 77-100 K. This suggests the presence of a RD with an activation energy in the range 150-200 meV, which overcompensates the double OTDs, even in the non-irradiated back-side. There are no traces of the deep levels associated with the OTDs, in either side of the samples. Indeed for such DLTS mode the typical thermal donors' peaks can be observed only at T< 80K [15]

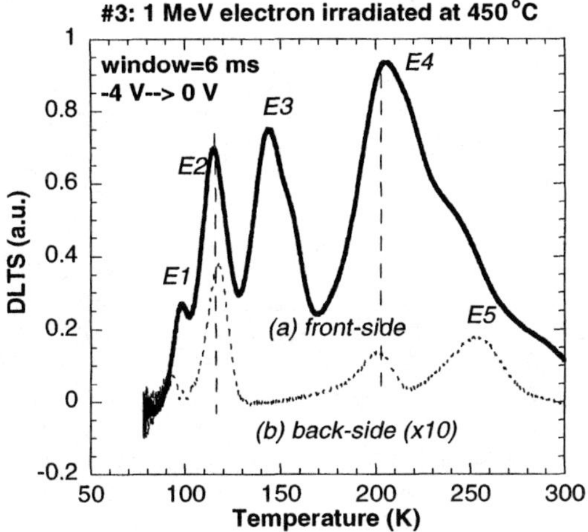

Fig. 4. DLT-spectrum from –4→ 0V for sample №3 after a 1 MeV electron irradiation at 450°C, corresponding with the irradiated front- and non-exposed back-side. The sensitivity is 10 times higher for b compared with spectrum a.

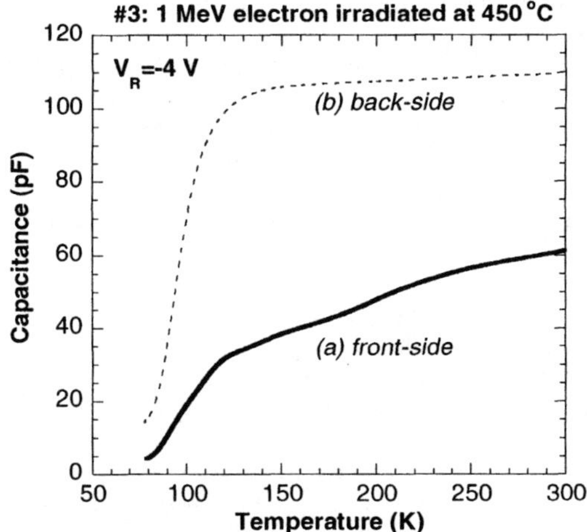

Fig. 5. C-T plots corresponding with the DLT-spectra of Fig. 4 and obtained on a sample irradiated by 1 MeV electrons for 2 h at 450°C. The reverse bias is –4 V. Complete carrier freeze-out occurs at ~80 K in both cases.

An Arrhenius analysis of some of the main peaks reveals an activation energy of 206 meV (E2), 280 meV (E3) and 302 meV (E4), respectively. Deep level E1 could correspond to the V-O or A center. It has been shown that after a 400°C irradiation with 5.6 MeV Si atoms, the A center is still the dominant radiation defect in n-type CZ silicon [14]. The levels E2 and E4 could be similar as the 0.22 eV and 0.47 eV levels of Ref. 14. In that paper, it was suggested that these traps are related to the divacancy (V_2), but shifted in energy to the higher side for the single acceptor state and to lower energy for the double acceptor state. The absence of a Poole-Frenkel shift in the present spectra also suggests an acceptor nature of the dominant RDs. As the divacancy normally anneals in the range 200-300°C, it was proposed by Lalita et al. [14] that these two levels belong to higher-order vacancy complexes and possibly constitute the single and double acceptor state of such a complex. In conflict with this interpretation is the absence of the 0.22 eV level in n-type Float-Zone (FZ) material irradiated at 330°C [13]. This suggests that there could be a correlation with oxygen for E3. The same applies in fact for the E3 level found here: it seems to be typical for CZ Si irradiated at high temperatures [13,14].

Therefore, an alternative picture could be that (most of) the RDs observed here are somehow related to stable V-O complexes. Support for this comes from infrared absorption measurements of silicon irradiated in the range 400-600°C [16], showing the dominance of VO_2 and VO_3 centers. In high-temperature electron irradiated FZ material, a level at E_c-0.36 eV has been ascribed to the V_2O_2 defect [13]. It has, furthermore, been shown in the past that upon annealing in the range 250-400°C the activation energy of the V_2 donor level shifts to higher values in irradiated p-type silicon [17]. It is believed that this increase in activation energy accompanies the transformation from V_2 to a more stable V_2O (or V_xO_y) complex. Note also in Fig. 4 that most of the peaks show clear shoulders, indicating the presence of other overlapping deep levels. It could mean that there is a whole family of related V_xO_y centers which gradually form at 450°C irradiation. Another candidate defect is the V_2O defect which has recently been identified in irradiated standard and oxygenated FZ silicon [18]. It could for example correspond to the E5 peak in Fig. 4.

An intriguing fact is the observation of some of the RDs in the back-side spectrum of Fig. 4 and the freeze-out at 77 K. As 1 MeV electrons are stopped in silicon in the first 1 mm or so, no direct creation of Frenkel pairs is occurring there. It implies that some of the primary (V and I) or even secondary radiation defects (V-O,...) are highly mobile at 450°C and create similar deep acceptors as in the damaged region, after being trapped. This long range migration may also explain the enhanced OTD formation if oxygen is involved in some of these mobile centers.

Another implication of the findings is that the proposed V_xO_y complexes may well serve as additional heterogeneous nucleation centers for OTDs (N_z term in (Eq.4)). So it is possible that the observed enhancement of the OTD rate is caused not only by an increase in D, but also by an increase of the N_z or the capture radius r.

In order to overcome the freeze-out problem illustrated by Fig. 5, it was decided to irradiate a second set of samples having a higher starting doping concentration, described in Table II. Here, only preliminary DLTS results, obtained on the irradiated side are reported, but the set of peaks is equal for all studied materials. This is illustrated by Fig. 6 showing DLTS spectra for the exposed side of the samples № 4 and 5. As can be seen, the same radiation peaks are observed, albeit with slightly different ratios and lower concentrations. Table III summarizes the generation rate of the observed RDs under 1 MeV electron exposure at 450°C (for the samples №4-6) and at 535°C (for the sample №7). From the comparison of Tables I and III it becomes clear that the generation rate of diverse RDs depends differently on the concentration ratio of oxygen and phosphor impurities. Apparently this determines the distinction of DLTS spectra for the different materials.

The following qualitative trends have been observed: comparing №7 and №6 one can conclude that increasing the irradiation temperature lowers the introduction rate of the main RDs. This is in line with the results of Ref. 13, where values in the range of a few times 10^{-2} cm^{-1} have been found in oxygen-lean FZ material at 360°C. This is about one decade higher than in Eq. (1) or for №1, which correspond to low-oxygen n-type CZ material, irradiated at 450°C. Increasing the oxygen content results in a lowering of the RD introduction rate, according to №4 and №5 in Table III. The same applies for a higher doping concentration (compare e.g. №6 with №5). Analyzing these data in more detail will lead to a better insight.

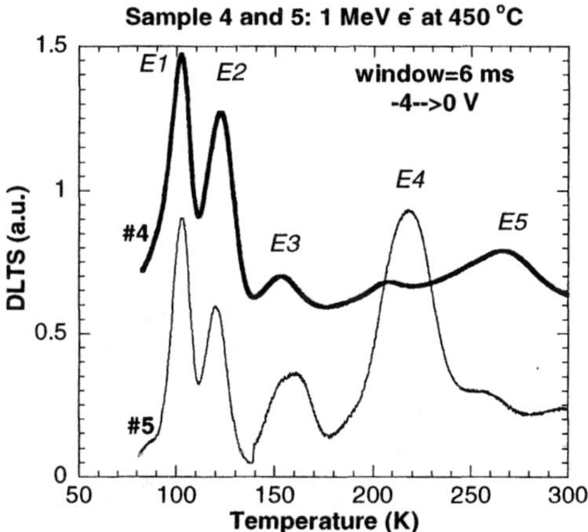

Fig. 6. DLTS-spectrum of sample № 4 and sample № 5 (irradiated side) corresponding with a pulse from –4 to 0 V and a rate window of 6 ms.

In order to derive the concentration of the OTDs, DLTS measurements down to 20-30 K are necessary on the samples of Table III. Based on the absence of a strong freeze-out in the 77 K range, it is in principle possible to reveal the second and first OTD level [15]. The next step will be to do the same analysis on the back-side. In addition, a new digital DLTS system has become available, which enables the extraction of the trap parameters in a single temperature scan. With this information, it should be possible to further refine the proposed model and to identify more firmly the observed RDs.

Table III. Introduction rate $dn_{RD}/d\Phi$ (cm^{-1}) of the main radiation defects in the samples described in Table II.

Level	№4	№5	№6	№7
E1	$1.6x10^{-3}$	$2x10^{-4}$	$5.3x10^{-4}$	$1.7x10^{-4}$
E2	$1.2x10^{-3}$	$1.4x10^{-4}$	$9x10^{-5}$	$1.4x10^{-5}$
E3	$6.3x10^{-5}$	$2.8x10^{-5}$		
E4	$1.4x10^{-4}$	$1x10^{-4}$	$6.7x10^{-4}$	

CONCLUSIONS

In spite of the rather qualitative character of the obtained results, it is possible to make two important conclusions:

1. For n-Si it is most likely the formation of deep radiation-induced acceptors that are stable at 450°C and 535°C and accordingly essentially more stable at room temperature than A-centers. The generation efficiency of these acceptors is sufficient for defect engineering applications requiring a control of the semiconductors properties instead of the commonly used approach with A-centers.
2. The electron irradiation accelerates significantly the OTD generation. This may be a manifestation of the radiation-assisted diffusion of oxygen in Si or of the formation of extra OTD nuclei during the irradiation.

REFERENCES

[1] V.V. Emtsev, Yu.N. Daluda, V.I. Shakhovtsov, V.L. Shyndich, V.B. Neimash, R. S. Antonenko, and K. Schmaltz, *Sov. Phys. Semicond.*, **24**, 232 (1990).

[2] P. Hazdra, V. Haslar, and M. Bartos, *Nucl. Instrum. Methods in Phys. Res. B*, **55**, 637 (1991).

[3] D.C. Schmidt, B.G. Svensson, J.L. Lindström, S. Godey, E. Ntsoenzok, J.F. Barbot, and C. Blanchard, *J. Appl. Phys.*, **85**, 3556 (1999).

[4] J.L. Lindström, L.I. Murin, T. Hallberg, V.P. Markevich, B.G. Svensson, M. Kleverman, and J. Hermansson, *Nucl. Instrum. Meth. in Phys. Res. B*, **186**, 121 (2002).

[5] W. Kaiser, H. Frisch, and H. Reis, *Phys. Rev.*, **112**, 1546 (1958).

[6] V.B. Neimash, V.M. Siratski, M.G. Sosnin, V.I. Shakhovtsov, and V.L. Shindich, *Sov. Phys. Semicond.*, **23**, 250 (1989).

[7] D. Wruck and P. Gaworzewski, *Phys. Stat. Sol.* (a), **56**, 557 (1979).

[8] A.N. Kabaldin, V.B. Neimash, V.I. Tsmots, V.V. Voronkov, and G.I. Voronkova, *Ukrainian Physical Journal*, **38**, 34 (1993).

[9] A.M. Kraitchinskii, V.B. Neimash, I.S. Rogutskii, and L.I. Shpinar, *Ukrain. Phys. Journ.*, **44**, 259 (1999).

[10] V.B. Neimash, A. Kraitchinskii, M. Kras'ko, O. Puzenko, C. Claeys, E. Simoen, B. Svensson, and A. Kuznetsov, *J. Electrochem. Soc.*, **147**, 2727 (2000).

[11] R.C. Newman, J.H. Tucker, and F.M. Livingston, *J. Phys. C: Sol. St. Phys.*, **16**, 151 (1983).

[12] A. Kraitchinskii, L. Mizruhin, N. Ostashko, V. Shakhovtsov, *Journal of Technical Physics*, **16**, 1180 (1988).

[13] J.-G. Xu, F. Lu and H.-H. Sun, *Phys. Rev. B*, **38**, 3395 (1988).

[14] J. Lalita, B.G. Svensson, and C. Jagadish, *Nucl. Instrum. Meth. in Phys. Res. B*, **96**, 210 (1995).

[15] A. Chantre, S.J. Pearton, L.C. Kimerling, K.O. Cummings, and W.C. Dautremont-Smith, *Appl.Phys.Lett.*, **50**, 513 (1987).

[16] J.L. Lindström, T. Hallberg, D. Åberg, B.G. Svensson, L.I. Murin, and V.P. Markevich, *Mat. Sci. Forum*, **258-263**, 367 (1997).

[17] M.-A. Trauwaert, J. Vanhellemont, H.E. Maes, A.-M. Van Bavel, G. Langouche, and P. Clauws, *Appl. Phys. Lett.*, **66**, 3056 (1995).

[18] I. Pintillie, E. Fretwurst, G. Lindström, and J. Stahl, *Appl. Phys. Lett.*, **81**, 165 (2002).

In Situ Observations of Point Defect Generation and Complexing during Electron Beam Irradiation of Nitrogen Doped Czochralski Silicon

N. Stoddard, A. Karoui, G. Duscher, A. Kvit and G. Rozgonyi*
Materials Science and Engineering Department,
North Carolina State University, Raleigh, NC, 27695
** also with Solid State Division,*
Oak Ridge National Laboratory, Oak Ridge, TN 37831

Abstract

Samples of nitrogen-doped and nitrogen-free Czochralski(CZ) silicon were observed *in situ* and *ex situ* after irradiation by a convergent electron beam in a transmission electron microscope (TEM). Short periods of exposure produced effects in the N-doped irradiated region that were absent under the same conditions in the N-free reference sample. The combination of analysis by conventional TEM, high resolution TEM and Z-Contrast STEM indicates that a ring rich in self-interstitials surrounds a central vacancy-dominant region containing small point defect clusters. An explanation is offered wherein the 200kV electrons create Frenkel pairs that immediately recombine in N-free CZ silicon, whereas in nitrogen-doped samples, the vacancies interact with nitrogen (or grown-in aggregates unique to N-CZ) to form stable complexes that subsequently either capture more vacancies to form faceted voids, or oxygen to form SiO_2 precipitate nuclei. Meanwhile, the interstitial silicon atoms are kinetically excited via electron collisions and diffuse out of the irradiated area, forming a dense interstitial-rich halo and leaving the central core with a high vacancy concentration. This study demonstrates that it is possible to use the TEM as a nanoscale laboratory in which to observe the formation of voids and precipitate nuclei from their component point defects at low temperature.

Introduction

The use of nitrogen doping in high purity silicon has been a topic of much research in the last few years.[1] Both Czochralski and Float Zone single crystal silicon benefit from the introduction of low concentrations ($< 10^{15}$ cm^{-3}) of nitrogen. Specifically, nitrogen-doped Czochralski silicon (N-CZ Si) has increased denuded zone integrity and size[2], along with a higher density of smaller oxygen precipitates in the bulk which provides improved gettering ability compared to nitrogen free CZ Si.[3] The oxygen precipitate growth rate and size are limited by oxygen diffusion which requires high temperatures, but the volume density of precipitates is independent of annealing temperature since they grow from nuclei that exist in the as-grown N-CZ Si at a high density[3]. Theoretical work[4] suggests that the split-interstitial N_2 complex is quite stable and tends to form vacancy complexes, either as VN_2 or, more prevalently, V_2N_2. These complexes tie up vacancies at high temperature and reduce the formation of voids by delaying their onset temperature during cooling.[5] Transmission electron microscopy (TEM) of oxygen precipitates in N-CZ has shown that two distinct octahedral amorphous phases exist. It has been suggested that the first, with a balanced N+O composition, transitions into a second oxygen rich phase as more oxygen accumulates.[6]

Within the TEM, conservation of momentum allows a 200keV electron to impart up to 20eV to a silicon atom[7]. Since it takes somewhere between 11 and 22eV to displace a silicon atom from its lattice position, displaced atoms will have little kinetic energy left from the collision. Subsequent electron collisions can effectively enable diffusion by imparting up to the full 20eV of pure kinetic energy to silicon interstitial atoms. In fact, atoms on interstitial sites will have a greater probability of electron scattering because of their exposed position in the Si lattice columns. Although interstitial atoms of both oxygen and silicon exhibit negligible diffusion below 400°C, they can easily move after collision with electrons. During point defect diffusion, it is possible for new defect complexes to form and pre-existing aggregates to grow, even at low temperature. In this report, we present the TEM observation of point defect agglomeration in N-CZ Si samples during exposure to a converged 200kV electron beam.

Materials and Methods

The samples examined were [100] oriented CZ wafers; one nitrogen-doped from Nippon Steel with $[N] = 5x10^{14}cm^{-3}$ (measured by FTIR), and two nitrogen-free reference samples. One N-free reference sample was from Wacker and had an unknown grown-in vacancy/interstitial prevalence, while the other, from MEMC, was grown at the 'perfect' Voronkov v/G ratio, such that neither intrinsic point defect was dominant[8]. The samples were polished to ~80 μm, dimple ground to <10μm, placed on a Mo ring and ion-milled at 10°. In a Topcon 002B TEM, all samples were aligned to the [100] zone axis and irradiated by a convergent 200kV e-beam from a LaB_6 thermionic filament chosen for its high electron flux. The converged beam diameter was ~400nm and irradiation was performed on different areas of the same foil for between 5 and 45 minutes. Analysis by conventional TEM methods was performed *in situ* in the Topcon microscope, and subsequently in a JEOL 2010F field emission scanning TEM capable of high resolution TEM (HRTEM), Electron Energy Loss Spectroscopy (EELS) and Z-Contrast imaging. Both diffraction patterns and the Fourier transforms of high-resolution images were examined for evidence of nanoscale second phase formation. The Z-Contrast imaging is particularly important in separating atomic number contrast from stress contrast, which is unavoidable in conventional modes of TEM. Brighter areas in Z-Contrast correspond to regions with higher average atomic number, higher spatial density of a single element, or that have a greater thickness[9]; whereas dark areas must be thinner or have a higher concentration of vacancies.

Results

Changes in the irradiated area of the N-CZ sample were evident after only five minutes of exposure to the convergent electron beam, with more dramatic effects occurring after twenty minutes, see Fig. 1(a). Note the grainy texture in the center of Fig. 1(a) which is surrounded by stress contrast lines, while the inset Kikuchi diffraction pattern verifies alignment with the [100] zone axis during irradiation. The Wacker N-free reference sample in Fig 1(b) is largely unchanged with no discernable grainy texture. Although a dark ring of wavy stress contrast lines exists, it passes through the irradiated area rather than emanating from it. At higher magnification, see Figure 2, more details of the changes within the irradiated volume of the N-CZ sample of Fig 1(a) are evident. Dark triangular areas suggest the presence of faceted voids. A STEM mode Z-contrast image, see Fig. 3, provides additional information about local modulations in the average atomic density while being free from interference due to stress contrast. The micrograph reveals a dark central region, corresponding to the size of the converged beam, surrounded by a bright ring. Because of the transparency of Z-contrast image analysis, we may interpret the dark central region to be vacancy-rich while the bright surrounding ring is interstitial rich.

The spatial separation of the Frenkel pair components can be attributed to the creation of self-interstitials by the incident electrons and the subsequent ejection/diffusion of those atoms from the irradiation region by further collisions. Note that spots of brighter contrast within the darker central region are discernible that may be clusters of lattice atoms or possibly small areas of a second phase with a higher density than the rest of the vacancy rich region. To examine the clusters more closely, HRTEM studies were performed that provided some evidence that these clusters are due to a second phase, such as areas with altered lattice periodicity, see Fig.4(a). The use of the Fourier transform, see Fig. 4(b), shows two smaller diffraction spots inside of and slightly off-axis from the normal (100) spots. There are also two streaks indicated by arrows on the perpendicular axis of the pattern. These observations provide evidence for point defect agglomeration and the presence of a second phase. The precipitate-like regions may be due to the presence of grown-in nitrogen-oxygen nuclei in the N-CZ sample; however, their known low volume density and small size make it more likely to be the result of processes stimulated by the irradiation, since they cannot be found outside the grainy contrast area. STEM chemical analysis via EELS has thus far been inconclusive on the question of local variations in nitrogen and oxygen concentration.

Discussion

These results demonstrate that Frenkel pair formation can be monitored during the interaction of 200kV electrons with silicon. In the N-free samples, the Frenkel constituents are likely to recombine immediately; however, in the presence of nitrogen, the vacancies (or interstitials) may also combine with nitrogen immediately after separation to form stable complexes of N-V or N-I. Subsequently, these complexes can capture either more vacancies to form faceted voids, or oxygen atoms to form SiO_2 precipitate nuclei. Our results support this model, providing evidence for the presence of both voids, see Fig. 2, and nuclei of a second phase, see Figs. 3 and 4. Meanwhile, the interstitial silicon atoms, energized via additional electron collisions, move either to the surface or out of the irradiated area where they form a dense interstitial-rich halo around the excess vacancy central core, as in Fig. 3. In the N-free reference samples, the lack of lattice disturbance indicates that there is no long-lasting separation of vacancies and Si self interstitials, and hence no significant diffusion or point defect mediated complex formation. Although there is some subtle strain contrast in the Wacker reference sample, even that is absent in the perfect MEMC sample. This difference in the behavior of the two reference samples can be explained as follows. Unlike the MEMC reference, the Wacker sample was grown under conditions such that an excess of either vacancies or interstitials is present on solidification. Whether the sample is inherently vacancy rich or interstitial rich, the excess point defect will diffuse under the electron irradiation, even if the lattice on the whole is not disturbed. The e-radiation induced diffusion creates a point defect concentration gradient that will introduce stress. This effect will be absent in the MEMC sample because of the perfect growing conditions and the resulting lack of a prevailing point defect.

The likelihood of oxygen precipitate formation will also be enhanced in the irradiated region simply because of its vacancy rich character and the stability of V-O complexes. We have indeed observed clusters in the vacancy rich region, see Fig. 2, and calculations indicate that there is enough oxygen in the excited volume to form significant aggregates, although we do not yet have direct evidence for the presence of oxygen. Alternatively, it is possible that these clusters do not involve oxygen at all, but are places where the radiation damaged silicon lattice exhibits local strain gradients. In any case, these aggregates must be somewhat stable under the electron beam in order to survive. If they involve oxygen, they should have less likelihood of being dissociated by the e-beam since SiO_2 has stronger bonds than Si, and 200kV electrons (which barely create Frenkel pairs in Si) may have insufficient energy to produce knock-on dissociation of an Si-O complex. In either case, further investigation is necessary to elucidate the situation. Modeling simulations will greatly enhance the interpretation of point defect behavior in such a vacancy rich lattice.

Finally, there is a somewhat confounding fact to consider with regard to the stated $5x10^{14}$ cm^{-3} nitrogen concentration. Namely, if one assumes a homogeneous distribution of nitrogen in the sample, then the dramatic differences we have observed between nitrogen doped and nitrogen free silicon are caused by only one or two atoms of nitrogen in the irradiated volume. This is conceptually problematic, and some alternatives must be considered. For example, it is possible that, over time, each nitrogen atom could be the seed of many extended defects. That is, once the evolution from the original complex towards a discrete void or precipitate has begun, the nucleating nitrogen atom detaches and is free to diffuse and complex with other vacancies. Obvious problems with confirming this hypothesis make it necessary to look elsewhere for additional possibilities. Although there is some evidence that the quoted nitrogen concentration measured by FTIR may be too small, it would be off by at most by a factor of 4.

Another proposal for the nitrogen stimulated observations is that there may be a sample preparation induced intensification of the nitrogen concentration in the thin TEM foil. Experiments are now in progress attempting to verify this type of phenomena. Finally, a completely different possibility is that the nitrogen itself may not be the prime factor in the *in situ* TEM experiments, since complexes may actually be inherited from nitrogen stimulated processes that were frozen in during ingot cooling. It is well known that the dynamics of oxygen precipitate and void formation at high temperatures are seriously impacted by the action of nitrogen[2,3]. The remnants of these high temperature effects, for example a V-O complex, may result in decidedly different room temperature behavior in N-free or N-doped samples

Conclusions

The creation and segregation of silicon Frenkel pairs under a converged electron beam at temperatures below 100C permits the observation of the nucleation and growth of extended defects such as voids, stacking faults, and oxygen precipitates in a controlled fashion. In situ TEM observation of this process is a unique tool because extended defect creation usually only occurs at elevated temperatures (> 650C) and often results in such low concentrations that the defects are quite difficult to find using electron microscopy. The role of nitrogen is critical, since the interaction of nitrogen with vacancies and/or oxygen creates large differences between N-free CZ Si and N-CZ Si during electron beam exposure. We have postulated that nitrogen acts to separate the Frenkel constituents creating N-V complexes which capture supersaturated interstitial oxygen initiating the nucleation of amorphous oxygen precipitates. Similarly, N-V complexes which capture additional vacancies leads to void formation. In the literature, electron irradiation phenomena are generally deemed destructive and problematic in nature. However, the results of this study indicate that we now have the possibility of having nitrogen, oxygen and silicon interstitials/vacancies interacting under the electron beam. Thus, the damage created in situ in the TEM moves from being destructive to constructive as new information on extended defects is generated. We can view this as a nanoscale laboratory where one can induce in specific regions of a sample point defect/impurity interactions at room temperature which normally require high temperature processing. The formation of voids and agglomerates of point defect stimulated impurity reactions in the TEM correspond to the formation of octahedral voids and oxygen precipitates during crystal growth and processing, but here the dynamics of their early stage formation can be studied directly and conveniently.

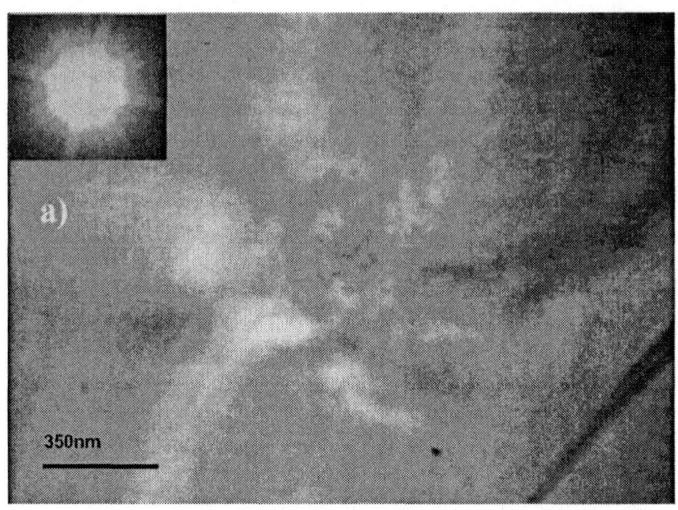

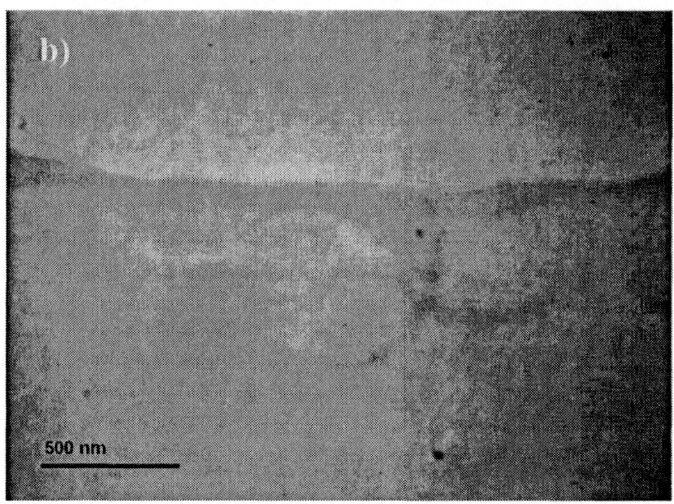

Fig. 1: Conventional TEM showing the entire irradiated areas after 20 min. exposure to a 200keV convergent beam. Note grainy texture in a) is not present in b). The inset Kikuchi pattern illustrates zone axis alignment.

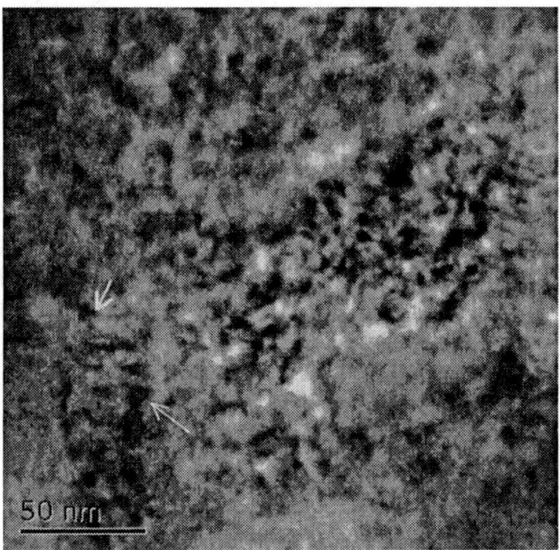

Figure 2: Close-up of irradiated area in N-doped sample of Fig. 1(a). Some faceted dark areas, indicated by arrows, suggest the presence faceted voids.

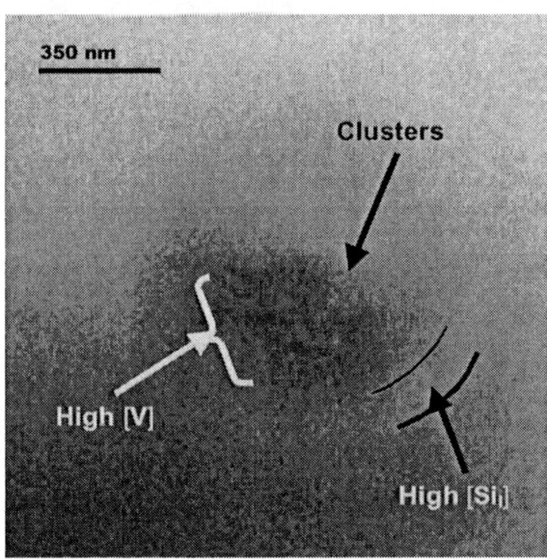

Figure 3: STEM Z-contrast image of N-CZ Si irradiated area, as in Fig. 1a. A central dark area is attributed to a large concentration of vacancies while the brighter ring surrounding it is due to excess self-interstitials. Some bright spots exist in the dark region, suggesting clustering.

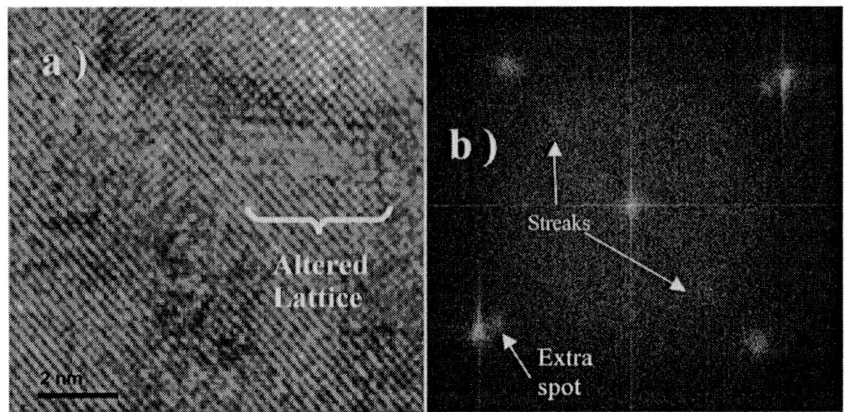

Figure 4: a) is a HRTEM image of the irradiated area. b) is a Fourier transform. The lattice has a different periodicity in the indicated part of the image, and the extra spots in the Fourier transform suggest the existence of a second phase.

References

[1] Rozgonyi, GA, ECS PV **2002-1**, 149-162 (2002)
[2] Tamatsuka M, Kobayashi N, Tobe S and Masui T, ECS PV **99-1**, 456-467 (1999)
[3] Nakai K, Inoue Y, Yokota H, Ikari A, Takahashi J, Tachikawa A, Kitahara K, Ohta Y and Ohashi W, *Journal of Applied Physics*. **89** (8), 4301-4309 (2001)
[4] Karoui A, Karoui FS, Rozgonyi GA et al, ECS PV **2002-2** 670-676 (2002)
[5] Kageshima H, Taguchi A, Wada K. *Applied Physics Letters* **76** (25), 3718-3720 (2000)
[6] Karoui A, Karoui FS, Kvit A, Rozgonyi G and Yang D. *Applied Physics Letters*, **80** (12) 2114-2116 (2002)
[7] David B. Williams and C. Barry Carter, *Transmission Electron Microscopy*. Plenum Press, NY, 1996
[8] Falster R. et al. ECS **PV 98-13**, 135 (1998)
[9] Pennycook SJ and Boatner LA, Nature **366** (1988) 565.

Acknowledgements

This work was supported by NREL contract grant AAT-2-31605-05 and NSF grant EEC-9726176. We thank them for their support.

Alternative Silicon Substrates and Structures

SMART-CUT®: A GENERIC LAYER TRANSFER SOLUTION FOR HIGH VOLUME SOI PRODUCTION

Christophe Maleville
SOITEC SA;
Parc Technologique Des Fontaines ; 38926 Crolles Cedex ; France
Christophe.Maleville@Soitec.fr

Smart-Cut® process is opening new ways by enabling transfer of monocrystalline films on any kind of substrates. Layer transfer feasibility has been already demonstrated for silicon, III-V compounds, Ge, ... and can be involved using various insulating (SiO2, Si3N4) or conductive (Wsi) layers. A wide range of substrates can be used from silicon to plastic, including transparent substrates such as quartz and glass. With the SOI experience, industrial compatibility of Smart-Cut® has been demonstrated. Unibond® SOI volume production is today running in Bernin fab, showing that splitting and bonding steps can be controlled at production levels, with high yields. Taking advantage of standard equipments flexibility, process has been successfully scaled up to 300mm. Most advanced 200mm processes were successfully transferred to 300mm, with products showing uniformity and defectivity results compatible with industry requirements for fully depleted applications.

Deposition techniques like chemical vapor deposition (CVD) offer to the semiconductor industry the initial flexibility to deposit thin films of key materials on many kinds of substrates. The homoepitaxy or heteroepitaxy techniques using CVD or Molecular Beam Epitaxy (MBE) add the flexibility to get a pure monocrystalline thin film but with a major limitation : the starting substrate has to be monocrystalline. The missing technology has been the one which allows the growth of a thin monocrystalline film on any kind of substrates (Figure 1). Hydrogen induced splitting of a given wafer (known today as Smart-Cut®), discovered at the LETI laboratory in 1991, provides this unique opportunity (1).

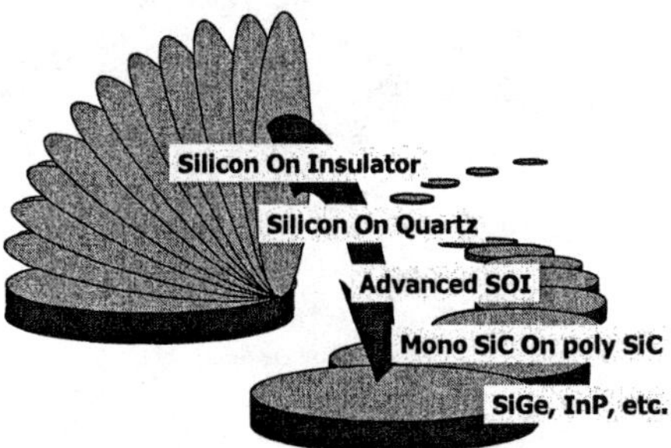

Figure 1: Schematic showing Smart-Cut process flexibility, allowing to transfer a thin monocrystalline film on any kind of substrates.

Therefore, a new technique is offered today to the semiconductor industry which enables the development of substrate solutions tailored to specific applications, for example the combination of high quality III-V thin films on Si substrate. This technique is in use in production today on a first application: silicon-on-insulator (SOI), which consists of a monocrystalline film of silicon, on a thin amorphous silicon dioxide layer, on top of a silicon wafer. A dedicated fab has been opened in 1998 for volume production of 4" to 8" SOI wafers. Based on high volume 8" experience, tool-set has been optimized, enabling high quality SOI volume production. SOI has been established as the material of choice for advanced microprocessor applications, with a transition from partially depleted to fully depleted operation for the sub-90nm node devices. It involves two major technological challenges for SOI material, uniformity control at the atomic level and 300mm availability.
In this paper we will describe the large range of applications addressed by Smart-Cut® process. New ultra thin 300mm Unibond® is described with properties which fulfill industry requirements in terms of uniformity, quality and availability.

SMART CUT®: AS A GENERIC TRANSFER SOLUTION

With the hydrogen induced splitting it is possible to combine the advantages of different materials and optimize the resulting substrate for manufacturing of III-V optoelectronic and microwave devices, with advantages like improved substrate availability, larger wafers and reduced costs, but also to enable new structures (3) and devices (Figure 2).

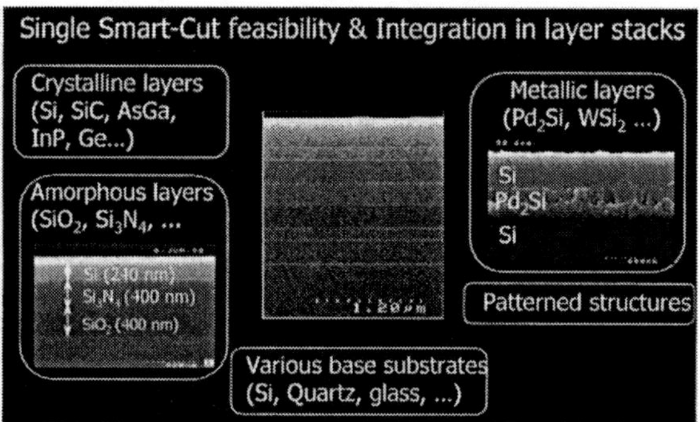

Figure 2: Single Smart-Cut feasibility and integration in layer stacks for a wide range of transferred materials and supports.

Heteroepitaxy of compound semiconductor materials on silicon for example, has been extensively investigated, but devices performances are generally limited by the high density of threading dislocation. Thus, Smart-Cut® technology, combining wafer bonding and thin film transfer, offers a new and unique opportunity for a major technological and industrial breakthrough in the fast growing compound semiconductor market. Recent breakthroughs have been achieved on silicon carbide, Silicon on Quartz and III-V compound semiconductors.

HIGH VOLUME SOI MANUFACTURING

SOI technology is evolving since almost 25 years, SOI material being used initially for niche application such hardened devices. In the last 5 years, following IBM example, main IC suppliers agreed that they have to design their device on SOI to achieve better performance in terms of speed and power consumption (2). Materials in general are coming back at the center of the performance race as it appears that scaling is not enough to continue to follow the Moore's law. In the run for faster microprocessors, fully depleted designs are now targeted for sub 90nm devices, requiring ultra thin SOI material. When operating in FD mode, transistor threshold voltage is directly linked to silicon thickness. Thickness uniformity is then becoming a critical parameter to guarantee Vt control. Looking to ITRS roadmap guidelines, one can see that in 2004 timeframe, 20 nm silicon layers, verifying ± 5% uniformity will be needed in high volume (3). SOI material manufacturability is the key to establish this new breakthrough in the silicon based industry.

200mm high volume manufacturing experience

Smart-Cut® process has been the enabling technology, allowing large volume production SOI wafers with consistent quality. Bernin facilities, opened in 1998, is now running at full capacity, with 900k wafers starts per year. 200mm experience is demonstrating production stability and Unibond® products are today qualified in production by customers in all diameters up to 8" and for applications ranging from thin films to thick films. As an example, figure3 shows defectivity chart for more than 50000 wafers, grouped by production week. When comparing defect counts with typical performance for CZ and high quality epi wafers, it can be seen that with less than 20 count per wafer for 160nm detection level, Unibond® product is close to the best silicon Epi wafer quality.

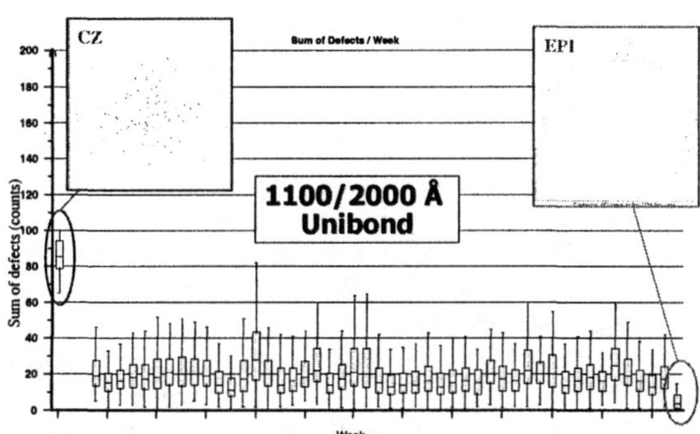

Figure 3: Sum of defects per wafer distribution grouped per production week for more than 50000 wafers, defects>0.16 μm. 1100/2000Å Unibond®. Typical CZ and Epi performance for the same threshold are given for comparison.

300mm SOI wafer manufacturing

Several IC suppliers are today in qualification and ramp-up of 300mm fab running SOI based advanced applications. As the Smart-Cut® process is based on standards equipments from the semiconductor industry, then transition from 200 to 300mm has been enabled by availability of 300mm generation equipments. 300mm Unibond® feasibility was demonstrated in 1997, samples built in 2000, and 300mm pilot line is operating in Bernin I since January 2002. Now in ramp-up, this first 300mm experience proved that there is no major limitation for 300mm Unibond®. As 300mm silicon wafers incorporate today the most advance silicon wafer engineering, the SOI quality and manufacturability has been even improved when going to this new diameter. Nanotopology improvements are particularly beneficial. Table 1 gives a comparison in terms of equipments type and configuration between 200 and 300mm lines.

From a manufacturing point of view, 300mm line has been set-up using all 200mm experience and is really close from 200mm line. The major change is in the carrier type since 12" is handled in FOUP for all equipments and all processing steps from incoming material to shipment.

Table 1: 200mm and 300mm tool-set comparison

Process step	200mm	300mm	Comment
Oxidation and HT anneal	Vertical furnace Double boat	Vertical furnace Double boat	Boat designs adapted to 300mm
Implantation	Batch 17 wafers	Batch 13 wafers	Medium current
Cleaning	Cassette less cleaner	Cassette less cleaner	Exact copy
Bonding	Automatic bonder	Automatic bonder	New machine generation
Polishing	Multiple head/platens	Multiple head /platens	Integrated cleaning for 300mm
Thickness measurement	Full wafer reflectometer	Full wafer reflectometer	>4000 pts per wafer in 300mm
Defectivity monitoring	SP1-DLS	SP1-DLS	Also used for mapping after HF revelation

Regarding high temperature treatment, boat designed for 300mm, reducing wafer stress is used. Oxide uniformity is typically same than for 200mm. For the implantation step, the same machine is used meaning that hydrogen beam is unchanged. A 12" wheel is installed instead of 8" wheel and handling system is also converted. From a throughput point of view, area to be scanned is 30% higher in 12" and wheel is 13 positions resulting in a throughput ratio of 50%. New source and beam line configurations are now involved allowing to reach 75 mA current when implanting at 20 keV, a typical energy for new products generation. Such conditions corresponds to 17 300mm wafers implanted per hour.

Concerning other processing steps, a few gaps can be noted with 8" lines. Because of FOUP operation, dry-in dry-out polishing machine is used, but End Point Detection by In-situ Reflectivity Monitoring is of course in place, lowering wafer to wafer variations for mean thickness.

ULTRATHIN PRODUCTS

New Unibond® process generations

When reviewing ITRS roadmap targets, several challenging items appear. Silicon film thickness and uniformity are the most aggressive, with values typically of 20nm and ± 5%, 6σ. These requirements lead to 10Å thickness accuracy on a 300mm wafer. As a general trend, it appears that such high uniformity needs to be guaranteed whatever the spatial wavelength of the measurement down to Å scale what is currently the domain of roughness measurement. The "nano-uniformity " will certainly be the key challenge rising metrology difficulty.

Figure 4 shows how Smart-Cut® technology can be extended to very thin films, both for silicon and oxide layers. 500 Å silicon films are already running in production while 200 Å SOI product is today in advanced prototyping. 100/200 Å Unibond® feasibility has been recently demonstrated.

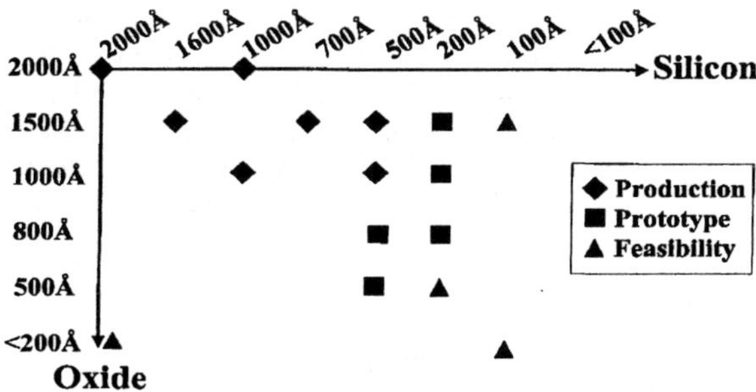

Figure 4: Smart-Cut® extendability chart showing product status towards production, depending on silicon and buried oxide layers thickness combinations.

Roadmap from table 2 has been defined to address ultra thin SOI layers for 65nm node and beyond. This roadmap is focusing on both 200 and 300 mm for ultra thin products with a full commonality for process and toolset technology.

Table 2: Unibond® ultrathin SOI roadmap

Time line	2000	2001	2002	2003	2004
Process generation	PD	FD	UT1	UT2	XUT
Silicon Layer Thickness (Å)	1000	500-700	200-700	200-300	200
Si unif 6σ (Å)	150	100	70	30	10
Si unif 6σ; all wrs all sites (%)	± 7 %	± 7 %	± 6 %	± 5 %	± 2.5 %
Roughness AFM(RMS)					
• 1x1 μm	1 Å	1 Å	1.5 Å	2 Å	2 Å
• 10x10 μm	1.5 Å	1.5 Å	3 Å	4 Å	5 Å
Box Layer Thickness (Å)	1500-2000	1000-1500	1000-1500	800-1500	800-1500
Box Layer TTV	± 4%	± 4%	± 3%	± 3%	± 3%

Ultrathin film processing in production

Present production is designed for ± 50Å with UT1 process generation. In this roadmap, uniformity is calculated at 6σ (Mean ±3σ) assuming that each wafer is measured using more than 1700 points in 200mm and 4000 points in 300mm. Then, plotting min-max values for each wafer (M-3σ and M+3σ), overall uniformity is obtained, all wafers, all sites (Figure 5). Uniformity is then driven by 2 parameters, wafer to wafer mean thickness variation and on-wafer sigma. For this typical 200mm example, mean thickness is controlled at ±15Å while typical on-wafer values are around 4.5Å.

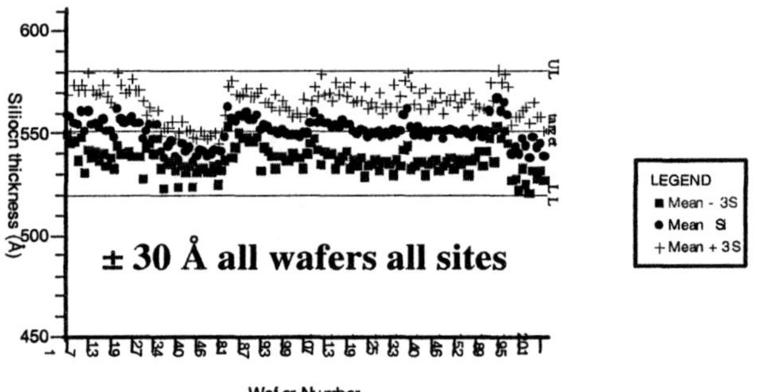

Figure 5: 200 mm UT1 overall thickness uniformity chart for a 550/1450Å product. For all sites on all wafers, thickness is within 550 ± 30Å, 6 sigma.

Uniformity can not be disconnected from surface roughness. Unibond® UT1 is offering the best combination regarding these 2 parameters. Both for 1x1 μm and 10x10μm AFM scans, we can see on AFM pictures from figure 6 that roughness is kept at very good levels, especially for small scans with less than 2Å RMS. Interface roughness exhibit also low values, typically less than 2Å RMS.

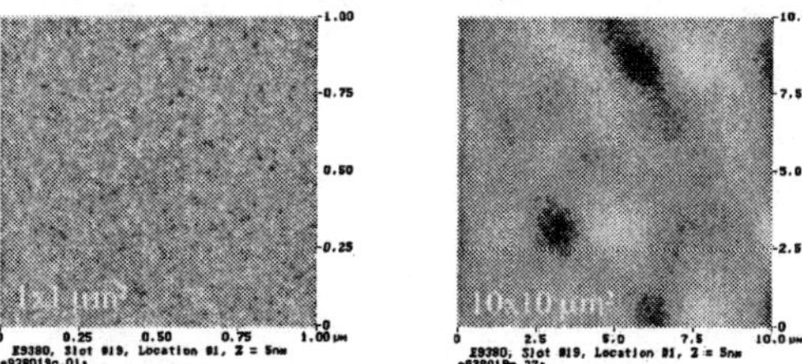

Figure 6: Surface roughness for 1x1 and 10x10 µm scans for UT1 products, respectively 1.8Å and 2.7Å RMS.

300mm UT1 process is in ramping-up phase and is capable of ±50Å total uniformity, 3mm edge exclusion. All characterization parameters including defect density, roughness, HF defect density, secco defects, metallic contamination, electrical properties are already exhibiting the same capability than for high volume 200mm.

Towards UT2 and XUT generations

In Unibond® process, final uniformity is defined by oxidation, implantation and polishing steps.

In new processes generations, oxidation and implantation conditions optimizations led to very uniform as-split off structures, approaching 1 Å uniformity (1 sigma) (Figure 7). UT1, UT2 and XUT process are based on such highly uniform (7Å range) SOI structure. Differences in these generations appear in the steps involved to erase surface roughness and obtain final SOI structure.

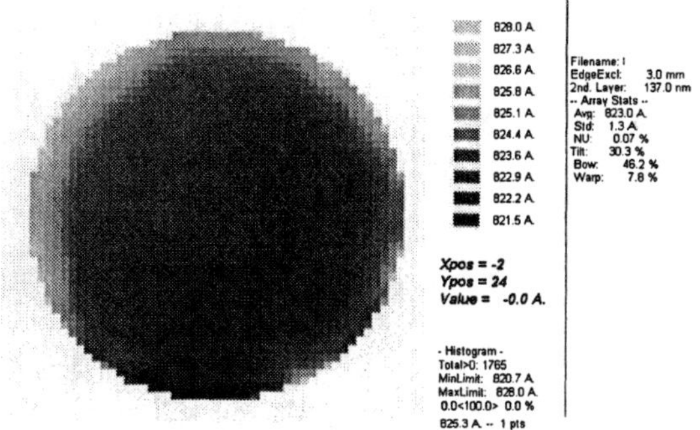

Figure 7: 200mm (as-split off) map for new UT generation processes showing 1.3Å uniformity (1σ) – Range is 7Å.

When comparing UT1 performance and ITRS targets, improvements need to be achieved going towards thinner and even more uniform films.

Regarding thickness, there is no limitation in the Smart-Cut® process for low energy implantation and 5 to 10 keV is a typical range. Then, 200Å to 500Å films are obtained and TEM cross-section (Figure 8) allows to verify that crystalline quality and interface sharpness is not degraded for these ultra thin films. This TEM picture also prove that, talking about nano-uniformity, outstanding performance can be achieved.

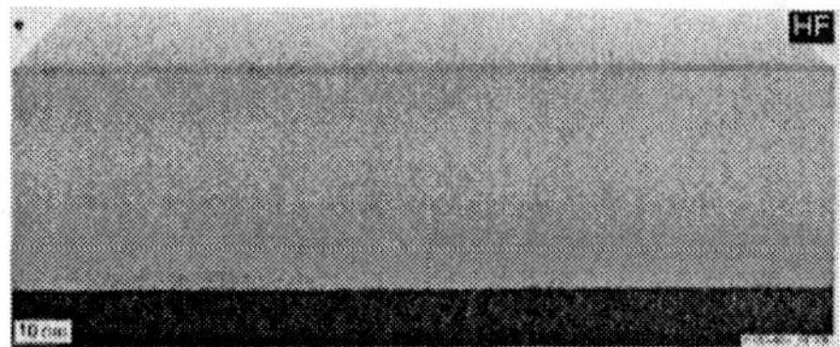

Figure 8: TEM cross section imaging from of top silicon layer from a UT2 generation 500/1500 Å 300mm Unibond®. Picture enhances very good silicon uniformity and sharp silicon/Box interface.

Regarding uniformity at wafer level, next figure (Figure. 9) is giving a typical thickness map for a UT2 300mm wafer, 500/1500Å at 3mm edge exclusion. Standard deviation lower than 5Å are achieved and combined with a ±5Å control of mean thickness value, wafer to wafer, UT2 is compatible with ±20Å overall uniformity, as shown on figure 10 for 200/1500Å product.

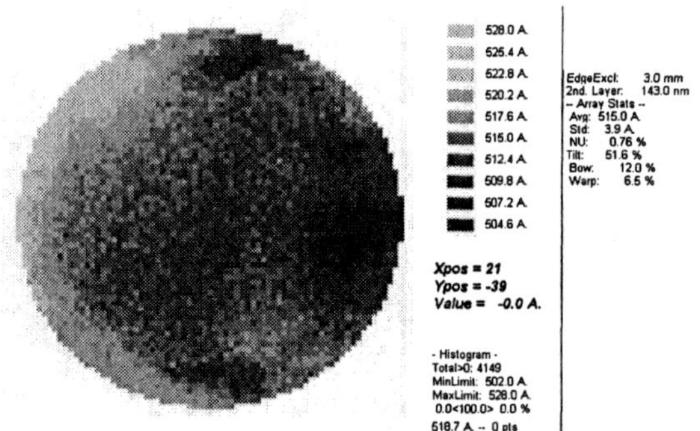

Figure 9: 300mm UT2 product thickness map. 500/1500Å, 3mm edge exclusion. Sigma is 3.9Å and range is ~25Å for 4149 points measured.

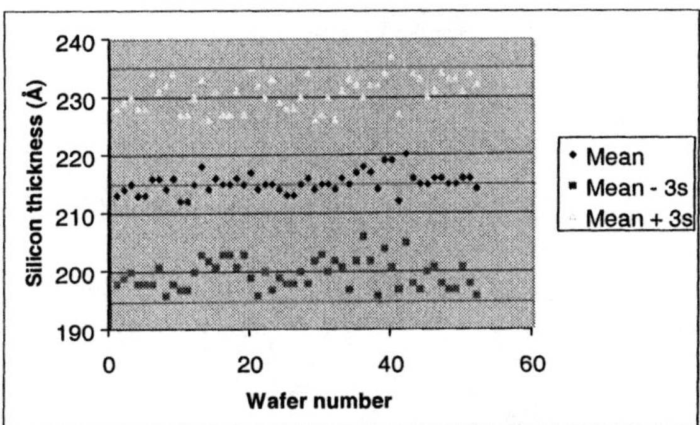

Figure 10: UT2 300 mm overall thickness performance preliminary results for 200/1500Å product. Total thickness variation is ± 20Å.

As discussed earlier, roughness and defectivity are key parameters to consider when talking about ultrathin products. When entering ultrathin films arena, reflectivity effects modifying laser scattering inspection of SOI structure is enhanced because optical absorption is decreased. Anyway, when tuning SP1 machine recipe, 0.3µm threshold can be achieved even on dark wafers (reflectivity <0.1). For that detection level, ultra thin UT2 Unibond® defect density (see example figure 11) is less than 0.2 /cm², confirmed by HF defect densities in the 0.05 /cm² range.

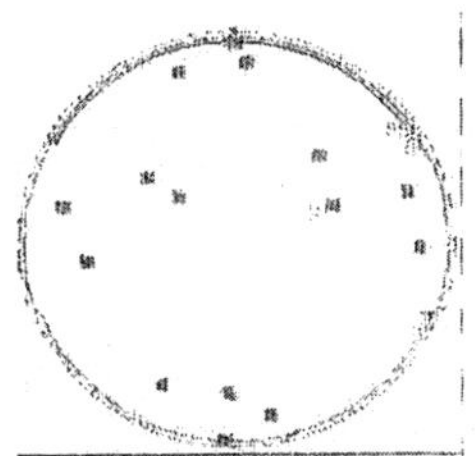

Figure 11: SP1 defect map on UT2 300mm 200/1500 Å Unibond®, oblique illumination @0.3µm.

Following recent publication (4), further improvements have been made in splitting and finishing steps allowing surface roughness to be respectively 1 and 3.5Å RMS at 1x1 and 10x10µm scans (Figure 12).

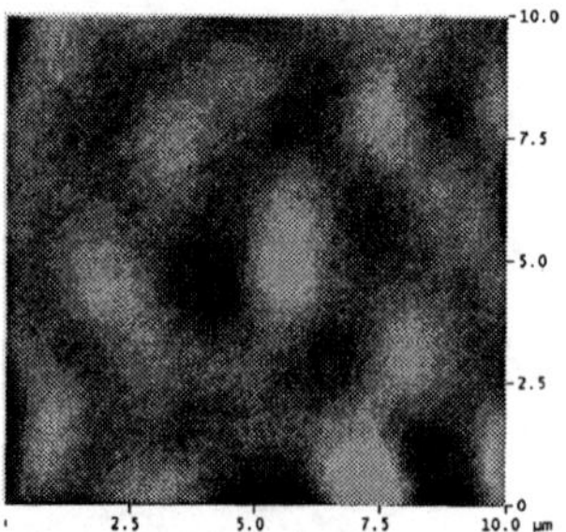

Figure 12: 10x10 µm AFM scan on UT2 300mm 200/1500 Å Unibond®. R= 3.5Å RMS

Last figure (figure 13) is imaging what will be XUT generation with first results obtained in 200mm, demonstrating that less than 2Å standard deviation is achievable.

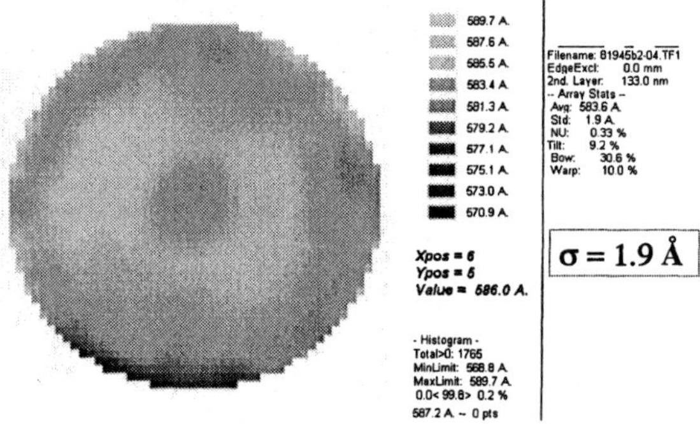

Figure 13: 200mm XUT product thickness map. 580/1500Å, 3mm edge exclusion. Sigma is 1.9Å and range is <20Å for 1765 points measured.

CONCLUSION

SOITEC has grown a large high volume production expertise in 200mm, allowing to drive product defectivity close to epi reference. No major change was introduced in equipment set and configuration when switching to 300mm, then same control and consistent high quality is achieved. New process strategy are allowing to improve layer uniformity compatible with IC supplier roadmaps for 90nm node and below. New Smart-Cut applications are ready for industrialization and prototyping ramp-up. The more promising are Silicon On Quartz and Strain Silicon On Insulator. SOQ and SSOI will increase in volume in the next years. SOQ will enable high speed application and high resolution on transparent substrates whereas SSOI is offering higher mobility for future generation high speed microprocessors.

REFERENCES

1. M. Bruel, Nuclear Instruments and Methods in Physics Research B, **108**, (1996), P. 313-319
2. www.soisolutions.com/news.html
3. http://public.itrs.net
4. C.Maleville et al., 2001 IEEE SOI conference, p.155.

Electrical and Structural Characterization of Silicon on Silicon Bonded Interfaces

P. McCann[1], J. McKeever[2], D. Nicholson[2], F. Ruddell[2], H.S. Gamble[2], W.A. Nevin[1]

[1]Analog Devices Belfast
5, Hannahstown Hill, Belfast, BT17 0LT, N. Ireland.
[2]Queen's University of Belfast, Ashby Buildings, Stranmillis Road, Belfast, BT7 1NN, N. Ireland.

ABSTRACT

In this work we have investigated the influence of bonding conditions, annealing temperature, and differences in crystal orientation and growth on the performance of silicon on silicon bonded substrates. Heavily doped substrates were bonded to intrinsic device layers to investigate PIN diode characteristics, and low resistivity handles and devices were bonded together to investigate the bonded interface. Samples were prepared by hydrophobic cleaning, and the joined pairs were annealed at temperatures between 600 °C and 1200 °C. The cleaning and bonding conditions were varied in order to study the effect of lag time and native oxide growth. The physical characteristics of the substrates were analyzed using SIMS, TEM and spreading resistance profiling and related to electrical performance. A range of techniques was used to highlight the electrical interface quality including I-V profiling of PN diodes, capacitance-voltage relationships of MOS capacitors, and surface charge analysis measurements. The bonding conditions and interstitial oxygen was shown to have an effect on device performance for low resistivity material. Under optimized conditions, the analysis shows that there is no adverse impact of the bonding interface on the junction characteristics. The silicon direct bonded substrate exhibits low leakage currents, low defect density, and controllable dopant distribution, which enables it to be used to replace thick epitaxy in manufacturing processes to reduce costs and improve quality.

INTRODUCTION

For semiconductor device manufacturers the silicon on silicon bonded wafer offers a cost effective alternative to thick epitaxial layers and inverse epi that have traditionally been used for applications such as high power devices. The use of silicon wafer bonding technology allows silicon substrates to be produced containing several layers of single crystal silicon. These layers can have doping ranges from 10,000 Ωcm to very highly doped implanted regions and can include combinations of different growth types and orientations, which is not possible using epitaxial technology. The bonding process gives a high wafer quality with low leakage, low warpage and a low defect density with a small thickness deviation in the substrate. In addition, the transition of dopant levels between layers can be tailored to be steep or soft depending on the annealing conditions used, to achieve the required dopant impurity profile. The epitaxial deposition step in RF attenuator and PT IGBT (Punch Through Integrated Gate Bipolar Transistor) manufacturing processes can be removed, increasing the device performance and yield and decreasing the complexity of the process. This reduces the capital cost associated with device manufacturing, allowing greater process flexibility. This process is also ideal for photodiodes, where silicon quality and thickness control are important for maximizing device efficiency.

Direct wafer bonding requires the starting substrates to be very flat, smooth, haze free and clean. The wafers must be free of particles larger than 0.3 μm and free of organic and metallic contamination. Two main categories of cleaning process can be used, hydrophilic and hydrophobic. Hydrophilic cleaning typically involves a RCA cleaning process, which results in a surface terminated with OH groups and water molecules, so that a thin native oxide of 2-20 Å is formed on the wafer surface. A very high temperature annealing step is required to diffuse the oxide into the silicon to produce a silicon to silicon bond(1). Hydrophobic cleaning involves the use of HF cleaning or exposure to an SF_6 plasma(2). The wafer surface is terminated with hydrogen and/or flourine atoms after cleaning, allowing silicon to silicon bonds to be realized at lower temperatures. During annealing of hydrophobic bonded substrates, the hydrogen is released from the mating surface to form bubbles, which appear as voids at the interface at temperatures between 300 °C and 900 °C(1). Hydrophobic wafer bonding was used in this work to allow silicon to silicon bonds at temperatures from 1000 °C upwards to be investigated.

The basic operation of RF attenuators, PT IGBT's and photodiodes can be analyzed using PIN diodes, but for a more detailed examination of the bonded interface lightly doped PN diodes are required. The performance of PIN diodes can be related to the parameters of dopant distribution, defect density and impurity levels at the bonded interface. In this work the structural and chemical characteristics were investigated using spreading resistance profiling (SRP), transmission electron microscopy (TEM) and secondary ion mass spectrometry (SIMS). The bonding interface was inspected for voids using a scanning acoustic microscope (SAM) and an infrared microscope. Electrical characterization was performed on PIN diodes and PN diodes using capacitance-voltage and current-voltage measurements. Surface charge analysis (SCA) was also used for electrical characterization of the bonded interface.

EXPERIMENTAL

This report covers material differences of growth, orientation, and dopant, and correlates the effects with anneal temperature and bonding conditions. Four different types of material were used, as shown in Table 1, labelled A-D. The handle wafers were prepared for bonding by using chemical-mechanical polishing to achieve a low surface roughness and good thickness variation. The wafers were cleaned pre-bond with an HF process to minimize the native oxide and provide a hydrophobic surface for the silicon to silicon bond. A high resistivity, n-type, FZ, <111> device (group A) was bonded to a low resistivity Cz handle (groups B and C). In addition, a lightly doped FZ, <100> device (group D) was bonded to a handle of the same group in order to conduct detailed electrical testing of the bonded interface. Each sample group was cleaned and bonded using two methods, either by cleaning in a wet deck, then bonding on an automated wafer bonding machine in ambient cleanroom atmosphere or by cleaning and bonding in-situ on an automated bonding tool. All the samples were immediately annealed in nitrogen at a range of temperatures from 600 °C upwards. The device wafers were then ground and polished to between 2 μm and 200 μm. Group A and B/C samples were thinned to 10 μm and 200 μm for PIN diode manufacturing, while Group D samples were ground and polished to 5 μm for PN diode testing. Samples from each set of substrates were thinned to 2 μm for high resolution SIMS and cross-sectional TEM. The samples were prepared for TEM by mechanical polishing and Ar^+ ion milling from the centre of the wafer. The analysis was performed in a JEOL 4000EX TEM at 400 keV. All TEM images shown are bright field. The SIMS profiles were obtained using a Cameca 4f spectrometer with a Cs^+ primary ion beam. Quantitation was obtained by reference to implant standards in silicon.

SCA measurements were made using a Semitest instrument on bonded samples with a 0.2 μm thermal oxide. The diodes were produced by diffusion of a 3 μm and 1 μm deep P+ layer for PIN diodes and PN diodes respectively with a 1mm diameter Aluminium gate. Electrical characterization of the diodes involved I-V profiling under forward and reverse bias voltages up to 100 V using a HP4145A parameter analyzer. MOS capacitors were also produced on substrates from group D to further test the bonded interface by C-V, C-t, pulsed C-V analysis, and Zerbst plots. All devices were manufactured and tested at the Northern Ireland Semiconductor Research Centre (NISRC) at Queen's University Belfast. Physical characterization also entailed Spreading Resistance Profiling (SRP).

RESULTS AND DISCUSSION

The PIN diode forward bias I-V analysis showed that the silicon direct bonded substrates have an ideality factor close to 1 for low level injection as shown in Figure 1. High level injection occurs above 0.7V voltage and the ideality factor is 2 under these conditions(3). The reverse breakdown was calculated at 1780 V and was not obtained with the characterization tools used; however the diode was fully depleted at –23 V. SCA measurements were taken from samples annealed at 1000 °C, 1100 °C and 1150 °C and Figure 2 shows a typical plot of charge versus depletion width with overshoot in inversion. The overshoot is due to the lifetime of the excess minority carriers generated by the applied light being longer than the dark cycle of the SCA measurement. The SCA works by illuminating the surface with modulated light, which generates an ac photocurrent, and applying a dc bias to keep the silicon in depletion or inversion. The ac signal is proportional to the depletion width and the applied light intensity, and from this the doping level, oxide charge, and density of interface traps are calculated. The induced charge is calculated from the probe capacitance and the applied voltage bias and plotted against the depletion width. The electrical characteristics of the PIN diode substrates at various anneal temperatures are listed in Table 2. The SCA is a qualitative measurement tool, which can able to examine differences in similar substrates. All the data in Table 2 is similar, indicating identical electrical performance for each annealing temperature.

The main areas of concern with silicon direct bonding for PIN diodes are the possible presence of oxide at the bonding interface and defects due to voids, or crystal dislocations. Inspection for interface voids using a high resolution scanning acoustic microscope showed none were present. Figure 3 shows a typical TEM image of the bonded interface showing an absence of crystal dislocations. Oxide islands can be reduced to a small percentage of the bonded interface using hydrophobic cleaning prior to the joining but are increased by the oxygen and dopant present in the Cz handles(4). Localized stress is apparent at the interface when joining wafers of the same orientation due to mismatch of the crystal plane(4); however no differences in the electrical characterization of the different substrates was found. This may be due to the diffusion of dopant from the handle into the device masking any effect.

In order to highlight the effects of the bonded interface, low resistivity (group D) bonded material was analyzed and compared to bulk silicon wafers of the same material. SCA measurements on group D substrates indicated no difference between the various annealing temperatures. Electrical characterization of PN diodes using I-V curves indicated good ideality factors and low leakage currents, comparable to PN diodes manufactured on the control bulk silicon substrates. C-V analysis indicated that there was no change in the dopant level as the device was biased through the bonded interface, and a resulting lifetime between 700-1000 μs was measured. Figure 4 correlates SRP data taken on bonded substrates to C-V measurements on the same material.

SRP data from substrates annealed at a range of temperatures indicated a strong influence of the bonding conditions on the resulting dopant profile. Time delay between HF cleaning and bonding results in the formation of a mono-layer of oxide on the surface. This oxide diffuses readily into float zone silicon and acts as a p-type dopant at the interface. In-situ hydrophobic cleaning and bonding gives the ability to control and minimize the growth of this oxide layer. Figure 5 depicts SRP data showing the difference between in-situ hydrophobic cleaning and hydrophobic cleaning with a time delay between cleaning and bonding. The increased dopant level results in an exponentially higher electric field required to bias the device. On the other hand SIMS data from in-situ cleaned substrates from group A and C indicated several orders of magnitude difference in the oxygen levels at the interface. Since both groups were cleaned using the same process an in-process variation in the oxygen level would not account for this difference. This indicates that supersaturated oxygen in Czochralski growth material contributes to a minimum level of 1e15 atoms/cm^3, irrespective of bonding conditions.

CONCLUSIONS

This work gives a physical and electrical characterization of the silicon direct bonded substrate and assesses its suitability as a replacement for thick epitaxial layers in device manufacture. The effects of cleaning conditions, annealing temperature, crystal growth method and crystal orientation were investigated, demonstrating that this process can produce high quality bonded substrates. The anneal temperature is limited by the formation of hydrogen bubbles below 900 °C, while SRP profiling was used to show that sharper transitions between dopant levels can be achieved than with comparable epitaxial layers. PIN diodes produced on bonded substrates showed ideal behaviour, allowing the advantages of thickness control, low defect density, and control of resistivity to be applied. The bonded interface was analysed in detail using high resistivity bonded substrates. PN diodes manufactured on this material indicated high quality silicon to silicon bonds with low leakage current and ideal characteristics. Capacitance-voltage measurements gave a further analysis of the interface and enabled a direct measurement of the dopant level using pulsed C-V. Correlation between SRP data and SIMS depth profiles on the same substrates indicated that the interfacial oxygen contributes to p-type doping at the bonded interface, and can be controlled by the bonding environment.

REFERENCES

(1). Q.-Y. Tong and U. Gösele, *Semiconductor Wafer Bonding: Science and Technology*, Wiley Interscience, New York (1999).

(2). M. Reiche, E. Hiller, D. Stolze, IEEE Sensors 2002 Conference, June 12-14, 2002, Orlando, Vol. 1, pp. 607-612.

(3) S.M. Sze, Physics of Semiconductor Devices, Wiley Interscience, New York (1969).

(4) P. McCann, S. Bryne, W.A. Nevin, *6th Int. Symp. on Wafer bonding*, San Francisco, Sept 2001.

ACKNOWLEDGEMENTS

The authors would like to thank Allan Pidduck, Alan Simmons and Dave Wallis of QinetiQ for TEM, SIMS and analysis.

Group	**Diameter**	**Type**	**Dopant**	**Resistivity**	**Orientation**	**Growth**
A	100 mm	n^-	Phosphorous	>2kohmcm	<111>	FZ
B	100 mm	n^+	Arsenic	<0.07 ohmcm	<100>	Cz
C	100 mm	n^+	Arsenic	<0.07 ohmcm	<111>	Cz
D	100 mm	n	Phosphorous	14-20 ohmcm	<100>	FZ

Table 1. Silicon Material properties

Anneal temperature °C	Dopant concentration Atoms/cm^3	Oxide charge q/cm^2	Dit no./(cm^2.eV)	Lifetime μs
1000	4.43e12	5.64e11	1.30e11	436
	3.98e12	5.85e11	1.40e11	453
1100	5.45e12	5.83e11	1.44e11	453
	3.79e12	5.96e11	1.40e11	444
	5.17e12	5.87e11	1.47e11	453
1200	5.46e12	5.75e11	1.42e11	508
	4.27e12	5.64e11	1.45e11	251
	4.65e12	5.89e11	1.34e11	621
Control	2.96e14	1.12e11	1.02e10	248

Table 2. SCA data for PIN substrates at different anneal temperature

Anneal temperature °C	Dopant concentration Atoms/cm^3	Oxide charge q/cm^2	Dit no./(cm^2.eV)	Lifetime µs
1000	2.03e14	1.43e11	4.72e10	78
	2.53e14	1.34e11	4.38e10	92
	2.49e14	1.36e11	3.99e10	109
	2.23e14	1.38e11	4.60e10	82
1100	3.15e14	1.26e11	4.67e10	103
	3.15e14	1.3e11	4.75e10	75
	2.88e14	1.22e11	4.51e10	111
1150	2.47e14	1.25e11	4.44e10	69
	2.74e14	1.29e11	4.57e10	54
	2.21e14	1.31e11	4.43e10	57
	3.00e14	1.3e11	4.75e10	76
1200	2.83e14	1.27e11	4.43e10	55
	2.83e14	1.27e11	4.43e10	69

Table 3. SCA data for PN substrates with different anneal temperatures

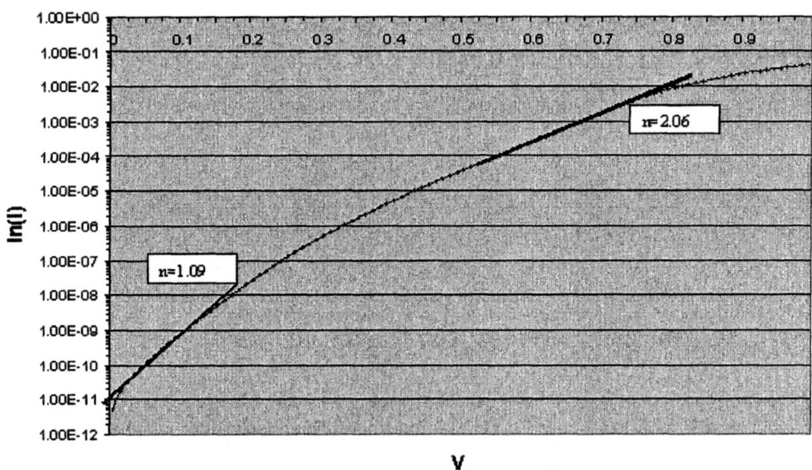

Figure 1. Ideality factors for PIN diodes on bonded substrates.

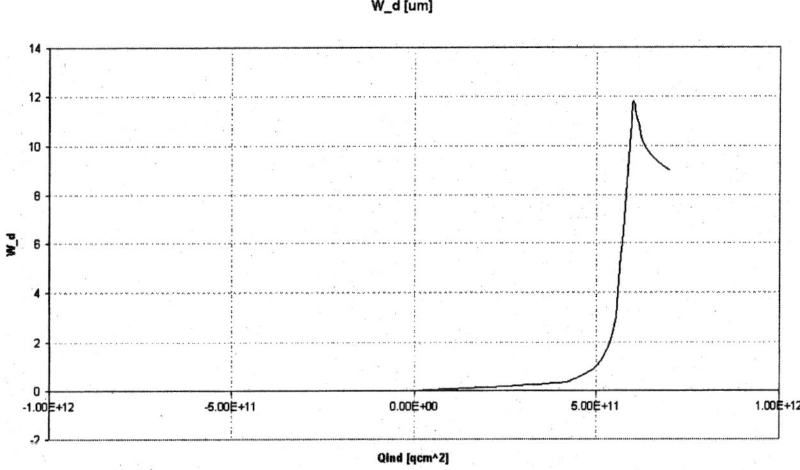

Figure 2 SCA plot for 8um n+/n- bonded substrates

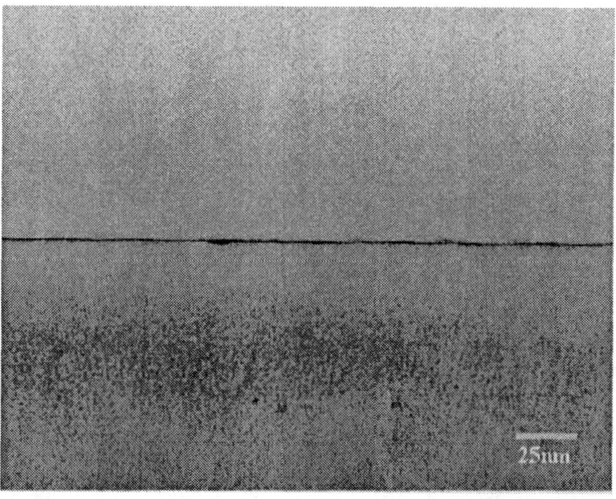

Figure 3 TEM image of the bonded interface

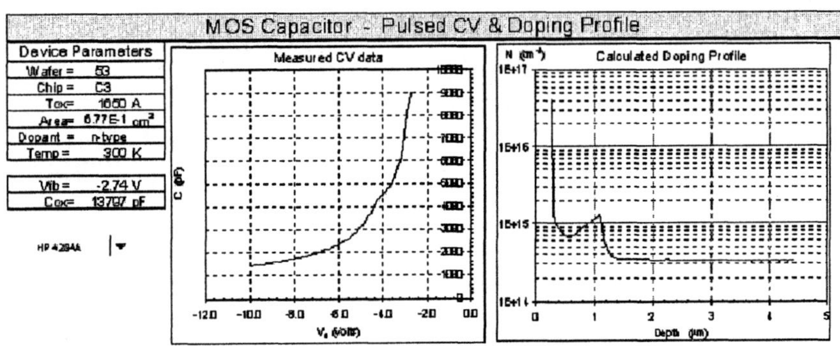

Figure 4. Pulsed C-V measurements on bonded substrates

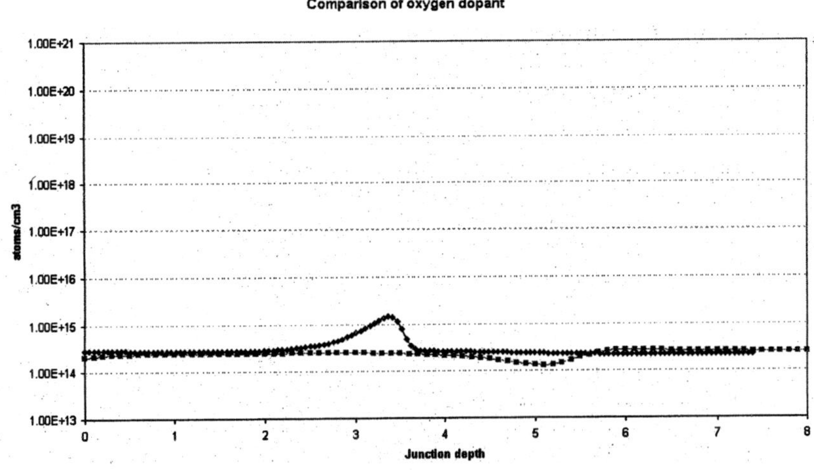

Figure 5 SRP indicating the effect of oxygen dopant.

EPITAXIAL WAFERS PREPARED BY HYDROPHOBIC WAFER BONDING

M. Reiche [1], E. Hiller [2], and D. Stolze [2]

[1] Max-Planck-Institut für Mikrostrukturphysik, Weinberg 2, D – 06120 Halle, Germany
[2] CIS Institut für Mikrosensorik gGmbH, K.-Zuse-Str. 14, D – 99099 Erfurt, Germany

ABSTRACT

Bonded hydrophobic wafer pairs (PEW wafers) are applied as alternative substrates for optical devices (pin-diodes and complex MOEMS). Important parameters of diodes are analysed at temperatures up to 250°C (dark current, photocurrent, CV-curves, rise time). Diodes produced on bonded hydrophobic wafer pairs show a similar or better behaviour than components prepared on the standard epitaxial wafers.

INTRODUCTION

Epitaxial wafers (epi-wafer) are increasingly important as substrates for numerous device technologies including CMOS, BICMOS, bipolar, and IGBT as well as optoelectronic applications (MEMS). The epitaxy process, however, limits the thickness and resistivity range of the deposited layers making that epi-wafer cannot be used for different applications. Furthermore, the epitaxial deposition is hardly to control especially for large diameter wafers resulting in thickness and dopant inhomogeneities, respectively, or an increasing defect density.

Semiconductor wafer direct bonding is an alternative technique for preparing (pseudo-) epitaxial wafers. Contrary to SOI fabrication, hydrophobic surfaces are brought into contact for this application. A Si/Si interface without any interfacial layers is formed. The process allows the combination of a large variety of different materials which cannot be realized by epitaxial deposition. Moreover, the production costs can be reduced significantly.

The present paper deals with a technique for preparing pseudo-epitaxial wafers (PE wafer) by semiconductor wafer direct bonding and their characterization. Instead of wet chemical procedures (HF dip) hydrophobic surfaces are produced by a treatment of both silicon wafers in a fluorine-containing plasma. After bonding and annealing one of the wafers (representing the layer) is thinned down to a few micrometers by mechanical grinding and subsequent polishing. This results in the formation of thick pseudo-epitaxial layers (d > 10 μm) which are important for optoelectronic applications. Thinner layers can also be prepared by using a modified smart-cut process.

SEMICONDUCTOR WAFER DIRECT BONDING

Widespread interest in modern wafer bonding techniques was generated by reports on silicon-silicon wafer bonding about 15 years ago [1-3]. Semiconductor wafer direct bonding (SWDB) requires wafers with a high degree of flatness, parallelism and smoothness. Also clean surfaces are necessary which are free of particulate, organic, and metallic contaminations. This is important because the surface cleanliness has a direct effect on both the structural and electrical properties of the bonding interface as well as on the resulting electrical properties of the bonded material. After cleaning an activation of the surfaces is required prior to bonding. Then the two mating wafers are brought together face to face in air at room temperature. The top wafer is floating on the other due to the presence of a thin cushion of air between both wafers. When an external pressure is applied onto a small part of the pair to push out the intermediate air, a bond is allowed to be formed by surface attraction forces between the wafers at this location. The bonding area spreads laterally over the entire surface of the wafers within a few seconds. It is generally assumed that the attraction forces between both surfaces are mainly caused by van der Waals interaction and hydrogen bridge bonds. Assuming typical conditions for SWDB (room temperature, atmospheric pressure, bonding of silicon wafers), the surfaces are covered with a thin, a few monolayers thick, water layer. If the surfaces are brought into contact, hydrogen bridges are generated between adsorbed water molecules. The interface structure is complex and may be analyzed by various methods such as multiple internal reflection spectroscopy. Different models of the atomic structure and bonding mechanisms are discussed in detail in [4]. The bonding energy of room-temperature bonded Si/Si or Si/SiO_2 wafer pairs is low. Surface energies (used as a measure for the bonding energy) of about 100 mJ/m^2 are reported. It has been found, however, that the surface energy of bonded wafer pairs increases with time at a given temperature, or by annealing at higher temperatures. There is a discontinuous increase of the surface energy of hydrophilic bonded wafer pairs with the annealing temperature. Typical temperature ranges with different slopes of the curve exist [5]. The representation of the surface energy data vs. the reciprocial temperature (Arrhenius plot) result in different activation energies for the different temperature ranges which are related to different atomic rearrangements of the interface. At higher temperatures ($T > 900°C$) the interfacial bonds are transformed into stable Si-O bonds over the entire interface area. The behaviour of the surface energy on the annealing temperature is quite different for hydrophobic bonded wafer pairs reflecting different hydrogen release processes at the interface. Here, Si-Si bonds exist at elevated temperatures. These two types of interfacial structure are shown in the high-resolution electron microscope images (Figure 1). The interfaces of hydrophilic bonded wafer pairs are atomically smooth, while interfaces of hydrophobic bonded wafer pairs are atomically rough. The interface represents a 1 to 2 nm thick defective layer. Most of the defects within the layer are of a rod-like morphology and oriented parallel to {111} lattice planes. They cannot be observed in TEM plan-view samples due to their 1-dimensional extension. Rod-like defects were shown to consist of silicon self-interstitials rather than of foreign phases composed of impurities detected by SIMS [6].

Bonded hydrophobic wafer pairs can prepared either by a HF dip or by a plasma treatment of the wafer surfaces prior bonding. Both treatments causes that oxide layers are removed and the surfaces are passivated by hydrogen. A HF dip, however, may cause an incomplete removal of the oxide layer or, at higher HF concentration, results in a increasing surface roughness. Treatments in a fluorine-containing plasma causes a more homogeneous removal of oxide layers. The surface structure is similar as for a HF dip.

Moreover, adding oxygen into the fluorine plasma results in an analogous surface structure, i.e. a complete removing of the native oxide, but a bonding behaviour similar to hydrophilic bonded wafer pairs [7,8].

EXPERIMENTAL

A CMOS process typically used for the preparation of the low-capacity and high-speed single photodiode SFD3.7 was applied [9]. The diode is characterized by their large spectral response range (400 – 1100 nm), the low dark current (I_d = 150 pA at T = 300K), and the low rise time of 4.7 ns. The standard material for the production process acts as reference (epitaxial wafers having an n^--layer (about 500 Ωcm) on an n-type substrate (ρ = 10 – 20 Ωcm, diameter 4 in.). The diodes posses an island-like structure produced by deep-trench etching in order to suppress the cross-talking in diode arrays. Front side contacts are used.

Czochralski-grown (CZ) silicon wafers (diameter 4 in., (100) orientation, n-type, ρ = 10 – 20 Ωcm) and high resistivity float-zone (FZ) wafers (diameter 4 in., (100) orientation, n-type, ρ = 3 - 5 kΩcm) were used for the preparation of bonded wafer pairs. The CZ-grown wafer acts as substrate in all cases, while the FZ material is used as top layer. All wafers were initially cleaned in RCA 1,2 solutions. Bonded hydrophobic wafer pairs were prepared by a SF_6/O_2 plasma treatment of the wafer surfaces. Plasma treatments were carried out in an ALCATEL MCM 100 system (reactive ion etching at 13.56 MHz) using a SF_6/18.9%O_2 plasma at a pressure of 1.4 Pa for 40 sec (rf power 80 W). The bonding experiments were carried out at room temperature and with water flushing in a CL 200 (Suess). After bonding the wafer pairs were subsequently annealed at 1050°C for 4 hours in an oxygen atmosphere. After that the top wafers were thinned down by grinding and polishing (CMP) to thicknesses between 5 μm and 80 μm. The procedure is explained elsewhere [10, 11].

The wafer pairs were analysed with respect to microscopic interface defects by infrared transmission microscopy immediately after bonding and annealing. The bonded wafer pairs were processed in the same CMOS runs as the epitaxial wafers. All diodes of the completely processed wafers were measured using the standard automatic test equipment. The results of the measurements of the dark current (I_d), the photocurrent (I_p), the rise time (t_r), and CV measurements are especially regarded for the interpretation. Furthermore, the temperature dependence of I_d was also measured on diodes separated and mounted in TO packages.

RESULTS AND DISCUSSION

An example of a bonded hydrophobic wafer pair is shown in Figure 2. There are no defects over the whole wafer detectable by infrared or acoustic microscopy. Furthermore, the corresponding thermography image proves a homogeneous current flow through the interface. Temperature inhomogeneities caused by interface defects are only visible in the rim area (Figure 3).

In order to characterize the bonded wafer pairs photodiodes were prepared on such wafers after thinning the top layer. Silicon photodetectors are used as sensing elements in optoelectronic integrated devices for numerous applications such as in the near-infrared region or as particle detectors for high energy physics experiments. The basic structure is a reversed biased pn-junction with an intermediate intrinsic layer (pin-diode). In this operation mode the space charge region is extended over the whole intrinsic layer (i-layer). Their thickness and quality strongly influence the efficiency of the diode (photocurrent).

The general dependences of the dark current (I_d) and photocurrent (I_p) on the layer thickness, doping, and temperature were calculated (SPICE simulations) for design simulation of the photodiodes. The main results can be summarized as followed:

1. The dark current increases by a factor of about 1.5 if the layer thickness increases from 5 μm to 50 μm. Thicker layers does not change I_d because the space charge region is extended over the whole layer at a reverse voltage of 2.5 V.
2. Increasing the temperature from room temperature to T = 200°C causes that the dark current increases about 4 orders of magnitude.
3. The dark current decreases as the dopant concentration of the layer increases.
4. The photocurrent increases with increasing thickness and doping of the layer.

The opposite behaviour of I_d and I_p causes that optimum parameters for the layer thickness and their doping concentration are needed. In addition, the large spectral range for application requires also an minimum thickness due to different carrier generation depths.

First electrical measurements were carried out on fully CMOS-processed wafers. An example of a bonded hydrophobic wafer pair is shown in Figure 4 after device processing. Diodes of different size have been prepared in this case. The SFD3.7 is located in the middle of the upper part of the wafer. Results of the measurements are summarized in Table 1. The dark current of individual pin-diodes prepared on the standard epitaxial material is about $9.5 \cdot 10^{-12}$ A at a reverse voltage of $U_C = 0.5$V and at room temperature. An increasing U_C causes an increasing dark current ($I_d = 1.08 \cdot 10^{-9}$A at $U_C = 50$V). The reason is the increasing width of the space charge region d_s with increasing U_C:

$$d_s = \left(\frac{2\varepsilon_H (U_D - U_C)(N_A + N_D)}{eN_A N_D} \right)^{1/2} \qquad (1)$$

In eq. (1) U_D means the diffusion voltage, N_A and N_D are the concentration of acceptors and donors, respectively, ε_H is the dielectric constant and $e = 1.6 \cdot 10^{-19}$ As is the elementary charge.

Diodes prepared on bonded hydrophobic wafer pairs (PE wafer) result in dark currents equivalent or lower than for diodes prepared on the standard epitaxial material (Table 1). The increase of I_d with increasing layer thickness corresponds to the results of SPICE simulation. Complete measurements of wafers proved, however, characteristic differences between both materials. Figure 5 shows the radial dependence of the dark current of diodes from the rim area of the wafer (at chip position 0) to the central portion. It can clearly be seen that higher values of I_d are obtained for chips near the wafer rim if standard epitaxial wafers are applied. On the other hand, I_d is constant over the wafer radius for the PE wafer. The different behaviour of the epitaxial wafer is caused by dopant inhomogeneities resulting from the epitaxial deposition process. Furthermore, there is also a wave-like behaviour of I_d which is caused by cross-talking to neighbouring (larger) diodes. The cross-talking is also suppressed for diodes on PE wafers.

The high-temperature behaviour of the dark current was also analysed on individual pin-diodes. Increasing the measurement temperature increases also the dark current (Figure 6). The slope of the curve measured for diodes on the epitaxial material, however, is higher so that I_d becomes similar or higher than for diodes on PE wafers even at T ≥ 140°C. This let us assume that thermally stimulated generation processes in the intrinsic layer (epitaxial layer) mainly contribute to the dark current. Carrier generation processes on the bonded interfaces appears to be less important. This interpretation is also confirmed by the fact that the differences of I_d increase at higher reverse voltages (which causes the extension of the space charge region up to the bonded interface). For instance, a dark current of $2.1 \cdot 10^{-5}$ A results for diodes on epitaxial wafers at T = 160°C (U_C = 30V), while $I_d = 1.1\cdot10^{-5}$ A is measured for diodes on PEW wafers under the same conditions.

Besides the dark current the photocurrent was also measured. The results are summarized in Table 1. Photocurrents measured on pin-diodes prepared on epitaxial wafers are $2.86\cdot10^{-6}$A (mean value) at a reverse voltage of 5 V. Equivalent values are also observed for diodes prepared on PE wafers. Here, the photocurrent is increasing as the thickness of the top layer increases (corresponding to an increasing thickness of the intrinsic layer). In addition, an analogous radial dependence of the photocurrent exists for diodes on epitaxial wafers as proved for the dark current. A similar behaviour is not found for diodes prepared on PE wafers.

Table 1: Results of measurements of the dark current (I_d) and photocurrent (I_p) of pin diodes (SFD3.7) prepared on standard epitaxial wafers and PE wafers (PEW). The data represent mean values of 90 diodes.

Wafer characteristics			Measurement at room temperature		
Substrate	Epitaxial/Device layer thickness	Epitaxial/Device layer resistivity	Id @ 5V	Id @ 50V	Ip @ 5V
Epitaxial wafer	30µm	> 500 Ωcm	7,85E-10	1,08E-09	2,86E-06
PEW	50µm	2,5...6 k Ωcm	3,86E-10	5,17E-10	9,13E-07
PEW	80µm	2,5...6 k Ωcm	4,14E-10	5,97E-10	1,16E-06
PEW	100µm	2,5...6 k Ωcm	5,67E-10	1,08E-09	1,25E-06

Measurements of the capacity of photodiodes as a function of the reverse voltage (CV-curves) are shown in Figure 7. Large size diodes OC 808 having a active area of 100 mm² were used for these analysis. The data for diodes prepared on epitaxial wafers proved the dependence of the capacity on the thickness of the epitaxial layer, i.e. the highest values of the capacity were obtained for the thinnest epitaxial layer (thickness 15 µm). This result agrees with the general assumption of the capacity of the space charge layer which is given as

$$C_s = A\frac{\varepsilon_H}{d_s} \qquad (2)$$

where A is the active area. Using eq. (1), the capacity can also be expressed as

$$C_s = A\left(\frac{e\varepsilon_H N_A N_D}{2(N_A + N_D)(U_D - U_C)}\right)^{1/2} \qquad (3)$$

This means that the capacity of the space charge region depends mainly on their depth. C_s decreases as the depth of the space charge region increases. If the depth is limited (by the thickness of the epitaxial layer) the capacity is constant. CV-curves were also measured on diodes prepared on bonded hydrophobic wafers. An analogous behaviour of the CV-curves was found if the same thickness of the top (intrinsic) layer is assumed (Figure 7b).

Furthermore, the rise time was measured for different pin-diodes prepared on epitaxial wafers and on bonded hydrophobic wafers (Table 2). The thickness of the epitaxial layer (or the top layer for the bonded wafer) was 50 µm. Diodes on epitaxial wafers result in rise times between 3.9 and 4.3 ns mean value 4.1 ns. Analogous values are measured also for the fall time. On the other hand, the rise time measured on the same diodes on PE wafers is between 3.5 ns and 3.6 ns mean value 3.5 ns). In addition, similar or lower rise and fall times are obtained for diodes prepared on PE wafer having thicker top layers (Table 2). Shorter rise times (as obtained for the bonded wafers) refer to a lower diffusion current contributing to the photocurrent. This means that diodes with higher switching frequencies can be realized on the bonded hydrophobic wafers instead of the conventional epitaxial material.

Table 2: Results of measurements of the rise and fall times for pin diodes prepared on standard epitaxial wafers and different PE wafers.

Conditions of the measurement: RL = 50 Ω, UR = 20 V, I = 857 nm, f = 10 kHz, impulse width: 70 ns. Backside contact.

	Rise Time (ns)					
	1	2	3	4	5	Mean Value
Epitaxial wafer (d = 50μm)	3.9	4.0	4.1	4.1	4.3	4.1
PE wafer (d=50μm)	3.5	3.5	3.5	3.6	3.5	3.5
PE wafer (d= 80μm)	3.2	3.2	3.2	3.2	3.2	3.2
PE wafer (d = 100 μm)	3.5	3.6		3.5	3.5	3.5
	Fall Time (ns)					
Epitaxial wafer (d = 50μm)	4.1	4.1	4.1	4.1	4.3	4.1
PE wafer (d=50μm)	3.5	3.5	3.6	3.6	3.6	3.6
PE wafer (d= 80μm)	3.2	3.3	3.3	3.3	3.3	3.3
PE wafer (d = 100 μm)	3.6	3.8		3.6	3.5	3.6

CONCLUSIONS

Bonded hydrophobic wafer pairs (PE wafers) are substrates comparable to conventional epitaxial wafers. The advantages of the bonded materials are

(i.) the flexibility for different material combinations which cannot be realized by epitaxy processes,

(ii.) the variability of the layer thickness, and

(iii.) the reduced production costs.

The preparation of different photodiodes (pin-diodes) demonstrates that important parameters of these sensitive devices are comparable or better by application of PE wafers. TEM and electrical measurements (thermography, dark current measurements) proved that the interface produced by hydrophobic wafer bonding does not act as generation source of carriers. There are no electrically active impurities on the interface. The application of single-crystalline wafers for the preparation of PE wafers result in defined and homogeneous top layers for optical devices. Defects in epitaxial layers such as dopant inhomogeneities does not exist. This increases the production yield and constant device parameters.

REFERENCES

1. M. Shimbo, K. Furukawa, K. Fukuda, and K. Tanzawa, *J. Appl. Phys.*, **60** , 2989 (1986)
2. J.B. Lasky, S.R. Stiffler, F.R. White, and J.R. Abernathy, *Proc. of the IEEE Internat. Electron. Dev. Meeting*, 684 , IEEE, Piscataway, NJ, 1985
3. J.B. Lasky, *Appl. Phys. Lett.*, **48**, 78 (1986)
4. Q.-Y. Tong and U. Gösele, *Semiconductor wafer bonding: science and technology*, p. 62, Wiley-Interscience, New York 1999
5. Q.-Y. Tong, E. Schmidt., U. Gösele, and M. Reiche, *Appl. Phys. Lett.*, **64**, 625 (1994)
6. M. Reiche, S. Hopfe, U. Gösele, H. Strutzberg, and Q.-Y. Tong, *Jpn. J. Appl. Phys.*, **35**, 2102 (1996)
7. M. Reiche, M. Wiegand, and V. Dragoi, in Semiconductor Wafer Bonding: Science, Technology, and Applications V, C.E. Hunt, H. Baumgart, U. Gösele, and T. Abe, Editors, PV 99-35, p. 292, The Electrochemical Society Proceedings Series, Pennington, NJ, (1999)
8. M. Reiche, U. Gösele, and M. Wiegand, Cryst. Res. Technol., **35**, 807 (2000)
9. CIS Institut für Mikrosensorik, Erfurt, product information SFD3.7 – single photodiode
10. U. Gösele and M. Reiche, in Proc. of the SEMICON Europa 2000 Silicon-On-Insulator Conference, SEMI Technical Publication (2000)
11. M. Reiche, K.H. Priewasser, E. Wittenzellner, P. Nauert, and W. Nadrag, in Semiconductor Wafer bonding: Science, Technology, and Applications V, C.E. Hunt, H. Baumgart, U. Gösele, and T. Abe, Editors, PV 99-35, p. 200, The Electrochemical Society Proceedings Series, Pennington, NJ, (1999)

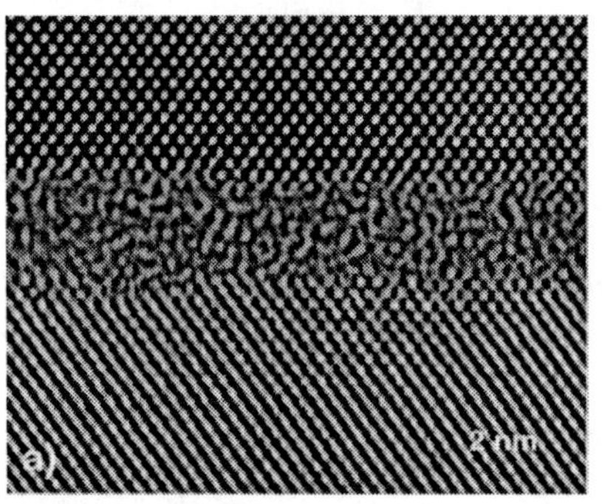

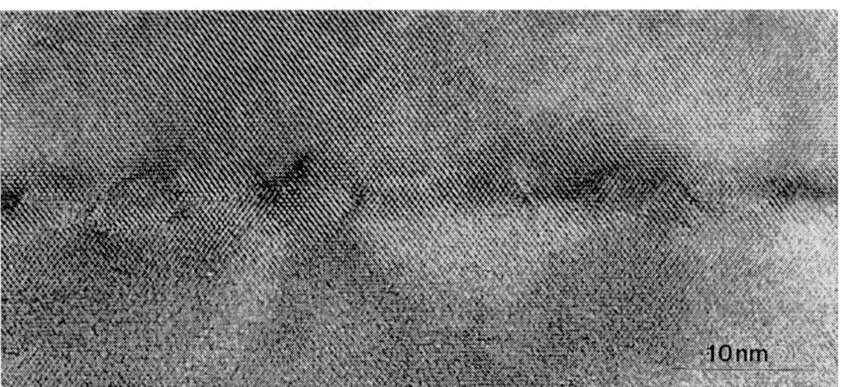

Figure 1: Cross-sectional high-resolution electron microscope images (XTEM) of interfaces of a hydrophilic bonded (top) and a hydrophobic bonded wafer pair (bottom).

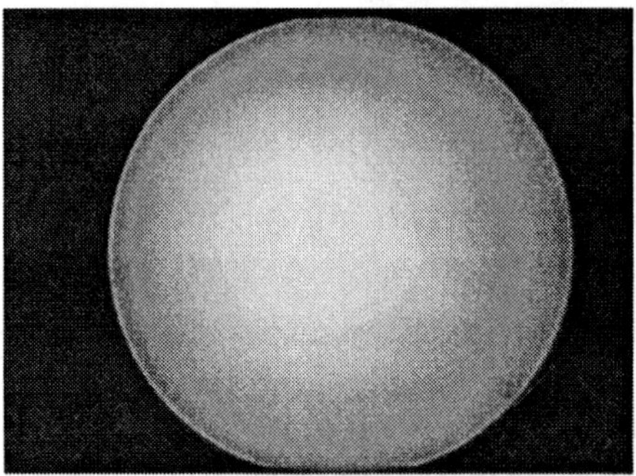

Figure 2: Infrared microscope image of a bonded hydrophobic wafer pair. A SF_6/O_2 plasma treatment was applied before bonding.

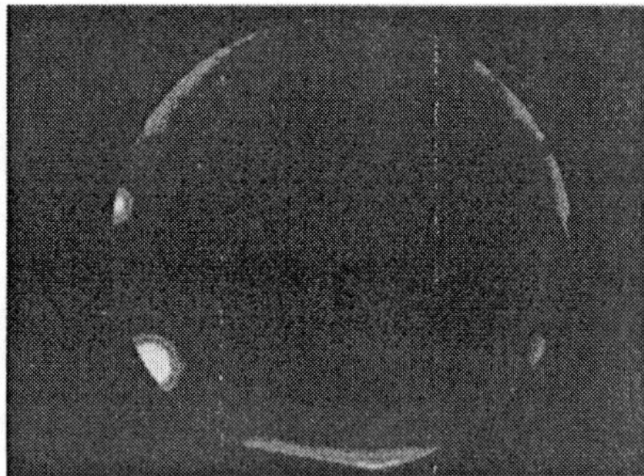

Figure 3: Thermography image of the wafer pair shown in Figure 2. The wafer diameter is 4 in.

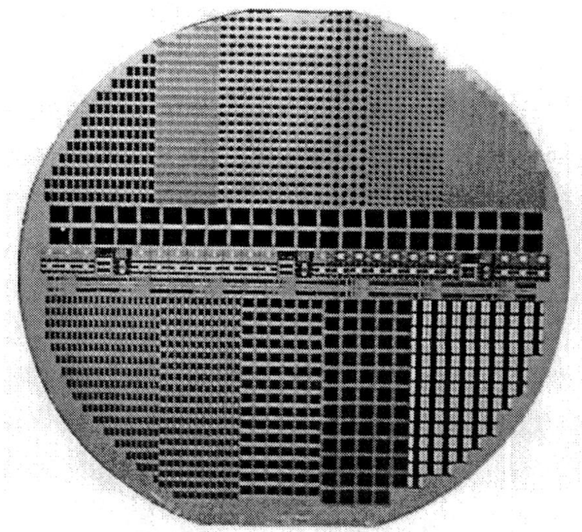

Figure 4: Image of a bonded hydrophobic wafer pair (PE wafer) after complete CMOS processing. Wafer diameter 4in.

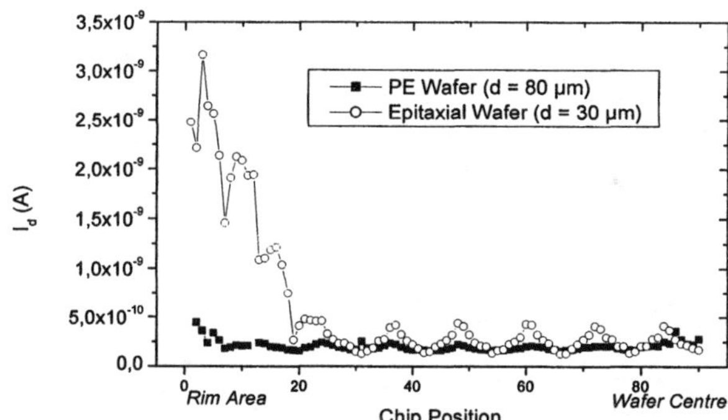

Figure 5: Radial dependence of the dark current (I_d) of pin-diodes prepared on a PEW wafer and a standard epitaxial wafer. The wafer diameter is 4 in. Measurements at a reverse voltage of 5V and at room temperature. A matrix of diodes of 15 lines (from the rim to the center) and 6 columns was measured.

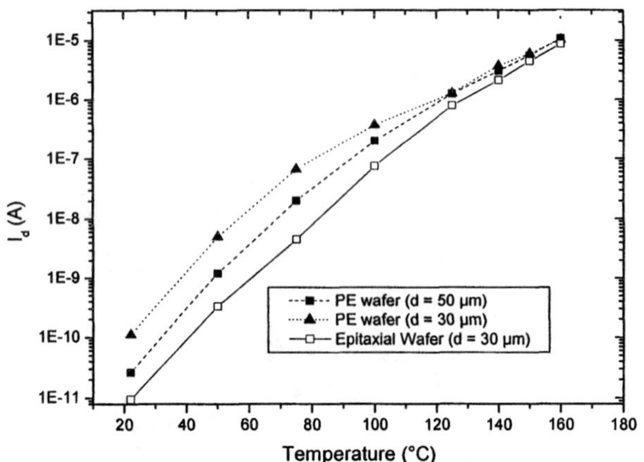

Figure 6: Measured dark currents of pin-diodes (active area 3.7 mm²) as a function of the measurement temperature for the different materials used. The reverse voltage is 0.5 V.

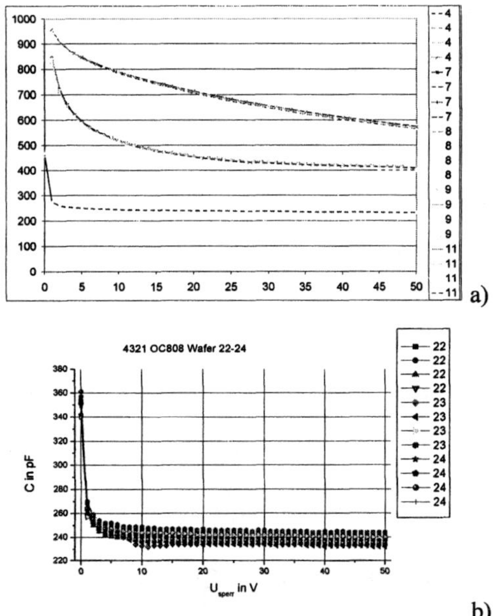

Figure 7: CV-curves of pin-diodes prepared on epitaxial wafers (a) having layers thicknesses of 15 µm (top), 30 µm, and 50 µm (bottom). b) CV-curves of pin-diodes prepared on bonded hydrophobic wafers (layer thickness 50 µm). The area of the diodes (OC 808) is 100mm².

FABRICATION OF 3-D STRUCTURE ON P-TYPE SILICON USING ELECTROCHEMICAL ETCHING

Shinichi Izuo, Fumio Saitoh, Hiroshi Ohji, Tatsuya Fukami,
Patrick J. French* and Kazuhiko Tsutsumi
Advanced R&D Center, Mitsubishi Electric Corporation
8-1-1 Tsukaguchi-honmachi, Amagasaki Hyogo 661-8661 Japan
*DIMES, EI/ITS, Mekelweg 4, Delft University of Technology 2628CD Delft

Abstract

We successfully demonstrated a 3-D structure in p-type silicon using one-step electrochemical etching. Macro porous silicon is employed for 3-D structure fabrication. We found that morphology of macro porous silicon is affected by silicon resistivity and composition of the electrolyte, and that sidewall width of silicon between trenches becomes thicker with higher silicon resistivity. Unetched portions along trenches were observed with high resistivity silicon. Unetched portions can be eliminated by increasing the water content of the electrolyte. From these results, we have proposed single-step ECE for a 3-D structured device using an epitaxial wafer consisting of a higher-resistivity of epi-layer and lower-resistivity of substrate layer. Trench formation in an epi-layer with higher resistivity and electropolishing in a bulk substrate with lower resistivity occurs. We successfully demonstrate an accelerometer structure using the electrochemical etching technique.

1. INTRODUCTION

Silicon micromachining techniques have been developed for high performance micro electromechanical devices. Three- dimensional silicon structure is fabricated by a combination of these techniques. However, the processes are sometimes complex and costly. Macro-porous silicon with high aspect ratios formed by electrochemical etching (ECE) was presented in the early 1990's[1], with research mainly focusing on macroporous silicon in n-type silicon [2-4]. To dissolve silicon, electric holes and HF molecules are necessary. Light illumination is needed to generate electric holes in n-type silicon. Several applications have been presented, such as a photonic crystal, an accelerometer and a particle separation device using n-type silicon [5-7]. Recently, macroporous formation in p-type silicon was presented by using an organic electrolyte [8]. Several groups have already presented macro-pore formations with high aspect ratios in p-type silicon using ECE [9-11]. The pore formation mechanism of organic electrolytes is under discussion [12,13]. However, macroporous formation in p-type silicon is attractive because it is simple to set up without a light source. Etching morphology, however, has been discussed only for pore formation, not for trench formation. In this work, we show influences of electrolyte composition and resistivity of silicon on etching properties and trench formation to fabricate 3-D structured devices.

2. EXPERIMENTAL

We used p-type (100) silicon substrates with mirror polished surfaces were used. Before ECE, initial V-grooves were formed by alkaline etching. A silicon dioxide layer was patterned by photolithography as a masking layer for alkaline etching. Silicon dissolution in ECE preferentially occurs at the initial grooves formed by alkaline etching, because electric holes, which are necessary for silicon dissolution, are collected at initial grooves by the concentration of electric field. Figure 1 shows a schematic view of etching equipment. Electrolyte was prepared by mixing dimethylformamide (DMF), deionised water and 50wt% hydrofluoric acid. All electrolytes, when mixed, contained 5wt% hydrofluoric acid. Tetrabutylammonium perchlorate was added in electrolyte to enhance electrical conduction. An anodic constant voltage was applied to silicon anodically relative to a platinum counter-electrode. The etched sample was observed by scanning electron microscopy, and a cross-sectional view was observed by cleaving the sample and polishing its surface.

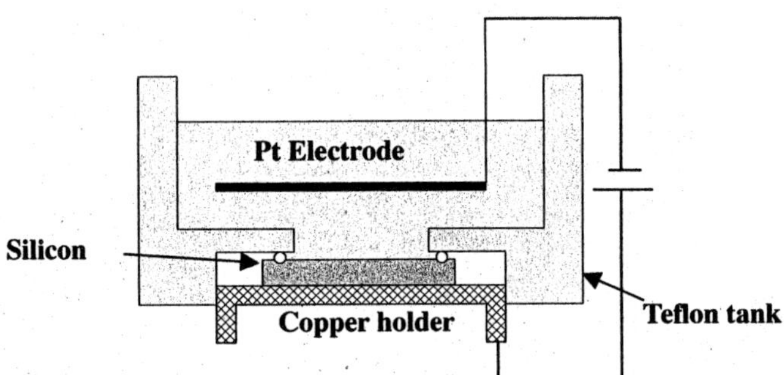

Fig.1 Schematic view of etching equipment

3. RESULTS AND DISCUSSION

3.1 Current-Voltage Characteristic

Figure 2 shows the current-voltage characteristic for three levels of silicon resistivity. The current peak is observed to shift toward the anode direction with higher silicon resistivity. The morphology of etched surfaces is classified into two regions separated by critical voltage at the current peak. Macro-porous silicon is formed below the critical voltage, whereas a flat surface is formed above the critical voltage that is called electropolishing. In the electropolishing region, a silicon dioxide layer is formed on the etched surface and chemically dissolved by a HF electrolyte. Constant current density above the critical voltage is limited by the dissolution rate of silicon dioxide; in the macroporous region, silicon is dissolved directly. Pore shape strongly depends on crystal orientation, and figure 3 shows the dependence of the etch rate on pore formation at the applied voltage and a silicon resistivity of 150 Ω · cm. Both the etch rate and pore diameter increase with increasing applied voltage. However, in the electropolishing region at 1.6 V, the etch rate decreases drastically and a flat surface with no pores is formed. Therefore, we selected the voltage just below the critical voltage to achieve trench formation.

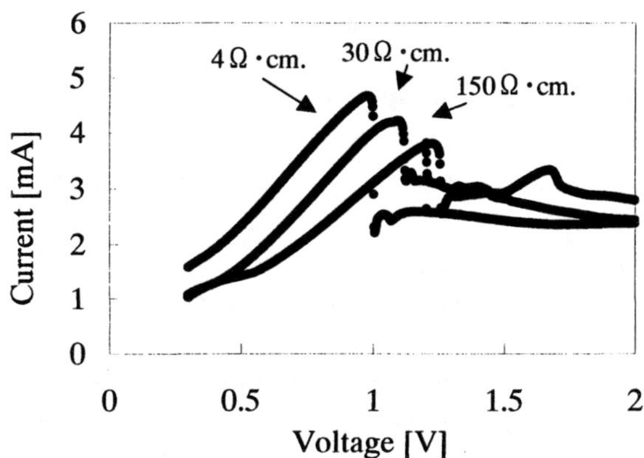

Figure.2 current-voltage characteristics for various silicon resistivities.

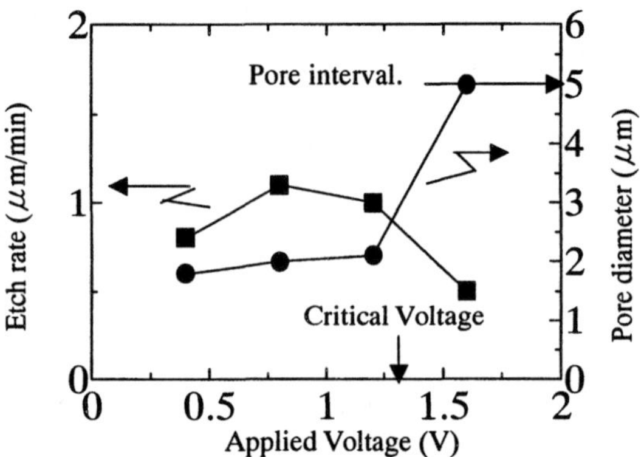

Figure.3 Dependence of applied voltage on etch rate and pore diameter.
This result is for pore structure, not trench structure.

3.2 Silicon Resistivity Effects

Figure 4 shows cross sectional views of etched samples for four levels of silicon resistivity. For 4.6Ω·cm silicon resisitvity, thin silicon sidewalls less than 1μm are observed as shown in Fig.4(b). For 0.08Ω·cm silicon resistivity, sidewall disappears, instead leaving a surface morphology similar to that produced by electropolishing. Sidewall width becomes thick with high silicon resistivity.

For trench formation, however, unetched portions along the initial V groove are observed at higher silicon resistivity, as shown in Fig. 4(d) and Fig. 5. This implies that starting points for etching are dotted along the initial V-groove. Therefore, it is necessary for trench formation that each pore along the V-groove is connected.

The dependence of sidewall width on silicon resistivity can be explained by the Space-Charge-Region (SCR) theory [12]. A space charge layer is formed in the silicon near the electrolyte interface due to the fermi-level difference between silicon and the electrolyte. In space charge layer, electric holes, which are necessary for silicon dissolution, are depleted; therefore, silicon sidewall width is equivalent to double the SCR thickness. Figure 6 shows the relationship between silicon resistivity and sidewall width. The curved line represents thickness twice that of depleted layer, calculated by the SCR model under a barrier height of 0.2 eV. Quantitatively, small inconsistencies exist between the SCR model and experimental data because barrier height changes according to applied voltage.

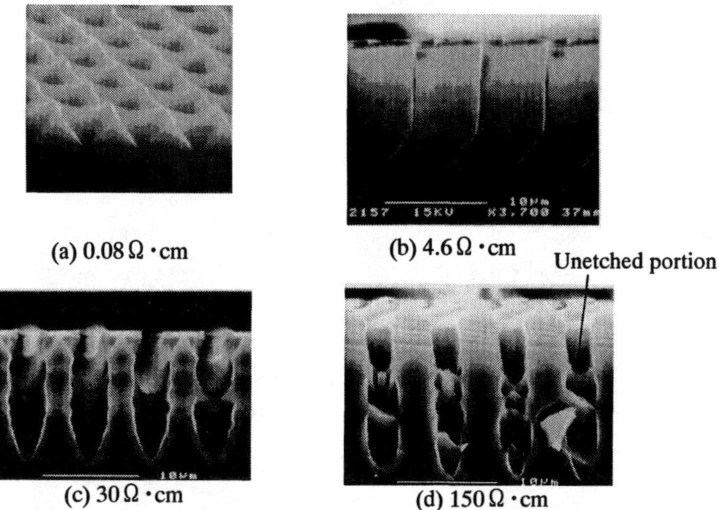

(a) 0.08 Ω・cm (b) 4.6 Ω・cm

(c) 30 Ω・cm (d) 150 Ω・cm

Fig. 4. Cross-sectional view of trench structures for various silicon resistivities.
Electrolyte consists of 90wt% DMF, 5wt% water, 5wt% HF

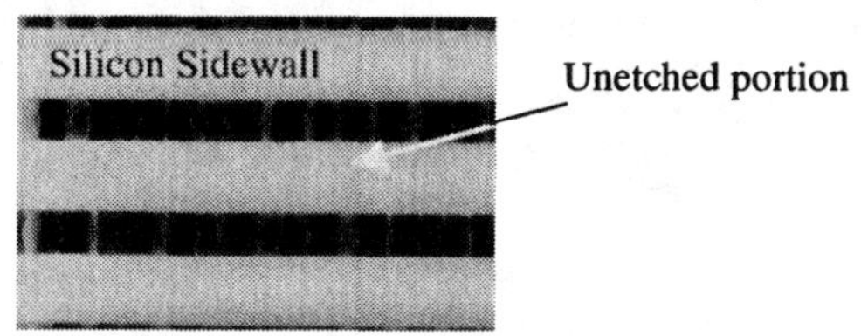

(d) plain view of etched sample with 150 Ω cm

Fig. 5. Plain view of trench structures for various silicon resistivities.
Electrolyte consists of 90wt% DMF, 5wt% water, 5wt% HF.

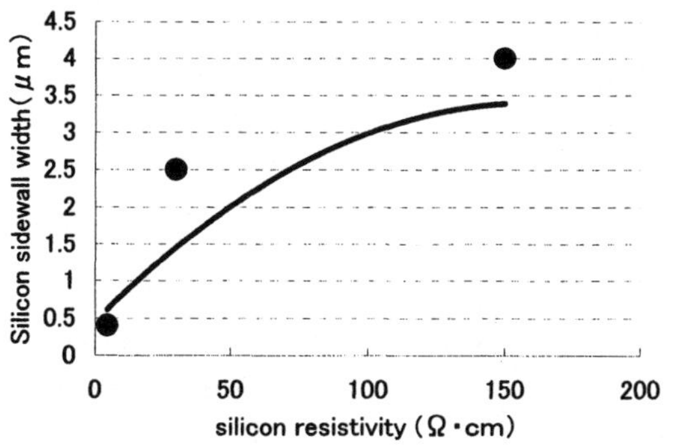

Fig.6 Relationship between silicon resistivity and sidewall width.
(Circles represent experimental data. The line is calculated from SCR model under a 0.2 eV barrier height)

3.3 Electrolyte Composition Effects

Figure 7 shows the cross-sectional view of an etched sample with 150Ω·cm resistivity in various composition of the electrolyte, specifically the water ratio; the silicon sidewall width decreases as the water ratio increases; moreover, Increasing the water ratio is effective in eliminating unetched portions along the V-groove. In addition, sidewall width slightly decreases with increasing water ratio. Water enhances the electropolishing reaction, which is an isotropic etching that makes sidewalls narrower, finally connecting each pore along the V-groove. Although increasing the water ratio has a positive effect for trench formation, it has a negative effect, in that the etch rate of silicon dioxide used as a mask layer increases. Therefore, there is a trade-off between trench formation and mask selectivity. For 70wt% H_2O based HF, 5000Å thick silicon dioxide is tolerable for 30 minutes etching.

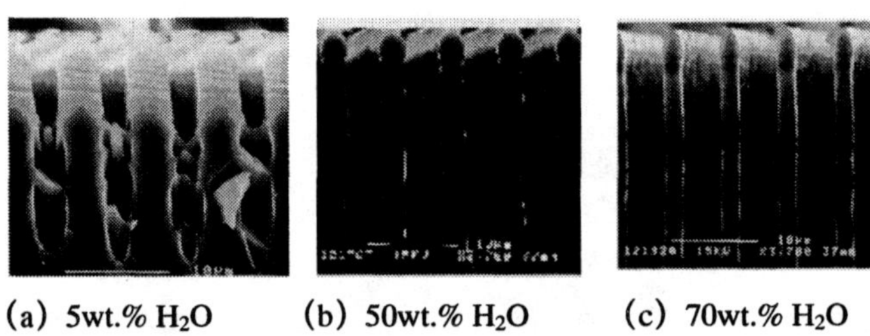

(a) 5wt.% H_2O (b) 50wt.% H_2O (c) 70wt.% H_2O

Fig. 7. Cross-sectional view of trench structures in p-type silicon with 150Ω·cm resistivity in a mixture of DMF and H_2O-based 5wt% HF electrolyte.

Figure 8 shows the dependence of the initial V-groove layout on etching morphology. The groove Intervals are set at 4μm, 7μm, 10μm and opening width is 2μm., The trench width increases as the groove intervals increases, independent of opening width. Whereas, silicon sidewall thickness become constant that depends on silicon resistivity. This fact supports the SCR model showing that silicon sidewalls are protected by a depletion layer.

3.3 Fabrication of 3-D Microstructure

As mentioned above, etching morphology with respect to trench formation and polishing is strongly affected by silicon resistivity, as shown in Fig. 4. Trench formation at high resistivity and polishing at lower resistivity is observed under same etching condition. Utilizing these results, we propose a single-step ECE for 3-D structured devices using epitaxial wafers.

Interval	30 Ω·cm	150 Ω·cm
4 μm	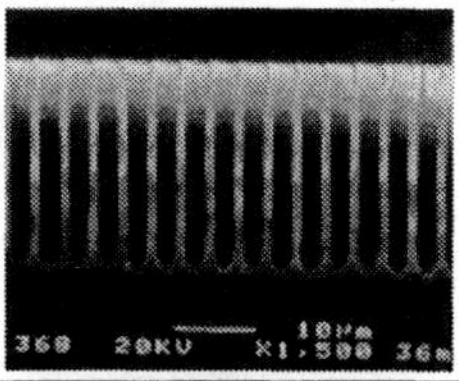	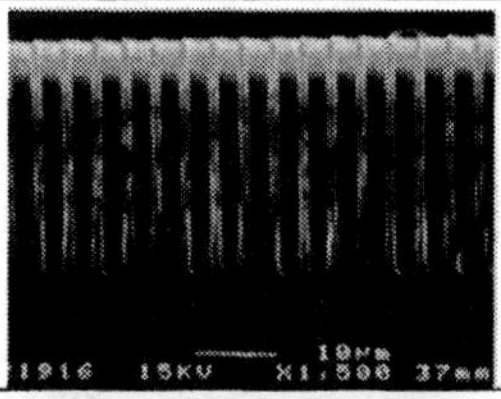
7 μm	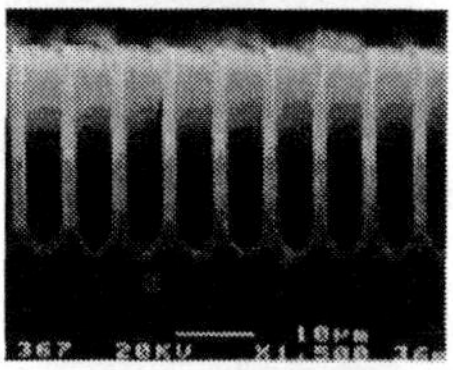	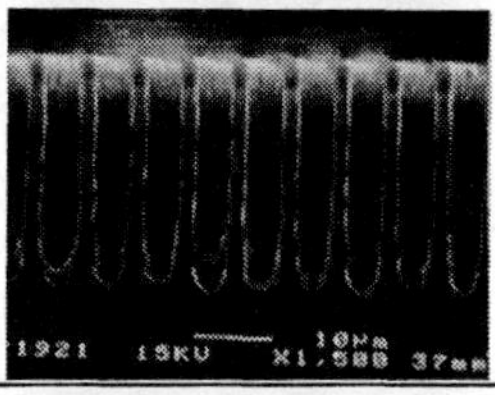
10 μm	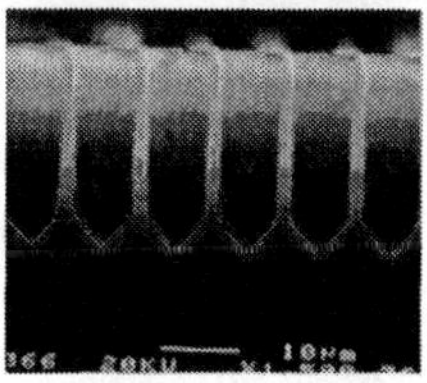	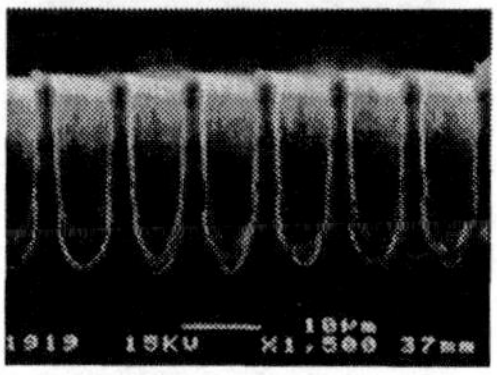

Fig. 8. Cross-sectional view of trench structures at various V-groove intervals. Electrolyte contains 25wt% DMF, 70wt% H_2O 5wt% HF.

The process sequence is shown in Fig 9.Before electrochemical etching, a V-groove is formed as initial pits. In the electrochemical etching process, trench with moderately thick walls is formed in an epitaxial layer with high resistivity. Following the tips of a trench reaching the bulk substrate, electropolishing will occur due to lower resistivity of the substrate material. Silicon sidewalls are protected during electropolishing due to depletion of electric holes in the silicon sidewall. Finally, each cavity under the trench connects, allowing freestanding structure to be fabricated as shown in Fig. 9(d). In this process, 3-D structure is achieved simply by one-step etching. We successively fabricated a 3-D comb structure for an accelerometer using this technique as shown in Fig 10. A moving mass with a comb structure is anchored by a beam spring, along with a directly anchored fixed comb, and the distance between the moving mass comb and the fixed comb changes with acceleration. The height of the freestanding structure, that is necessary to precisely control for spring's rigidity, is equal to the epitaxial layer thickness, which can be controlled precisely.

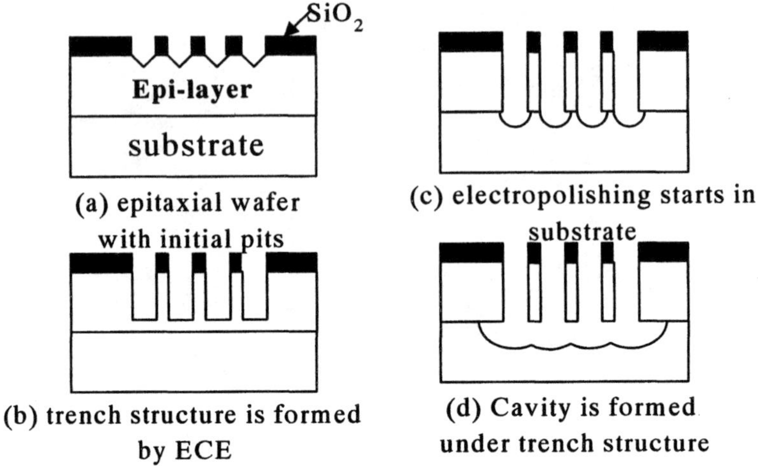

(a) epitaxial wafer with initial pits

(b) trench structure is formed by ECE

(c) electropolishing starts in substrate

(d) Cavity is formed under trench structure

Fig. 9. Process sequence for fabrication of 3-D structure by electrochemical etching

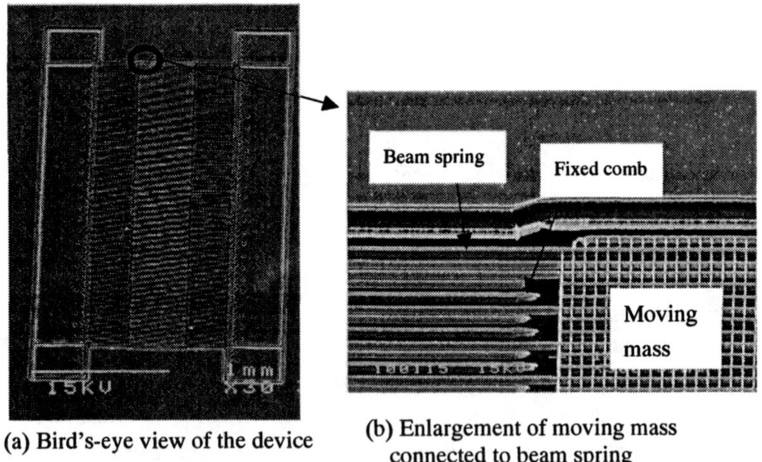

(a) Bird's-eye view of the device

(b) Enlargement of moving mass connected to beam spring

Fig. 10. SEM photograph of fabricated accelerometer structure.

4. CONCLUSIONS

In this research we have shown trench formation using an ECE technique. A higher water ratio enhances lateral etching, which prevents unetched portions along the trench. Silicon resistivity influences sidewall width. This result was extended to one-step silicon micromachining for a 3-D structure. We have successfully demonstrated an accelerometer structure using this electrochemical etching technique.

ACKNOWLEDGEMENT

This work was partly supported by Millennium Project from the Ministry of Education, Culture, Sports, Science and Technology of Japan (12408).

REFERENCES

1. V. Lehmann and H. Foll, J.Electrochem. Soc., **137**, 653 (1990).
2. M. Hejjo, *et al*, J. Electrochem. Soc., **147**, 627 (2000).
3. K. Grioras, *et al*, J. Micromech. Microeng., **11**, 371(2001).
4. J. E. A. M. van den Meerakker *et al.* J.Electrochem. Soc., 147, 2757(2000)
5. J. Shilling, *et al*, Appl. Phys. Lett., **78**(9), 1180 (2001)
6. H. Ohji, P.J.French and K.Tsutsumi, J. Micromech. Microeng., **10**, 440 (2000)
7. C. Kettner, P Reimann, P. Hanggi and F. Muller, Phys. Rev., E, **61**(1), 312(2000).
8. E. K. Propst and P.A. Kohl, J.Electrochem. Soc., **141**, 1006 (1994).
9. M. Christophersen, J. Carstensen, A. Feuerhake and H. Foll, Mater. Sci. Eng., B69-70, 194 (2000).
10. K.J. Chao, S.C. Kao, C.M. Yang, M.S. Hseu and T.G. Tsai, Electrochem. Solid-State Lett., **3**(10), 489 (2000).
11. H. Ohji, P.J.French and K.Tsutsumi, Tech. Digest, The 10th Int. Conference Solid state Sensors and Actuators, Sendai, Japan, pp 1086-1089 (1999).
12. V. Lehmann and S. Ronnebeck, J. Electrochem. Soc., 146, 2968 (1999)
13. R.B. Wehrsphon, F. Ozanam and J.N. Chazalviel, J. Electrochem. Soc., **146**, 3309 (1999).

AUTHORS INDEX

SUBJECT INDEX

Z